Role and Impact of Tourism in Peacebuilding and Conflict Transformation

Jorge Tavares da Silva
GOVCOPP, University of Aveiro, Portugal

Zélia Breda
GOVCOPP, University of Aveiro, Portugal

Fabio Carbone
Centre for Trust, Peace, and Social Relations, Coventry University, UK & GOV-COPP, University of Aveiro, Portugal

IGI Global
PUBLISHER of TIMELY KNOWLEDGE

A volume in the Advances in Hospitality, Tourism, and the Services Industry (AHTSI) Book Series

Published in the United States of America by
IGI Global
Business Science Reference (an imprint of IGI Global)
701 E. Chocolate Avenue
Hershey PA, USA 17033
Tel: 717-533-8845
Fax: 717-533-8661
E-mail: cust@igi-global.com
Web site: http://www.igi-global.com

Library of Congress Cataloging-in-Publication Data

Names: Silva, Jorge Tavares da, editor. | Breda, Zélia, 1975- editor. | Carbone, Fabio, 1978- editor.
Title: Role and impact of tourism in peacebuilding and conflict transformation / Jorge Tavares da Silva, Zélia Breda, Fabio Carbone, editors.
Description: Hershey, PA : Business Science Reference, 2020. | Includes bibliographical references and index. | Summary: "This book provides an approach to peace through tourism by presenting a theoretical framework of tourism dynamics in international relations, as well as a set of peacebuilding case studies that illustrate the role of tourism in violent or critical scenarios of conflict"-- Provided by publisher.
Identifiers: LCCN 2020004163 (print) | LCCN 2020004164 (ebook) | ISBN 9781799850533 (hardcover) | ISBN 9781799854081 (paperback) | ISBN 9781799850540 (ebook)
Subjects: LCSH: Tourism--Social aspects. | Tourism--Political aspects. | Peace-building.
Classification: LCC G156.5.S63 R65 2020 (print) | LCC G156.5.S63 (ebook) | DDC 306.4/819--dc23
LC record available at https://lccn.loc.gov/2020004163
LC ebook record available at https://lccn.loc.gov/2020004164

This book is published in the IGI Global book series Advances in Hospitality, Tourism, and the Services Industry (AHTSI) (ISSN: 2475-6547; eISSN: 2475-6555)

British Cataloguing in Publication Data
A Cataloguing in Publication record for this book is available from the British Library.

For electronic access to this publication, please contact: eresources@igi-global.com.

Advances in Hospitality, Tourism, and the Services Industry (AHTSI) Book Series

Maximiliano Korstanje
University of Palermo, Argentina

ISSN:2475-6547
EISSN:2475-6555

Mission

Globally, the hospitality, travel, tourism, and services industries generate a significant percentage of revenue and represent a large portion of the business world. Even in tough economic times, these industries thrive as individuals continue to spend on leisure and recreation activities as well as services.

The Advances in Hospitality, Tourism, and the Services Industry (AHTSI) book series offers diverse publications relating to the management, promotion, and profitability of the leisure, recreation, and services industries. Highlighting current research pertaining to various topics within the realm of hospitality, travel, tourism, and services management, the titles found within the AHTSI book series are pertinent to the research and professional needs of managers, business practitioners, researchers, and upper-level students studying in the field.

Coverage

- Service Management
- Casino Management
- International Tourism
- Service Design
- Hotel Management
- Destination Marketing and Management
- Tourism and the Environment
- Service Training
- Health and Wellness Tourism
- Travel Agency Management

IGI Global is currently accepting manuscripts for publication within this series. To submit a proposal for a volume in this series, please contact our Acquisition Editors at Acquisitions@igi-global.com or visit: http://www.igi-global.com/publish/.

The Advances in Hospitality, Tourism, and the Services Industry (AHTSI) Book Series (ISSN 2475-6547) is published by IGI Global, 701 E. Chocolate Avenue, Hershey, PA 17033-1240, USA, www.igi-global.com. This series is composed of titles available for purchase individually; each title is edited to be contextually exclusive from any other title within the series. For pricing and ordering information please visit http://www.igi-global.com/book-series/advances-hospitality-tourism-services-industry/121014. Postmaster: Send all address changes to above address.

Titles in this Series

For a list of additional titles in this series, please visit: http://www.igi-global.com/book-series/advances-hospitality-tourism-services-industry/121014

Strategies for Promoting Sustainable Hospitality and Tourism Services
Maximiliano Emanuel Korstanje (University of Palermo, Argentina) Babu George (Fort Hays State University, USA) and Alexandru-Mircea Nedelea (Stefan cel Mare University of Suceava, Roania)
Business Science Reference • © 2020 • 243pp • H/C (ISBN: 9781799843306) • US $195.00

Handbook of Research on Resident and Tourist Perspectives on Travel Destinations
Patrícia Pinto (Faculty of Economics and Research Centre for Tourism, Sustainability, and Well-Being (CinTurs), University of Algarve, Portugal) and Manuela Guerreiro (Faculty of Economics and Research Centre for Tourism, Sustainability, and Well-Being (CinTurs), University of Algarve, Portugal)
Business Science Reference • © 2020 • 400pp • H/C (ISBN: 9781799831563) • US $285.00

Global Niche Tourism and Opportunities for the Medical Tourism Industry
Manjeet Singh (Central University of Jammu, India) and Subbaraman Kumaran (Alliance University Bangalore, India)
Business Science Reference • © 2020 • 300pp • H/C (ISBN: 9781799834274) • US $215.00

Cases on Tour Guide Practices for Alternative Tourism
Gulsun Yildirim (Recep Tayyip Erdogan University, Turkey) Ozlem Ozbek (Bandirma Onyedi Eylul University, Turkey) Ceyhun Caglar Kilinc (Akdeniz University, Turkey) and Abdullah Tarinc (Akdeniz University, Turkey)
Business Science Reference • © 2020 • 350pp • H/C (ISBN: 9781799837251) • US $245.00

Accelerating Knowledge Sharing, Creativity, and Innovation Through Business Tourism
Muhammad Waseem Bari (Government College University, Faisalabad, Pakistan) Sadia Shaheen (Government College University, Faisalabad, Pakistan) and Meng Fanchen (Beijing Institute of Technology, China)
Business Science Reference • © 2020 • 351pp • H/C (ISBN: 9781799831426) • US $225.00

Contemporary Management Approaches to the Global Hospitality and Tourism Industry
Abraham Pius (Arden University, UK) Husam H. Alharahsheh (University of Wales Trinity Saint David, UK) and Adenike A. Adesanmi (University of East London, UK)
Business Science Reference • © 2020 • 290pp • H/C (ISBN: 9781799822042) • US $225.00

Global Entrepreneurial Trends in the Tourism and Hospitality Industry
Debasish Batabyal (Amity University, Kolkata, India)
Business Science Reference • © 2020 • 300pp • H/C (ISBN: 9781799826033) • US $225.00

IGI Global

701 East Chocolate Avenue, Hershey, PA 17033, USA
Tel: 717-533-8845 x100 • Fax: 717-533-8661
E-Mail: cust@igi-global.com • www.igi-global.com

Editorial Advisory Board

Table of Contents

Foreword xvi

Preface xviii

Acknowledgment xxiv

Section 1
Tourism, World Peace, and Conflict Resolution

Chapter 1
Peace: A Roadmap for Heritage and Tourism 1
Fabio Carbone, Coventry University, UK
Luiz Oosterbeek, Polytechnic Institute of Tomar, Portugal

Chapter 2
Prospects of Tourism for Peace: Is Tourism Part of the Problem or the Solution? 16
Maximiliano E. Korstanje, University of Palermo, Argentina

Chapter 3
Catalytic Effect of Tourism in Peacebuilding: Sustainability and Peace Through Tourism 29
Sweety Jamgade, Ramaiah University of Applied Sciences, India

Chapter 4
Tourism and Peacebuilding From a Holistic Approach: The Case of Trails for Peace 46
Silvia Aulet, University of Girona, Spain
Edgar Tarrés, University of Girona, Spain

Chapter 5
Tourism and the Social Reintegration of Colombian Ex-Combatants 66
Mónica Guasca, University of Leuven (KU Leuven), Belgium
Anne Marie Van Broeck, University of Leuven (KU Leuven), Belgium
Dominique Vanneste, University of Leuven (KU Leuven), Belgium

Chapter 6
The Contribution of Tourism to Peacebuilding Processes: The Case of African Peace Parks............ 87
Ana Margarida Teixeira, University of Coimbra, Portugal
Paula Duarte Lopes, University of Coimbra, Portugal

Chapter 7
Post-War Tourism in the Reconciliation Process of New Zealand Vietnam War Veterans................ 108
Maria Dorsey, University of Otago, USA

Section 2
International Relations and Tourism

Chapter 8
Tourism as an Agent of Peace and Reconciliation in Cross-Strait Relations 136
Jorge Tavares da Silva, University of Aveiro, Portugal
Zélia Breda, University of Aveiro, Portugal

Chapter 9
Tourism as an Instrument of Cultural De-bordering and Economic Connectivity: The Case of Macau... 148
Francisco José Leandro, City University of Macau, China

Chapter 10
EU Tourism Contributing to Peace and Unity? A Policy Rather Left for the Market!....................... 164
Dina Sebastião, University of Coimbra, Portugal
Alina Stoica, University of Oradea, Romania

Chapter 11
Contexts of Political (Un)Stability in Asia and the Chinese Instrumentalisation of Tourism............ 183
Jiawei Xing, University of Aveiro, Portugal
Zélia Breda, University of Aveiro, Portugal
Jorge Tavares da Silva, University of Aveiro, Portugal

Section 3
Human Interactions and Sustainable Development

Chapter 12
Hostels in Hostile Territory: Tourist Spaces of Transformative Dialogue in the Israeli-Palestinian Context... 195
Jack Shepherd, Mid Sweden University, Sweden
Daniel Laven, Mid Sweden University, Sweden

Chapter 13
Can Events Facilitate Intercultural Understanding? A Critical Investigation at Llangollen International Musical Eisteddfod 218
Karen Davies, Cardiff Metropolitan University, UK
Caroline Ritchie, Cardiff Metropolitan University, UK

Chapter 14
Factors Affecting the Impact of Tourism on Attitude Change: A Qualitative Research 238
Sedat Çelik, Şırnak University, Turkey

Chapter 15
Tourism, Citizenship, and Education: A Virtuous Triangle for Peacebuilding 255
Adalberto Dias de Carvalho, Higher Institute of Business and Tourism (ISCET), Portugal

Chapter 16
Academic Mobility: Bridging Tourism and Peace Education 275
Anabela Monteiro, Universidade Europeia, Portugal
Sofia Lopes, Universidade Europeia, Portugal
Fabio Carbone, Coventry University, UK

Chapter 17
Genetic Tourism: Key to Multicultural Understanding and Peacebuilding? 302
Carla Villagra Álvarez, Laboratory of Multidisciplinary Research in Tourism (LMRT), University of Girona, Spain
Neus Crous-Costa, Laboratory of Multidisciplinary Research in Tourism (LMRT), University of Girona, Spain

Chapter 18
The Role of Visiting Professionals in Peacebuilding 320
Katerina Antoniou, University of Central Lancashire, Cyprus

Compilation of References 342

About the Contributors 393

Index 400

Detailed Table of Contents

Foreword ... xvi

Preface ... xviii

Acknowledgment ... xxiv

Section 1
Tourism, World Peace, and Conflict Resolution

Chapter 1
Peace: A Roadmap for Heritage and Tourism ... 1
Fabio Carbone, Coventry University, UK
Luiz Oosterbeek, Polytechnic Institute of Tomar, Portugal

On 25 September 2015, the United Nations General Assembly adopted an Agenda for Sustainable Development establishing 17 Sustainable Development Goals (SDGs) to be achieved by 2030, on which states, civil society, and the private sector are called to contribute. In this context, tourism, which is based on billions of encounters between people of diverse cultural backgrounds, is hailed as the industry that can foster tolerance and multicultural and interfaith understanding, laying the groundwork for more peaceful societies. This chapter critically analyses the current discourse on tourism and peace, exploring opportunities and limitations. The authors affirm the urgency for a more complex approach by informing the debate with concepts, theories, and ideas coming from areas different than tourism (cultural anthropology, history, cultural heritage, cultural diplomacy, peace and conflict studies, peace education and sustainability science). The authors focus mainly on the role of cultural heritage and its management as the central hinge in the association between tourism and peace.

Chapter 2
Prospects of Tourism for Peace: Is Tourism Part of the Problem or the Solution? ... 16
Maximiliano E. Korstanje, University of Palermo, Argentina

Over the years, philosophers have debated the problem of war and peace. Is violence inherently enrooted in our minds? Or simply are we educated in a violent society? Is war a social malady to be eradicated or part of our nature? The myth of Cain and Abel gives some hints on this, but, of course, it does not suffice to explain the complexity of the conflict in the societal order. Humans often are slaves of negative emotions such as rage, fear, greed, and envy. Nonetheless, as Immanuel Kant imagined, a durable and perpetual peace can be internationally achieved, where a tacit agreement is convened among nations.

In so doing, they should share a common-grounded constitution and be subject to the law of federation. The figure of hospitality is vital to weave an international pact of cooperation and non-aggression. Here one might question if tourism is part of this panacea, as Kant in his days envisaged.

Chapter 3
Catalytic Effect of Tourism in Peacebuilding: Sustainability and Peace Through Tourism................. 29
Sweety Jamgade, Ramaiah University of Applied Sciences, India

Peace is the state of calmness attained when human beings of diverse culture, perceptions, and beliefs understand and respect the diversity of other humans in their surroundings. The understanding and respect are gathered through travel to places, observing and appreciating the culture, living style of the natives. It has been discussed in many tourism summits, conferences, forums, and symposiums about the contribution of tourism towards peacebuilding in the world. Many scholars have also investigated the impact of tourism on creating and maintaining peace in selected conflicted areas. The purpose of this chapter is to examine the significance of tourism in sustainable peacebuilding across the globe through a systematic literature review approach to accomplish the objectives of the study. It presents the concept of peace tourism, the Global Peace Index's domain of militarisation, safety and security in society, national-international conflicts, global leaders, analysis of some countries in tourism development, and sustainability through peace tourism.

Chapter 4
Tourism and Peacebuilding From a Holistic Approach: The Case of Trails for Peace 46
Silvia Aulet, University of Girona, Spain
Edgar Tarrés, University of Girona, Spain

Sánchez Sánchez and Fernández Herreira raise a reflection on the three dimensions of peace: form a personal dimension and the spiritual sphere, from a social dimension, and form an environmental-natural dimension. Tourism can play a fundamental role as a tool for establishing ties with oneself and with other people to contribute to the development of peace. In this chapter, authors propose to study the different dimensions of the concept of peace by studying the project of Trails for Peace, a project born as a non-profit social entrepreneurial project that integrates different routes around the world that have a special significance for their relationship with some conflict areas. Trails help travelers to know the natural areas and to build inner peace through various spiritual activities. The project helps to promote the welfare of people by involving local communities to foster contact between visitors and tourists and sustainable development. This chapter aims to analyse the impact the project has on the areas where these paths have been developed.

Chapter 5
Tourism and the Social Reintegration of Colombian Ex-Combatants... 66
Mónica Guasca, University of Leuven (KU Leuven), Belgium
Anne Marie Van Broeck, University of Leuven (KU Leuven), Belgium
Dominique Vanneste, University of Leuven (KU Leuven), Belgium

Research on the role of tourism in post-conflict scenarios has suggested that it contributes to peace and reconciliation by providing common ground for antagonists to meet. However, not enough attention has hitherto been given to cases where one of the opposing parties consist of former combatants. This chapter draws on academic literature on disarmament, demobilization, and reintegration (DDR) to discuss how

Colombian demobilised guerrillas are currently using tourism to assist their reintegration process. Based on document analysis, semi-structured interviews, and non-participant observation, it is argued that tourism can contribute to the social reintegration of ex-combatants by making them visible and active agents in the process and supporting alternative settings where reintegration can take place. Ultimately, the chapter posits that tourism can assist non-conventional approaches to reintegration, which might increase the likelihood of ensuring a successful transition from war to peace.

Chapter 6
The Contribution of Tourism to Peacebuilding Processes: The Case of African Peace Parks............ 87
Ana Margarida Teixeira, University of Coimbra, Portugal
Paula Duarte Lopes, University of Coimbra, Portugal

The concern for environmental protection within an increasingly globalised international system has led to the planning of Transfrontier Conservation Areas, among which Peace Parks aim to achieve or maintain peace across borders. Based on the growing importance that tourism practice has for the sustainability of Peace Parks, this chapter reflects on how the tourism activity developed within Peace Parks can contribute to peacebuilding processes. This dynamic is addressed by analysing a case study, focused on the Great Limpopo Transfrontier Park. It shows that tourism has the potential to positively contribute to sustainability, creating better living conditions for the local population and, consequently, promoting the achievement of peace. However, it also shows that tourism may have the opposite effect if national interests dominate, if there is insufficient consultation of local communities' interests or if sensitive border and land-use issues are not overcome.

Chapter 7
Post-War Tourism in the Reconciliation Process of New Zealand Vietnam War Veterans 108
Maria Dorsey, University of Otago, USA

Tourism has the potential to act as a positive force in reconciliation efforts between countries. The basis of tourism in facilitating reconciliation is premised on people coming into contact with one another in non-adversarial settings, which support a higher probability that positive effects can result from this contact. The investigation on post-war tourism and its role in moving the reconciliation process forward has been limited. Since the Vietnam War ended, there has been a growing phenomenon of Vietnam War veterans returning to visit Vietnam. This chapter examines the impact of New Zealand Vietnam veterans' visits to post-war Vietnam on the reconciliation process with the Vietnamese and with self.

Section 2
International Relations and Tourism

Chapter 8
Tourism as an Agent of Peace and Reconciliation in Cross-Strait Relations 136
Jorge Tavares da Silva, University of Aveiro, Portugal
Zélia Breda, University of Aveiro, Portugal

There is a non-violent conflict over Taiwan's sovereignty, between the People's Republic of China (PRC) and the Republic of China (ROC). For PRC, this division cannot persist forever and does not exclude a possible military solution. While political divisions remain, the population on both sides of the strait interact, existing sociocultural and economic dynamics. These are usually interpreted as people-to-people dynamics, in which individuals act as peace agents or citizen diplomats. Tourism is a good example of this phenomenon, considering the increasing visitor flows between both sides. This dynamism sometimes pressures the political power to transform the conflict, but also acts as a throwing weapon in times of hostility. After 2016, the political landscape in Taiwan changed, and tourism became one of the sectors involved in political tensions. This chapter explores several dimensions of tourism in this conflict, particularly its role in peace and reconciliation between Mainland China and Taiwan, but also its vulnerabilities regarding high-level bilateral relations.

Chapter 9
Tourism as an Instrument of Cultural De-bordering and Economic Connectivity: The Case of Macau.......... 148
Francisco José Leandro, City University of Macau, China

Since the retrocession of Macau to the People's Republic of China (hereafter referred to as China) in 1999, the territory has strengthened interpersonal relations related to trade and commercial engagement between China and the Portuguese-speaking countries (PSC). Moreover, the tourism industry has served as a catalyst in promoting human social capital, technology transfer, and economic security. Therefore, the main research question stands as follows: to what extent has the Macau Special Administrative Region (MSAR) of the People's Republic of China delivered a sustainable contribution to the relations between China and PSC, using the tourism industry as apparatuses of cultural and economic de-bordering? The author argues in favour of a positive correlation between the role of MSAR and the tourism industry as an instrument of economic advancement. Moreover, China's Belt and Road initiative has put forward new opportunities to MSAR, based on the concept of people-to-people (P2P) exchange and the physical integration into the Greater Bay Area (GBA).

Chapter 10
EU Tourism Contributing to Peace and Unity? A Policy Rather Left for the Market!.......... 164
Dina Sebastião, University of Coimbra, Portugal
Alina Stoica, University of Oradea, Romania

This chapter analyses the EU tourism policy regarding tourism sensitive to peace. It relies on the medieval, modern, and contemporary enlightenment philosophy of building a lasting peace in Europe, which were foundational ideas of European integration and keep being a reflex in its current values. Although the EU has been witnessing the longest period of this territory with peace, it is not taken for granted, and Euroscepticism and nationalism have been growing in Europe. The chapter assesses the conception of tourism as an intervening policy for the EU to contain nationalism, intolerance, and state conflict in Europe, using the theoretical framework of tourism sensitive to peace. It is concluded that Europe lacks an immaterial vision for tourism, as it is confined to the market functionality.

Chapter 11
Contexts of Political (Un)Stability in Asia and the Chinese Instrumentalisation of Tourism.......... 183
Jiawei Xing, University of Aveiro, Portugal
Zélia Breda, University of Aveiro, Portugal
Jorge Tavares da Silva, University of Aveiro, Portugal

Tourism is one of the most vulnerable sectors in the face of economic, political, environmental, and social changes. The study regarding the relationship between political phenomena and the dynamics of international relations and tourism is still at an infancy stage. The authors refer, in particular, to contexts of crisis, catastrophes, and conflicts in their relationship with tourism indicators. This is the purpose of this work, focusing on the Chinese case due to its particularities. In other words, they want to see how some regional political phenomena—the Asian context—are reflected in the development of Chinese tourism. They find that there is an immediate correspondence between the number of Chinese tourists and the contexts of political instability, but in which tourist flows are very dependent on government guidelines. Although subject to market dictates, they often obey geopolitical imperatives.

Section 3
Human Interactions and Sustainable Development

Chapter 12
Hostels in Hostile Territory: Tourist Spaces of Transformative Dialogue in the Israeli-Palestinian Context 195
Jack Shepherd, Mid Sweden University, Sweden
Daniel Laven, Mid Sweden University, Sweden

This chapter looks at the role youth hostels play in challenging some of the destructive narratives of the Israeli-Palestinian context, narratives revolving around Othering, and demonisation of the Other. To understand the role of hostels in this regard, the authors use the concept of transformative dialogue to demonstrate how hostels assist in the formation of partnerships between previously hostile groups, and also how hostels provide a tourism forum for self-expression and exposure to divergent narratives of space. The chapter is a comparative case study using qualitative data gathered at 14 hostels over a period of six years. The findings of the chapter stress the remarkable achievement of youth hostels in the region in challenging hegemonic discourses of separation and abjection, despite the enormous difficulty in doing so.

Chapter 13
Can Events Facilitate Intercultural Understanding? A Critical Investigation at Llangollen International Musical Eisteddfod 218
Karen Davies, Cardiff Metropolitan University, UK
Caroline Ritchie, Cardiff Metropolitan University, UK

The founding philosophy of many cultural events established after the Second World War was to enhance the dynamics of peace through supporting and developing multicultural understanding. Over 50 years after their establishment, this chapter investigates the potential of such iconic events to achieve this aim and contribute to the concept of peace through tourism, based on a longitudinal ethnographic case study of Llangollen International Musical Eisteddfod. The results show that this aim can be achieved by such events if they provide enough time and space for participants (performers and audiences) to interact. However, the study also identifies current cultural, political, and fiscal challenges in providing these temporal and physical spaces.

Chapter 14
Factors Affecting the Impact of Tourism on Attitude Change: A Qualitative Research 238
Sedat Çelik, Şırnak University, Turkey

Tourism plays a vital role in getting to know and understand each other. In this regard, the main purpose of this research chapter is to understand the relationship between tourism and attitude change and the factors affecting this relationship. Firstly, the role of tourism in changing attitudes is discussed within the framework of Allport's Contact Hypothesis, and then qualitative research results are given. The phenomenology research design was used in the research, and the interview method was preferred in obtaining data. Six open-ended questions were asked to 12 tourists, who came to Şırnak, determined by the purposeful sampling method, with a semi-structured interview form. The answers given by the participants were determined by the descriptive analysis method. The research reveals that tourism is critical in eliminating problems between societies, opening social communication channels, reducing prejudices, and making existing attitudes more positive.

Chapter 15
Tourism, Citizenship, and Education: A Virtuous Triangle for Peacebuilding 255
Adalberto Dias de Carvalho, Higher Institute of Business and Tourism (ISCET), Portugal

Tourism is not necessarily an activity that contributes to the promotion of peace. Well-known testimonies and criticisms point out their adverse effects and consequent rejection, namely in terms of intensive occupation of territories and tampering with local cultures and identities. These harmful impacts undermine the sustainability of tourism, in its complexity (which imply a harmonious but fragile system of interdependent variables, considering the existence of open and universal hospitality) and multidimensionality (because its variables are of a different, but coherent, nature). In this context, sustainability, as necessary and vulnerable, can be easily threatened by immediate economic interests and by significant gaps in civic awareness and the exercise of citizenship by all or part of the tourism protagonists and responsible. Thus, the importance of an education that takes into account the perspective of universal solidarity, which privileges, among others, the contributions of John Dewey's "cultural criticism" and of interactive constructivism.

Chapter 16
Academic Mobility: Bridging Tourism and Peace Education 275
Anabela Monteiro, Universidade Europeia, Portugal
Sofia Lopes, Universidade Europeia, Portugal
Fabio Carbone, Coventry University, UK

Education has played an influential role over time in economic, social, and cultural development and, hence, in the formation of individuals. The considerable increase in academic mobility, in recent decades, shows that it is an essential tool in students' education. Due to its specific characteristics, academic mobility has a strong link to multiculturalism, and one of its missions is to be a transmitter of values and is an important element for educational tourism. In this sense, this chapter intends to demonstrate the pedagogical potential of academic mobility, its individual and social effects in terms of peace education, and to explore the relationship between academic mobility and tourism for an environment of intercultural and peace sharing.

Chapter 17
Genetic Tourism: Key to Multicultural Understanding and Peacebuilding? 302
Carla Villagra Álvarez, Laboratory of Multidisciplinary Research in Tourism (LMRT), University of Girona, Spain
Neus Crous-Costa, Laboratory of Multidisciplinary Research in Tourism (LMRT), University of Girona, Spain

This chapter introduces the birth of a new tourism modality, genetic tourism, by comparing it with genealogical tourism. Once this comparison was carried out, it was possible to determine the characteristics of genetic tourism and to study its possible positive effects concerning cultural sensitivity and transculturality, related to multicultural understanding and peacebuilding. The exploratory study was done by analysing travel blogs and online news related to genetic tourism, which results were used to identify a series of topics related to cultural sensitivity and transculturality. Among these topics, the author highlights the predisposition of the studied subjects to define themselves as citizens of the world, enabling people-to-people interactions and the belief of a positive relationship between genetic tourism, tolerance, cultural ethnorelativism, and reductions of negative stereotypes.

Chapter 18
The Role of Visiting Professionals in Peacebuilding 320
Katerina Antoniou, University of Central Lancashire, Cyprus

Seeking to inform the peace-through-tourism scholarly debate, this chapter revisits this complex relationship from the angle of peacebuilding professionals. International peacebuilding settings habitually welcome peace professionals as visiting contributors, consultants, and freelancers. These visiting peacebuilders are characterised by a form of geographical hybridity that identifies them as a distinct audience in international peacebuilding and a niche audience of international travellers. In order to examine their role and contribution in international peacebuilding, the present study employs a qualitative methodological approach that combines historical narrative and field interviews to the case of the Cyprus peacebuilding discourse. This chapter argues that visiting peacebuilders have a significant role in formulating optimal encounters between peacebuilder audiences and in increasing the effectiveness of international peacebuilding projects.

Compilation of References 342

About the Contributors 393

Index 400

Foreword

An ancient Roman adage states "*Si vis pacem, para bellum*" (If you want peace, prepare for war). In the past 3,400 years, we have been at peace for 268 of them or a mere 8% of recorded history.

On November 11 2018, the world commemorated the end of the First World War (WWI) – "the war to end all wars". Yet, since the end of WWI, more than 100 million persons have lost their lives as a result of wars – many of them innocent women and children.

Perhaps it is time to realise "*Si vis pacem, para pacem*" (If we want peace, we must prepare for peace). And no industry has the potential to contribute to peace more than the global tourism industry. The book *Role and Impact of Tourism in Peacebuilding and Conflict Transformation* provides us with a timely navigational star to lead us in this direction, by providing both theoretical frameworks from a range of different disciplines, as well as concrete case studies.

As this book prepares for publication, the world is experiencing the worst pandemic in 100 years. SARS-CoV-2, which led to the COVID-19 pandemic, has infected more than 6.2 million persons in the world, and more than 376,000 persons have died from the virus. Before the pandemic is over, deaths resulting from it will likely equal or exceed the number of persons who have died in major war zones since the terrorist attacks of September 11, 2001 (Afghanistan, Pakistan, Iraq) – a total of 480,000, including civilians and other non-combatants.

Tourism is one of the hardest-hit industries by the COVID-19 outbreak. Airlines, hotels, and cruise lines are experiencing severe losses as travel restrictions have been put in place in virtually every country. Job losses have surged, and many small and medium-sized enterprises (SMEs) are at risk. The pandemic will likely have a longer-lasting effect on international tourism compared to other industries, bringing an abrupt end to a 10-year period of sustained growth since the 2009 financial crisis.

The Chinese word for crisis (危机 – wéijī) is composed of two characters, one representing danger and the other opportunity. While COVID-19 currently, and in the near future, represents a crisis for the global tourism industry, perhaps it also presents a time to pause and re-think the future development of tourism – its future development, particularly, in terms of its contribution to the 17 Sustainable Development Goals (SDGs), to be achieved by 2030 and issues such as 'overtourism'.

The Institute for Economics and Peace estimated that the global cost of violence to the global economy in 2018 was approximately US$ 10 trillion, 11.2% gross world product or $1,853 for every person in the world. Imagine if this amount could be reduced by re-thinking tourism in a manner that contributed to:

- Peace and tranquillity within ourselves;
- Peace with others, from our neighbours next door to our neighbours in the global village;
- Collaboration among nations;

- Peace with nature and our common home – planet Earth;
- Peace with past generations – by which we honour our respective cultures, heritage and achievements of our forefathers;
- Peace with future generations – through sustainable lifestyles and practices;
- Peace with our Creator – by practising the universal principle of all faiths and humanists: do unto others as we would have them do unto ourselves; bringing us full circle to 'peace and tranquillity within ourselves'.

Imagine if the amount saved from the cost of violence could then:

- improve the lives of the remaining 600 million men, women and children in the world still living in extreme poverty;
- provide clean water to 780 million persons who never get a refreshing drink of clean water;
- save the lives of more than 3.4 million persons who die each year from waterborne diseases, making it the leading cause of disease and death around the world – most of those deaths are young children, about 4,000 a day;
- provide improved sanitation to 2.5 billion people (more than 35% of the world's population).

COVID-19 gives us an opportunity to pause and re-think the issue of 'overtourism' – mainly an issue in Europe, which currently receives 51% of international arrivals. At the same time, Europe, particularly Western Europe, seriously exceeds its ecological capacity. Sub-Saharan Africa and South America currently receive less than 5% and 3% of international arrivals, respectively, and each also has an excess of ecological capacity. Re-thinking tourism would emphasise increasing travel to both these regions, which also would contribute to enhancing socio-economic development in these regions and reducing global poverty.

COVID-19 gives us an opportunity to pause and re-think the important roles that Cultural and Sports Tourism, Adventure Tourism, Pro-Poor Tourism, Volunteer Tourism, Hosteling and other newly emerging forms of tourism can play towards achieving peace; re-thinking tourism so that it promotes international understanding; collaboration among nations; being socially and environmentally sustainable; enhancing cultures and local heritage; reducing poverty; and helping to heal the wounds of conflict.

COVID-19 gives us an opportunity to pause and re-think the role that tourism can play in helping us to realise that we are "One Global Family", sharing one common home – planet Earth. A planet that is currently facing global challenges, such as climate change – the depletion of resources – and species' extinction. An opportunity to realise that these global challenges require that we all come together – as people, societies, nations – to find global solutions and work together to implement those solutions. The book *Role and Impact of Tourism in Peacebuilding and Conflict Transformation* provides us with a beginning road map towards this end.

Louis D'Amore
International Institute for Peace Through Tourism (IIPT), USA

Preface

The book *Role and Impact of Tourism in Peacebuilding and Conflict Transformation* fits into a moment of operational uncertainty and theoretical (re-)elaboration on the nature of the tourist phenomenon within a society marked by geopolitical turmoil and reduced international security, as well as rapid changes at the global level and even a pandemic (the COVID-19 pandemic, during 2020) that is currently posing the sector in front of new challenges. Thus, although the conceptual association between tourism and peace is hardly new, today it is more contemporary and appropriate than ever to reflect on it with renewed vigour and through a more complex approach.

Heraclitus (535 BC-475 BC) argued that war is the mother of all things. Hegel (1770-1831), not only shared Heraclitus's assertion, but added that war, as a moment of the antithesis of society, is beneficial to humans, as only through conflicts freedom and dignity are fully recognised. Even the "enlightened" Kant (1724-1804), author of the "Perpetual peace", takes as its starting point the indisputable observation that the states are constantly at war or in any case in a situation of fragile peace with each other, being conflict part and parcel of their nature. Nevertheless, the idea that conflicts are not inevitable, however utopian it may seem, continues alive. The theoretical-practical search for the best way to create peace is itself in constant evolution. Conflict transformation mobilises human intervention in order to redirect the consequences of conflict towards reduced violence, altered relationships, understanding, redistribution, structural adjustments, and work which will help to account for root causes (Thiessen, 2017). In other terms, conflict transformation attempts to make explicit, and then reshape, the social structures and dynamics behind the conflicts, representing, therefore, a process of engaging with and transforming the relationships, interests, discourses and, if necessary, the very constitution of society that supports the continuation of violent conflict.

Such an approach is evidently based, among other things, on the promotion of mutual knowledge and formal, non-formal and informal education for peace. Since the link between tourism and these issues is more than immediate, the sector has embraced this challenge both from a practical and a theoretical point of view. The United Nations World Tourism Organization (UNWTO) unmistakable defines tourism as "a catalyst for peace and development", also providing in dedicated publications an overview of tourism as a tool in peacebuilding and reconciliation efforts (Wohlmuther & Wintersteiner, 2014).

Wilson (2012), who is one of the world's leading biologists, points out that human evolution has been defined by conflict, and affirms without any doubt that war is embedded in our very nature. Even Kant actually started from this assumption to promote the firm conviction of the need for international laws that guarantee world peace (as human nature itself would not eventually be enough for this purpose). One can thus say that the search for international peace and security is, therefore, one of the results of human reason's evolution. Coherently, the constitution of UNESCO states that "since wars begin in the minds of men, it is in the minds of men that the defences of peace must be constructed" (UNESCO, 2020). Today, the achievement of "peace and fair institutions" is among the Sustainable Development Goals of the United Nations (UN). It is, therefore, a shared responsibility to overcome our nature in order to reach, with our reason, a more just and peaceful world. Furthermore, today our responsibility is amplified by the fact that the compression of time and space, as well as the technological modernisation and transport revolution, offered a new awareness of the world as a whole (Robertson, 1992).

In this context, tourism, based as it is on culture and identities, on the enhancement of diversity and, at the same time, on international encounters (albeit the "tyranny of time" (Bank, 1997) in tourism practice represents an obstacle and limitation in this sense), can play a significant role in the construction of a culture of peace. According to the World Travel & Tourism Council (WTTC, 2016, p. 2), "countries with more sustainable and open tourism industries tend to be more peaceful", and "both conflict-affected and non-conflict-affected countries show a correlation between the Tourism Index and higher positive peace, indicating that higher scores on the Tourism Index correlate with higher levels of positive peace". Coherently, scholars followed the debate on the conceptual association between tourism and peace, making it part the general research agenda in tourism studies. The literature on this subject is already extensive, indeed. However, the need for an upgrade on the approach is clear, namely through the application of a more complex approach to the subject, looking beyond the tourism studies' theoretical horizon and drawing on models from areas other than tourism.

This book proposes more mature perspectives about the discourse on "tourism and peace" in the contemporary world, by including works that reflect influences from areas of study that range from international relations to cultural heritage management, from cultural diplomacy to peace and conflict studies, and so on. The book associates tourist practices with peacebuilding, in particular with the third generation approach for making peace, that is, conflict transformation. The latter "move beyond the pursuit of 'conflict resolution' – an approach which prioritises the de-escalation of conflict dynamics and the diffusion of crises – and it will focus instead on deeper issues and underlying patterns which, if addressed, can progress far beyond the temporary relief of unjust conditions" (Thiessen, 2017, p. 8). By doing so, the book contains matters that normally are not included in the debate about tourism and peace, such as the difference potential of tourism within positive and negative peace (Galtung, 1996), and the role of tourism in the reduction or eradication of what Galtung (1990) defined as cultural violence.

The book offers new angles of understanding through case studies and best practices that could help students, academic faculty and researchers, whose work and/or research are related to tourism, international relations, international politics, peace studies and conflict resolution, to deal in real case scenarios. In addition, it will be an important tool for practitioners or mediators in order to better understand tourism as a vehicle of global understanding and peace in the contemporary world. Students will also benefit from case studies that will showcase contemporary issues in different geographical locations around the world.

The book is composed of three main sections: "Tourism, World Peace, and Conflict Resolution", "International Relations and Tourism", and "Human Interactions and Sustainable Development".

SECTION 1: TOURISM, WORLD PEACE, AND CONFLICT RESOLUTION

Chapter 1, "Peace: A Roadmap for Heritage and Tourism," alerts to the need for a more mature and complex approach to "tourism and peace", by informing the debate with concepts, theories and ideas coming from areas different than tourism (cultural anthropology, history, cultural heritage, cultural diplomacy, peace and conflict studies, peace education and sustainability science). In this sense, the chapter explores, in a theoretical and practical way, the role of cultural heritage and its management within the conceptual association between tourism and peace, proposing cultural institutions as the keystone of such architecture.

The following chapter, Chapter 2, "Prospects of Tourism for Peace: Is Tourism Part of the Problem or the Solution?" provides the readers with an interesting excursus on the nature of Man and human conflicts. Afterwards, it suggests an interesting link between tourism and hospitality and the Kantian concept of hospitality contained in the remarkable work "Perpetual Peace". Intentionally, the authors' approach to the topic consists of never taking for granted that tourism could contribute to the creation of a culture of peace.

Chapter 3, "Catalytic Effect of Tourism in Peacebuilding: Sustainability Through Peace Tourism," is of great interest not only because it contains the state-of-the-art of the debate on tourism and peace, but also because it then puts tourism in relation to various topics related to the field of peace and conflict studies, such as the Global Peace Index's domain of militarisation, safety and security in society, national-international conflicts, global leaders. Finally, the author links her complex discourse with the idea of sustainability.

The case study presented in Chapter 4, "Tourism and Peacebuilding from a Holistic Approach: The Case of Trails for Peace," leads us to explore different dimensions of the very concept of peace. Trails for peace, a project that integrates different routes around the world, some of them related to conflict areas, aims to create and/or strengthen inner peace through travel experiences that put visitors in contact with the authenticity of local communities. The chapter analyses the impact of this approach in the regions where the project is operating, particularly from the point of view of promoting a culture of peace.

Another case study, but with a totally different work perspective, is described in Chapter 5, "Tourism and the Social Reintegration of Colombian Ex-Combatants". The chapter draws on academic literature on Disarmament, Demobilisation and Reintegration (DDR), focusing on the case of Colombian demobilised *guerrillas.* Based on the discussion of the results of the empirical work, the authors argue that tourism is actually contributing to ex-combatants' social reintegration, increasing the likelihood of a successful transition from war to peace.

Chapter 6, "The Contribution of Tourism to Peacebuilding Processes: The Case of African Peace Parks," reflects on how the tourism activity developed within Peace Parks can contribute to peacebuilding processes. However, through the specific study of the Great Limpopo Transfrontier Park (Zimbabwe, Mozambique, South Africa), the authors of this chapter illustrate both positive and negative impacts of the project.

Chapter 7, "Post-War Tourism in the Reconciliation Process of New Zealand Vietnam War Veterans," proposes another interesting and relevant perspective: the association between tourism and reconciliation in post-conflict areas, being reconciliation one of the fundamental aspect of peacebuilding. The authors consider the growing phenomenon of Vietnam War veterans returning to visit Vietnam, examining the impact of New Zealand Vietnam veterans' visits to post-war Vietnam on the reconciliation process with the Vietnamese and with themselves.

SECTION 2: INTERNATIONAL RELATIONS AND TOURISM

In the context of the association between tourism and international relations, Chapter 8, "Tourism as an Agent of Peace and Reconciliation in Cross-Strait Relations," opens the section with the interesting study on the complex and delicate situation of Taiwan, disputed between the People's Republic of China (PRC) and the Republic of China (ROC). The so-called "Taiwan Strait Issue" – or, from a Taiwanese perspective, the "Mainland Issue" – is a result of the Chinese Civil War and no armistice or peace treaty has been signed to date, so this continues being a latent conflict. The authors analyse the informal sociocultural interaction and dynamics of people from both sides of the Strait, particularly those related to tourism activities, taking into account the increasing visitor flows. The chapter explores, in particular, the political context after 2016, when President Tsai Ing-wen from DPP takes power in Taiwan. The new framework has been having consequences on tourist flows in the Taiwan Strait. Xi Jinping's authoritarian Mainland China strongly influences this process according to the political trends on the island, more or less adjusted with his own interests. The relationship between the intensity of the conflict and tourism flows is an excellent example of analysis in the context of international relations.

Chapter 9, "Tourism as an Instrument of Cultural De-Bordering and Economic Connectivity: The Case of Macau," explores the complex geopolitical consequences of the re-established Chinese sovereignty over Macau in 1999, in particular through the prism of tourism. The author, for example, highlights the role of tourism in the renovation of ancient bridges between Macau and Portugal (of which Macau was a colony), but also new bridges, namely between Portuguese speaking countries and China, through Macau. The author argues in favour of a positive correlation between the role of Macau and its tourism industry as an instrument of economic development and bridge between Asia and Europe.

Chapter 10, "EU Tourism Contributing to Peace and Unity? A Policy Rather Left for the Market!" introduces one more new topic: the contemplation of the idea of tourism as a vehicle for building a culture of peace within the supranational policies of tourism development. In the light of the results of their research, the authors assert the lack of a European vision of tourism that goes beyond market issues and, through their statement, they warn of the need to review the current political approach to tourism.

The link between geopolitics and tourism is explored in Chapter 11, "Contexts of Political (Un)Stability in Asia and the Chinese Instrumentalisation of Tourism". Authors start by recognising the lack of studies regarding the relationship between political phenomena, international relations and tourism. Focusing on the Chinese case, the chapter analyses how some regional political phenomena are reflected in the development of Chinese tourism. Evidence show, with regards to the case study, that tourism dynamics, although subject to market dictates, often obey geopolitical imperatives.

SECTION 3: HUMAN INTERACTIONS AND SUSTAINABLE DEVELOPMENT

Chapter 12, "Hostels in Hostile Territory: Tourist Spaces of Transformative Dialogue in the Israeli-Palestinian Context," examines the role that youth hostels play in challenging some of the destructive narratives of the Israeli-Palestinian context, narratives revolving around "Othering and demonisation of the Other. The authors use the concept of transformative dialogue to demonstrate how hostels assist in the formation of partnerships between diverse groups also by providing a tourism forum for self-expression and exposure to divergent narratives of space. Based on empirical evidence, the chapter stresses the remarkable achievement of youth hostels in the region in challenging hegemonic discourses of separation and abjection.

Chapter 13, "Can Events Facilitate Intercultural Understanding? A Critical Investigation at Llangollen International Musical Eisteddfod," illustrates the results of a longitudinal ethnographic case study to analyse the potential of this event to contribute to the concept of peace through tourism. The founding philosophy of many cultural events established after the Second World War is to enhance the dynamics of peace through supporting and developing multicultural understanding. The important conclusion of this work is that this aim can only be achieved if event organisers provide enough time and space for participants (performers and audiences) to interact. The study also identifies current cultural, political and fiscal challenges in providing these temporal and physical spaces.

Chapter 14, "Factors Affecting Tourism and Attitude Change: A Qualitative Research," explores, in the context of the debate on tourism and peace, the relationship between tourism and attitude change, and the factors affecting this relationship. The role of tourism in changing attitudes is discussed within the framework of Allport's Contact Hypothesis, and then the results of a study carried out in Şırnak (Turkey) are discussed. The importance of this chapter lies in the fact that a large part of the arguments in favour of tourism as a vehicle for peace is precisely the contact between visitors and the local community.

Chapter 15, "Tourism, Citizenship and Education: A Virtuous Triangle for Peacebuilding," opens by pointing out testimonies and criticisms against the association of tourism and peace, namely highlighting the harmful impacts that undermine the sustainability of tourism. Sustainability can be, thus, easily threatened by immediate economic interests and by significant gaps in civic awareness and the exercise of citizenship by all, or part of, tourism stakeholders. Therefore, the author alerts for the role of education in promoting universal solidarity as a solid basis for sustainable tourism development.

Also Chapter 16, "Academic Mobility: Bridging Tourism and Peace Education," explores the association between formal, informal and non-formal education and tourism for the creation of a global culture of peace. Namely, the authors investigate the considerable increase in academic mobility, in recent decades, and its strong link to intercultural capacities. This chapter demonstrates the pedagogical potential of educational tourism, namely academic mobility, its individual and social effects in terms of peace education, and explores the relationship between academic mobility and tourism for an environment of intercultural and peace sharing.

Chapter 17, "Genetic Tourism: Key to Multicultural Understanding and Peacebuilding?" proposes a brand new concept, that of genetic tourism, by comparison with genealogical tourism. Once this comparison was carried out, it was possible to determine the characteristics of genetic tourism and to study its possible positive effects concerning cultural sensitivity and transculturality, related to multicultural understanding and peacebuilding. The author highlights the predisposition of the studied subjects to define themselves as citizens of the world, enabling people-to-people interactions and the belief of a positive relationship between genetic tourism, tolerance, cultural ethnorelativism and reductions of negative stereotypes. This chapter is one more example of the degree of complexity that the debate on tourism and peace can reach.

Chapter 18, "The Role of Visiting Professionals in Peacebuilding," revisits this intricate relationship from the angle of peacebuilding professionals. International peacebuilding settings habitually welcome peace professionals as visiting contributors, consultants and freelancers. The chapter argues that visiting peacebuilders have a significant role in formulating optimal encounters between peacebuilder audiences and in increasing the effectiveness of international peacebuilding projects. This chapter puts in direct contact tourism and peacebuilding, not just theoretically, focusing on the behaviours and attitudes of peacebuilding operators travelling around the world.

The book *Role and Impact of Tourism in Peacebuilding and Conflict Transformation* definitively establishes the association between tourism and the most sophisticated form of peacebuilding in the history, which, according to Laderach (1997), move beyond "traditional" diplomacy – which often emphasises top-level leaders and short term perspectives – towards a more holistic approach. In doing so, the book is a considerable and necessary theoretical breakthrough in the debate about tourism and peace, with enormous potential for real-world impacts. It indeed provides clues for future research, but also insights for those involved in the tourism sector, as well as mediators and peacemakers, national and supranational agents dedicated to peacebuilding. For everyone who dreams of a more united and peaceful world.

REFERENCES

Bank, R. (1997). *The tyranny of time: When 24 hours is not enough.* Wipf & Stock.

Galtung, J. (1990). Cultural violence. *Journal of Peace Research, 27*(3), 291–305. doi:10.1177/0022343390027003005

Galtung, J. (1996). *Peace by peaceful means: Peace and conflict, development and civilisation.* PRIO.

Laderach, J. P. (1997). *Building peace: Sustainable reconciliation in divided societies.* United States Institute of Peace Press.

Robertson, R. (1992). *Globalisation: Social theory and global culture theory, culture & society.* Sage Publications.

Thiessen, C. (2017). The evolution of conflict transformation: Theory and practice in Israel and the occupied Palestinian territory. In A. Özerdem, C. Thiessen, & M. Qassoum (Eds.), *Conflict transformation and the Palestinians: The dynamics of peace and justice under occupation* (pp. 3–19). Routledge.

UNESCO. (2020). *UNESCO: Building peace in the minds of men and women.* Retrieved from https://en.unesco.org/70years/building_peace

Wilson, E. O. (2012). *Is war inevitable?* Discover. Retrieved from https://www.discovermagazine.com/planet-earth/is-war-inevitable

Wohlmuther, C., & Wintersteiner, W. (2014). *International handbook on tourism and peace.* Drava.

WTTC. (2016). *Tourism as a driver of peace.* World Travel & Tourism Council.

Acknowledgment

We would like to acknowledge the assistance of all academics and researchers who contributed to the successful completion of the book, in particular, the authors and the colleagues who took part in the review process. Without their support, this publication would not have been possible to bring to light.

We also would like to express our sincere gratitude to Louis D'Amore for his insightful foreword.

Jorge Tavares da Silva
University of Aveiro, Portugal & University of Coimbra, Portugal

Zélia Breda
University of Aveiro, Portugal

Fabio Carbone
Coventry University, UK

Section 1

Tourism, World Peace, and Conflict Resolution

Chapter 1
Peace:
A Roadmap for Heritage and Tourism

Fabio Carbone
https://orcid.org/0000-0002-6711-4892
Coventry University, UK

Luiz Oosterbeek
https://orcid.org/0000-0003-3303-5958
Polytechnic Institute of Tomar, Portugal

ABSTRACT

On 25 September 2015, the United Nations General Assembly adopted an Agenda for Sustainable Development establishing 17 Sustainable Development Goals (SDGs) to be achieved by 2030, on which states, civil society, and the private sector are called to contribute. In this context, tourism, which is based on billions of encounters between people of diverse cultural backgrounds, is hailed as the industry that can foster tolerance and multicultural and interfaith understanding, laying the groundwork for more peaceful societies. This chapter critically analyses the current discourse on tourism and peace, exploring opportunities and limitations. The authors affirm the urgency for a more complex approach by informing the debate with concepts, theories, and ideas coming from areas different than tourism (cultural anthropology, history, cultural heritage, cultural diplomacy, peace and conflict studies, peace education and sustainability science). The authors focus mainly on the role of cultural heritage and its management as the central hinge in the association between tourism and peace.

INTRODUCTION

Strive,

at individual and collective level,

for a better understanding of the world surrounding us and

DOI: 10.4018/978-1-7998-5053-3.ch001

act without hesitation,

because History will not forgive

the apathy or inconsistency of the commitment to a better future.

This chapter critically presents the evolution of the academic and non-academic discourse relating to the conceptual association between tourism and peace and in particular tourism and peacebuilding. The first section provides a background of the work by adopting a complex approach. The authors highlight in this first section some current socio-cultural traits to be considered in the context of the study, as well as some contemporary geopolitical aspects. Afterwards, the very concept of peacebuilding is presented and, finally, the last subsection focuses on the state of the art of the debate on the relationship between tourism and peace, namely tourism and peacebuilding, by highlighting issues, controversies, theoretical and practical limitations of the current discourse.

The second section contains the main focus of the chapter and, in response to the need highlighted in the first part of the chapter to raise the level of complexity of the debate, the authors propose the integration of models and ideas from areas other than tourism. In particular, reference is made to studies on the management of cultural heritage and their social value. Cultural heritage and its management and its association with tourism are presented as the keystone in the conceptual architecture binding together tourism and peace.

BACKGROUND

This section presents the background of the present study by contextualising it in the socioeconomic and sociocultural current situation (at least with regards to Western society) and through some geopolitical considerations. Afterwards, the main theme is introduced with a brief note on the meaning of peacebuilding. For space limitations and to avoid going off-topic some very specific aspects such as the debate on liberal peace, its internal components and the often-ignored tensions between them (Richmond, 2006) are not included here. Once defined this broad context, the analysis of the current debate on the link between tourism and peacebuilding will represent the focus of the last subsection.

Current days present us with challenges that were unthinkable just a few decades ago. The utopia of a new world order in the name of a peaceful interdependence of peoples and nations – favoured by the rapid process of globalisation – has given rise to several and often conflicting needs. From a socioeconomic point of view, for instance, half of the world's 736 million impoverished people lived in just five countries, and 85% of them (629 million) in Sub-Saharan Africa (Figure 1).

Figure 1. Share of poor people in the world by region or country, 2015
Source: PovcalNet, the World Bank. Retrieved from https://www.worldbank.org/en/news/feature/2019/12/20/year-in-review-2019-in-charts.print

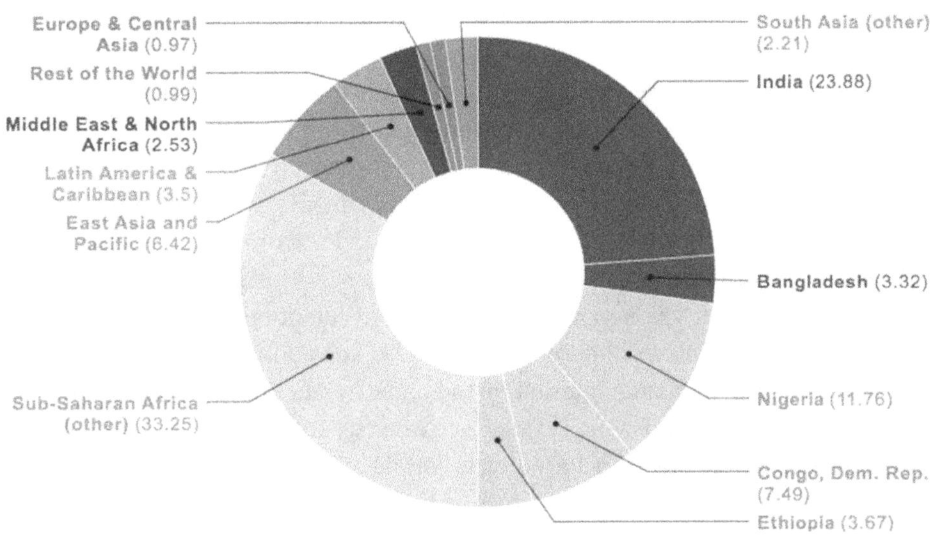

While inequalities at global level remain dramatic, cultural tensions also arise, as a result of growing mobility driven by various factors (from tourism to refuge) and unaccompanied by adequate monitoring and education for cultural diversity appreciation.

Figure 2. Refugee population by country or territory of origin (thousands)
Source: World Development Indicators (SM.POP.REFG.OR), the World Bank. Retrieved from https://www.worldbank.org/en/news/feature/2019/12/20/year-in-review-2019-in-charts.print

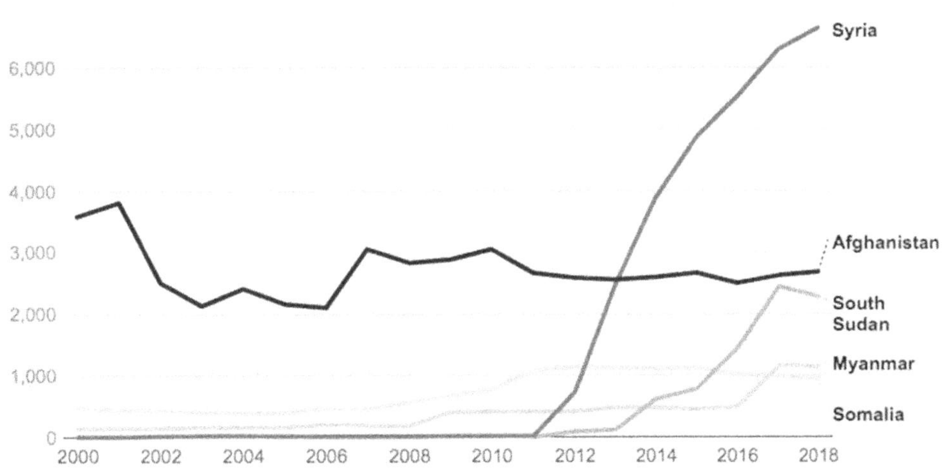

Furthermore, the number of people seeking international protection outside of their country of origin has increased by 70% since 2011. About 85% of refugees are hosted by developing countries. In 2018, 67% of refugees came from 5 countries: Syrian Arab Republic, Afghanistan, South Sudan, Myanmar, and Somalia (Figure 2). And more than 4.6 million people left Venezuela between 2016 and November 2019, primarily for Colombia, Peru, and Ecuador.

Many processes and concerns are still left on the table: is the Middle East moving towards new balances or new disasters? What results will bring the new confrontation between the United States, Russia and China? Which will be the evolution and impact of their respective internal cultural diversity? From a social point of view, what is going to happen to Western primacy, which after a devastating economic crisis is now being dramatically undermined by the repeated terrorist attacks that threaten its model of coexistence based on democracy, freedom, equality and social justice? Has reaction to the dramatic migratory phenomena of recent years made people aware of latent ethnocentrism and a weak sense of identity, leading toward potential xenophobia, racism? Has substantial apathy towards the suffering of others made all part of a post-humanity, guided by the banality of evil (Arendt, 2000)? In this sense, the emergence of a toxic political agenda – as defined by Amnesty International – made of propaganda and populism, is typical of our ages, and it is based on a simple and bold assumption to spread their hostile messages: the public does not care about the truth, what they want to hear is something that is simple and seems to solve their problems. It does not matter if it corresponds to the facts or not, if one just keeps saying it again and again and again, sooner or later, people will start to believe it. This dynamic is then related to a further paradox, that is, the incredible advances in technology and science on the one hand, and, on the other, the widespread functional illiteracy, emotional illiteracy and nihilism of Western youth? Is it a set of transitory phenomena or a structural reality?

The ground for fake news and tension building is the same: ignorance resulting from alienation. Whereas alienation was originally perceived as, primarily, the depriving of material resources, throughout last century it proved to be, above all, a depriving of knowledge and is becoming, in the dawn of the third millennium, a depriving of cognitive skills (Oosterbeek, 2019).

"We are viewers still unconscious and uncertain of a major historical breakthrough", says French sociologist Yves Mény (2019), and all the above aspects compose this age of uncertainty. How to deal with its challenges? This condition of uncertainty and instability should actually appeal to our greater responsibility and action. Striving for better analysis and a greater understanding of what surrounds us, linking the past, the present and the future should represent our first responsibility individually as well as collectively. Another misunderstanding typical of our times, an illusion, therefore a trap, is the confidence in the progress of techniques as being the main lever for human alleviation. While one cannot deny the important benefits brought about by technical advances in all fields, it would be necessary to pause to reflect on the fact that, in any case, a technique, in general, and technology, in particular, do not provide, by themselves, future horizons or perspectives to human beings, which require, instead, considerations on ethics and reflection. Techniques as an end in themselves are empty containers, with no possible meaning; this is their advantage, but also their key limitation: it is human agency, informed by cultural values, that assigns meaning to the use, or refusal, of techniques.

There is, thus, a call to rediscover the ethics and values behind human things: history, politics, coexistence between peoples. In this sense, we are convinced that at the base of all human matters, there must be the purpose of improving the condition of the human being himself, recovering the Kantian philosophy by which man, guided by culturally agreed values, must be the end of any action, instead of a mere means. This reflection leads us to a paradigm shift and a perspective that changes (or at least should change) each person's approach to any activity, professional and non-professional and at any level: local, regional, national or international and global. If all areas of human activities are involved, then, what is, in particular, our role as cultural heritage managers and tourism operators in creating a sustainable and peaceful future?

ONTOLOGICAL NOTES: THE VERY NATURE OF PEACEBUILDING

It is good (or at least it would be) to look with a certain scepticism at any practical or theoretical work that concerns peace and/or peacebuilding in which a brief clarification does not appear on what those terms actually mean. In the present case, the authors of this chapter took care of introducing the concept of peacebuilding with the aim to clearly identify where tourism can be conceptually connected with the process of construction of peace.

The theoretical and, subsequently, practical development of the concept of peacebuilding took place, as often happens, thanks to an intense intellectual/scientific debate on a revisited meaning of peace – to be contextualised mostly in the cold-war period (1947-1991) – and to the resulting reflections in the creation of supranational procedures and recommendations. United Nations (UN) officially adopted the term *peacebuilding* in 1992, when the Secretary-General Boutros Boutros-Ghali released a report entitled "An Agenda for Peace: Preventive Diplomacy, Peacemaking and Peace-keeping" (Boutros-Ghali, 1992). Section II of the report, paragraph 20, provides the definitions of the first three concepts:

- (Preventive diplomacy, seeking to resolve disputes before violence breaks outs);
- Peace-making, the "action to bring hostile parties to agreement";
- Peacekeeping, the operation of "deployment of a United Nations presence in the field, hitherto with the consent of all the parties concerned, normally involving United Nations military and/or police personnel and frequently civilians as well".

But the same author added a new, complementary approach to achieving peace, in paragraph 21, by arguing the need for:

- Post-conflict peacebuilding, the "action to identify and support structures which will tend to strengthen and solidify peace in order to avoid a relapse into conflict" (Boutros-Ghali, 1992).

As mentioned, this innovation was firmly based on previous intellectual, scientific debates. It is indeed important to take into account the considerations made by Galtung (1976) years before the elaboration of the report. Peacemaking and peacekeeping operations are required to halt conflicts and preserve peace once it is attained, but they are based on a *dissociative* approach, which seeks to keep antagonists separate - with the help of third parties if necessary.

On the contrary, based on the understanding of peace as a structure, Galtung presents peacebuilding as based on an *associative* approach. Peacebuilding indeed seeks to achieve peace by identifying the structures "that remove [the] causes of wars and offer alternatives to war in situations where war might occur" (Galtung, 1976, p. 298). Observe the reader that the most characteristic aspect of the concept of peacebuilding is the idea of building peace, and not just maintaining or preserving it, in order to avoid the recurrence of new conflicts after the end of armed conflict. And it is exactly this conceptual and practical aspect the associative approach - that allows establishing a link between tourism and peace. However, it is legitimate at this point to wonder how tourism becomes one of the instruments for building peace?

TOURISM AND PEACEBUILDING. LIMITATIONS AND POTENTIALS OF THE CURRENT DEBATE

Tourism is considered a global force for good, as it provides societies with opportunities for boosting economies, societies and cultures worldwide. The sector is responsible for 1 in 10 jobs worldwide, 10.4% of the global GPD and 30% of services. According to the major trend among tourism scholars' discourse, beyond the economic implications, tourism has positive repercussions on societies as it leads to the direct encounter of different peoples and cultures. UNWTO says tourism "can consolidate peace in post-conflict societies" (UNWTO, 2020).

Although these assumptions may occur, such an assessment seems to us to be insufficient. On the one hand, tourism also entails the danger of aggravating cultural tensions and triggering conflicts, namely when it is run mainly as an extractive business, with no care for residents' cultural habits and traditions. The focus on "good" as a primarily economic and social gain is a naïf approach, since goodness is always a result of valuing, i.e., is anchored on values, attitudes and agency. On the other hand, the main positive impact of tourism for preventing conflicts and building peace is that, among all humans, tourists tend to be better equipped not only to recognise but to appreciate cultural diversity, and to actually aim at preserving it. In this sense, the interaction between tourists and residents tends to be a real process of co-construction of a common ground of cultural relations within a new sociocultural matrix (Oosterbeek, 2017), beyond the soliloquies of ethnocentric communities.

On the other hand, one of the main scholars in peace and conflict studies, John Paul Lederach, makes the following consideration: "Peacebuilding is a complex task. It is [...] an overwhelming challenge. How, really, do we get whole societies wrapped in histories of violence that date back generations to move toward a newly defined horizon?" (Lederach, 2005, p. 31). So how can we associate the idea of tourist activity effectively with such a complex concept as that of peacebuilding?

Several world leaders – John F. Kennedy, Ghandi, Bill Clinton, among others – have recognised the importance of the travel and tourism industry as a driver of peace, security and understanding. In 1988, even anticipating the concept of sustainability launched a few years later by the United Nations, the concept of "Tourism: industry of Peace" was introduced by Louis D'Amore (1988), setting the stage for the creation of the International Institute for Peace through Tourism (IIPT). Soon an intense debate also started at the academic level. In an article published by the Annals of Tourism Research at the end of the '90, Smith (1998) talked about *post-conflict tourism*, focusing on tourist activities after the Second World War. In this article, he states that "World War II was a landmark in tourism: it created technology for rapid and efficient global mobility; generated the motivations for mass tourism; and, because of the human horror of nuclear holocausts, it stimulated new policies that stimulated contact between peoples, also through tourism " (Smith, 1998, p. 204). Since then, the topic has been approached from different perspectives, such as transitional justice and the (re) construction of the State (Novelli, Morgan, & Nibigira, 2012), the commodification for tourist purposes of former war sites (Causevic & Lynch, 2011; Packer, Ballantyne, & Uzzell, 2019), (re)construction of contested national identities through post-conflict tourism (Zhang, Xiao, Morgan, & Ly, 2018; J. J. Zhang, 2017), and even the process of creating the tourist destination brand after an armed conflict (Shirley, 2018). On the other side, Selwitz (1996) explores the benefits of peace for tourism and hospitality.

Shortly some ideas took hold, such as:

- tourism contributes to more peaceful societies;
- countries with more sustainable tourism are more likely to enjoy higher levels of positive peace in the future, as the more sustainable the country's tourism sector, the lower the level of violence;
- in conflict-affected countries, tourism represents a vehicle for socioeconomic recovery (Burnett, 1990), having a high level of resilience to violence and conflicts, even in the presence of terrorist attacks directed at tourists.

Based on these convictions, and within the scope of the United Nations Sustainable Development Goals (SDGs), the World Tourism Organisation (UNWTO, 2020) declares:

tourism can make a significant contribution, especially in the context of Goal 16 - JUSTICE, PEACE AND STRONG INSTITUTIONS. Tourism is based on billions of meetings between people from different cultural backgrounds. As such, the sector can promote tolerance and multicultural and interfaith understanding, laying the foundation for more peaceful societies. Tourism, which benefits and involves local communities, can also consolidate peace in post-conflict societies.

Figure 3. SDG, Objective 16
Source: WTO. Retrieved April 1 2019, from: http://tourism4sdgs.org/sdg-16-peace-justice-institutions/

More sceptical authors expressed their reservations about the real capacity of tourism to create peace as early as the late eighties (Brown, 1989) affirming that it is rather a tourism that benefits from peace (Litvin, 1998). These approaches, however, had no followers or follow-up. It is instead necessary to make important considerations in the light of the existing literature. While it is true that both international agencies and academics have given important inputs to the debate on tourism as a vehicle of peace so far, on the other hand, this debate should be reviewed today and raised to a higher level of complexity. Without any doubt the first inputs to the debate are still to be considered as particularly valid intuitions. Nonetheless, they appear somewhat naive nowadays. If it is true that even the very idea of *peace* is not unanimous, and "where there should be research agendas there are silences and assumptions" (Richmond, 2005, p. 6) a research agenda should also be established in the context of the association between tourism and peace. Deeper and theoretically more consistent integration between disciplines (tourism studies, cultural studies, peace and conflict studies and international relations, among others) would be the first step towards more solid arguments (and theoretical models) within the debate on tourism and peace.

In particular, the authors of the present chapter aim to investigate the link between cultural heritage and tourism development in promoting a culture of peace (Carbone, 2018a, 2018b, 2019) by articulating the field of tourism with theories from areas such as cultural heritage management and heritage studies, peace and conflict studies, cultural diplomacy and international relations.

CULTURAL HERITAGE AND TOURISM. THE COMMON PATH TO PEACE

The concept of tourism as a peacebuilding activity is fascinating. However, as aforementioned, a deep revision of the current discourse should be made, taking steps towards the emancipation from the initial intuition, developing the discourse and taking it to other levels of complexity. For instance, the idea itself of "peace industry" should be, in our opinion, reviewed. An *industry* is an entity that produces something in series, and it does not seem to be the correct approach when talking about the creation of peace through tourism. It is not just a rhetorical exercise to propose to change the term "industry with craftmanship of peace". "Blessed are the artisans of peace", wrote the apostle Matthew (5.3). Craftsmanship presupposes endeavour, commitment ability and skills (all ideas that the term *industry* does not suggest). Peace is indeed a construction that must be modelled. It does not exist in nature. Here, then, is the implication of this first consideration, which at first glance is purely rhetorical. Creating a culture of peace requires endeavour, ability, originality and the wise use of different arts and abilities. What is the role of cultural heritage in this context?

Cultural Heritage in Today's Society

There is a widespread understanding that the role of cultural heritage (and subsequently that of cultural institutions such as museums) has changed, at least partially. This understanding has led some to consider that the nature of museums must change, standing in the sphere of information dissemination in line with new emerging societal values, using essentially new media and attracting more audiences. Understanding needs is a permanent negotiation between contexts perceived in the short term and integrated trends in the long term.

The difference between contexts and trends is certainly in the value attributed to each one. But a bigger difference lies in the fact that contexts imply values that immediately and effectively condition human options (very restricted to reactive actions). At the same time, trends integrate values built in the long run and, therefore, more strongly conditioned by human views of the future, more or less informed by their understanding of the past. Over the past three centuries, this duality has produced divergent approaches to adaptive relationships between human beings and the environment, emphasising human will and agency (as in Nietzsche or Marx) or highlighting environmental or other extra-human restrictions (as in Hegel or Darwin). In any of the aforementioned approaches, the keyword is "process", that is, the "envelope" in which needs are expressed and human strategies are executed.

The "envelope" that led to the current context of perception of the "urgency of reforms" and the "climate emergency" started to be designed in the post-World War II, through a combination of two main factors: the scientific advances that, from the 1950s onwards, led to optimism in a "problem-oriented strategy"; and the social and economic transformations that, from the mid-1960s onwards, raised growing doubts about the capacity of existing political management models to deal with the newly perceived needs, related to the expansion of individual rights.

The past seven decades have therefore expressed a peculiar trend, which has tended to structure a long-term process not through a set of long-term global goals, or utopias, but through a series of unique short-term goals. This trend is interesting, since, despite the growing anxiety due to the incapacity of a problem-solving agenda to cope with human dilemmas, public policies and strategies pursued in the same road, reinforcing concerns and disputes (Oosterbeek, 2016).

The virtualisation of behaviour, the expansion of short-term agendas, ageing and the shift to market-oriented "STEM" education (Science, Technology, Engineering and Mathematics) are components of this same trend, interdependent and feeding each other in a destructive "great acceleration". Public debate and dispute tend to be replaced by totalitarian, "well-intentioned" and simplified views, fuelled by "unique perspectives", with no room for reflection and interaction. The reaction against the collectivist totalitarianism of the first half of the last century, generated a kaleidoscope of singular agendas, each trying to impose itself on the others. Expressions of this process are the crisis of nation-states (which, after a cycle of confederative inclination, seem to give way to networks of cities, opening up a new space for insecurity among them), the naive pulverisation of identity discourses (which feed xenophobic drives) or the clamour for emergencies (simplifying the contradictions between them).

The notion of heritage has expanded, now integrating entire landscapes, to accompany this spread of interests. However, it can consolidate itself only as an expression of this pulverisation (and, in this perspective, assuming as a product intended for specific consumer markets) or assume itself in the secular logic and formation of elites (which, in a democratic society implies the global opening to society), not treating its users as customers of a product, but as potential leaders who must learn to co-construct interpretive mechanisms and complex action.

"Archaeology is partly the discovery of the treasures of the past, partly the meticulous work of the scientific analyst, partly the exercise of creative imagination" (Renfrew & Bahn, 2012). This fascinating definition does not reflect, however – at least not explicitly – the complexity of the role nowadays attributed to cultural heritage within society and the shared commitment to creating a sustainable and peaceful future. In this sense, efforts are being made by an increasing number of academics to promote a new role of cultural heritage within society, as demonstrated for instance by the works presented at the XVI World Congress of the International Union of Prehistoric and Protohistoric Sciences titled Archaeology, Societies and Environments in Africa (Oosterbeek, Camara, & Martins, 2014). On their side, supranational institutions are going in the same direction, and the "Faro Convention on the value of cultural heritage for society" is one of the most recent documents aimed at revising the traditional paradigm on the value of cultural heritage, including archaeology. This Convention is based on the following ideas:

- knowledge and the fruition of cultural heritage are part of the citizen's right to participate in cultural life, as defined in the Universal Declaration of Human Rights;
- cultural heritage is a resource for human development, valuing cultural diversity and promoting intercultural dialogue
- cultural assets are part of a model of economic development based on the principles of sustainable use of resources (Council-of-Europe, 2005).

Nevertheless, the association between cultural heritage and peace, the potential contribution that cultural heritage could bring to peacebuilding, was largely ignored (Walters, Laven, & Davis, 2017). But, literature is emerging in this direction, although not without controversy, about the new responsibilities attributed to cultural heritage professionals and their "emerging role" (Stone, 2017) in the promotion of intercultural dialogue for peace.

The Emerging Trend: Bridging Heritage, Tourism and Peace

The 8th International Conference on Peace Museums "The role of peace museums in preventing war and promoting remembrance, historical truth and reconciliation", held in No Gun Ri, (Korea) in 2014, and convened under the umbrella jointly by the International Network of Museums for Peace and the International Peace Foundation of the NGR, is representative of this new trend whereby supranational organisations such as UNESCO invite curators and cultural heritage managers to embrace this new challenge. In this perspective, ICOM's attempt to change the definition of a museum is also very recent, proposing a description that better reflects the responsibility to promote intercultural dialogue and global understanding. In turn, the failure of this attempt made during the last ICOM conference, in Kyoto in August 2019, represents the extremely controversial nature of these new conceptual proposals.

In line with this emerging trend, a few years ago a conceptual model was formulated, called "Paideia approach to cultural heritage management for the development of tourism and local communities" (Carbone, 2010; Carbone, Oosterbeek, & Costa, 2012). This model established for the first time a link between cultural heritage management and tourism as a driver of peace, also including the capacity of enhancing the social value of cultural assets among the dimensions of quality in cultural heritage management (Carbone, 2016; Carbone, Oosterbeek, & Costa, 2013). The arguments underlying the discourse on tourism, cultural heritage and peace are, thus, better defined, more concrete and visible. The initial simplicity - although characteristic of a great intuition - that characterised the origin of the debate on "Tourism and Peace", thus leaves room for greater complexity and concreteness. The process whereby tourism could actually contribute to global understanding and peace is presented in Figure 4.

Figure 4. Dynamics inherent in the association between cultural heritage management and tourism development in view of the effective contribution to human development, intercultural dialogue and the culture of peace.

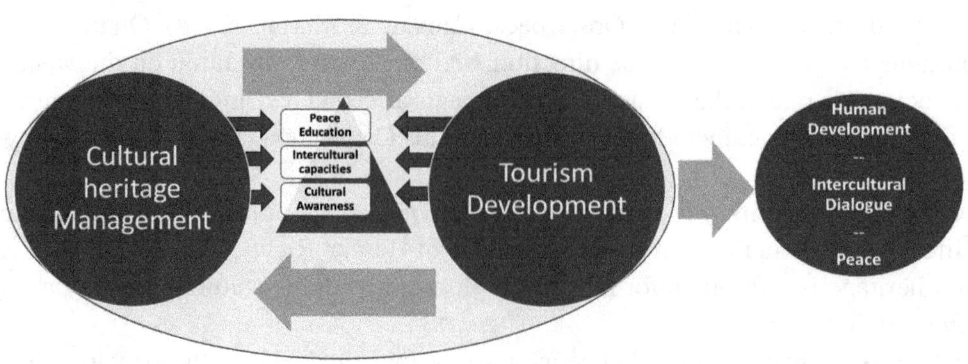

Figure 4 displays the key elements and phases. The association between cultural heritage management and tourism development is the starting point of the process. This association goes beyond the traditional, neoliberal view of cultural heritage as a mere component of the tourism product and supply. On the contrary, the cultural asset becomes a resource and vehicle for local development based on socioeconomic development and territorial identity and branding, as well as experiential learning and education (Timothy, 2018). In particular, cultural heritage management in the context of its association with tourism development aims, in this conceptual proposal, to boost cultural awareness and promote intercultural capacities. In this sense, the "paideia approach to cultural heritage management and tourism development" (Carbone et al., 2013) represents the keystone of the conceptual architecture binding together tourism and peace. Such an approach, is inspired by Socratic philosophical ideal of Paideia (5th century BC), related to one of the most famous Oracle of Delphi's mottos, thus adopted by Socrates: *know thyself!* (γνῶθι σεαutόν). The "self-knowledge" is the (pre)condition for each individual to establish a peaceful relation with the *other*. From the application of this concept to cultural heritage management, as shown in the figure (left), the objective of the association between cultural heritage and tourism is a process of sociocultural based on pedagogical practices aiming at:

1. The promotion of cultural awareness (*know thyself!*), by promoting greater public participation, including through a stronger connection with formal, informal and non-formal education;
2. The promotion of intercultural capacities and the debate on inclusion and pluralism, to be put into practice through the opportunities of contact with other cultures, represented also (but not exclusively) by tourists.
3. Cultural awareness and intercultural capacities converge thus into peace education practices.

In this sense, Cultural heritage and its association with tourism should have an important role, particularly within peace education. A considerable development in the field of peace education – that caters early child care to adult – reflects growth in social concern of this past century about horrific forms of violence – ecocide, genocide, modern warfare, ethnic hatred, racism, sexual abuse and domestic violence – and according to Harris (2004) it is mainly based on five aspects: (i) international education, (ii) human rights education,(iii) development education, (iv) environmental education, and (v) conflict resolution education.

Cultural heritage and its association with tourism can thus make an excellent contribution to further development of peace education, by pursuing a process of co-construction, in line with UNESCO's guidelines on sustainability science. Heritage sites not only effectively become "space for 'social complicity' and meaning construction" (Valecillo, 2009), but also the very place where the link between tourism and peace materialise. The process so far exposed, strongly based on the proposal by Carbone et al. (2013), ideally trigs the second part of the model (figure 4, on the right), by representing the root for human development, intercultural dialogue and peace. Ideally, if this process would be applied to all destinations, it would trigger a virtuous circle, by sustaining a complex chain of events that reinforce themselves through a feedback loop (Carbone, 2017). Heritage and tourism thus converge in a common path, taking part of that comprehensive, integrative, and strategic approach to the creation of a culture of peace.

CONCLUSION

This chapter contains a series of reflections on the current debate on the possible contribution of tourism to the creation and promotion of a culture of peace. Despite the optimistic - and sometimes naive - impulses that characterised the origins of the debate, the modalities and conditions necessary for tourism to become an effective vehicle for peace are still unclear and the debate remains vague. Empirical evidence contradicts - at least partially - the belief according to which the mere contact between tourists and the local population automatically result in a greater mutual understanding.

A new, advanced approach, in theory, and in practice, is therefore needed. In theory, it is necessary to raise the level of complexity of the discourse concerning the association between tourism and the creation of peace. From a practical point of view, it is necessary to implement this complexity, considering that a greater mutual understanding and the effective promotion of intercultural dialogue through tourism necessarily require preparation of both tourists and residents, based on the strengthens of cultural awareness, on the one hand, and the reinforcement of intercultural capacities, on the other, as well as on a reviewed setting of the destination and the entire tourist experience, more focused on the sociocultural and human factor and less on the merely economic aspect.

In sum, for tourism to effectively being a vehicle of peace, some precondition exists. Firstly, it is necessary to free tourism from the neoliberal economicist yoke, by valuing the sociocultural aspects. In this sense, another precondition for tourism to lead to intercultural dialogue and an effective contribution to peaceful societies, it is the reconsideration of cultural heritage in the context of tourism development, that is, consider it no longer as a mere component of the tourist supply but a key element for sociocultural growth towards human development and peace. A conceptual approach that can be put in practice by becoming part of the standard operating procedures within the process of planning, development and management of each tourist destination. In this process heritage and tourism thus converge in a common path, taking part of that comprehensive, integrative, and strategic approach to the creation of peace.

REFERENCES

Arendt, H. (2000). *Eichmann em Jerusalém*. Companhia das Letras.

Boutros-Ghali, B. (1992). *An agenda for peace: Preventive diplomacy, peacemaking and peacekeeping*. United Nations-Security Council. doi:10.1177/004711789201100302

Brown, F. (1989). Is tourism really a peacemaker? *Tourism Management*, *10*(4), 270–271. doi:10.1016/0261-5177(89)90001-0

Burnett, G. W. (1990). On the nature of peace in relationship to tourism: Three cases. *Tourism Review*, *45*(1), 2–7.

Carbone, F. (2010). Valorização turística do património arqueológico em prol da sociedade: O Paideia approach. *Boletim ICOM*, *2*(11), 2–10.

Carbone, F. (2016). Cultural heritage quality management: Analysis of archaeological heritage managers' perception. *European Journal of Tourism Research*, *14*, 114–118.

Carbone, F. (2017). International tourism and cultural diplomacy: A new conceptual approach towards global mutual understanding and peace through tourism. *Tourism. An International Interdisciplinary Journal*, *65*(1), 61–74.

Carbone, F. (2018a). A highest purpose of archaeology: Towards a future worthy of our children. *Global Journal of Archaeology and Antropology*, *6*(4). Advance online publication. doi:10.19080/GJAA.2018.06.555693

Carbone, F. (2018b). Post-multicultural challenges for cultural heritage managers and museums in the age of migrations. *Museum Management and Curatorship*, 1–22.

Carbone, F. (2019). Tourism and peace: a tool for global integration and peace. In L. Oosterbeek & L. Caron (Eds.), *Resilience and Transformation in the territories of low demographic density: integrated methodologies of human and social sciences for integrated cultural landscape management* (pp. 149–165). Instituto Terra e Memoria.

Carbone, F., Oosterbeek, L., & Costa, C. (2012). Paideia approach: Bridging tourism and community development on the basis of heritage enhancement. *Journal of Tourism and Development*, *17/18*, 165–166.

Carbone, F., Oosterbeek, L., & Costa, C. (2013). Paideia approach for heritage management: The tourist enhancement of archaeological heritage on behalf of local communities. *Pasos - Journal of Tourism and Cultural Heritage, 2*(2), 285-295.

Causevic, S., & Lynch, P. (2011). Phoenix tourism: Post-conflict tourism role. *Annals of Tourism Research*, *38*(3), 780–800. doi:10.1016/j.annals.2010.12.004

Council-of-Europe. (2005). *Council of Europe Framework Convention on the Value of Cultural Heritage for Society*.

D'Amore, L. (1988). Tourism: The world's peace industry. *Journal of Travel Research*, *27*(1), 35–40. doi:10.1177/004728758802700107

Galtung, J. (1976). Three approaches to peace: Peacekeeping, peacemaking, and peacebuilding. In J. Galtung (Ed.), *peace, war and defence: Essays in peace research* (pp. 282–304). Christian Ejlers.

Harris, I. M. (2004). Peace education theory. *Journal of Peace Education*, *1*(1), 5–20. doi:10.1080/1740020032000178276

Lederach, J. P. (2005). *The moral imagination: The art and soul of building peace*. Oxford Press. doi:10.1093/0195174542.001.0001

Litvin, S. (1998). Tourism: The world's peace industry? *Journal of Travel Research*, *37*(1), 63–66. doi:10.1177/004728759803700108

Mény, Y. (2019). *Popolo ma non troppo. Il malinteso democratico*. Il Mulino.

Novelli, M., Morgan, N., & Nibigira, C. (2012). Tourism in a post-conflict situation of fragility. *Annals of Tourism Research*, *39*(3), 1446–1469. doi:10.1016/j.annals.2012.03.003

Oosterbeek, L. (2016). Becoming human. New approaches for uncertain times. In L. Oosterbeek, M. Quagliuolo, & L. Caron (Eds.), Sustainability dilemmas. Transdisciplinary contributions to integrated cultural landscape management (Vol. 38-39, pp. 85-107). ITM.

Oosterbeek, L. (2017). Kóios and Phoïbe: knowledge through sociocultural matrices, in the framework of cultural integrated landscape management and sustainability science. In L. Oosterbeek, B. Werlen, & L. Caron (Eds.), *Sociocultural matrices: Transdisciplinary contributions to integrated cultural landscape management* (pp. 45–64). ITM.

Oosterbeek, L. (2019). Persona: From unrest to adaptation. *Plastir,* (56), 1-15.

Oosterbeek, L., Camara, A., & Martins, C. (2014). *Archaeology, societies and environments in Africa.* In *Proceedings of the XVI World Congress of the International Union of Prehistoric and Protohistoric Sciences.* Oxford: Archaeopress. 10.30861/9781407312972

Packer, J., Ballantyne, R., & Uzzell, D. (2019). Interpreting war heritage: Impacts of Anzac museum and battlefield visits on Australians' understanding of national identity. *Annals of Tourism Research, 76*, 105–116. doi:10.1016/j.annals.2019.03.012

Renfrew, C., & Bahn, P. (2012). *Archaeology. Theories, methods and practice* (6th ed.). Thames & Hudson.

Richmond, O. P. (2005). *The transformation of peace*. Palgrave. doi:10.1057/9780230505070

Richmond, O. P. (2006). The problem of peace: Understanding the 'liberal peace'. *Conflict Security and Development*, *6*(3), 291–314. doi:10.1080/14678800600933480

Selwitz, R. (1996). Peace brings tourists to Northern Ireland. *Hotel and Motel Management, 211*(7), 8, 40.

Shirley, G. (2018). The branding of post-conflict tourism destinations: Theoretical reflections and case studies. In E. Wylie, N. Andreas, & G. Jesse Hession (Eds.), *The tourism–disaster–conflict nexus* (Vol. 19, pp. 119–139). Emerald Publishing Limited. doi:10.1108/S2040-726220180000019007

Smith, V. L. (1998). War and tourism. An American ethnography. *Annals of Tourism Research, 25*(1), 202–227. doi:10.1016/S0160-7383(97)00086-8

Stone, P. (2017). Preface. In D. Laven, D. Walters, & P. Davis (Eds.), Heritage and peacebuilding (pp. xii-xiv). Boydell & Brewer.

Timothy, D. J. (2018). Making sense of heritage tourism: Research trends in a maturing field of study. *Tourism Management Perspectives*, *25*, 177–180. doi:10.1016/j.tmp.2017.11.018

UNWTO. (2020). *SDG 16 – Peace justice and strong institutions*. Retrieved April 1, 2010, from http://tourism4sdgs.org/sdg-16-peace-justice-institutions/

Valecillo, Z. (2009). ¿Cómo acercar los bienes patrimoniales a los ciudadanos? Educación Patrimonial, un campo emergente en la gestión del patrimonio cultural. *PASOS: Revista de Turismo y Patrimonio Cultural, 7.*

Walters, D., Laven, D., & Davis, P. (2017). Introduction. In D. Laven, D. Walters, & P. Davis (Eds.), Heritage and peacebuilding (pp. 1-4). Boydell & Brewer.

Zhang, C. X., Xiao, H., Morgan, N., & Ly, T. P. (2018). Politics of memories: Identity construction in museums. *Annals of Tourism Research*, *73*, 116–130. doi:10.1016/j.annals.2018.09.011

Zhang, J. J. (2017). Rethinking 'heritage' in post-conflict tourism. *Annals of Tourism Research*, *66*(Supplement C), 194–196. doi:10.1016/j.annals.2017.06.005

Chapter 2
Prospects of Tourism for Peace:
Is Tourism Part of the Problem or the Solution?

Maximiliano E. Korstanje
https://orcid.org/0000-0002-5149-1669
University of Palermo, Argentina

ABSTRACT

Over the years, philosophers have debated the problem of war and peace. Is violence inherently enrooted in our minds? Or simply are we educated in a violent society? Is war a social malady to be eradicated or part of our nature? The myth of Cain and Abel gives some hints on this, but, of course, it does not suffice to explain the complexity of the conflict in the societal order. Humans often are slaves of negative emotions such as rage, fear, greed, and envy. Nonetheless, as Immanuel Kant imagined, a durable and perpetual peace can be internationally achieved, where a tacit agreement is convened among nations. In so doing, they should share a common-grounded constitution and be subject to the law of federation. The figure of hospitality is vital to weave an international pact of cooperation and non-aggression. Here one might question if tourism is part of this panacea, as Kant in his days envisaged.

INTRODUCTION

Over the years, philosophers debated the problem of war and peace. Is violence inherently enrooted in our mind, or are we simply educated in a violent society? Is war a social malady to be eradicated or part of our nature?

Some scholars have agreed that hospitality and peace are inextricably intertwined. One of the pioneering philosophers in focusing on the role of hospitality as an instrument of understanding was Immanuel Kant, who in his days, imagined, a durable and perpetual peace can be internationally achieved whether a tacit agreement is convened among nations. In so doing, they should share a common-grounded constitution and be subject to the law of federation. The figure of hospitality is vital to weave an international

DOI: 10.4018/978-1-7998-5053-3.ch002

pact of cooperation and no-aggression (Kant, 2015). Here one might question, is tourism part of this panacea that Kant envisaged in his days?

The dominant discourse within the constellations of tourism research punctuates that the tourism industry emerged in the days after the industrial revolution because of the combination of multiple and complex factors. The political and economic stability not only ensured the acceleration of technological breakthrough but also laid the foundations for the rise and expansion of tourism. In this context, peace was the precondition for the well-functioning of the tourist system (Moufakkir & Kelly, 2010; Sharpley, 2018). Not only peace but also democracy played a leading role in the configuration and consolidation of the tourism industry worldwide (Sandell, 2005; Burns & Novelli, 2007). It is safe to say, at least for these theorists, that undemocratic governments face serious limitations and challenges to promote tourism simply because the necessary economic conditions, which are inherited in the global capitalism, are not given. In this vein, tourism, democracy and capitalism seem to be inevitably entwined. Sasha Pack in a seminal book not only contradicted this thesis but explained clearly how undemocratic governments such Franco's Spain or Nazi Germany have successfully developed the tourism industry to boost the domestic consumption. Centred on the case of Spain, Pack holds the thesis that tourism has historically and successfully flourished in undemocratic conditions or post-conflict destinations. Pack does not deny that tourism needs stable and peaceful conditions, but her well-documented case shows clearly how tourism can be assertively promoted by bloody dictatorships. This moot point begs some pungent questions such as is tourism an efficient instrument to promote peace?. What is the evidence that sustains this thesis? Is tourism part of the problem or the solution to placate violence?

The present chapter pursues four main goals. Firstly it interrogates on the nature of peace in its connection in what Elias dubbed as a civilizatory process. In a second stage, the section reviews publications, where some voices have recently alerted on the complex nature of tourism which under some conditions far from promoting peace, accelerates a process of rivalry and conflict. In third, the specialised literature that discusses post-conflict destinations, as well as the needs of stimulating a dark consumption is widely assessed in the sections of this chapter. Last but not least, some remarks are posed revolving around the agenda for the years to come. This section aims at answering to what extent tourism will be work as a mechanism towards peace or conflict.

PRELIMINARY DEBATE

The fact is that, after all, peace played a crucial role within the constellations of sociology –as well as other social sciences- and government. At a closer look, the founding parents of sociology and anthropology believed that the process of civilisation brought the man from the state of nature back. In so doing, he overly renounced to his atavistic drive to violence. In Hobbesian terms, when lay-people abandons their wildness, law and order are widely embraced and respected. In the eighteenth and nineteenth centuries, this ideological position led European writers to think that aborigines were living in a state of nature while the (European) civilising process should domesticate the native disposing them from their bloody past (Hamowy, 1990; Ivison, 2004). The Hobbesian theory was particularly important after WWII ended, where millions lost their lives. One of the first authoritative voices in questioning such an assertion, doubtless, was Norbert Elias who innovatively acknowledged that the cycles of peace and warfare are ambiguously repeated, fluctuated and accumulated in the line of history. In his influential book "The Civilising Process", Elias (1978) toys with the belief that violence was an essential part of medieval civilisation and its passage to the industrial society. The shame and repugnance, disposed by the rule of etiquette, was imposed to a deep sentiment of self-restraint in order for the self to be finally domesticated (Elias, 1978); but far from disappearing violence is still encapsulated in the core of modern civilisation. The archetype of "the noble savage", which was questioned by scholars like Hobbes, Rousseau or Locke, became not only conducive to the colonial exploitation but also created a false image of the "non-western Other", which lasts to date. Still further, Miguel Angel Centeno proffers an interesting analysis of the influence of war in the efficiency in the economic budget. Simply put, those nations which went through "total wars" (like WWI and WWII) developed more financial resources to control their budget, laying the foundations towards a stage of economic abundance. These examples apply for Japan, the US, England, France and Germany (only to name few). Rather, the Latin American economies which never took part in these events failed to regulate their domestic markets. As a result of this, they were pressed to bid for foreign loans while their economic situations gradually aggravated. To put the same in bluntly, Centeno argues convincingly that war is at the foreground at the configuration of society (Centeno, 2002).

TOURISM AND PEACE

From its inception, the tourism industry was understood as an activity which not only promoted peace in the tourist-receiving countries but also coordinated economic resources to boost local economies (Crouch & Ritchie, 1999). However, this growth has some limits. At a first glimpse, tourism seems to be strictly sensitive to disruptive events as conflicts, riots, and even warfare. The security of tourists is a key factor in the preservation of tourism consumption. Nobody wants to travel to a destination where its sense of security is threatened (George, 2003). In some respect, scholars strongly believed that tourism opened the doors to a more competitive atmosphere that consolidates finally democratic institutions (Lea, 2006; Richards & Hall, 2003). What does happen with those countries where tourism has not flourished?

Over time, some voices questioned the fact that development and tourism would be directly proportional. As de Kadt (1979), puts it, some under-developing nations not only failed to adopt tourism as a primary source of profits but also even aggravated their economic situation once development-program failed. This happened simply because those countries with a historical background based on slavery or interethnic conflict are culturally unable to develop a sustainable tourism industry (de Kadt, 1979). This assumption is far from being empirically validated but what it is important not to lose the sight of the fact that more than four decades elapsed after de Kadt´s publication, but his seminal book holds fair and actual even to date. There is a dearth of new publications which focus on tourism as a vehicle towards peace and institutional stability (Litvin, 1998, Var & Ap, 1998; D'Amore, 1998; Cho, 2007). All these authors concentrate efforts and time in debating to what extent tourism is a vital force for peace, or simply it needs from stable conditions to flourish and survive (Leslie, 1996). As Litvin (1998) puts it, tourism is often contemplated as a beneficiary vehicle towards peace even in conditions where the distribution of wealth is successfully achieved.

Moufakkir and Kelly (2010) call the attention to gather empirical study-cases to find a coherent answer to the question of conflict. Per their viewpoint, peace should be defined as something more complex than the absence of war. In context, a durable climate of deprivations or even worse, poverty gradually leads to war. The quest for peace and security seem to be invariably intertwined. However, here the problem becomes more acute. Since the notion of security is culturally negotiated and individually perceived, there are no clear indicators to affirm a country is secure while another is not. As noted, this slippery matter deserves attention and further discussion because the verifiable evidence is not conclusive. Other voices claim that tourism – far from being an instrument of peace – activates long-dormant prejudices and conflicts. In a seminal book “Ethnicity Inc”, Comaroff and Comaroff (2009) interrogate furtherly the nature of tourism in Africa as the precondition towards ethnic cleansing. As they hold, tourism revives historical conflicts when the nation-state imposes heavy taxes on historically-relegated ethnicities. Though tourism helps in poverty relief at a short-run, in the threshold of time it creates a situation of rivalry between stakeholders. Quite aside from this controversy, what this chapter looks to debate is the uncanny intersection of tourism and conflict. We start from the premise that the best ways to understand how tourism promotes peace are related to the study of conflict as a social phenomenon strongly associated with our human nature. To here we have debated preliminarily the role of tourism as an agent of transformation that boosts local economies while reaffirming the democratic institutions, but less attention was paid to understand in what way, that is, does tourism promote peace?

For some reason which is very hard to precise here, in tourism research, scholars adopted an economic-based viewpoint that led them to demonise conflict and war as social maladies and serious threats for the well-functioning of the tourism industry. For Barlow (1988), or Rovelstad (1988) (forerunners in the field), tourism showed to be an ambassador towards international understanding and –at the best- peace. To wit, Var and Ap (1998) place this hypothesis under the critical lens of scrutiny. They hold that there is little evidence that proves tourism instils or undermines the intra-societal levels of violence. To fill this gap, they conducted empirical research-oriented to explain the potential of tourism for a genuine process of peace and cooperation. The sample was designed by combining participants who work in the tourism industry and others from other industries. The outcomes are frustrating and evident in some respect (as an ecological fallacy) peace is stressed as the tug of war or at least the precondition for the rise of tourism by the interviewees. One of the problems of this investigation lies in the fact they echo what the professionals of tourism indeed believe though little evidence is given to test that tourism promotes stability, as they anticipate. In this vein, Pratt and Liu (2015) hold that those travellers in contact with other cultures are more open and develop more firm ties with locals. Particularly, this assumption suggests that tourism paves the ways for political stability which is a condition for peace. However, the current research appears to be descriptive and theoretical, leaving little space to empirical research. As they noted, in spite of the volume of publication, there is a serious dispute to what extent tourism promotes peace or conflict.

Methodologically speaking, Pratt and Liu (2015) employ the Global Peace Index (GPI), which is formed by 22 qualitative and quantitative indicators measuring not only the levels of perceived safety or security but also the degree of militarisation. They finally conclude that though tourism is important for peace, it is not determinant. Peace, anyway, seems to be more relevant for medium-income destinations. Here two problems come to us. On the one hand, the notion of peace or conflict is very broad to be empirically investigated, leaving the issue inexpugnable even for the well-trained eye. Although terms such as prejudice, racism, or simply tourist-phobia can be framed as valid variables for future research, no less true is that they share different causalities. On another, racism can be kept in secrecy or covert depending on the involving interests. Besides, the subordinated role of hosts respecting to guests impedes overt expressions of racism or xenophobia, at least in the world of work. Desk-staff can be seriously punished or dismissed whether they express hostility to their clients because of their religion, race or nationality. We have conducted early research to note that after more than 20 years, Chileans and Argentineans kept the same degree of hostility than days after the Beagle conflict. The findings revealed that though Argentinean workers – in the tourism sector – had a bad image of Chilean Tourists who visited the country favoured by a currency exchange after the end of convertibility system in 2002. Beyond the paradigm in the specialised literature, Argentineans do not express their prejudices and hostility against tourists, though qualitative research unveiled them. This suggests two important hypotheses. First and most important, some themes should be approached following qualitative methods, above all, when they are morally condemned. Secondly, the tourism industry – sometimes – instils violence and conflict even after more than three decades. The historical background in geopolitics predicts or explains part of the hostility between tourists and locals. In this token, Salazar (2006) did an important step affirming that war – not necessarily – represents a serious drawback for the expansion of tourism. The industry proved to be resilient, while conflict an important stimulus for the tourism recovery.

On the other hand, certain acts of terrorism are specifically targeted at tourism. In conflict zones, tourists can be targeted because they are viewed as ambassadors for their countries, as soft targets, and often because of their "symbolic value as indirect representatives of hostile or unsympathetic governments". In countries like Burma, tourism brings international recognition and fosters an illusion of peace while providing foreign exchange to pay for arms which strengthen the military junta. In cases such as this, tourism fortifies the undemocratic regime whose members may benefit personally and politically from any increase in arrivals. Although controversial, a tourism boycott is believed to help diminish the access to such rewards and erode the foundations of the government, advancing the necessary political changes to establish peace. (Salazar, 2006, pp. 325-326)

In sum, low-politics has few probabilities of developing a peaceful society through tourism than high-politics, as Salazar (2006) adheres. In its complexity, one might speculate that tourism appears to be a fertile ground to the competitiveness among stakeholders. The degree of conflict or rivalry depends exclusively on how these relations are founded as well as the history behind. The same remarks are found in one of his landmark books, *Envisioning Eden: mobilising imaginaries in tourism and beyond.* With a strong focus in the figure of conflict in Southeast Asia, Salazar starts from the premise the mobility theory shows some limitations in the ways the "Other" is represented – if not possessed. Based on a sentiment of ethnocentrism, Europeans enthrals the theory of mobilities as ideals lines of democracy and superiority over other cultures. To what extent we live in mobile cultures is one of the paradigms Salazar (2006) confronts. It is impossible to imagine tourism without violence simply because it recreates social relations once a border is crossed. Echoing the anthropological legacy which validated that conflict and multicultural landscape are inevitably entwined, Salazar deciphers that tourism makes grounds when tourists trespass borders creating discrepancies and rivalries while violence –through the articulation of stereotypes- restore the balance. This point begs some important questions we shall approach in the next section: is conflict inherently enrooted in the essence of tourism?, why is the industry under attacks of different risks such as climate change or terrorism?, is tourism a resilient activity?.

TOURISM IN POST-CONFLICT DESTINATIONS

With the benefits of hindsight, some specialists assert that the turn of the century opened the doorsteps to new global risks ranging from terrorism and political violence to natural disasters without mentioning the outbreaks of viruses. Without any doubt, the 9/11 inaugurated a new epoch of anxiety and fears where security and safety were key factors of global governance and sustainability (Korstanje & Olsen, 2011; Korstanje & Tarlow, 2012; Mansfeld & Pizam, 2006; Tarlow, 2014). Having this said, the notion of security in international tourist destinations has been radically shifted (Laws, 2005; Timothy, 2006) and situated as the urgency for policymakers and even governments. The theory of risk perception, which was originally coined in psychology four decades ago, has been adopted as the tug of war of tourism-management. This theory devoted efforts in identifying those psychological dangers that potentially may place the destination, or the tourist system, in jeopardy. Based on a preemptive logic, which means that specialists need to propose a rational model to prevent disasters, proponents of risk perception theory enumerated a set of fruitful steps to enhance security as well as reducing the level of conflicts at the involving destinations (Blake & Sinclair, 2003; Ritchie, 2004; Glaesser, 2006). It is noteworthy to mention that this academic wave – which surfaced since 2001 – had serious problems to explain the upsurge of a radicalised activity in developed countries as Belgium, France, England and the US. To some extent, just after 2012, the emergence of terrorist cells like ISIS or *Boko Haram* certainly prompted a serious crisis for theorists of risk perception and tourism-management. Let us remind readers that France was the epicentre of combined bloody attacks from January to November of 2015. All these coordinated blows targeted tourist destinations as well as leisure-hot spots. The same happened in other nations such as England, the US, Spain and Belgium, which went through different attacks in their territories, respectively. For this reason, risk perception theory and risk- management as sub-disciplines fell in discredit since 2012 (Seraphin, Butcher & Korstanje, 2017). What terrorism revealed seems to be the impossibility of the precautionary principle, all together to the quantitative methods, to predict potential threats and understand the complex nature of the conflict. Over recent years, some scholars started to question not only the failure of risk perception in making more secure tourist destinations but also the needs of forging an all-encompassing model that helps policymakers dealing with risks (Walters & Mair, 2012; Chew & Jahari, 2014; Portoff & Neal, 2014). In perspective, these academicians accepted conflict as part of society, and not as a radical malady to be effaced. In view of this, the precautionary logic has little opportunities to avoid the disaster lest by the articulation of coherent conceptual corpora that help in the mitigation of its effects for the local community. The *post-disaster and post-conflict destination theories*, in this way, emphasise the urgency to enhance the resiliency of tourist destinations (Seraphin, Korstanje & Gowreesunkar 2019). In a prefatory introduction to the book "Tourism and Hospitality in Conflict-ridden Destination", Isaac, Cakmak and Butler (2019) recognise that the difficult intersections between conflict and tourism lead scholars to interrogate to the dominant discourses in tourism fields mainly tended to ponder security and safety as unquestionable mandates.

Studying the complex intersections prevents the hegemonic discourses of global security for completing itself, stabilising its boundaries and securing a totalised presence. These hegemonic discourses create a separation between socio-political conflicts and tourism through recurrent reminders of the necessity of safety, and security in tourism. (Isaac, Cakmak, & Butler, 2019, p. 1).

From this citation, naturally, two questions arise. The question of whether conflict varies in the time says little about the accuracy behind the event. This means that there is a type of heritage of conflict which rememorises the event but does not explain its previous background. The causalities of disaster often tend to be misinterpreted, fabricated or even forgotten to protect the interests of politicians. This moot-point was particularly criticised by Joy Sather-Wagstaff and Rodanthi Tzanelli in their respective works. As Sather-Wagstaff (2016) observes, tragic events wake up a much deeper natural human solidarity, which in usual conditions transcends ideologies, borders and nationalities. With basis on the 9/11 and Ground Zero, she eloquently argues that the US received global support after the attacks that whipped New York, but it changed once the politicians ideologically manipulated the message. The archetype of dark heritage exhibits a fabricated story articulated to protect the interests of the status quo. Secondly, if tourism is a resilient industry, we run the risk to commoditise conflict as a form of attraction. When this happens, conflict perpetuates as a source of the attraction itself. Butler (2019) tries to give a clear answer to these two above questions. He distinguishes three clear-cut stages: pre-conflict tourism, during-conflict tourism and post-conflict tourism. For the sake of clarity, pre-conflict tourism starts from an imagined future where the danger has not taken a room. Precisely, the problem lies in the fact that, as he cites, "conflict is not always predictable, particularly in terms of the opening of hostilities between protagonists, and in some cases, tourists anticipating visiting a peaceful destination may get an unpleasant surprise if conflict breaks out while they are present" (Butler, 2019, p. 14). Instead, during-conflict destination signals the psychological motivations as well as the deliberate decision to travel towards destinations while the conflict is underway. This seems to be the case of war-tourism, a new segment where travellers experience the monstrosity of war in the battlefront. Finally, post-conflict destination corresponds with a decision of visiting a zone which has been devastated by wars or a disaster. Unlike the other two typologies, post-conflict destinations include the re-memorisation of past-events where human suffering occupied a central position. A clear example of this type associates to dark or thana-tourism, which often promotes the visit to medieval castles, demilitarise zones, or concentration camps (to new a few of them). The particular interests of dark tourists look to imagine their own death through the lens of others. Each tragedy leaves a moral lesson, and dark tourists want to hear it. Although Butler does not provide with a clear basis to what extent post-conflict destinations foster peace –because it is out of his scope-, no less true is that he sheds light on post-disaster consumption as an emerging horizon for research in peace studies. Complementarily, Tzanelli and Korstanje (2019) alert on the risks of consuming demilitarised areas – emulating an anti-terrorist fantasy. Hundreds of visitors choose Israel (the illegal West Bank Settlement) to train in demilitarised zones as the Israeli Defense Forces (IDF). The scheduled activities encompass stabbing attacks, a sniper tournament, and dog attacks giving a taste of what Israeli forces often experience. The idea of *being there* plays a leading role in the configuration of these types of leisure-spots. As Tzanelli and Korstanje (2019) write, West bank settlement often attracts people interested in military-tourism, a post-modern interplay between war and fantasy. Basically, these types of new consumptions emulate the long-dormant ideology of colonial oppression in the former centuries, which punctuate that after all atrocities the army forces are doing the correct to protect us:

Fantasy terror camps both obey the logic of domination and act as the apotheosis of consumer capitalism, turning the murder to fellow humans into an anodyne spectacle. As part of an infotainment industry specialising in blends of morbidity with patriotic values, terror camps dictate to tourists that they can have fun with impunity. One may argue that contemporary terrorism, tourism and the value of hospitality are structurally intertwined with histories of European colonisation. Historically, Europeans used tourism and the value of hospitality to legitimate expansion, by fashioning civilised travelers-colonists as protectors and educators of the noble savages.. Nowadays in globalised contexts of terrorist attacks, such structures of hegemony are reinterpreted, placing the law of hospitality in danger". (Tzanelli & Korstanje, 2019, p. 76)

The above-noted excerpt reminds two relevant things on post-conflict tourism. On the one hand, some ethical concerns should be placed as the main entry in this debate. The introduction of ethics in the post-conflict consumption assists policymakers to plan sustainable programs aimed at protecting the individual rights of locals. Otherwise, the local agent runs the risk to be commoditised as a part of the spectacle. Secondly, some doubts are cast to what extent post-conflict destinations create safer destinations while curbing conflict. Comaroff and Comaroff (2009) have conducted an auto-ethnography in the US and Africa, which is self-explanatory. During decades, aboriginal communities (organised in the forms of nations for the American case) have been historically pushed away from the promises and benefits of capitalism and democracy. As a result of this, aboriginal nations were fraught of social maladies as alcoholism, violence and poverty. In the 90s decade, the idea of adopting tourism for poverty relief seduced many of the aboriginal communities. Gradually, the industry of heritage becomes a source of profits for the capital-owners and a source of prosperity and employment for the community. As the previous situation of backwardness and pauperism is finally reversed, the nation-state decides to impose further taxes on the gains the community earns. Unilaterally or negotiated, this decision meets a great resistance in local community paving the ways for a tensional climate of conflict and discontent. Although the case of American aboriginal nations is pretty different to Africa, where ethnic cleansing and civil wars are every-day events, the tourism industry recreates a long-dormant climate of rivalry, which lest controlled, can usher the country into war. This was the case of some African nations where inter-ethnic disputes were never solved. What is more than important to discuss is to what extent tourism is the vehicle to peace, or sometimes its main obstacle (Comaroff & Comaroff, 2009).

CHALLENGES OF TOURISM FOR THE NEXT CENTURY

Despite the criticism exerted on the study of tourism as a promoter of peace, there is a dearth of examples, and works, where tourism succeeded in bringing stability to affected regions like Northern Ireland or Rwanda. As the previous backdrop, this section enumerates a set of real concerns and challenges tourism will go through in the next decades. As Pernecky (2010) infers, the philosophical dilemmas of tourism have been systematically overlooked by specialists of all pundits. Since they come primarily from managerial disciplines, they ignored the significant contribution to existentialism (particularly Martin Heidegger) to the role of tourism to placate social conflict. The term being in the world (dasein), which was proper of Heidegger, sets the pace to "The being of Tourism". To be more exact, we live in a world where natural beings circulate, for example, men, but this does not mean, these men are tourists since they born. The notion of what a tourist means is culturally and historically created and internalised. Even it changes according to time. Almost clear concepts as poverty, peace or injustice, as Pernecky (2010) adheres, are understood in different ways by different actors geographically located in different landscapes. This suggests that what it is experienced is previously determined by the biographical background of the observer. A Palestinian Tour guide, naturally, has its own conception of geography than a Dutch geography teacher. Their cosmologies are based on opposing experiences which lead to see the world in a different way. While travelling, we are being transformed by the "Other" who interpellate us. Although well-skilled to understand the reality, tourists interpret the visited landscapes according to their stereotypes, hopes and fears. Tourism, above all, engenders a relational logic between hosts and guests which not always ends in conflictive situations. The Cold War has divided the world into two. The capitalist nations were automatically fitted against the communist ones. Of course, the Korean Peninsula was not an exception. With basis on this, Kim and Prideaux (2003) examine the tensions between the South and North Korea and the role of tourism in the disarticulation of hostility. The Hyundai Corporation was legally allowed to build a tourist resort in North Korea, but for some reasons South Korean tourists remained indifferent to this resort. Although the project received considerable financial support, it has not been prospering because of political and ideological issues. Kim and Prideaux (2003) conclude that even if tourism can be considered a successful vehicle towards peace, in some conditions it triggers conflict. By this way, Korstanje and George (2012) investigated the case of Malvinas-Falkland an Islands archipelago in dispute between Great Britain and Argentina. Per their viewpoints, though Argentineans and their social imaginary claim sovereignty over the Islands they manifest no intention to travel there. At a first look, this does not happen because of the hostility of the isleños (kelpers) but also because Malvinas-Falkland is symbolically deemed as a sanctuary which should not be tainted by the tourism industry. These observations remind not only host-guest conflicts are given by micro-social interaction but also the importance of ideology as a push-pull factor. In this respect, for tourism to flourish, as David Leslie shows, political stability should be durable in the time (Leslie 1996). Causevic & Lynch (2013) conduct a participant observation to describe how the tourism industry enhances or undermines political stability in post-conflict destinations like Bosnia and Herzegovina (former Yugoslavia). They hold the thesis that governance and the political administration sometimes fail to achieve cooperation between divided communities, but no less true seems to be that in B&H the industry leads the ways in achieving cooperation in other economic sectors.

Last but not least, Movelli, Morgan and Nibigira (2012) lament that civil unrest seriously affects the tourist destination image while declining the economies of poorest nations. Merging the cosmologies of two westerners and one African researcher, they offer a multicultural answer to the problem of conflict in the Third World. Almost 60% of developing countries have experienced a type of civil war or conflict in their territories. In a situation of the fragility of this calibre tourism would be a valid toolkit to boost local economies. The nations which have developed considerable resources in improving their infrastructure have further opportunities and probabilities of developing tourism than those nations which have serious infrastructure deficiencies. Security and safety, in this vein, play a leading role in the formation of an agreement for peace between two communities or nations in dispute. Centred on the study case of Burundi, Novelli, Morgan and Nibigira (2012) accept that this country seems to be a fertile ground to understand the uncanny connection of tourism development and conflict since the country was subject to several civil wars over decades. The outcome cannot be extrapolated to other countries or context outside Burundi, but very well it serves to enhance the current understanding of the issue. With this in mind, it is important to appreciate the political obstacles towards the formulation of coherent plans and programs in tourism development. One of the aspects that impede tourism development associates to the lack of planning to organise communities, in view of the durable state of conflict or rivalry among districts. This leads invariably to the absence of a strategic program to stimulate tourism. The typical situation of fragility can be very well explained not only by the lack of interests in the community but also by the poor understanding of the opportunities of the sector by the side of politicians. The second element to take into consideration is the higher levels of uncertainty which are proper of post-conflict destinations. Without a clear vision and whipped by years of wars, the ways and transport infrastructure in Burundi still remains as a great obstacle to the adoption of tourism as a primary source of economic production. All these factors combined results in the lack of interests of foreign investors, or what is worse, the real-estate speculation in the case they are really interested. The wounds left by war – unless regulated by a transitional justice like the case of Rwanda – can be agitated even for the inclusion of the first tour operators in the country. This is in consonance with the original observations of Comaroff and Comaroff (2009). However, to some extent, tourism may be a potential instrument to create the necessary synergy to create a shared memory of the conflict, which helps in reducing conflict and rivalry. Needless to say, the challenges are many, but the opportunities to learn from the "Kantian peace" through the stability that tourism promotes should not be disregarded. Still further, the increase of terrorist attacks as well as other global risks such disasters is changing the nature of tourism. Today, many western tourists use virtual technology to visit places otherwise would be inexpugnable to them. However, this is a much deep-seated issue which needs further investigation.

CONCLUSION

The present chapter focused on the importance, challenges and advantage of tourism by providing the stable condition of reproduction that lead the community to a durable peace. In the first section, we preliminary debated the first steps of the sociology of conflict to interpret the correct reasons and historical background that dispose people to war and violence. In so doing, we mention the pro and cons in considering tourism as a vehicle towards peace and democracy. Complementarily, the second section explains the needs of adopting a new epistemological method to study post-conflict landscapes in an ever-globalised world. Finally, a collection of ideas and study-cases in good practices are shown to the readers to understand this slippery matter. Basically, though Comaroff and Comaroff´s (2009) thesis is correct respecting to the opportunities of tourism to pacify a region, no less true is that tourism should be considered a key piece in the orchestration of sustainable practices, policies and regulations that assist communities to recover after a civil war or, at best, to a state of a conflict prolonged in the threshold of time.

REFERENCES

Barlow, M. (1988). Tourism, peace, and conflict: A geographer's perspective. *Tourism-a Vital Force For Peace, 108*.

Blake, A., & Sinclair, M. T. (2003). Tourism crisis management: US response to September 11. *Annals of Tourism Research*, *30*(4), 813–832. doi:10.1016/S0160-7383(03)00056-2

Burns, P. M., & Novelli, M. (Eds.). (2007). *Tourism and politics*. Routledge. doi:10.4324/9780080467184

Butler, R. (2019). Tourism and conflict: A framework for examining risk versus satisfaction. In R. Isaac, E. Cakmak, & R. Butler (Eds.), *Tourism and hospitality in conflict-ridden destinations* (pp. 13–24). Routledge. doi:10.4324/9780429463235-2

Causevic, S., & Lynch, P. (2013). Political (in) stability and its influence on tourism development. *Tourism Management*, *34*, 145–157. doi:10.1016/j.tourman.2012.04.006

Centeno, M. A. (2002). *Blood and debt: War and the nation-state in Latin America*. Park University Press, Penn State Press.

Chew, E. Y. T., & Jahari, S. A. (2014). Destination image as a mediator between perceived risks and revisit intention: A case of post-disaster Japan. *Tourism Management, 40*, 382–393. doi:10.1016/j.tourman.2013.07.008

Cho, M. (2007). A re-examination of tourism and peace: The case of the Mt. Gumgang tourism development on the Korean Peninsula. *Tourism Management, 28*(2), 556–569. doi:10.1016/j.tourman.2006.04.019 PMID:32287720

Comaroff, J. L., & Comaroff, J. (2009). *Ethnicity, Inc.* University of Chicago Press. doi:10.7208/chicago/9780226114736.001.0001

Crouch, G. I., & Ritchie, J. B. (1999). Tourism, competitiveness, and societal prosperity. *Journal of Business Research, 44*(3), 137–152. doi:10.1016/S0148-2963(97)00196-3

D'Amore, L. J. (1988). Tourism: A vital force for peace. *Tourism Management, 9*(2), 151–154. doi:10.1016/0261-5177(88)90025-8

De Kadt, E. (1979). *Tourism: Passport to development: Perspectives on the social and cultural effects of tourism in developing countries*. World Bank Press.

Elias, N. (1978). *The civilising process* (Vol. 1). Blackwell.

George, R. (2003). Tourist's perceptions of safety and security while visiting Cape Town. *Tourism Management, 24*(5), 575–585. doi:10.1016/S0261-5177(03)00003-7

Glaesser, D. (2006). *Crisis management in the tourism industry*. Routledge. doi:10.4324/9780080464596

Hamowy, R. (1990). Cato's Letters, John Locke, and the republican paradigm. *History of Political Thought, 11*(2), 273–294.

Isaac, R., Cakmak, E., & Butler, R. (2019). Introduction. In R. Isaac, E. Cakmak, & R. Butler (Eds.), *Tourism and hospitality in conflict-ridden destinations* (pp. 11–12). Routledge. doi:10.4324/9780429463235-1

Ivison, D. (2004). Locke, liberalism and empire. In *The Philosophy of John Locke* (pp. 99–118). Routledge.

Kant, I. (2015). *On perpetual peace*. Broadview Press.

Kim, S. S., & Prideaux, B. (2003). Tourism, peace, politics and ideology: Impacts of the Mt. Gumgang tour project in the Korean Peninsula. *Tourism Management, 24*(6), 675–685. doi:10.1016/S0261-5177(03)00047-5

Korstanje, M. E. (2011). Influence of history in the encounter of guests and hosts. *Anatolia, 22*(2), 282–285. doi:10.1080/13032917.2011.597991

Korstanje, M. E., & George, B. P. (2012). Falklands/Malvinas: A re-examination of the relationship between sacralisation and tourism development. *Current Issues in Tourism, 15*(3), 153–165. doi:10.1080/13683500.2011.569535

Korstanje, M. E., & Olsen, D. H. (2011). The discourse of risk in horror movies post 9/11: Hospitality and hostility in perspective. *International Journal of Tourism Anthropology, 1*(3), 304–317. doi:10.1504/IJTA.2011.043712

Korstanje, M. E., & Tarlow, P. (2012). Being lost: Tourism, risk and vulnerability in the post-'9/11'entertainment industry. *Journal of Tourism and Cultural Change*, *10*(1), 22–33. doi:10.1080/14766825.2011.639455

Laws, E. (2005). *Tourism crises: Management responses and theoretical insight*. Psychology Press.

Lea, J. (2006). *Tourism and development in the Third World*. Routledge. doi:10.4324/9780203040393

Leslie, D. (1996). Northern Ireland, tourism and peace. *Tourism Management*, *17*(1), 51–55. doi:10.1016/0261-5177(96)83161-X

Litvin, S. W. (1998). Tourism: The world's peace industry? *Journal of Travel Research*, *37*(1), 63–66. doi:10.1177/004728759803700108

Mansfeld, Y., & Pizam, A. (Eds.). (2006). *Tourism, security and safety*. Routledge.

Moufakkir, O., & Kelly, I. (Eds.). (2010). *Tourism, progress and peace*. CABI. doi:10.1079/9781845936778.0000

Novelli, M., Morgan, N., & Nibigira, C. (2012). Tourism in a post-conflict situation of fragility. *Annals of Tourism Research*, *39*(3), 1446–1469. doi:10.1016/j.annals.2012.03.003

Pack, S. (2006). *Tourism and dictatorship: Europe's peaceful invasion of Franco's Spain*. Springer. doi:10.1057/9780230601161

Pernecky, T. (2010). The being of tourism. *The Journal of Tourism and Peace Research*, *1*(1), 1–15.

Pottorff, S. M., & Neal, D. D. M. (1994). Marketing implications for post-disaster tourism destinations. *Journal of Travel & Tourism Marketing*, *3*(1), 115–122. doi:10.1300/J073v03n01_08

Pratt, S., & Liu, A. (2016). Does tourism really lead to peace? A global view. *International Journal of Tourism Research*, *18*(1), 82–90. doi:10.1002/jtr.2035

Richards, G., & Hall, D. (Eds.). (2003). *Tourism and sustainable community development* (Vol. 7). Psychology Press.

Ritchie, B. W. (2004). Chaos, crises and disasters: A strategic approach to crisis management in the tourism industry. *Tourism Management*, *25*(6), 669–683. doi:10.1016/j.tourman.2003.09.004

Rovelstad, J. M. (1988). *World awareness and perception search*. Paper presented at the first Global Conference Tourism: A vital force for peace. Vancouver, Canada.

Salazar, N. B. (2006). Building a 'culture of peace' through tourism: Reflexive and analytical notes and queries. *Universitas Humanistica*, (62), 319–336.

Salazar, N. B. (2010). *Envisioning Eden: Mobilising imaginaries in tourism and beyond* (Vol. 31). Berghahn Books.

Sandell, K. (2005). Access, tourism and democracy: A conceptual framework and the non-establishment of a proposed national park in Sweden. *Scandinavian Journal of Hospitality and Tourism*, *5*(1), 63–75. doi:10.1080/15022250510014291

Sather-Wagstaff, J. (2016). *Heritage that hurts: Tourists in the memoryscapes of September 11*. Routledge. doi:10.4324/9781315427539

Séraphin, H., Butcher, J., & Korstanje, M. (2017). Challenging the negative images of Haiti at a pre-visit stage using visual online learning materials. *Journal of Policy Research in Tourism, Leisure & Events*, *9*(2), 169–181. doi:10.1080/19407963.2016.1261146

Seraphin, H., Korstanje, M., & Gowreesunkar, V. G. (2019). *Post-disaster and post-conflict tourism: Toward a new management approach*. Apple Academic Press.

Sharpley, R. (2018). *Tourism, tourists and society*. Routledge. doi:10.4324/9781315210407

Tarlow, P. (2014). *Tourism security: strategies for effectively managing travel risk and safety*. Elsevier.

Timothy, D. J. (2006). Safety and security issues in tourism. In *Tourism Management Dynamics* (pp. 43–51). Routledge. doi:10.1016/B978-0-7506-6378-6.50013-5

Tzanelli, R. (2016). *Thanatourism and cinematic representations of risk: Screening the end of tourism*. Routledge. doi:10.4324/9781315624105

Tzanelli, R., & Korstanje, M. E. (2019). 6 On killing the 'toured object. In R. Isaac, E. Cakmak, & R. Butler (Eds.), *Tourism and hospitality in conflict-ridden destinations* (pp. 71–83). Routledge. doi:10.4324/9780429463235-6

Var, T., & Ap, J. (1998). Tourism and world peace. In W. Theobald (Ed.), *Global tourism* (pp. 45–47). Routledge.

Walters, G., & Mair, J. (2012). The effectiveness of post-disaster recovery marketing messages: The case of the 2009 Australian bushfires. *Journal of Travel & Tourism Marketing*, *29*(1), 87–103. doi:10.1080/10548408.2012.638565

KEY TERMS AND DEFINITIONS

Dark Consumption: It is understood to the fascination for visiting spaces of mass death, war, pain, suffering or genocide. The term often associates to dark-tourism or Thana-tourism.

Peace: It is conceptually associated with a state of friendship and harmony where hostility is placated.

Post-Conflict Destination: It is a new sub-discipline of tourism research which studies in what ways, destinations are recycled, commercialised and sold just after an episode of violence or after years of war.

Prosperity: It means with good fortune or flourishing of a person, nation or group in view of wealth or social status.

Tourism: It consists in the act of travelling away from home less than six months and more than 24 hours. Today it is one of the largest and prosperous industries in the world.

Chapter 3
Catalytic Effect of Tourism in Peacebuilding:
Sustainability and Peace Through Tourism

Sweety Jamgade
https://orcid.org/0000-0002-6109-8801
Ramaiah University of Applied Sciences, India

ABSTRACT

Peace is the state of calmness attained when human beings of diverse culture, perceptions, and beliefs understand and respect the diversity of other humans in their surroundings. The understanding and respect are gathered through travel to places, observing and appreciating the culture, living style of the natives. It has been discussed in many tourism summits, conferences, forums, and symposiums about the contribution of tourism towards peacebuilding in the world. Many scholars have also investigated the impact of tourism on creating and maintaining peace in selected conflicted areas. The purpose of this chapter is to examine the significance of tourism in sustainable peacebuilding across the globe through a systematic literature review approach to accomplish the objectives of the study. It presents the concept of peace tourism, the Global Peace Index's domain of militarisation, safety and security in society, national-international conflicts, global leaders, analysis of some countries in tourism development, and sustainability through peace tourism.

INTRODUCTION

Tourism: A catalyst to peace

The world is a book and those who do not travel read only a page.

Saint. Augustine

DOI: 10.4018/978-1-7998-5053-3.ch003

This quote envisaged the catalytic effect of tourism as a book of reading and learning cultures through travel. Travellers experience the cultures and diversity of other regions. Cultural experience creates an intellectual, moral and spiritual understanding, the broad-mindedness to accept a human as they are and thus leading to peace through travel. Tourism not only flourishes in peaceful environments but may also contribute to the achievement of peace and transformation in the life of the individual, the community, the nation, and international relations (Lankford, Grybovych, & Lankford, 2008). International travelling promotes understanding and trust among people from different backgrounds. This line of thinking has brought up to the surface, the concept of peace through tourism (Jimenez, & te Kloeze, 2014). Tourists play an important role in the promotion of peace, many a time they being as peace messengers. Promotion of peace through tourists is more impactful than any of the tourism organisations' meetings and forum discussions. In many developing countries, peace is promoted between local inhabitants, as tourism creates job opportunities and is a major factor of household income to them for their satisfaction (Gajdošík, Sokolová, Gajdošíková, & Pompurová, 2019). Sustainability of peace is ensured when travelling across the region and globe becomes calmer and easy with hassle-free border formalities and less of conflicts which are created due to regional crisis, lack of safety and social injustice. Right cultural understanding through travel for tourism and not for migration purposes can reduce national and international conflicts. Global tourism is not only the world's biggest business, but the world's biggest business for peace. It forces people to understand each other, to see another point of view other their own, to see their differences and to see their sameness (Malley, 2002). Thailand's visa exemption to selected 64 countries and bilateral agreements between countries is a good indication of the catalytic effect of tourism; which is the primary economic contributor in Thailand's GDP. It is noted that Bangkok ranks in the top 20 most visited cities in the world (Hedrick-Wong, & Choong, 2014). Strengthening of trust in the travel business is an important criterion to have peace and sustainable tourism worldwide. According to Yin and Zhao (2006), a lot of businesses are emerging from the fact that people know each other. With the strengthening of trust, the competitive level of Regional Tourism Alliance (RTAs) could be greatly improved.

According to Litvin (1988), tourism is undoubtedly a beneficiary of peace, but as tourism is never successful in the absence of peace, it cannot, therefore, be a generator of peace. This relationship between tourism and peace is always debated. Research on the contributions of tourism towards peacebuilding is no longer a nascent concept. Several studies by individuals and associations have been conducted since the 19th century; however, most are either hypothetically based and lack precision. There were many professional meetings, forums held worldwide under the theme of "tourism and peace", to discuss the catalytic effects of tourism and resolve regional issues. Some of the prominent meetings, conferences with highlighting declarations are recorded below.

Lists of Some World Conference/ Meetings on Tourism and Peace

1929: Inaugural meeting of British Travel and Holidays Association with the theme of 'Travel for Peace'.

1967: UN had its International Tourism Year with a slogan of 'Tourism: Passport to Peace'.

1975: Conference on Security and Cooperation in Europe (CSCE), which led on to the Organization for Security and Co-operation in Europe (OSCE) founding document, the 'Helsinki Final Act', mentioned tourism as an important means of achieving increased corporation and better understanding.

1980: Manila declaration by the World Tourism Organization highlighted the fact that tourism could contribute to the establishment of a new international economic order, and could be a vital force for world peace.

1987: A conference on 'Tourism- a Passport to Peace' in Shannon.

1988: Global Conference on 'Tourism- a Vital Force for Peace' in Vancouver resulted in a first publication on 'Peace Through Tourism' by International Institute for Peace through Tourism.

2000: Conference in Amman, Jordan brought an official document 'The Amman Declaration of Peace Through Tourism' (Wohlmuther, & Wintersteiner, 2014).

2001: UN General Assembly endorsed the UNWTO Code of Ethics with Tourism as a force for peace (All Regions, 2012).

2002: 21st Century African Agenda for Peace and Poverty Reduction through Tourism – Nelspruit, Mpumalanga Province, South Africa

2003: Tanzania Action Plan – Dar es Salaam, Tanzania was the 2nd IIPT African Conference on Peace through Tourism on the theme of "Community Tourism: Gateway to Poverty Reduction", where it was confirmed that global tourism could be a vital force for peace, poverty reduction and the realisation of the Millennium Goals of the United Nations and sustainable development (International Institute for Peace through Tourism, 2003).

2011: Lusaka Declaration on Sustainable Tourism Development, Climate Change, and Peace– Lusaka, Zambia (International Institute for Peace through Tourism, n.d.c).

2012: UNWTO Conference on Sustainable Development held in Rio de Janeiro declared International Day of Peace to be dedicated to "Sustainable Peace for a Sustainable Future" (UNWTO, 2012).

2016: UNWTO Conference in 2016 on 'Tourism, a catalyst for development, peace and reconciliation' held in Passikudah, Sri Lanka concluded that 'Tourism can play a key role in building peace and supporting reconciliation processes' (UNWTO, 2016).

2017: UN Conference on 'The Question of Jerusalem' held in Baku concluded that tourism is the main driver for cities' economy, international identification and commercialisation with new tourism products and religious tourism as instrumental to peace (United Nations, 2017).

2019: United Nations General Assembly, 106 Plenary Session declares 2021 the International Year of Peace and Trust to constitute the means of mobilising the efforts of the international community to promote peace and trust among nations based on, inter alliances, political dialogue, mutual understanding, and cooperation, to build sustainable peace, solidarity and harmony (United Nations, 2019).

With the aim to study the significance of tourism in sustainable peacebuilding across the globe, the above- stated conference declarations were analysed in detail. To create awareness about the concept of peace tourism and peace tours initiated by various organisations, a literature analysis was conducted. Case analyses of some conflict-affected areas in the Asia Pacific region were noted. The Global Peace Index and the role of world peace leaders and activists were analysed to have a better understanding of indicators, donors in peace development.

Global Peace Index and World Peace Leaders

The Global Peace Index (GPI) indicates the state of peacefulness of the assessed list of countries conducted by the Institute for Economics and Peace every year. GPI is measured under the three thematic domains of the degree of militarisation in countries, level of societal safety and security and extent of ongoing domestic and international conflicts using 23 qualitative and quantitative indicators from respected sources of high relevance. According to the 13th edition of the GPI report (Institute for Economics and Peace, 2019), which ranked 163 nations based on the level of peacefulness, the average country score of global peacefulness improved by 0.09% with 86 countries improving and 76 showing deteriorations. Iceland remains the most peaceful country in the world, a position it has held since 2008, and the other top four countries being New Zealand, Austria, Portugal and Denmark. The remarkable country is Bhutan which has recorded the largest improvement of any country in the top 20, rising to 43 places in the last 12 years.

Afghanistan is now the least peaceful country in the world, replacing Syria, which is now the second least peaceful. South Sudan, Yemen (first time listed in the least level index) and Iraq comprise the remaining five least peaceful countries. The greatest increase in peacefulness can be seen in four regions out of nine least index regions, which are Russia and Eurasia, followed by the Middle East and North Africa (MENA). In MENA, the number of deaths from conflict has declined, owing to the de-escalation of violence in Ukraine and Syria, respectively, as exhibited in figure 1.

Figure 1. Regional GPI Results 2019
Source: Vision of Humanity (2019)

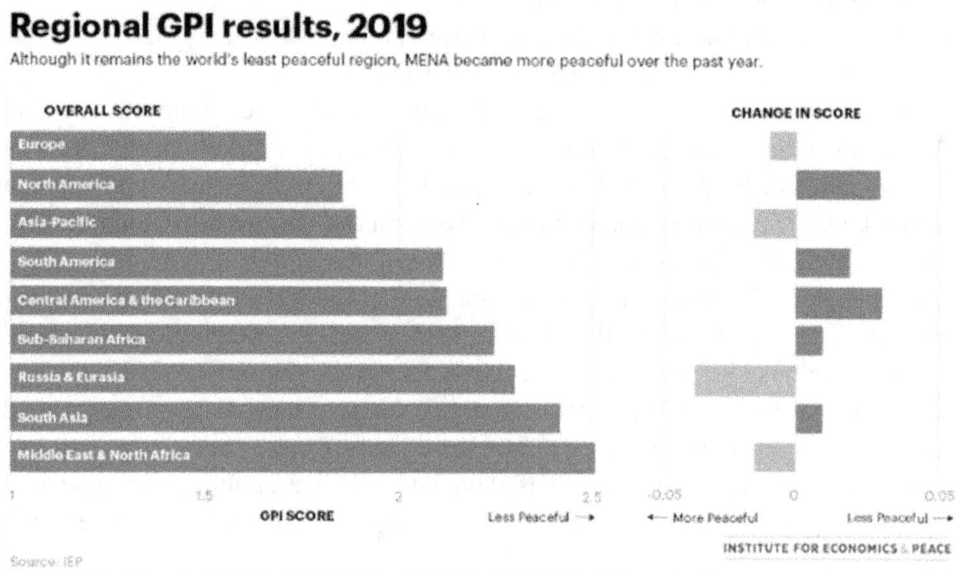

According to the GPI report, there is a strong correlation between perceptions of peacefulness and actual peacefulness as measured by the GPI. Both men and women in more peaceful countries are more likely to report that they feel safe walking alone at night than people in less peaceful countries. There is also a higher level of trust in police in more peaceful societies. On the other side, violence continues to have a significant impact on economic performance around the globe, as shown in table 1.

Table 1. Average economic cost of violence

Violence	Average economic cost of violence (percentage)
Countries affected by violence	35% of GDP
Countries least affected by violence	3.3% of GDP

Source: Institute of Economics and Peace, 2019

Whereas the year 2018 and the previous four years, GPI reports were very alarming with the findings of 0.27% deteriorations in the global level of peace due to the conflicts, tensions, and crisis being unresolved. The eleven-year trend since 2008 in peacefulness indicates the deteriorations of global peace by 3.78% stressing that improvement is less and deteriorations in the domain of ongoing conflict and safety and security have increased. The indicator with the most widespread deterioration globally was the terrorism impact indicator. Just over 63 per cent of countries recorded increased levels of terrorist activity (Institute of Economics and Peace, 2019).

The ongoing domestic and international conflicts are majorly due to the resource crises, civilisation issues, and misunderstandings due to rigid beliefs of racism and religion leading to rigidity in visa regulations thus restricting the book of travel to spread the message of peace. Hospitality has been adversely impacted because of terrorism, war, and restricted travel.

Many proactive world leaders with rational beliefs have foreseen the deterioration of humanity and strived to bring peace for humankind. These peace leaders understood the requisite of the peace movement by travelling to places and being empathetic. They started their journey of peace for some cause and improvement of global peace. Their profound contribution is remarkable, which supplements tourism and peace relations. Some of the most admired leaders who were politicians, spiritual leaders, humanitarian and activists, who have sacrificed their lives in peace journey, are listed below with their contributions.

Mahatma Gandhi (1869-1948): He was the messenger of nonviolence in the world and father of the independence movement in India. His travel and stay in South Africa backed his political, ethical view against discrimination and brought the leader out of him. His quote below proposed the catalytic effect of travel to peace.

I have watched the cultures of all lands blow around my house and other winds have blown the seeds of peace, for travel is the language of peace. Mahatma Gandhi (Litvin, 1988)

Martin Luther King Jr. (1929-1968): He was the leader of the civil rights movement, known for progressing civil rights through civil disobedience and nonviolence inspired by Christian beliefs and Mahatma Gandhi. His march on Washington for freedom and jobs and his speech "I have a dream" inspired many. The below passage from his speech indicates the impulse he has for peace in the world through equality.

I have a dream that my four little children will one day live in a nation where they will not be judged by the color of their skin, but by the content of their character. Martin Luther King Jr. (Gates, & Appiah, 1999)

Nelson Mandela (1918-2013): He was a leader who made a significant contribution to peace in the South African region. Although imprisoned for fighting apartheid, he was released and became the first democratically elected black president.

Pope Benedict XV (1854-1922): Throughout World War I, he advocated peace, opposed aerial warfare and promoted humanitarian initiatives for the prisoners of the war, their children and missing people.

Gautama Buddha (563-483 BC): He was born a prince in India, but he disowned the comforts of the palace to seek enlightenment. After attaining Nirvana, he spent many years travelling across northern regions of India and southern Nepal, teaching his philosophy of inner peace, detachment and how to attain liberation from earthly suffering.

Jesus of Nazareth (2BC – 7 AD): Jesus' travels and acts to Jerusalem taught a radical philosophy of love and forgiveness. These teachings formed the basis of Christianity

Blessed are the peacemakers, for they shall be called the children of God. Jesus of Nazareth (Pettinger, 2013).

Sustainability and Peace Through Tourism

Tourists are conflict-sensitive and response positively to peace. Peace and sustainability, measured as the indicator of expansion, are vulnerable due to numerous conflicts – social, economic, political, cultural and environmental. Tourism holds the key to the promotion of peace and sustainability (Sharma, Charak, & Kumar, 2018). Upadhayaya, Khatiwada, Upreti, and Sapkota (2013) and Upadhayaya, Muller-Boker and Sharma (2011) explored the relationship between tourism, conflict, and peace in Nepal. They stated that tourism could be an effective tool in coping with and responding to conflict and can be a mediator for peace by assisting in the development of a supportive environment. They believed that the sustainability of tourism is highly dependent on peace, security and safety. According to the United Nations World Tourism Organization (UNWTO), Sustainable Development Goal (SDG) 16 is on "Peace, justice and strong institutions", being more than ever close to the vision of inclusive and universal tourism, peaceful tourism that strives for justice. It focuses on to promote peaceful and inclusive societies for sustainable development, provide access to justice for all and build effective, accountable and inclusive institutions at all levels (Lim, n.d.).

Peace Tourism: A Need

Peace is the state of attainment of calmness and a sense of understandings with oneself, people, past generations, future generations and nature. Peace tourism is travelling to places to attain an understanding of people, learning their culture, history and environment by visiting peace memorials, museums and parks and attending peace programs to eradicate poverty, social injustice and conserve the environment. Thus the other types of tourism namely ecotourism, sustainable tourism, responsible tourism, heritage tourism, cultural tourism, green tourism, open tourism and new such categories of tourism can be associated as subsets of peace tourism and Sustainable tourism.

Promoting a culture of peace through tourism exposes the ideas of how to achieve "dialogue" for peace and promote ideals of open tourism. Peace tourism then becomes concerned with the processes of communication and interaction rather than voyeuristic descriptions of other worlds (Urry, 1990). Respect for the difference lies in the processes of cross-cultural communication. Underlying these processes is the language of nonviolence, which implies knowledge and understanding of comparative and conflicting values. It includes education about economic, political and sociocultural circumstances of specific regions to encourage a sensitivity to respect for cultural differences (Blanchard, & Desbiolles, 2013). So, peace tourism can play a vital role in minimising the root causes of violence, which are inevitable in its present state majorly by creating human understanding. It can act as a facilitator of sustainable development. According to Ward (2009), peace tourism is a holistic, multi-dimensional phenomenon that intersects natural, social, economic, political and spiritual systems. Peace tourism then is certainly relevant to citizens of nations involved in terrorism, war, genocide, drug or human trafficking, arms sales or any systematic armed violence against other nations. According to the International Institute for Peace's (IIPT) vision, the notion of peace through tourism encompasses poverty alleviation, international understanding, preservation of heritage, protection of the environment, and sustainability (Jimenez, & Jan te Kloeze, 2014). All these notions are the elemental criteria to have sustainability of any place and resources (Shin, 2008). The extent to which a country is developed correspondingly indicates its relationship to peace tourism development through various donors. The three influential factors of peace tourism are the nation's stage of development, cultural or ethnic diversity and its global status – the power of leadership. All these factors are closely associated with the level of peace tourism a nation has due to the contributors. Table 2 represents such association of a country's development to its cultural/ethnic diversity, its global status and, peace tourism contributors/donors.

Table 2. Development of nations and peace tourism

Nation stage (Development)	Cultural/ethnic diversity	Global status (Power)	Peace tourism (Contributors)
Early	· Homogeneous · Limited migration, single citizenship	Limited	· Emergency relief · NGOs, individual · Initiatives/philanthropies
Middle	· Mostly homogeneous · Ease of travel	Middle	· Education travel courses · Ecotourism
Mature	· Heterogeneous or multi-citizenship · Freedom of travel	Established	· People-to-People cooperatives and environmental focus

Source: Ward, 2009

The above table indicates that if a country is in the early stages of development, it will have limited cultural diversity and more homogeneity of the ethnic group due to limited tourism and global status. Peace tourism in such nations will be through emergency relief activities, non-governmental organisations (NGOs) and individual philanthropic activities or initiatives. Nations in the middle stage of development have "middle global status" with a mostly homogenous group of people, but the ease of travel exists. The peace contributions in such nations are mostly through education, awareness, travel institutes with courses and conservation is in the form of ecotourism principles. Mature developed nations have heterogeneous people having multi-citizenship indicating the freedom of travel. Developed nations are the key leaders with established global status and have people to people relationship building with environmental focus as peace initiatives.

Sustainability Through Tourism

The global significance of sustainability is through peacebuilding. Open and sustainable tourism can be a force for peace, with causal mechanisms operating both at the individual and state level (D'Amore, 1988). Open tourism or international openness which is one of the criteria in the Travel and Tourism Competitiveness Index (TTCI) refers to the ease and extent to which tourism flows within and between countries allowing the exchange of money, ideas, and cultures ensuring safety and security measures. Ease and extent of travel are associated with the visa regulations and safety and security measures for the tourists by the host nations. Open tourism can broaden minds and opinions so that individuals may become more informed and tolerant of their fellow human beings (World Travel and Tourism Council, 2016). Hypothetically open tourism/international openness may lead to peace tourism.

Sustaining peace should be about stimulating and facilitating the capacity of societies to self-organise, so that they can increase their ability to absorb and adapt to stress, to the degree necessary to sustain peace (de Coning, 2016). Militarisation may hinder international openness; in borderlands, tourism regularly functions as a strategic policy tool to strengthen regional images, shape identity narratives and facilitate cross-border interactions (Scott, 2013; Timothy, 2001). The Eurogio Karelia (EK) formation provides an illuminating example of the role of similarities and dissimilarities with cross-border mobility and border region-building. Its founders emphasised three fundamental similarities between the Finnish and Russian partner regions: a common border, a common natural landscape and a common set of cultural and cooperation traditions. On the other hand, both the ofðcial EK documents and various expert evaluations also mentioned several dissimilarities, from language to availability and quality of services. The Finnish–Russian borderland presents an interesting case where familiarity does not mean similarity. After 20 years of cross-border interaction, familiarity has undoubtedly increased in Finnish-Russian border. In addition, despite its deteriorated state, the Finnish cultural and historical heritage still attracts Finns to visit Sortavala. (Izotov, & Laine, 2013). Iraqi Kurdistan is an oasis of peace in a desert full of conñict. The imposition of arbitrary borders to place Iraqi Kurdistan to Iraq by British Colonial officials and suppression of Kurdish identities and their rights led to the rebellion of people and government. The political conflicts have added to the deterioration of the peace in Kurdistan (Romano, 2006).

Studies of cross-border tourism have generally approached borders as barriers that limit travelling and the growth of tourism (Wachowiak, 2006). One of the world's most visible, significant, and conflicting border landscapes is the Shankill ¤ Falls divide in Belfast, Northern Ireland. This border fence or wall is a signature feature of the cultural landscape that sharply divides Catholic and Protestant communities, especially in the western part of the city. Community efforts are being implemented to reduce fear and hostilities, the use of large colourful murals depicting the history and political views of both traditions is a way of marking territory attracting Belfast Peace line tours (Brunn, Bryne, McNamara,& Egan, 2010). Houaphan Province, in Laos, is becoming again a transcultural and transnational crossroads that it was in the late 19th century. Cross-border tourism promotes these flows and interactions of the local with the global and increases a deeper understanding. The travel of Western and Japanese development experts bring in additional transnational influences through the transfer of knowledge, legal systems, and universal discourses of human rights (Tappe, 2011). The more the border travel, the more will be a deeper understanding of human and nation issues and sustainable development.

Peace Tours and Their Promotion

The organisation of peace tours by various associations to promote peace has contributed to some extent in the activity of peace tourism as a product and creating awareness. Some of the noted peace tours are described below.

Peace Rally

The protracted liberation wars in Africa had affected the continent's peace and eventually its tourism. The violent land clashes in Laikipia County have threatened the communities living. Thus Kenya leaders are promoting peace amateur rally by driving 180 km to conflict bandit attack areas since the year 2017 according to KTN News. Africa, if settled with these issues, can bring in peace and development.

Peace Boat

Since its founding in Tokyo, the organisation has launched 25 international voyages of passenger ships. They form a floating university of sorts that provides humanitarian aids at their various stops and holds conferences related to global peacekeeping. It is granted a Special Consultative Status by the United Nations and hence submits reports and proposals to them regularly (Vashishtha, 2016). The Peace Boat team believes that travel can be a tool for positive social and political change, and seeks to create and implement best practices in responsible travel leading to sustainability. Socio-political considerations rather than commercial interests largely determine the choice of destinations. Peace Boat seeks to create awareness and action to effect positive social and political change through global educational programmes, responsible travel, cooperation projects and advocacy. These activities are carried out on a partnership basis with other civil society organisations and communities in Japan, Northeast Asia, and around the world (Lim, n.d.).

Reality Tours

Encouraging people-to-people ties, this idea motivated the US-based international human rights organisation Global Exchange to organise travel that can be educational and positively influence international affairs with relationship building is regarded as essential to this transformation. Reality Tours initiated in 1988 are meant to educate participants about how people, individually and collectively, contribute to global problems, and, then, to suggest ways in which they can contribute to positive change locally and internationally (Reality Tours, n.d.).

Peace Museums, Memorials and Organisations in the World

The peace museums, memorials movements were initiated mostly before and after the World War II, built on the battleground of wars to show solidarity to the people who lost their lives, the brutality of wars, and a message of peace to curb such human instigated conflicts. The vacillation is whether the memorial sites are constituents of war tourism or peace tourism or both. There are programs organised by various associations to create awareness and promote peace tourism. Some of the popular peace museums and memorials contributing to peace tourism are:

Hiroshima Peace Memorial, Nagasaki Peace Park and Museum

In August 1945, the United States dropped two atomic bombs on the cities of Hiroshima and Nagasaki, Japan. A decade later, the Hiroshima Peace Memorial Park and Museum were opened, experiencing increased tourist footfalls, mostly with the purpose to know about the history and pay tribute. As recorded, nearly two million people visited in the year 2016 and mounting every year. The Nagasaki Peace Park is filled with statues, commemorative plaques and remains of the bomb drop. It is located only 200 metres from the bomb site. The Hiroshima Peace Memorial Park is designed in such a way that the park's axis points to the atomic bomb dome, which is kept as a memorial. It has the Children's Peace Monument, which was dedicated to all the children who became victims of Hiroshima's atomic bomb.

Okinawa Prefectural Peace Memorial Museum

The Battle of Okinawa was one of the bloodiest of the Second World War, and the inhabitants faced not only the American invasion but also the tactics of the Japanese army, who viewed the islands as expendable in order to slow invading forces from reaching mainland Japan. More than 4,00,000 visitors visit Okinawa Prefectural Peace Memorial Museum in Itoman city annually. Many school excursions are arranged to make students learn about the importance of peace. The peace park storytellers or *kata-ribe*, who are the Second World War survivors, have the critical task of mediating the message to both domestic and international tourists (Hashimoto & David, 2019).

Dayton International Peace Museum

The founders of Dayton International Peace Museum were Lisa Wolters, Ralph and Christine Dull, J. Frederick Arment and Steve Fryburg. It spreads knowledge on the methods of bringing peace in a non-violent way. Its mission is to influence the culture of peace at multiple levels from local to international level. This museum is housed in Isaac Pollack House and got a place in National Register of Historic Places and a member of International Network of Peace Museum (Dayton International Peace Museum, n.d.).

The Tehran Peace Museum

The Tehran Peace Museum is situated in the construction donated by the Theran Municipality, located in the Historic City Park. It organises and holds various peace programs on international humanitarian law and advocacy of peace (Tehran Peace Museum, n.d.).

Yser Tower

The Yser Tower memorial is a peace memorial and is located next to Belgian Yser river in the DiksMuide city. It was created to respect of soldiers that were killed during World War I on the Yser. It signifies appeal for Nooit meer Oorlog (No more War) and is written in four different languages. It is known as the highest peace monument in the entire Europe (De Izertoren, n.d.).

International Institute for Peace through Tourism (IIPT)

Louis D'Amore founded IIPT in 1986, the International Year of Peace in New York, with a vision of travel and tourism becoming the world's first global peace industry. Its emblem inscribes "travel promotes peace" (International Institute for Peace through Tourism, n.d.a). It organises international conferences and global summit in different regions with a focus on actual case studies that demonstrate and promotes international understanding; cooperation among nations; improved quality of the environment; cultural enhancement and the preservation of heritage; poverty reduction; reconciliation and healing wounds of conflicts; and through these initiatives, helping to bring about a peaceful and sustainable world.

A primary goal of IIPT is to mobilise the travel and tourism industry as a leading force for poverty reduction. It initiated poverty reduction projects in four countries of the Caribbean and Central America and documented code of ethics and guidelines for sustainable tourism. The major initiative of Global Peace Parks Project with a goal of 2,000 peace parks by 2020 commemorating the Centenary of World War I and its theme "No More War" is commendable (International Institute for Peace through Tourism, n.d.b).

The second flagship project of IIPT's 30th Anniversary Year is the "IIPT Travel For Peace Campaign" that encourages every traveller to be an "Ambassador for Peace" by practising the IIPT Credo of the Peaceful Traveler and the ethic inherent in the iconic Padre Johnson portrait of the Global Family reminding us that we are "One Earth One Family." Travel agents, tour operators, hotels, airlines and other sectors of the travel and tourism industry are being invited to become charter members of the campaign and to encourage their clients to be "Ambassadors for Peace" (International Institute for Peace through Tourism, n.d.c).

United Nations World Tourism Organisation (UNWTO)

It was formed in 1974, is a leading international organisation in the field of tourism to promote tourism as a driver of economic growth, inclusive development and environmental sustainability and offers leadership and support to the sector in advancing knowledge and tourism policies worldwide. It encourages the implementation of a global code of ethics for tourism and achieving Sustainable Development Goals (UNWTO, n.d.).

The International Network of Museums for Peace (INMP)

It is a non-profit organisation, established since 1992 and associated with the United Nations Department of Public Information in New York. It aims to build a global culture of peace by strengthening the work of museums for peace, peace gardens and other peace-related sites, institutions (International Peace Bureau, n.d.).

CASE STUDIES

World War-Affected Hiroshima and Nagasaki in Maintaining Peace

Hiroshima and Nagasaki are described as an interesting, educational and emotional example of "dark tourism," "grief tourism," "battlefield/ war tourism," "educational tourism," or "peace tourism" (Ehrlich, 2014). For many years, numerous research works have been conducted on Hiroshima and Nagasaki. Yoshida, Bui and Lee (2016) explored the extent to which these sites of atomic bomb explosions have been memorialised and embedded in tourism promotion for educational purposes. It reveals the themes of tourism education and war peace are intertwined with the historical context of the two cities. As a tourist destination, conservative Hiroshima centres on the remembrance of the atomic tragedy and the traditional or nostalgic aspects of Japanese culture, whereas relatively liberal Nagasaki presents itself as a melting pot of cultures. Both cities retain their central role of publicly commemorating the victims of the atomic bomb for the Japanese.

Conflicted Kashmir and its Dependency on Tourism for Peacebuilding

Kashmir despite a major tourist destination, the spread of terrorism, insurgency, social-political unrest and militarisation especially during the last two decades have hindered the smooth growth of the tourism industry (Chauhan, & Khanna, 2009). There is a plunge in economy and tourism activity and an increase in national and international conflicts. Enforced calm and discourse around tourism, rather than achieving a political resolution are just adding fuel to the conflicts (Junaid, 2012). Being the paradise of India's natural beauty and tourism, it has suffered socially and economically. The revocation of article 370 (special status) and declaration of Jammu and Kashmir as a union territory in August 2019 by the government might change the face of Kashmir, its impact and assumptions can be analysed in due course of time. The travel industry hopes that this new status might bring fortunes (Banerjee, 2019).

Morocco in Maintaining Peace

Amid widespread regional conflicts and insecurity, the Kingdom of Morocco has succeeded in preserving its stability. The threats of terrorism are creating domestic conflicts and insecurity. To deal with such a daunting challenge, Morocco's comprehensive security strategy combines a wide range of policies that link the improvement of the socio-economic situation with the preventative capacity to anticipate the risks of terrorism and the operational aspects of counter-terrorism work (Alaoui, 2017). It has a vision of peace in Africa and is acting on it, its membership in the African Union Peace and Council will help in asserting peace (The North Africa Post, 2018). The Global Peace of Index of Morocco's is 90th and 7th most peaceful in the Middle East and North Arica (MENA) region. It has also made significant improvements in tourism with its ranking as 66th out of 140 countries in the 2019 Travel and Tourism Competitiveness Index.

SOLUTIONS AND RECOMMENDATIONS

Findings of the study of Stephen and Liu (2016) indicate that peace is most important to tourism in medium-income destinations, but still important for high-income nations. It is advisable considering the sustainability element of the world: high, medium and low-income nations should cohesively work for peacebuilding and cultural exchange programs to broaden the understanding and relationship buildings to eliminate conflicts. Peace initiatives should majorly focus on the transmission of social justice to all ethnic groups, which are a solution to many conflicts. Cultural and heritage tourism could also be integrated to promote peace tourism as a product. A holistic approach to promoting peace tourism as a product should be incorporated to ensure sustainability.

Relationship building and alliances with tourism service providers are recommended for the regional promotion of peace tourism. Emphasising the economic and sustainability effects of tourism and peace is imperative to delegate responsibility to the tourism industry for developing and promoting peace tourism as a product. Peace education programs having a peace tour as activities should be formulated by tourism bodies and implemented in the primary education system for the children. Cultural-exchange programs should be a mandated course at the higher education system to avoid human conflicts arising due to cultural differences.

FUTURE RESEARCH DIRECTIONS

Many debate that tourism is not the generator of peace rather beneficiary of peace, thus it is required to study the relationship of tourism and peace with precision by the officiating bodies and organisations. There is a high requirement to identify and address challenges emerging through the benefits of peace tourism. The trend of visiting warfare zone, battlefields and war memorials are termed as war tourism whereas the same elements are the constituent of peace tourism. Hence, there is a need to study these emerging trends of tourism and clearly state the working definitions with specifications of activity and purposes. Future implications could be to define pragmatically the code of conduct for peace tourism in-order to regulate sustainability as a tourism product and attain improvement in all domains of peacefulness.

CONCLUSION

Travel and tourism are extremely sensitive to any political or social unrest. They require a climate of peace if they are to stimulate the social, economic and cultural development of society. Therefore, lasting peace is a prerequisite for the continuing increase of travel for tourism and not for migration purposes. Border regulations and Open tourism/ International openness play an important role in travel which eventually leads to peacebuilding by the understanding of cultures and makes a mind. Countries having a high state of peacefulness are having high ranks in travel and tourism competitiveness index and vice-versa, thus betoken its relationship. The continuing growths of air travel and cross border movements have contributed towards the growth of tourism. According to the Institute for Economics and Peace (2019), countries with very high levels of peace, on an average, have achieved over three times higher GDP *per capita* growth compared to the least peaceful countries, for the last 60 years. Mature and middle level developed countries are the beneficiaries of the impact of peace tourism mostly initiated by peace activists, leaders and organisations at all levels. Tourism stimulates international understanding and benefits any nation's GDP through its trade collaborations, foreign exchange earnings (FEE) and revenue through the utility of services. It also conserves and protects the ecosystem through ecotourism, culture through cultural tourism, heritage through heritage tourism, sustainable development through sustainable tourism, and, therefore, contributes to the peace and sustainability of any nation. Collectively, all forms of tourism have an enormous catalytic effect in peacebuilding, for that reason peace tourism should be promoted worldwide.

The latest report of Global Peace Index indicates the deterioration of the level of peacefulness globally due to international and national conflicts which majorly occurs due to resource sharing or crisis, social injustices, political or instigated individual interest, terrorism, etc. which is a crucial deterrent to the development of peace tourism. Many leaders and organisations have initiated and accepted the responsibility on their shoulders to create awareness about the disastrous effects of war and endurance of peace to achieve sustainability through peace tourism, by arranging peace programs, conferences, summits for decision making, peace tours, fundraising activity and so on.

REFERENCES

Alaoui, B. A. (2017). Morocco's security strategy: Preventing terrorism and countering extremism. *European View*, *16*(1), 103–120. doi:10.100712290-017-0449-3

Banerjee, T. (2019). One good tourist brings 10 more: ' Travel industry hopes J&K's new status will change fortunes. *ET Online*. Retrieved March 24, 2020, from: https://economictimes.indiatimes.com/magazines/panache/one-good-tourist-brings-10-more-travel-industry-hopes-jks-new-status-will-change-fortunes/articleshow/70586412.cms

Blanchard, L.-A., & Desbiolles, H. F. (Eds.). (2013). *Peace through tourism: Promoting human security through international citizenship*. Routledge. doi:10.4324/9780203370384

Brunn, D. S., Bryne, S., McNamara, L., & Egan, A. (2010). Belfast landscapes: From religious schism to conflict tourism. *Focus on Geography*, *53*(3), 81–123. doi:10.1111/j.1949-8535.2010.00011.x

Chauhan, V., & Khanna, S. (2009). Tourism: A tool for crafting peace process in Kashmir, J&K, India. *Tourismos*, *4*(2), 69–89.

D'Amore, L. J. (1988). Tourism: A vital force for peace. *Tourism Management*, *9*(2), 151–154. doi:10.1016/0261-5177(88)90025-8

Dayton International Peace Museum. (n.d.). *About us: A brief history of America's only peace museum*. Retrieved January 5, 2020, from: https://daytonpeacemuseum.org/about/

de Coning, C. (2016). From peacebuilding to sustaining peace: Implications of complexity for resilience and sustainability. *Resilience*, *4*(3), 166–181. doi:10.1080/21693293.2016.1153773

De Izertoren. (n.d.). *The Yser Tower*. Retrieved January 7, 2020, from: http://www.museumaandeijzer.be/ijzertoren/en/ijzertoren-diksmuide/yser-tower

Ehrlich, S. R. (2014). *Hiroshima, Japan was devastated by an atomic bomb but tourists can't keep away*. CNN Press.

Gajdošík, T., Sokolová, J., Gajdošíková, Z., & Pompurová, K. (2019). *Peace promotion through volunteer tourism. Marketing peace for social transformation and global prosperity*. IGI Global.

Gates, H. L., & Appiah, A. (1999). *Africana: The encyclopedia of the African and African American experience*. Basic Civitas Books.

Hashimoto, A., & David, J. T. (2019). Contested geopolitical messages for tourists at the Okinawa Peace Park and memorials. *Japan Forum*, *31*(3), 378–407. doi:10.1080/09555803.2018.1451354

Hedrick-Wong, Y., & Choong, D. (2014). *MasterCard 2014 Global Destination Cities Index*. MasterCard.

Institute for Economics and Peace. (2019). *Global Peace Index 2019: Measuring peace in a complex world, Sydney, June 2019*. Retrieved December 20, 2019, from: http://visionofhumanity.org/reports

International Institute for Peace through Tourism. (2003). *Tanzania Action Plan*. Retrieved March 16, 2020, from: https://www.iipt.org/conference/africanconference2003/presentations/TanzaniaActionPlan.pdf

International Institute for Peace through Tourism. (n.d.a). *About us: Mission statement*. Retrieved November 29, 2019, from: https://peacetourism.org/about-iipt/

International Institute for Peace through Tourism. (n.d.b). *Global Peace Parks Project*. Retrieved November 29, 2019, from: https://peacetourism.org/peace-parks-project/

International Institute for Peace through Tourism. (n.d.c). *Marketing peace through tourism: A 35 year history of IIPT*. Retrieved November 29, 2019, from: https://peacetourism.org/marketing-peace-through-tourism//

International Peace Bureau. (n.d.). *International Network of Museums for Peace*. Retrieved December 15, 2019, from: http://www.ipb.org/members/international-network-of-museums-for-peace/

Izotov, A., & Laine, J. (2013). Constructing (un)familiarity: Role of tourism in identity and region building at the Finnish-Russian Border. *European Planning Studies*, *21*(1), 95–110. doi:10.1080/09654313.2012.716241

Jimenez, C., & te Kloeze, J. (2014). Analysing the peace through tourism concept: The challenge for educators. *Sociology and Anthropology (Alhambra, Calif.)*, *2*(3), 63–70. doi:10.13189a.2014.020301

Junaid, M. (2012). *Peace tourism and political games in Kashmir*. Retrieved January 10, 2020, from: https://www.aljazeera.com/amp/indepth/opinion/2012/07/201272612535508711.html

Lankford, V. S., Grybovych, O., & Lankford, K. J. (2008). Context for peace and community tourism: A worldview framework. *World Leisure Journal*, *50*(3), 199–208. doi:10.1080/04419057.2008.9674553

Lim, Y. M. (n.d.). Imagine peace translated from Korean by Jessica Lee. *Transforming Tourism*. Retrieved March 15, 2020, from: http://www.transforming-tourism.org/goal-16b-peace.html

Litvin, W. S. (1988). Tourism: The world's peace industry? *Journal of Travel Research*, *37*(1), 63–66. doi:10.1177/004728759803700108

Malley, M. (2002). Letters/perspective. *Hotel & Motel Management*, 8.

Pettinger, T. (2018). *People who made a contribution to world peace*. Retrieved December 12, 2019, from: https://www.biographyonline.net/people/world-peace.html

Reality Tours. (n.d.). *Reality tours*. Retrieved January 3, 2020, from: https://globalexchange.org/realitytours/

Romano, D. (2006). *The Kurdish National Movement: Opportunity, mobilisation, and identity*. Cambridge University Press. doi:10.1017/CBO9780511616440

Scott, J. W. (2013). Constructing familiarity in Finnish–Russian Karelia: Shifting uses of history and the re-interpretation of regions. *European Planning Studies*, *21*(1), 75–92. doi:10.1080/09654313.2012.716240

Sharma, P., Charak, N. C., & Kumar, R. (2018). Sustainable tourism development and peace: A local community approach. *Journal of Hospitality Application & Research*, *13*(1), 37–56.

Shin, Y. (2008). Peace and tourism vs. tourism and peace: Conceptual Issues. *International Journal of Hospitality and Tourism Systems*, *1*(1), 25–35.

Stephen, P., & Liu, A. (2016). Does tourism really lead to peace? A global view. *International Journal of Tourism Research, 18*(1), 82–90. doi:10.1002/jtr.2035

Tappe, O. (2011). Memory, tourism, and development: Changing sociocultural configurations and upland-lowland relations in Houaphan Province, Lao PD. *Journal of Social Issues in Southeast Asia, 26*(2), 174–195. doi:10.1355j26-2b

Tehran Peace Museum. (n.d.). *Home*. Retrieved December 20, 2019, from: http://www.tehranpeacemuseum.org/index.php/en/

The North Africa Post. (2018). *Morocco's membership in Peace & Security Council boosts peace, stability in Africa*. Retrieved from https://northafricapost.com/23101-moroccos-membership-peace-security-council-boosts-peace-stability-africa.html

Timothy, J. D. (2001). Tourism planning in borderlands. In *Tourism and political boundaries* (pp. 133–135). Routledge.

United Nations. (2017). *International Conference on the Question of Jerusalem.* Retrieved March 18, 2020, from: https://www.un.org/unispal/wp-content/uploads/2017/07/Baku-Report-Final.pdf

United Nations. (2019). *A/RES/73/338 Seventy-third session Agenda item 15*. Resolution adopted by the General Assembly on September 12, 2019. Retrieved January 4, 2020, from: https://undocs.org/en/A/RES/73/338

UNWTO. (2012). *International day of peace: Tourism as a force for development and peace*. Retrieved from https://www.unwto.org/archive/global/news/2012-09-21/international-day-peace-tourism-force-development-and-peace-0

UNWTO. (2016). *Tourism: A catalyst for peace and development*. Retrieved January 2, 2020, from: https://www.unwto.org/archive/asia/press-release/2016-07-14/tourism-catalyst-peace-and-development

UNWTO. (n.d.). *About UNWTO*. Retrieved November 29, 2019, from: https://www.unwto.org/who-we-are

Upadhayaya, P. K., Khatiwada, S. S., Upreti, B. R., & Sapkota, T. (2013). *Codes of conduct for peace responsive tourism in Pokhara: A manual*. Pokhara Tourism Council, Nepal Tourism Board, NCCR North-South.

Upadhayaya, P. K., Muller-Boker, U., & Sharma, S. R. (2011). Tourism amidst armed conflict: Consequences, copings, and creativity for peacebuilding through tourism in Nepal. *The Journal of Tourism and Peace Research, 1*(2), 22–40.

Urry, J. (2005). *The tourist gaze* (2nd ed.). Sage Publications.

Vashishtha, P. (2016). *20 world organisations that strive to establish world peace*. Retrieved November 29, 2019, from: https://listsurge.com/20-world-organizations-strive-establish-world-peace/

Vision of Humanity. (2019). *Global peace index*. Retrieved March 23, 2020, from: http://visionofhumanity.org/indexes/global-peace-index/

Wachowiak, H. (Ed.). (2006). *Tourism and borders: Contemporary issues, policies and international research*. Ashgate.

Ward, E. V. (2009). Conñicts of interest: Plasticity of peace tourism and the 21st century nation. In J. Harris (Ed.), *The nation in the global era: Conflict and transformation* (pp. 304–316). Brill., doi:10.1163/ej.9789004176904.i-476.109

Wohlmuther, C., & Wintersteiner, W. (2014). *International handbook on tourism and peace*. Retrieved January 5, 2020, from: https://www.e-unwto.org/doi/pdf/10.18111/9783854357131

World Travel and Tourism Council. (2016). *Tourism as a driver of peace: Quantitative analysis on the link between peace and tourism*. World Travel and Tourism Council.

Yin, M., & Zhao, S.-Z. (2006). Research on a dynamic model of trust-building within regional tourism alliances evidence from China. *Chinese Economy*, *39*(6), 5–18. doi:10.2753/CES1097-1475390601

Yoshida, K., Bui, H. T., & Lee, T. J. (2016). Does tourism illuminate the darkness of Hiroshima and Nagasaki. *Journal of Destination Marketing & Management*, *5*(4), 333–340. doi:10.1016/j.jdmm.2016.06.003

KEY TERMS AND DEFINITIONS

Cross-Border Tourism: Provision of hassle-free movement of people in conflict-affected regions to create deeper understanding and relationship-building.

Deteriorations: The state of a nation having a low level of global peacefulness due to high militarisation, less of societal safety and security and more of ongoing conflicts.

Global Peace Index: An indicator to measure the state of the peacefulness of assessed countries in the domains of the degree of militarisation, level of societal safety and security, and extent of ongoing domestic and international conflicts.

International and National Conflicts: Indifferences and misunderstandings due to social injustice, resources crisis, political unrest leading to militarisation and societal safety and security concerns.

International Openness/Open Tourism: The state of free travel to nations due to the provisions of high safety and security, hassle-free visa and border regulations to travellers.

Militarisation: It is the indicator of military expenditures on the number of armed forces per 1, 00,000 citizens, export and import of nuclear and heavy weapons, overall a process of readiness for war and conflicts.

Peace Parks: Parks curated as a symbolism of endurance of peace and sustainability of culture and ecosystem.

Peacemakers: Leaders, humanitarian, activists, organisation and individuals working to eliminate social injustice and conflicts, conservation of human rights and ecosystem for the attainment of peace and sustainability of the society.

Chapter 4
Tourism and Peacebuilding From a Holistic Approach:
The Case of Trails for Peace

Silvia Aulet
https://orcid.org/0000-0003-4022-6568
University of Girona, Spain

Edgar Tarrés
University of Girona, Spain

ABSTRACT

Sánchez Sánchez and Fernández Herreira raise a reflection on the three dimensions of peace: form a personal dimension and the spiritual sphere, from a social dimension, and form an environmental-natural dimension. Tourism can play a fundamental role as a tool for establishing ties with oneself and with other people to contribute to the development of peace. In this chapter, authors propose to study the different dimensions of the concept of peace by studying the project of Trails for Peace, a project born as a non-profit social entrepreneurial project that integrates different routes around the world that have a special significance for their relationship with some conflict areas. Trails help travelers to know the natural areas and to build inner peace through various spiritual activities. The project helps to promote the welfare of people by involving local communities to foster contact between visitors and tourists and sustainable development. This chapter aims to analyse the impact the project has on the areas where these paths have been developed.

INTRODUCTION

Rodríguez Rojo (1995, p. 46) understands peace as "the non-violent regulation of the conflict (personal, sociopolitical, environmental) to positively build justice or a New International Order that implies analytical, critical, creative and global awareness". Bauman (2007, p. 13) says that if human beings want peace, they must worry about justice. Peace must be understood from an integrative and transformative

DOI: 10.4018/978-1-7998-5053-3.ch004

perspective and must have a holistic vision of reality. Peace is not living without conflict; having conflicts is natural in human nature; what is not natural is the absence of justice and the absence of individual commitment. Peace, or the aspiration for peace, work for peace, should seek the implementation of fundamental values in the human person, such as (Ruiz Monroy, n.d.), autonomy, charity, freedom, equality, justice, forgiveness, commitment or recognition of the other.

Sánchez Sánchez and Fernández Herreira (1996) raise a reflection on the three dimensions of peace: in the peaceful, harmonious relationship of man with himself and with others; in the relation intra and inter groups, institutions, peoples, states (at the level of social structures) and in the harmonious relation of the man with nature. We can refer to three dimensions of peace: personal, social and environmental.

Tourism can play a fundamental role as a tool for establishing ties with oneself and with the others in order to contribute to the development of peace. In this chapter, we propose to study the different dimensions of the concept of peace by studying the proposal for Trails for Peace.

Through this chapter, authors want to present this project and analyze the impact it has on some of the areas where these paths have been developed. Gandhi claims that peace is the way, not the goal or the ending point. Peace is the way, the commitment, the action, the conviction of each person to walk (Fisas, 2002, 17).

The chapter is structured in four main sections (apart from this introduction). In the first section, the concept of peace will be presented based on the interconnection of the three dimensions mentioned above. This section aims to present the philosophical approach of the authors concerning the area studied. In the second section, the case study will be presented, describing the origin of Trails for Peace, the different paths that exist already and how they function. In the third section, an analysis of the impacts of Trails for Peace regarding peacebuilding will be done. Finally, the last section is the conclusion, reflecting on how the case study presented can foster peacebuilding through tourism.

PHILOSOPHICAL BASES; THE CONCEPT OF PEACE

As mentioned in the introduction, peace will be addressed from three perspectives, according to Sánchez Sánchez and Fernández Herreira (1996), peace from a personal dimension, from a social dimension and from an environmental dimension.

Before the three dimensions are explored, a brief definition of the concept of peace will be presented. Rodríguez Rojo (1995, 46) understands peace as "the non-violent regulation of conflict (personal, socio-political, environmental) in order to build positive justice or a new International Order that involves analytical, critical, creative and global awareness." Peace should be understood from an integrative and transformative perspective and must be approached with a holistic vision of reality. Thus, Mayor Oreja says that

Peace in the world can only begin with peace with yourself and then peace with all the organizations in which you participate more actively and directly. Peace is the result of the sum of personal attitudes, based on the search for truth, from freedom. (quoted in El Mundo, 2009)

Peacebuilding should seek to implement fundamental values in human beings. Banda and Tarradellas (1991, 99-100) make six statements or proposals that are key to understanding the concept of peace. The first one is that where there are people, there are conflicts. It is not the conflict that is avoidable but the violent solution of it. Despite the instinct of aggression innate in the human personality, work must be done to foster the rejection of violence as a way to deal with conflict. The second one refers to peacebuilding as a historical process. Peace is not a circumstantial and concrete claim that can be achieved in one concrete moment; peace is not an episode. It has all the characteristics of a peculiar and unprecedented historical process, establishing a culture for peace must be the result of the creative work of different generations were peace education play a fundamental role in the construction of the future.

Thirdly, it is not possible to live without accepting a certain risk. The obsessive eagerness to eliminate risks eventually creates new ones. When a reasonable share of risk is not accepted, people may fall into irrational actions that create new risks and generate a chronic state of anxiety. Bauman (2007, 16) refers to today's society as one who feels "miserable and vulnerable, overwhelmed by (and perhaps subordinated to) forces that neither control nor fully understand; a population terrified of their own helplessness and obsessed with the effectiveness of their borders and the security of the population that inhabits them".

The fourth statement is that peace is not free. Peace has a price that must be paid in two ways: with personal, solitary work, done with one's interiority; and with social work, with solidarity, with society. The active presence in society of people who have internalized peace-generating values, people who have made serious personal choices in this field, is needed.

Strategies for peace must be created. There are issues for which there is still no answer. Further research on peacebuilding is needed and should be encouraged.

Last, the world cannot be understood without understanding the concept of interdependence. The world has become small but humankind has not grown big. Panikkar was one of the great philosophers who also defended the concept of interdependence, there is nothing completely separate, and dualities do not exist (Pigem, 2007).

Several institutions have worked on tourism as a peacebuilding tool, as the International Institute for Peace through Tourism (IIPT) born in 1986 or the World Tourism Organization (UNWTO) and research on this topic has been promoted since the early days of tourism research (Wohlmuther and Wintsersteiner, 2014). Farmaki (2019) remarks that, while peace is a fundamental issue for tourism, the role of tourism to peace can be questioned in some cases because of the negative effects it may have on the developing areas. For becoming a tool for peacebuilding, tourism should consider political, economic and social dimensions (Farmaki, 2017).

Actual research tend to see peace as a complex reality not only considered as the absence of war and violence. As Haessly (2010, 5) states "peace researchers describe peace as the presence of some desirable conditions in society, including integration, justice, harmony, equity, freedom, wholeness, the promotion of the dignity of each person and the wellbeing of all of a society's citizens".

In order to contribute to peacebuilding, tourism should:

- Contribute to development and poverty reduction by offering job opportunities to local people, including youngsters and women, and supporting the development of local businesses. Tourism industry can also support other complementary activities such as agriculture and handicraft (Moufakkir & Kelly, 2010; Wohlmuther & Wintsersteiner, 2014; Becken & Carmignani, 2016)
- Help to protect cultural landscapes and traditions, to preserve tangible and intangible heritage and to make local communities proud of their identities while promoting intercultural exchanges among hosts and visitors (Moufakkir and Kelly, 2010; Wohlmuther & Wintsersteiner, 2014; Becken & Carmignani, 2016)

- Help to the sustainable development of the regions and encourage local communities to be more aware of natural environment and sustainability, protecting landscape, fauna and flora (Moufakkir & Kelly, 2010; Wohlmuther & Wintsersteiner, 2014; Becken & Carmignani, 2016)
- Promote a culture of peace and eliminate the conditions that lead to conflict by involving local community in the decision-making process (Wohlmuther & Wintsersteiner, 2014; Becken & Carmignani, 2016).
- Protect spiritual traditions of local communities, including tangible (sacred spaces, build heritage) and intangible (rituals and traditions) heritage. Special attention is payed to pilgrimage trails (Wohlmuther & Wintsersteiner, 2014; Becken & Carmignani, 2016; McIntosh & Harman, 2017).

As seen, the links between tourism and peace are not related to only one dimension but affect the economic, social and environmental dimensions. Quoting Wohlmuther and Wintsersteiner (2014) a complex understanding of tourism and peace is needed, "we have to study tourism in the broader context of human mobility while developing a concept of peace that goes far beyond the political dimension or negative peace (absence of war)" (Wohlmuther & Wintsersteiner, 2014, p. 19). In the following sections, the different dimensions of peace mentioned above are going to be explored in order to apply them into the case study.

The Personal Dimension: Spirituality

The first dimension to explore is peace from a personal and spiritual dimension, being well with oneself. Panikkar (2004, p. 126) states that living without outer peace is difficult; living without inner peace is impossible.

Spirituality is a concept that remains a term difficult to explain due to the great diversity of approaches adopted towards it. The word spirit comes from the Latin word spiritualitas, related to the Greek word pneuma meaning breath; spiritus refers to the breath, the air. The word spirit also has several meanings, but in general they refer to a substance or an incorporeal being, not a material one. It can also refer to an internal principle of people's resolutions (to have a spirit to do something, to be poor in spirit) or to an impulse according to which we can act (spirit of rebellion, the spirit of sacrifice).

The Western dualistic vision of spirituality (spirit versus body/material) is complemented by a trinitarian vision in which the three dimensions of human beings converge: body, psyche and spirit (Salom, 2006, 33). The idea of spirituality varies across disciplines and different cultural and religious traditions but, essentially, refers to the individual search for the meaning of life (Willson et al., 2003). Spiritual consciousness is natural and universal among humans. Spiritual pursuit has become a dominant feature of late-twentieth-century life. It seems that while participation in religious institutions and practices are diminishing, the search for spirituality is growing rapidly through the search for a huge variety of beliefs and practices like feng shui, yoga or retreats among others (Sharpley & Jepson, 2011; Aulet, 2018).

As mentioned by Willson et al. (2013), spirituality affects higher levels of Maslow's Hierarchy of Needs. According to Maslow (1963), healthy people have met their basic needs: physiological, security, surrender, love, respect, and self - esteem so that they are motivated by trends that lead to self-fulfilment (quoted in Sánchez Sánchez and Fernández Herreira, 1996, 42). Maslow sees self-fulfilment as personal growth and development of one's potential; Pigem adds that self-fulfilment is a key element that should be the key to our personal and collective life:

An economy in line with life must focus not on the growth of monetary goods but the growth of people and life: it must be an economy of self-fulfilment. (Pigem, 2015, p. 122)

Thus, it is widely accepted that the search for spirituality today is closely linked to the pursuit of wellbeing understood as a multidimensional concept that encompasses many aspects of human life (Bimonte & Faralla, 2012).

Spirituality can also be described as interconnectedness or connection, which refers to the search for a harmonious or unity relationship with oneself, "others" (including other people, animals, Earth, and nature) and / or God / higher power (Willson et al., 2013, p. 153). Spirituality is defined as "experiencing a meaningful connection with our nuclei, other human beings, the world and / or a greater power, expressed through our reflections, narratives and actions" (Schulz, 2005, p. 4).

Panikkar, on numerous occasions, also refers to the need for personal inner growth, the need to develop contemplative and meditative capacities, as superficiality is one of the fundamental problems in our society. Thus, he says,

The man has two movements: the centrifuge and the centripetus, outward, which is love, and the centripetal, inward, which is meditation. The great modern dichotomy brings into play the polarity between knowledge and love. Both are bound together: one cannot know without loving, and one cannot love without knowledge. (Bassas, 2008, p. 253)

This reflection is what leads us to talk about this outward movement in the next section, which is also understood as a relationship with others.

The Social Dimension: Intercultural Dialogue

Panikkar (2004, 137) states that only forgiveness, reconciliation and ongoing dialogue lead to peace. As mentioned above, the reality must be holistically perceived and the culture of participation and commitment to the social and natural environment must be promoted. An important element to keep in mind is solidarity, understood as the supreme ethic of survival.

Pigem (2010) mentions values such as tolerance, equality, solidarity and freedom. These values (along with others) are highly prized. Pigem (2010), following Aristotle, reflects on the "value of use" and the "value of change" and states that not everything useful is considered valuable by the economy. It uses the paradox that Adam Smith already mentioned about water; water is very useful, but its purchasing power is very low, unlike diamonds.

This paradox has much to do with the unsustainability of the contemporary economy. Like stocks, the trading of ethical values also fluctuates, and it does so very quickly in times of uncertainty like ours. What values must rise in our culture and our imagination to achieve a sustainable society? (Pigem, 2010, p. 15).

The values, therefore, affect not only the economic or environmental, but also the social aspects; what is the change that must take place?

The Earth Charter (2012-2016) calls for the need for a shared vision of the core values that can guide the behavior of individuals, organizations, businesses, governments and transnational institutions towards greater sustainability. In this sense, dedicate Chapter III to social and economic justice and chapter IV to democracy, nonviolence and peace. Especially point 16 refers to:

Promote a culture of tolerance, nonviolence, and peace.

a) Encourage and support mutual understanding, solidarity, and cooperation among all peoples, within, and among nations.
b) Implement comprehensive strategies to prevent violent conflict and use collaborative problem solving to manage and resolve environmental conflicts and other disputes.
c) Demilitarize national security systems to the level of a non-provocative defense posture, and convert military resources to peaceful purposes, including ecological restoration.
d) Eliminate nuclear, biological, and toxic weapons and other weapons of mass destruction.
e) Ensure that the use of orbital and outer space supports environmental protection and peace.
f) Recognize that peace is the wholeness created by right relationships with oneself, other persons, other cultures, other life, Earth, and the larger whole of which all are a part.

Peace in relation to people contemplates tolerance and respect. However, the encounter and dialogue between people from different backgrounds should also be encouraged. Panikkar states that there is no peace without interculturality, he argues that if there is a possible solution to the turmoil of the contemporary world, it cannot come from a single religion, tradition, philosophy or culture, but must be constructed in collaboration between different traditions, religions, cultures and philosophies of the world (Boada, 2004, 23).

According to UNESCO (2019), interculturality is understood as all those evolutionary relationships between different cultural groups, the equitable interaction of different cultures and the possibility of creating new-shared expressions. Panikkar (2004) strongly emphasizes the idea of equitable interaction, since he argues that intercultural dialogue is often based on a situation of supremacy on the part of some of the participants. Often the processes of interaction between diverse cultures occur in settings where part of the participants are immigrants who dialogue with residents; this situation does not have to generate a subordination or the thought that one culture or idea is above the others or that it has the absolute truth. This can also be applied to tourism. Authors like Palmer (1994), Hall and Tucker (2004) or d'Hauteserre (2004), among others, explore human relations in the field of tourism, stating that in some cases unequal relations between residents and tourists occur. In this case, intercultural dialogue becomes almost impossible.

Some of the challenges facing an intercultural society, as proposed by the School of Culture of Peace of the Diputació de Barcelona (2010), are to guarantee of rights, both in the sense of avoiding discrimination and of generating the false idea of favoritism, and to guarantee equality between all groups by promoting the participation of all of them in the decision-making process.

UNESCO has a specific program, Spiritual Convergence and Intercultural Dialogue, which seeks to foster and promote dialogue between different religions and spiritual traditions as a tool to promote peace and respect for religious and cultural pluralism. Dialogue requires communication and interpretation.

It is not enough for my Quechua speaker to speak Spanish and for me to understand their language, it is also necessary for me (or anyone else) to know Quechua so that we can both confront and verify our respective translations, here we are referring to a knowledge of the spirit of a language and not just its vocabulary. (Panikar, 2004, p. 58)

For dialogue to be fruitful, inter-cultural dialogue must be articulated with intra-cultural dialogue; each tradition (or each person) must dialogue with itself in order to lay the groundwork for true dialogue. Dialogue is based on exploring the identity of the other, establishing an exchange to enrich one's own beliefs or traditions and becoming aware of the value and richness of diversity (Regull, 2007).

The Natural Dimension, Sustainability

Berger (1975, p. 42) described current development as "the growing condition of uprooting. The creation of uprooting is both a consequence of the ecological destruction of the Earth and of breaking the cultural and spiritual links of the population with the Earth. [...] Ecological destruction is, in essence, the destruction of Earth as to a spiritual and ecological home".

Environmental degradation constitutes a threat to the development of today's society as we know it, and therefore, a barrier to the aspirations of a more just, united and, ultimately, peaceful world.

Another of the challenges facing contemporary society is sustainability. Sustainable development is the one that ensures the needs of the present without compromising the ability of future generations to confront their own needs, as defined in the Bruntland Report in 1987.

It is essential to manage resources so that they can regenerate themselves as they are used. The three principles are:

1. Ecological sustainability, which ensures that development is compatible with the maintenance of essential ecological processes, biological diversity and biological resources;
2. Cultural and social sustainability, which ensures that development increases people's control over their lives and maintains their values and culture as strengths of community identity;
3. Economic sustainability that ensures development is economically efficient and resource managed so that they will endure for generations to come.

Since the publication of the Bruntland Report, little progress has been made on the road to sustainability. The Earth Charter (2012-2016) states that we are facing a critical moment in history, with a future with great challenges and risks.

The resilience of the community of life and the wellbeing of humanity depend upon preserving a healthy biosphere with all its ecological systems, a rich variety of plants and animals, fertile soils, pure waters, and clean air. The global environment with its finite resources is a common concern of all peoples. The protection of Earth's vitality, diversity, and beauty is a sacred trust. (Earth Charter, Preamble, 2012-2016)

Numerous agencies have raised their voices of concern about the current situation and the uncertainty of the future. In this context, in 2015, the UN approves the 2030 Agenda on Sustainable Development with 17 Sustainable Development Goals. These goals represent the way forward for a sustainable future for all. They interrelate with one another and incorporate global challenges such as poverty, inequality, climate, environmental degradation, prosperity, peace and justice.

Pigem's (2010) vision of sustainability is also holistic, the essence of sustainability is the fact that everything is related to everything, and therefore sustainability cannot be a simple speciality. Issues related to sustainability must emerge from academia and politics to root for culture and personal decisions. Thus, Pigem demands a change of values to move towards a true sustainable development.

Without this change of values, he path to sustainability is really difficult. We are the heirs of a materialistic vision that sees the Earth as a mere storehouse of resources and not as the Mother Earth of traditional indigenous cultures.

Several authors have explored the connections between peacebuilding and sustainability. Vandana Shiva (2004) states that in tribal and agrarian cultures, the cultural and religious identity derives from the land. Earth is the soul of society; it embodies the spiritual home of most cultures. She has defended the importance of seeds, as she says that everything starts being a seed. She fights against globalization and defending ideas like food sovereignty.

Schmitz et al. (2012) propose a reflection on sustainability based on a holistic and inclusive model where the fields of peace and conflicts studies and of economics are highlighted. Swain and Krampe (2011) state that environmental stress is one main catalyst that creates societal insecurities that may create conflicts. As mentioned above, the Sustainable Development Goals are aligned with these ideas as well, taking as bases the interconnected theories, being an example of new attitudes that are arising in coherence with the intuition of interdependence.

Fragmentation has led to our uprooting and ecological devastation. Roszack (1985) affirmed that the needs of the planet are the needs of the person and the rights of the person are the rights of the planet. But although this is a trend (and a challenge) is not new. In 1854, the Sioux chief Seattle sent a letter to the President of the United States remembering the importance of caring about the environment:

This we know: the Earth does not belong to man, man belongs to the Earth. All things are connected like the blood that unites us all. Man did not weave the web of life, he is merely a strand in it. Whatever he does to the web, he does to himself.

Chief Seattle

Following this idea of interconnection or interdependence, we can see that the three dimensions of peace are all connected. In the following section, the proposal of Trails for Peace will be presented in order to by analyzed according to this three-dimension perspective.

THE PROPOSAL OF TRAILS FOR PEACE

The project Trails for Peace is a Mindful Travel Destinations project that was born as a non-profit social entrepreneurial project. The project integrates different routes around the world that have a special significance for their relationship with some conflict, for the defence of the rights of the weakest, for their natural beauty, for their commitment to sustainability or their spiritual meaning. Trails for peace is a global project which looks for peace through hiking in special paths along the planet. Routes are designed in order to promote better awareness of peace in both the destination and the people walking through it.

Designing the routes where old war conflicts aroused in the past, is also important to promote peace in the world, consciousness, and tools to get a better life: meditation, silence, sharing stories of locals, nature and time for oneself. The main idea is to bring travellers to know natural places and authentic destinations, as well as to find inner peace through activities such as forest bathing and mindfulness. These experiences are thought to bring balance among body, mind and soul. A coherence, very useful nowadays, to face the ominous results of our new lifestyle and the growing nature deficit.

In a moment where wellbeing and transformative tourism is growing and gaining followers (Reisinger, 2013) this project is also looking to generate wealth to local communities through tourism experiences. The mission is to promote wellbeing to both visitors and locals.

The first Trail for Peace was set in the Sierra de Guadarrama near Madrid (Spain), in the Embalse de La Jarosa. The second one is about twenty kilometres out of Bogotá, in Colombia. Two new projects are about to come shortly, one in the isle of Menorca, and the other in the Cap de Creus Natural Park, in the municipality of Roses, north Costa Brava.

Trails for Peace in Spain

The first project of Trails for Peace was build in the Embalse de La Jarosa, about fifty kilometres from the capital of Spain, Madrid, in the middle of a Natural Park.

The Spanish Civil War occurred between the years 1936 to 1939, originated by a climate of internal turmoil due to a difficult economic, political, and social situation. In 1876, General Primo de Rivera carried out a coup d 'état by which he assumed the government of Spain, establishing a dictatorship that would last until his resignation in 1931. At this moment, the Second Republic was proclaimed, a very politically unstable period in which during five years 13 different governments followed one another. This period is characterized by a strong confrontation between parties of the right and the left (Rodriguez, 2011).

In the elections of 1936, the left-wing parties obtained power, and the President, Manual Azaña had a devastating attitude against estates the Church, the army, and the aristocracy. This attitude led to an increase in the differences between left and right, generating strikes and violence on the street. The trigger for the military uprising led by General Francisco Franco was the murder of Calvo Sotelo, a right-wing leader (Rojo, 2017).

This was the beginning of a bloody civil war between the two fronts, the Republicans (left-wing) and the National Army (right-wing). The war had multiple facets such as class struggle, religious war, and confrontation of nationalisms, among others. Both sides committed serious crimes such as disappearances of persons, extrajudicial courts, looting, etc.

Officially, the war ended in April 1939 with the appointment of General Franco as Spanish Head of State, but the post-war was also a very tough period both socially, politically, and economically and some of the wounds are still open.

As an example, it was not until 2007 that the government approved the Historical Memory Law to recognize and expand the rights of those who suffered persecution or violence during the civil war and the dictatorship, not without generating some discomfort. This situation was repeated in 2018 when the government decided to move the remains of General Franco outside the mausoleum where he was, generating a great public debate and raising opinions against it by some political parties.

The Embalse de la Jarosa, in Guadarrama, has the footprint of the Spanish Civil War as the path runs the old route where, during the war, both fronts Republicans and the National Army fight each other.

The Guadarrama trail takes advantage of the itinerary of the Trenches Route, already created in 2012 by the Guadarrama City Council, which is part of the network of thematic trails in the town. The main aim of the project is to transmit a message of peace combining walking, silence, and nature.

The project was launched in June 2018. An organized walk was proposed to present the project and establish the signposting along the path as well as an explanatory placard at the car park at the beginning of the route.

The inauguration day a speech about the project, the route, and the Civil War heritage and landscape was done to all the attendants. After the speech, and during the walk, mindfulness activities and forest baths took place. People attending include authorities, local people, and families. Activities were planned also taking into account children and were related to emotional intelligence and the importance of working with inner and world peace. This project has been developed by Mindful Travel Destinations, Koan Consulting, and Aulas en Calma working together with the City Council of Guadarrama with the support of other institutions.

After the inauguration, the municipality of Guadarrama is in charge of maintaining the roads and signalling so that visitors can walk individually. As mentioned before, the placard placed at the beginning of the walk remind visitors of the history of the place and the importance of walking mindfully. Guided walks with activities such as the one that took place on the opening day are also organized eventually.

A new Trail for Peace in Spain is now developing in Roses (Northeast of Spain). It is located in front of the sea, at the entrance of the Cap de Creus Natural Park. During the Spanish Civil War, there was a military site from where to control the ships entering the Bay of Roses; it was a strategic point. Some years later, after the war finished, the place was converted into military barracks used for military service. There is a coast path following the entire coast from Roses to Cadaqués. An amazing walking all over the cliffs, tiny rocky beaches, breathtaking sunrises...

The municipality of Roses and the Natural Park are now working together to design the Trail for Peace. This trail was supposed to be opened in June 2020, but due to the Covid-19 pandemic, the inauguration has been postponed. During 2019, while the Trail for Peace in Roses was being prepared the municipality of Roses started to organize mindful activities: forest bathing and mindfulness experiences as part of this project. The results have been very positive. In the activities proposed there was people from abroad and inhabitants of Roses. Although usually it is strange to see foreigners mixed with locals in the activities, everything relating wellbeing is an exception.

Trails for Peace in Colombia

The second Trail for Peace, opened in 2018 in Colombia, in Choachi. Colombia gained its independence from Spain in 1810. At this moment, political parties with multiple interests began to form, as well as new non-regularized revolutionary groups. The two main parties were the Conservative and the Liberal. The first protected the interests of the wealthy class; while the second, was presented as a reformist alternative and in defence of the interests of merchants and less favoured groups in society. The fight between both forces was evidenced through 54 civil wars that divided the population, pitting families, friends, and neighbours against each other (CIDOB, 2014).

During the presidential elections of 1948 the presidential candidate Jorge Eliécer Gaitán was killed. People went outside the streets to complain, overwhelmed with fury, desolation and sadness, looting buildings, burning public and private property, violating constitutional laws, and filling the city of Bogotá with violence. This event is known as "El Bogotazo" which ends with the coup d'etat of General Gustavo Rojas Pinilla in 1953 (CIDOB, 2014).

In 1956 the conservative party and the liberal party signed the Benidorm Pact in Spain, an agreement that established a system of political alternation between both forces. This pact was valid until 1974 and was defended as a return to the democratic system after the dictatorship. However, the pact included only the two traditional parties, causing the increase of opposition movements and the origin of the contemporary guerrillas. Among the most notorious guerrillas are the Fuerzas Armadas Revolucionarias de Colombia, Ejército del Pueblo (FARC-EP), the Ejército de Liberación Nacional (ELN), the Ejército Popular de Liberación (EPL) and the Movimiento Diecinueve de Abril (M-19). Thus, one of the characteristics of the conflict in Colombia is the plurality of actors (Menchero Sanchez, 2018).

In 2012, the Colombian government publicly announced talks with the FARC-EP, one of the main guerrilla agents, to start a negotiation process to find a solution to the armed conflict. After more than 50 years of conflict, in 2016 a "Final Agreement for the Termination of the Conflict and the Construction of a Lasting Peace" was signed in Havana. That same year a popular plebiscite is called to ratify the agreement, but the victory of the "no" leaves the process in a situation that forces the process to be renegotiated. After several summits between the government and the promoters of the "No" and introducing modifications to the agreement, at the end of the same year the FARC re-signed the agreement (Wola, n.d.).

Finally, in June 2017, the FARC finished the total delivery of the weapons, leaving behind a path full of devastation and violence. According to the Historical Memory Center, the conflict caused the death of 40,787 combatants and 177,307 civilians, to which are added about 25,000 disappeared, 27,000 kidnapped and 150,000 murdered (Centro Nacional de Memoria Historica, 2012).

The Trail for Peace was placed in Choachi, in the department of Cundinamarca, 40 quilometres from Bogota. It is an area of leafy vegetation, with abundant hills, ravines and rivers. According to Bonilla (2018), this is an ideal place due to its proximity to Bogota, and because suffered the conflict as in 1995 it was taken over by the FARC. It was considered a strategic place because of the road that connected different areas from the south with Bogota.

The Trail for Peace was inaugurated in October 2018. It started with one of the responsible for Tourism and Peace, who was a former military woman. The promoters of the project, together with the Mindful Travel Destinations are the municipality of Choachi and the association for ecotourism, Ventana de la Luna.

On the day of the inauguration, many different people attended the event, including representatives of the government, people living in the area, and representatives of different associations. They all walked around the trail mindfully. Contemplating in silence the breathtaking views. Sharing tea and food, and awakening very high and positive emotions.

In this trail, the same as in the first one of Madrid, there is an explanatory placard informing about the trail, the project, and the history of the place. Unlike the Madrid trail, the Colombia trail is not prepared to be walked individually but accompanied by a guide. The Choachi trail develops in the middle of the Aguadulce path in an environment of surprising nature and with a spectacular landscape that leads the traveller to enter a territory full of history, gigantic plants, streams, and ascents up to 3000m. The guide tells stories to the visitors related to the traditions of the community, as well as the recent history (Bonilla, 2018). The aim is to generate empathy with the community, with the environment, with care for the environment, with other walkers, and with the message of peace towards the world. Some food prepared by the host community is also provided during the walk.

In this process, members of the community get involved as guides, others prepare the pastries, others start to provide accommodation services, and there is also a house where additional food products are sold.

TRAILS FOR PEACE, TOURISM AND PEACEBUILDING, IMPACTS AND CHALLENGES

The proposal of Trails for Peace aims to foster peacebuilding in the three dimensions mentioned above, and it is a clear example of tourism development. The tourist proposal of Trails for Peace could be labelled under different typologies of tourism: mindful tourism, responsible tourism, wellness tourism, sustainable tourism, that are generally associated to peacebuilding (Isaac, 2014). It is difficult to assess, a priori, the impacts of tourism development on destinations and also the impacts different activities have on peacebuilding. There have been several attempts to establish a system of indicators for assessing peacebuilding, but, as noted by MacGinty (2013), most of them present deficiencies. The main causes are (MacGinty, 2013):

- The lack of agreement among different actors and institutions on the definition of peace, most of them focusing mainly on the idea of negative peace proposed by Galtung (1969) related to the absence of war but leaving aside other dimensions.
- Most of the indicators try to assess specific projects when peacebuilding is a long-term process.
- Many of the indicators are originated top-down
- The results tend to be presented as statistics and are published in governmental and official sources. A more qualitative approach such be needed and all populations (especially those living in conflict areas) should have access to the data as well.
- The indicators proposed tent to offer a global view and do not present the differences within and between communities.

Hinds & Carter (2015) also refer to a lack of accurate data, in many cases, as a barrier to elaborate indicators. Among the different institutions that have attempted to provide a list of indicators to assess peacebuilding, it can be mentioned the Everyday Peace Indicator Project, the International Dialogue on Peacebuilding and State-building, the Globally-Accepted Indicators (GAIN) for peacebuilding or the Global Peace Index or, even, the indicators for assessing the Sustainable Development Goals (United Nations, 2017). On the other hand, some authors have explored how can tourism contribute to peace, focusing, in this case, on three main areas: intercultural understanding and education, economic stimulus and development, and cooperation for development (Becker & Carmigani, 2016).

In the case of Trails for Peace, trying to assess the impacts is difficult for two main reasons. The first one is the lack of data referring to the number of visitors. The second one is the difference between the different places where the paths are located, as the context is different in each case.

For assessing the peacebuilding process in the case of Trails for Peace in-deep interviews have been made. Three interviews have been done with the promoters of Trails for Peace (one for Colombia and two for Spain). The information collected has been organized according to the three dimensions of peace mentioned in the literature review. Table 1 shows a comparison of the results obtained for both the path in Colombia and the path in Spain.

Table 1. Trails for Peace and Peacebuilding

	Spain	Colombia
Personal dimension	• Mindful activities are organized that help visitors to work the dimension of spirituality and internal growth, but only eventually	• The guide proposes visitors places were to rest and meditate while walking • The nature and landscape offer the possibility to connect with oneself while walking
Social dimension	• The history of the place is explained using interpretation tools, like placards, brochures and the forthcoming app. • Visitors are mainly local, it's important from the perspective of the Spanish Civil War and healing old hurts • In the case of Roses, international visitors get in touch with local people in the activities and are aware of the conflict as well • Increase of tolerance (peacebuilding)	• The guide is always from the local community, and normally visitors are international or from other areas of the country, providing the chance to get in touch with local traditions, believes, etc. • The guide explain the history • Food and accommodation is also provided by local communities • Increase of tolerance (peacebuilding) • Increase of income for local population thanks to tourism than decreases the risk of conflicts
Environmental dimension	• The two trails in Spain are developed in Protected Natural Areas	• The landscape is spectacular. The tourist development is based on the concept of sustainable and community tourism.

Source: The authors

The first dimension presented is the personal dimension or the one that connects with spirituality and inner peace. As mentioned before, one of the main aims of the project is to make people more aware when they walk, by participating in activities such as meditations, mindfulness, or forest baths.

Developing Trails for Peace means creating a place where to organize such mindful travel experiences. This is a way to innovate in tourism; bringing new ways to enrich the products, for which tourists will travel to the destination, while at the same time, enriching the lives of those enjoying them. A useful win-win proposal with clear positive results for society.

In both cases, different activities are proposed in the trails. In the case of Spain following a calendar, in the case of Colombia, having a guide for the tour that has to be contacted through the association promoting ecotourism. The activities focus on making people aware of the place where they are. When people start to walk along a Trail for Peace path, they must be aware that long ago, somebody had to pace the same road under very different circumstances, under a context of war. The fear or panic that such unknown people suffered can be perceived. They should walk in silence; mindfully. Whatever activity they perform, must be approached with humility, feeling that nature as well. Coming back to the present, they should realize how important is spreading peace all over the world. Global changes come from individuals. Activities such as yoga or mindfulness can help in developing oneself and encourage having a holistic approach that can foster changes in daily lives.

When an activity is performed in a specific place, the energy used in the activity is easy to perceive. It is not weird that people feel sad or fearful in places where there have been killings or suffering. The sites catch the energy of what is done or happened in there, so people can perceive this. It is easy to see when people can measure the frequencies of the sites. Low frequencies relate to energy dealing with suffering, violence, fear... and we can change them by performing some other kind of activities, relating to peace, love, and positive emotions. Such beautiful activities, done, again and again, lots of time, manage to transform the energy or the sensations that history has left along the landscape. The results are something that can be measured. The site changes and we change ourselves with it.

Reisinger (2003) explores the different dimensions of transformational tourism; Sharpley (2009), Aulet (2018), or Collins-Kreiner (2019) have studied the connections between spiritual tourism, pilgrimages, and nature. The road has always been believed as a journey. A special journey leading to our inner self. A spiritual path to connect human beings with nature and with the Universe, as a unity. Therefore, a road leading to peace must be a road that unifies everything: body and mind, human being and nature, human being and universe. Trails for peace recommend mindfulness techniques to go along such beautiful paths, so this way visitors will improve health, wellbeing, creativity and will develop the necessary tools to transform life to bring peace inside and around oneself: peace of mind. Lee et al. (2012) try to define the idea of peace of mind and propose a list of words related to peace and harmony that include peaceful, serene, and calm, at ease, comfortable, harmonious, and balanced, that were mentioned by one of the interviewees when asked for feedback among the participants of some of the activities organized.

When people leave the activity they leave with a great sensation of wellbeing. Even those who had a bad experience, because they realize that they have or had an emotion which didn't let them enjoy as they should. In the final circle, when everybody shares what they felt, those having such bad experience they understand, and then they change. The power of consciousness. When we perform activities where Silence, Nature, Awareness, and good feelings are involved, something moves inside. We can feel lots of beautiful sensations which bring peace inside; what we need to live with balance and harmony. (personal communication)

The second dimension to explore is the social dimension. The experience not only should transform visitors but communities.

The increase in cultural exchanges between the resident population and visitors is positive as it can help to maintain local traditions and preserving cultural heritage. Tourism can also support local small businesses, improve local quality of life, and foster positive social changes in terms of greater tolerance and wellbeing.

In the case of Trails for Peace is difficult to assess because there are very few data available on the number of visitors and their activities in the destination. From the interviews done, we can see significant differences in the different trails.

In the case of the Trail for Peace in Guadarrama, visitors and participants are local, there are no international visitors. According to the interviewee, this is a positive point as in the case of the Spanish Civil War there are still "many opened hurts" at the national level. According to the promoters, the two new trails that are being prepared (in Roses and Menorca) probably will have more presence of international tourists as they are located in developed tourism destinations.

Regarding the path in Colombia, visitors come from other parts of Colombia and also, but less, from other countries. As Farmaki, Khalilzadeh and Altinay (2019) state, safety and security issues are important issues related to demotivation, tourists look for low-risk destinations. Peace is necessary for tourism, but, in the case of Colombia, tourism is a key element of social, cultural, and environmental development, that can help reduce inequalities and differences caused by the conflict (Menchero Sanchez, 2018).

In the case of Colombia, as mentioned, local communities provide most of the services, being a base for the development of community tourism, which is considered to be a positive element as it helps communities to improve their quality of life and to reduce poverty. Authors refer to the enhancement of local communities, thanks to tourism (Suansri, 2003; Goodwin & Santilli, 2009).

"Peace through tourism stands on the belief that cultural understanding can promote attitude change and thus facilitate peace" (Sharma, Charak and Kumar, 2018, 41). Promoting the encounter of visitors with locals can help increase tolerance. As mentioned before, for a real intercultural dialogue is important to see each other as an equal and not to establish hierarchical relationships. The inter-cultural dialogue aims to make visitors aware of the values of the place they visit, including the history of the conflict and the importance of peace (Kang and Moscardo, 2006; Shamsub and Lebel, 2012).

Environmental issues are related to the third dimension. Tourism and peace is not only related to understanding between people and communities or poverty alleviation but also sustainability (Jimenez & Kloeze, 2014).

In the case of Trails for Peace, they all take place in the natural environment. According to Mallarach (2012) each natural site has also intangible values and protected landscapes usually have spiritual meanings as well. Many natural sites consist of physical elements (tangible) and intangible elements. The physical elements can be the landscape, a fountain, a cave, a mountain, etc. which will have a spiritual significance for local communities and visitors. Visitors are there to find the spirit of the site, to understand its meaning, and to be changed in some way. This forms part of the visitor experience.

Shackley (2001) refers to the physical impacts of tourism and mentions theft of objects, vandalism, pollution (acoustic, visual ...), damage ... and classifies them according to whether they are accidental or deliberate. Deliberate physical impacts can be prevented through visitor education and awareness. One of the risks that the promoting institutions may face is vandalism. Bhati and Pearce (2016) state that vandalism has historical roots and is an ongoing problem in many tourist destinations. How to prevent visitors from damaging the facilities and infrastructures provided especially in unattended areas, is one of the challenges. The facilities provided along Trails for Peace are just related to signposting, but to develop tourist products, more facilities should be added, like interpretation tools to transmit the meaning of the trail or rest areas.

A key concept is carrying capacity, which can be physical or spiritual (or perceived, in the words of Urry, 1990). Carrying capacity is essential for protection and sustainable development. It refers to the maximum use that can be made of any place without harming the resources, without reducing the satisfaction of the visitors or hurting the society, economy, or culture of the region. Boundaries are difficult to quantify, but they are essential and must be respected to maintain quality and visitor satisfaction.

In all the existing paths, numbers of participants are still low, and this is a positive aspect in the sense that it is a guarantee of not overpassing the carrying capacity of the place.

One of the challenges of Trails for Peace is how to promote this initiative and make people more aware of the three dimensions. One of the weaknesses of the project is communication. Even great efforts have been made in all destinations to arrange the path and have the trail available for walking; there is a lack of information in some cases on the values to be promoted. Mindful Travel Destinations is developing a project with the company Play&Go to develop a mobile application to transmit these values to all visitors. This app will generate a community of travellers and walkers for peace while providing geolocated routes with information on the points of interest. The app will allow users to leave messages or thoughts while doing the route that will be available to other users, and this will contribute to the dissemination of the values of Trails for Peace.

CONCLUSION

The project of Trails for Peace aims to transform people trough walking by making them aware of the environment, the path, the history and the local communities around the trail. Peacebuilding is the consequence of transforming people by making them aware of the space surrounding them. The three dimensions of peace are present in this proposal.

The inner dimension is related to walking, being aware of the place, participating in mindful activities. Authors like Alzamora Salom (2006) also refer to the spirituality of tourism. Tourism is practised during the no-working time, providing time for oneself, which is one of the basis for personal development.

The intercultural dimension is explored but putting in touch visitors with local communities, making to know each other and fostering tolerance and respect.

Finally, yet importantly, the third dimension is reached by connecting people with the natural environment where the trails are located.

This project is promoted by an NGO, and one of the difficulties is the lack of resources to develop services and facilities and the difficulty to get in touch with all possible stakeholders. Even this, the project has expanded, and it is supposed to grow with the two new trails for peace forthcoming. The commitment of local authorities is a key element in the development of this proposal as they are the ones in charge of arranging the path, promoting it, involving local communities and so on. It is important to establish a strong commitment to those in charge of the trails, especially regarding the development and management of the trails. They should be fully aware of the impacts of their activities and try to minimize negative impacts.

Nevertheless, more research is needed. In this chapter, authors have focused on the development of the project and the impact it has, especially on the destinations. Future research should focus on visitors, for example, and carry surveys or interviews to determine how in deep they have been transformed by the experience and to see how this can foster peacebuilding.

This project bets on a new way of understanding tourism as a way to change society. There is no alternative. Contemporary lifestyle is guilty of many new pathologies related to stress, and anxiety, for example. The challenge is not only to fight climate change and to be more sustainable but to promote changes in the behaviour of people. If changes are not done, the future looks dreadful. Wellbeing should be one the axis of a way we live. The challenge is complex and rather difficult; however, we, as human beings, we must always try to find our Trail for Peace.

REFERENCES

Alzamora Salom, A. (2006). La espiritualidad del turismo (3rd ed.). Parròquia de Santa Ponça.

Aulet, S. (2018). Spiritual Tourism in Europe: the SPIRIT-Youth project. In R. Butler & W. Suntikul (Eds.), *Tourism and Religion. Issues and implications* (pp. 200–219). Channel View Publications.

Banda i Tarradellas, A. (1991). *Educació per la pau*. Biblioteca Cultural Barcanova.

Bassas, A. (2008). *El matí amb Raimon Panikkar*. Proa.

Bauman, Z. (2007). *Tiempos líquidos: vivir en una época de incertidumbre*. Ed. Ensanyo TusQuets.

Becken, S., & Carmignani, F. (2016). Does tourism lead to peace? *Annals of Tourism Research, 61*, 63–79. doi:10.1016/j.annals.2016.09.002

Berger, P. L. (1975). *Rumor de ángeles: La sociedad moderna y el descubrimiento de lo sobrenatural.* Herder.

Bhati, A., & Pearce, P. (2016). Vandalism and tourism settings: An integrative review. *Tourism Management, 57*, 91–105. doi:10.1016/j.tourman.2016.05.005

Bimonte, S., & Faralla, V. (2012). Tourist types and happiness a comparative study in Maremma, Italy. *Annals of Tourism Research, 39*(4), 1929–1950. doi:10.1016/j.annals.2012.05.026

Boada, I. (Ed.). (2004). *La filosofia intercultural de Raimon Panikkar*. Pòrtic.

Bonilla, J. (2018). *Un sendero para la construcción de paz en Choachí*. Blogs - El Tiempo. http://blogs.eltiempo.com/felicidad-desarrollo-turismo/2018/11/27/sendero-la-construccion-paz-choachi/

Brundtland, I. (1987) *Informe: Nuestro futuro común*. Comisión Mundial sobre Medio Ambiente y Desarrollo.

Centro Nacional de Memoria Histórica. (2012). *¡Basta ya! Colombia: Memorias de guerra y dignidad.* Retrieved April 2020, from http://www.centrodememoriahistorica.gov.co/micrositios/informeGeneral/descargas.html

Charter, E. (2012-2016). *Earth Charter*. Recovered from: https://earthcharter.org/

Collins-Kreiner, N. (2019). Pilgrimage tourism-past, present and future rejuvenation: A perspective article. *Tourism Review, 75*(1), 145–148. doi:10.1108/TR-04-2019-0130

d'Hauteserre, A. M. (2004). Postcolonialism, colonialism, and tourism. A companion to Tourism. In A. Lew, M. Hall, & A. Willimas (Eds.), *A companion to tourism* (pp. 235–245). Bkacwell Publishing. doi:10.1002/9780470752272.ch19

de Cultura de Pau, E. (2010). *Idees per a construir la pau. Diplomàcia municipal i altres iniciatives dels governs locals per construir la pau.* Xarxa de Municipis de la Diputació de Barcelona. Recovered from https://escolapau.uab.cat/municipisipau/municipis/idees2_cultura_de_pau.pdf

Farmaki, A. (2017). The tourism and peace nexus. *Tourism Management, 59*, 528–540. doi:10.1016/j.tourman.2016.09.012

Farmaki, A. (2019). Visitor-host encounters in post-conflict destinations. The case of Cyprus. In R. Isaac, E. Çakmak, & R. Butler (Eds.), *Tourism and Hospitality in Conflict-Ridden Destinations*. Routlege. doi:10.4324/9780429463235-17

Farmaki, A., Khalilzadeh, J., & Altinay, L. (2019). Travel motivation and demotivation within politically unstable nations. *Tourism Management Perspectives, 29*, 118–130. doi:10.1016/j.tmp.2018.11.004

Fisas, V. (2002). *La paz es possible. Una agenda para la paz del siglo XXI*. Barcelona: de Bolsillo Plaza y Janés and Intermón Oxfam.

Fundació, C. I. D. O. B. (2014). *El proceso de paz con las FARC-EP*. Retrieved April 2020 from https://www.cidob.org/publicaciones/documentacion/dossiers/dossier_proceso_de_paz_en_colombia/dossier_proceso_de_paz_en_colombia/el_proceso_de_paz_con_las_farc_ep

Galtung, J. (1969). Violence, peace, and peace research. *Journal of Peace Research*, *6*(3), 167–191. doi:10.1177/002234336900600301

Goodwin, H., Santilli, R. (2009). Community-based tourism: A success. *ICRT Occasional Paper*, *11*(1), 37.

Haessly, J. (2010). Tourism and a culture of peace. In O. Moufakkir & I. Kelly (Eds.), *Tourism, progress and peace* (pp. 1–16). CAB International. doi:10.1079/9781845936778.0001

Hall, M., & Tucker, L. (Eds.). (2004). *Tourism and postcolonialism*. Routledge. doi:10.4324/9780203392270

Hinds, R., & Carter, B. (2015). Indicators for conflict, stability, security, justice and peacebuilding. In *GSDRC6 Helpdesk Research Report* (Vol. 1230). University of Birmingham.

Isaac, R. (2014). Responsible Tourism and Development in the Context of Peace-Building. In C. Wohlmuther & W. Wintsersteiner (Eds.), *The International Handbook on Tourism and Peace* (pp. 87–100). Centre for Peace Research and Peace Eduaction of the Klagenfurt Univeriesty and UNWTO.

Jimenez, C., & Kloeze, J. (2014). Analyzing the peace through tourism concept: The challenge for educators. *Sociology and Anthropology (Alhambra, Calif.)*, *2*(3), 63–70.

Kang, M., & Moscardo, G. (2006). Exploring cross-cultural differences in attitudes towards responsible tourist behaviour: A comparison of Korean, British and Australian tourists. *Asia Pacific Journal of Tourism Research*, *11*(4), 303–320. doi:10.1080/10941660600931143

Lee, Y. C., Lin, Y. C., Huang, C. L., & Fredrickson, B. L. (2013). The construct and measurement of peace of mind. *Journal of Happiness Studies*, *14*(2), 571–590. doi:10.100710902-012-9343-5

Mac Ginty, R. (2013). Indicators+: A proposal for everyday peace indicators. *Evaluation and Program Planning*, *36*(1), 56–63. doi:10.1016/j.evalprogplan.2012.07.001 PMID:22868180

Mallarach, J. M. (Ed.). (2012). *Spiritual Values of Protected Areas of Europe: Workshop 2011: Viim, Alemanya*. Federal Agency for Nature Conservation.

Maslow, A. H. (1971). *The farther reaches of human nature*. Viking.

McIntosh, I., & Harman, L. (2017). *The many voices of pilgrimage and reconciliation*. CAB International. doi:10.1079/9781786393265.0000

Menchero Sánchez, M. (2018). Colombia en posconflicto: ¿un turismo para la paz o paz para el turismo? *Revista Iberoamericana de Filosofía. Política y Humanidades*, *39*, 415–438.

Mindful Travel Destinations. (2018). *Trails for Peace*. Retrived from: https://mindfultraveldestinations.com/trails-for-peace-2/

Moufakkir, O., & Kelly, I. (Eds.). (2010). *Tourism, progress and peace*. CAB International. doi:10.1079/9781845936778.0000

Mundo, E. (2009). *40 definiciones por la paz*. Recoverd from: http://fund-culturadepaz.org/doc/40_Definiciones_PAZ.pdf

Palmer, C. A. (1994). Tourism and colonialism: The experience of the Bahamas. *Annals of Tourism Research*, *21*(4), 792–811. doi:10.1016/0160-7383(94)90084-1

Panikkar, R. (2004). *Pau i interculturalitat, una reflexió filosòfica*. Proa.

Pigem, J. (2007). *El pensament de Raimon Panikkar. Interdependència, pluralisme, interculturalitat*. Institut d'Estudis Catalans.

Pigem, J. (2010). *Qüestió de valors. Del consumisme a la sostenibilitat*. Institut del Territori.

Pigem, J. (2015). *Intel·ligència Vital. Una visió postmaterialista de la vida i la consciència*. Kairós.

Reisinger, Y. (Ed.). (2013). *Transformational tourism: Tourist perspectives*. CAB International. doi:10.1079/9781780642093.0000

Ribera Regull, R. (2007). *El Diàleg interreligiós*. Fragmenta.

Rodríguez, P. L. M. (2011). *Los orígenes de la guerra civil española* (Vol. 154). Encuentro.

Rodríguez Rojo, M. (1995). *La educación para la paz y el interculturalismo como tema transversal*. Vilassar de Mar: Oikos-tau.

Rojo, V. (2017). *Historia de la guerra civil española*. RBA Libros.

Roszak, T. (1985). *Person / Planet. Towards a new ecological paradigm*. Kairós.

Ruiz Monroy, H. E. (n.d.). *Cultura de paz. Un compromiso*. Retrived from: https://www.monografias.com/trabajos93/cultura-paz-compromiso/cultura-paz-compromiso.shtml

Sánchez Sánchez, A., & Fernández Herreira, A. (Eds.). (1996). Dimensiones de la educación para la paz. Teoría y experiencia. Granada: Eirene.

Schmitz, C. L., Matyók, T., Sloan, L. M., & James, C. (2012). The relationship between social work and environmental sustainability: Implications for interdisciplinary practice. *International Journal of Social Welfare*, *21*(3), 278–286. doi:10.1111/j.1468-2397.2011.00855.x

Schulz, E. K. (2005). The meaning of spirituality for individuals with disabilities. *Disability and Rehabilitation*, *27*(21), 1283–12195. doi:10.1080/09638280500076319 PMID:16298931

Seattle, C. (1985). *Letter to the President of the United States*. Recovered from Carolina State University: http://www.csun.edu/~vcpsy00h/seattle.htm

Shackley, M. (2001). *Managing sacred sites: service provision*. Continuum.

Shamsub, H., & Lebel, L. (2012). Identifying tourists with sustainable behaviour: A study of international tourists to Thailand. *Journal of Environmental Management & Tourism*, *3*(1). Advance online publication. doi:10.2478/v10260-012-0003-z

Sharma, P., Charak, N. S., & Kumar, R. (2018). Sustainaible tourism development and peace: A local community approach. *Journal of Hospitality Application & Reserach*, *13*(1), 36–56.

Sharpley, R. (2009). Tourism, religion and spirituality. In T. Jamal & M. Robinson (Eds.), *The SAGE handbook of tourism studies* (pp. 237–253). SAGE. doi:10.4135/9780857021076.n14

Sharpley, R., & Jepson, D. (2011). Rural tourism: A spiritual experience? *Annals of Tourism Research, 38*(1), 52–71. doi:10.1016/j.annals.2010.05.002

Shiva, V. (2006). *Earth democracy: Justice, sustainability and peace*. Zed Books.

Suansri, P. (2003). *Community based tourism handbook*. Responsible Ecological Social Tour-REST.

Swain, A., & Krampe, F. (2011). Stability and sustainability in peace building: Priority area for warfare ecology. In *Warfare Ecology* (pp. 199–210). Springer. doi:10.1007/978-94-007-1214-0_14

UNESCO. (2019). *Diversidad de las expresiones culturales*. Retrieved from: https://es.unesco.org/creativity/

United Nations. (2017). *Global indicator framework for the Sustainable Development Goals and targets of the 2030 Agenda for Sustainable Development Goals and targets (from the 2030 Agenda for Sustainable Development) Indicators*. UN.

Urry, J. (1990). *The tourist gaze: Leisure and travel in contemporary societies*. Sage.

Willson, G. B., McIntosh, A. J., & Zahra, A. L. (2013). Tourism and spirituality: A phenomenological analysis. *Annals of Tourism Research, 42*, 150–168. doi:10.1016/j.annals.2013.01.016

Witersteiner, W., & Wohlmuther, C. (2014). Peace Sensitive Tourism: how Tourism Can Contribute to Peace. In C. Wohlmuther & W. Wintsersteiner (Eds.), *The International Handbook on Tourism and Peace* (pp. 36–61). Centre for Peace Research and Peace Eduaction of the Klagenfurt Univeristy and UNWTO.

Wohlmuther, C., & Wintsersteiner, W. (2014). Tourism and Peace – an emerging field of action and Research. In C. Wohlmuther & W. Wintsersteiner (Eds.), *The International Handbook on Tourism and Peace* (pp. 17–28). Centre for Peace Research and Peace Eduaction of the Klagenfurt Univeristy and UNWTO.

WOLA - Washington Office on Latin America. (n.d.). *El proceso de paz en Colombia*. Retrieved Arpil 2020, from https://www.wola.org/es/programa/colombia/el-proceso-de-paz-en-colombia/

Chapter 5
Tourism and the Social Reintegration of Colombian Ex-Combatants

Mónica Guasca
https://orcid.org/0000-0002-2679-262X
University of Leuven (KU Leuven), Belgium

Anne Marie Van Broeck
https://orcid.org/0000-0003-3770-1807
University of Leuven (KU Leuven), Belgium

Dominique Vanneste
University of Leuven (KU Leuven), Belgium

ABSTRACT

Research on the role of tourism in post-conflict scenarios has suggested that it contributes to peace and reconciliation by providing common ground for antagonists to meet. However, not enough attention has hitherto been given to cases where one of the opposing parties consist of former combatants. This chapter draws on academic literature on disarmament, demobilization, and reintegration (DDR) to discuss how Colombian demobilised guerrillas are currently using tourism to assist their reintegration process. Based on document analysis, semi-structured interviews, and non-participant observation, it is argued that tourism can contribute to the social reintegration of ex-combatants by making them visible and active agents in the process and supporting alternative settings where reintegration can take place. Ultimately, the chapter posits that tourism can assist non-conventional approaches to reintegration, which might increase the likelihood of ensuring a successful transition from war to peace.

DOI: 10.4018/978-1-7998-5053-3.ch005

INTRODUCTION

Peace and conflict scholars are increasingly drawing attention towards the need to research peace from a micro-level perspective in war-shattered societies. Traditionally, scholars have given priority to macro-level analysis focusing, amongst others, on peace treaties, national policies or international interventions for peacekeeping and peacebuilding (Autesserre, 2014; Lederach 1997). However, the lack of understanding of how peace is built and maintained on the ground might explain to a large extent why many post-conflict transitions fail (Richmond & Mitchell, 2011). Researchers interested in unravelling 'peacebuilding from below', recognise that no lasting peace can be achieved if local realities are neglected (Mac Ginty & Firchow, 2016). Hence, they have focused on studying grassroots peace initiatives and local agency, and are advancing debates on the link between territorial configurations and peace to understand those everyday contexts (Björkdahl & Buckley-Zistel, 2016; Donais, 2009; Paffenholz, 2015). Through their work, the locals and their spaces are now considered essential for the study of peace.

Tourism studies are particularly well-positioned to shed light on peace outcomes in micro-contexts, assessing how different actors use tourism to transform territories marked by violence and influence transversal processes of societal healing. The research line that addresses post-conflict tourism and reconciliation is of particular interest. Inquiries in this field has mostly yielded positive results, concluding that tourism can provide common ground for opposing parties to meet and work together in the aftermath of violence, as shown for the cases of Northern Ireland (Anson, 1999; Simone-Charteris & Boyd, 2010), Bosnia and Herzegovina (Causevic & Lynch, 2013), Rwanda (Alluri, 2009), and Cyprus (Sonmez & Apostolopoulos, 2000). Yet, except for Northern Ireland, little is known about the background of those antagonists and their participation in violent acts during war-time. Specifically, the study of post-conflict tourism has not placed enough attention on a type of actor that played a key part in the conflict and is thus crucial for peace: the former combatant.

The idea to use tourism as a means in assisting the reintegration of fighters has been more seriously considered by national and international institutions, governments and ex-combatants themselves. The United Nations, for example, has encouraged the inclusion of ex-combatants as ecotourism guides or forest rangers taking into account their extensive knowledge of certain terrains, as has been done in Mozambique, Indonesia and Côte d'Ivoire (United Nations et al., 2013). Van Broeck, Guasca and Vanneste (2019) have pointed at several initiatives undertaken by Colombian ex-rebels, highlighting their increased involvement in tourism projects. In Guatemala and El Salvador, the ex-fighters also undertook their own tourism projects (Herrador-Valencia & Riera, 2016; Janzen, 2014), while in Namibia, the government employed their counterparts for wildlife protection and tourism promotion (McMullin, 2013 a; Metsola, 2006). Despite the multiple cases to examine, this line of inquiry remains underdevelopped.

This chapter sheds light on the role of post-conflict tourism in ex-combatant reintegration, and by doing so, contributes to advance the debate on the tourism-peace nexus. We draw on the Colombian case to explore, through qualitative research, how tourism is currently being used by demobilised guerrillas from the Revolutionary Armed Forces of Colombia (FARC - *Fuerzas Armadas Revolucionarias de Colombia*) in their reintegration process, after signing the peace agreement with the government in 2016. The chapter also engages with the existing research on Disarmament, Demobilisation and Reintegration (DDR) of ex-combatants, which provides a framework to understand the challenges that necessarily arise when trying to bring rebels back to society. Moreover, in line with the understanding of peace as being rooted in local realities, we pay special attention to the agency of ex-combatants and the spaces where their reintegration is taking place.

The chapter begins with a conceptual framework on the DDR approach, highlighting the stage of reintegration, which is our main focus. Various obstacles to reintegration found in the literature are presented. The next section introduces the case of Colombia as a country trying to emerge from a destructive war after the demobilisation of its largest guerrilla group. Then, we describe our research methods and give an overview of the ways ex-combatants are approaching tourism as a strategy for social reintegration. The fourth section discusses the implications of tourism development in these contexts, relating the findings to the academic debates on reintegration and peace at the micro-level. We conclude with reflections on the Colombian case, on the relationship between peace and tourism in contexts or armed insurgency, and with suggestions for future research.

BACKGROUND

The Disarmament, Demobilisation and Reintegration Approach to Peacebuilding

The Disarmament, Demobilisation and Reintegration (DDR) approach aims at ensuring the transition of former fighters into civilian life. It has been widely used in countries emerging from violent conflict, and it is often presented as a crucial component for post-conflict reconstruction (Muggah, 2009). According to the Integrated DDR Standards of the United Nations – a central player in sponsoring peacekeeping interventions worldwide – DDR is usually a result of peace negotiations when fighters agree to lay down arms and renounce to the use of violence as a political strategy. The process starts with the disarmament stage, which involves the collection and disposal of small arms, heavy weapons, ammunition and explosives. It is usually followed by the demobilisation stage when the group's structure is officially dismantled, and combatants are located in temporary areas to acquire civilian status before moving to other communities. This stage comprises a reinsertion phase, during which demobilised fighters receive assistance to attend immediate needs. Such assistance might include cash allowances, in-kind benefits like food supplies, provisional shelter, medical attention, and workshops about their rights and duties as part of society. The ultimate goal is to successfully complete the reintegration stage, which implies that ex-combatants find alternative livelihoods and support networks beyond their war-related contacts and do not consider taking up arms again (United Nations, 2014).

Despite the importance given to DDR programmes in war-to-peace transitions, the approach has been subject of substantial criticism, which points at the indiscriminate and standardised deployment of such programmes in the absence of clear evidence of their success (Humphreys & Weinstein, 2007; Muggah, 2005). Since 1989, when the first official DDR operation was launched in Namibia, more than sixty DDR interventions have been undertaken in several countries emerging from war, the majority of them in Sub-Saharan Africa (Muggah, 2009). However, successful interventions have been generally equated to the completion of the disarmament and demobilisation stages, adopting a minimalist perspective focused entirely on security and ignoring long-term commitments for reintegration and development outcomes (Berdal, 1996; Muggah, 2005). As a result, the well-being of fighters after demobilisation had received little attention compared to the relevance ascribed to them when they still held a weapon, which is worrisome so far as demobilised fighters who failed to reintegrate could easily become spoilers of the peace process (Stedman, 1997).

No Reintegration without Social Reintegration

Scholars agree that reintegration is the weakest link of the chain (Theidon, 2007). It is broadly defined as a long-term process in which ex-combatants return to their communities of origin or choice, and they and their families join the economic, social and political life of the civilian population (United Nations, 2014). Its multi-dimensional character implies that focussing primarily on providing employment and income to the former combatants, although imperative, is not enough to achieve a complete reintegration (Bowd & Özerdem, 2013; Özerdem, 2012). Certainly, employed ex-combatants can still be socially isolated and excluded from the community life they try to join, thus reducing the possibilities for reconciliation, personal and societal healing, and lasting peace. Therefore, the acceptance of ex-combatants and their families by receiving communities and society at large is a crucial piece in the reintegration scheme (Özerdem, 2012; Porto & Alden, 2016).

The social reintegration of ex-combatants is particularly complex in protracted conflicts where societies that have suffered in the scourge of war exhibit deep social fragmentations (Barakat & Chard, 2002). Rejection and stigmatisation of demobilised rebels are expected, especially when the groups they belonged to committed human right violations against the civil society. Former fighters are normally considered to be naturally violent, unreliable, and psychologically unstable, a threat to security that could trigger a new wave of bloodshed at any given moment (Mashike, 2004; McMullin, 2013 a; Mcmullin, 2013 b; Metsola, 2006). In Mozambique, for instance, ex-combatants were portrayed in the media as “creatures with a feeble and weak psychic structure, who have lost, many times, self-confidence and self-control” (Lundin, 1998, p. 114). This tendency to treat ex-combatants as a homogenous group, by invariably applying the negative image that is associated with the category of the ex-combatant to all group members, ignores the multiple individualities, motivations to enrol and experiences in war (Theidon, 2007).

Furthermore, when reintegration is detached from general development frameworks and takes place in contexts of pervasive poverty, as is the case in many war-shattered societies, it can create new conflicts between ex-combatants and receiving communities (Kingma, 2002). This was the case of Liberia, where strong resentment spread across civilians as they perceived that the cash allowances for demobilised combatants were a reward for their misbehaviour while the national unemployment rate reached over 85% and the rest of the population slipped further into poverty (Jennings, 2007; Willibald, 2006). In those scenarios of economic hardship and social hostility, it is difficult to experience an improvement in livelihoods compared to war times, as ‘re-marginalisation and not reintegration is the natural outcome awaiting most ex-combatants’ (McMullin, 2013a, p. 9).

The widespread stigmatisation also has considerable influence in ex-combatants’ decision on where to reintegrate (de Vries & Wiegink, 2011). One option is to return to their communities of origin or a community where they have support networks, running the risk that their past affiliation with a rebel group will be exposed and lead to social exclusion. What is more, sometimes returning to home towns is not even possible. Such can be the case when rebels were forced to commit atrocities against their own communities, as in the case of child soldiers in West Africa (Bowd & Özerdem, 2013). An alternative is to move to communities where they can hide their involvement in war and avoid the weight of stigmatisation. Reintegration is then associated with the idea of breaking with the past and considered a success when ex-combatants’ identities are erased, becoming invisible to society (Bowd & Özerdem, 2013; Mcmullin, 2013b; Rodríguez López et al., 2015). However, evidence has shown that hiding identities to avoid exclusion can bring affective implications like anxiety, depression, or shame, and behavioural implications like social avoidance or impaired relationship functioning (Goffman, 1963; Pachankis,

2007). A third option is to stay together with fellow combatants and do not 'return' to any community, having each other's support and protection to facilitate the transition (Pugel, 2006). Yet this option is generally not supported for fear of easier remobilisation or the worry of complete isolation (De Vries & Wiegink, 2011; Theidon, 2007).

The stigmatisation of former fighters and the excessive focus on their limitations rules out the possibility of recognising their agency in the reintegration process (Rodríguez López et al., 2015). Because of their devalued identities, demobilised rebels are discursively framed as a problem to be solved, rather than partners to contribute to joint solutions. In contrast, actively engaging ex-combatants and focusing on their strengths has proven to be a successful strategy (Bowd & Özerdem, 2013). This was the case of Northern Ireland, where former fighters have been at the forefront of conflict transformation and cross-community initiatives, being facilitators of the peace process rather than obstacles, which favoured their acceptance in society (Clubb, 2014; Dwyer, 2012; McEvoy & Shirlow, 2009; Rolston, 2007). In Colombia, demobilised guerrillas have also found ways to fight discrimination and be visible agents in their social reintegration, as discussed later in the chapter.

THE TOURISM - REINTEGRATION NEXUS

The DDR process with the Revolutionary Armed Forces of Colombia

For more than six decades, Colombia has been the battleground for armed confrontation between the government, far-left guerrillas and far-right paramilitary groups. It is the longest conflict in the hemisphere and one of immense magnitude. According to Grupo de Memoria Histórica (2013), the number of direct victims reaches almost nine million, most of them civilians. More than 50,000 people were forcibly disappeared, a figure that surpasses those of military dictatorships like in Argentina or Chile, all the while Colombia was considered a democracy. In addition, the widespread violence led to massive waves of displacement, generating the second largest population of internally displaced persons in the world (7.5 million people).

A major actor in such confrontation was the Revolutionary Armed Forces of Colombia (FARC for its acronym in Spanish), the oldest and largest guerrilla group in the country and in the western hemisphere. Founded by peasants in 1964 and guided by a Marxist-Leninist ideology, the FARC embraced in its beginnings the struggle of rural communities. Its revolution aimed at achieving greater equality and social justice, for which they received considerable support from various segments of the society, who saw in them the opportunity to put an end to a history of oppression and exclusion. However, as they acquired more power as an armed group, their initial ideologies got increasingly blurred. In need of funding to maintain several warring factions and expand their presence throughout the country, the FARC got involved in acts like kidnapping, extortion and drug trafficking (Pécaut, 2008). It conducted multiple attacks against the civilian population, which tarnished its reputation and undermined the credibility of its political-ethical discourse.

By the 1990s, the Colombian society increasingly started to identify the FARC more with a terrorist group than with an insurgent guerrilla (Echandía, 2006). Several failed attempts towards peace at negotiations and the widespread fear about the FARC taking control over the majority of the territory paved the way for Álvaro Uribe to the presidency. Uribe promised his voters to defeat the guerrilla with an aggressive display of military power, harming the FARC severely during his term (2002-2010). With the change of government, the FARC leaders decided to accept the invitation of the new president, Juan Manuel Santos, to restart peace dialogues. In 2016, after four years of negotiations in Cuba, the national government and the FARC leaders signed a peace agreement opening the doors for the disarmament, demobilisation and reintegration of FARC guerrillas.

In 2017, 6,804 FARC combatants handed in their weapons to the United Nations Verification Mission, which were melted down and transformed into floor tiles of an art gallery (Fundación Ideas para la Paz, 2017; Parkin, 2018). The total number of FARC members who demobilised (including those not owning weapons) was about 13,000 (Agencia para la Reincorporación y la Normalización, 2019). To help them cover basic needs after demobilisation, the government gave each ex-combatant an initial financial aid of two million Colombian pesos (approximately 550 euros) and committed to provide a monthly income of around 200 euros (corresponding to the 90% of the Colombian minimum wage) until December 2019, while ex-fighters would undertake and consolidate entrepreneurial projects or find a stable employment.

Taking into consideration that 96% of ex-combatants expressed being more inclined to have own ventures than being employed (Martinez & Lefebvre, 2019), the peace agreement also foresaw the creation of a multilateral fund to help kick-start small enterprises. The ex-combatants were given the possibility to formulate projects and apply for up to eight million pesos per applicant (about 2000 euros) to pursue individual or collective projects. A committee composed of FARC and government representatives would then evaluate their feasibility and approved projects would receive the requested seed capital. Nonetheless, there are many barriers to obtain approval and most of the projects undertaken by ex-combatants continue to run without resources from the multilateral fund. According to the United Nations Verification Mission in Colombia - UNVMC (2019b), 24 collective and 705 individual projects have received funds through this mechanism, involving only around 18% of the total demobilised population.

For the demobilisation stage, the ex-combatants were congregated in 26 zones distributed in different parts of the country, where they could start the reinsertion phase and have an adjustment period before moving to other communities. The locations of the zones, currently known as Territorial Training and Reincorporation Spaces (ETCRs – *Espacios Territoriales de Capacitación y Reincorporación*), were chosen taking into account the areas of influence of the FARC. The government leased lands in private properties, forest and indigenous reserves to temporarily locate the ETCR, and gave them legal status for two years (until August 2019). However, after this period, 26% of the ex-combatants still reside in an ETCR together with their families (Garzón, Prada, Silva & Zárate, 2019). Over time it became clear that not all demobilised fighters saw these spaces as transitory. The main issue is that ex-combatants cannot permanently settle in some of these spaces, either because the land does not belong to them nor the government or because it is not allowed by law to have permanent human settlements, as in the case of natural reserves. For the moment, the ETCRs are still supported by the government until August 2020, while looking for solutions to solve the concerns over land availability (UNVMC, 2019b).

The reintegration stage also presents significant challenges, notably concerning the social inclusion of ex-fighters. In a context of protracted conflict, the bitterness of war grows profound roots and inflicts wounds difficult to heal. As the FARC transformed their approach to the civilian population from persuasion to intimidation, the resentment towards their organisation, their actions and their members grew stronger (Grupo de Memoria Histórica, 2013). Even if the Colombian society longed for peace, a large segment of the population found it difficult to accept that FARC members would be able to demobilise without penalisation, interpreting this as impunity. This was one of the main reasons why in a plebiscite in 2016, 50.2% of the voters rejected the peace agreement. The government and the FARC came back to the negotiation table to address the demands of the critics and finally signed the agreement. Although the plebiscite did not impede the disarmament and demobilisation of the FARC, it made evident the immense challenges that were waiting ahead for the social reintegration of ex-combatants into a polarised society.

Colombia is no exception when it comes to the stigmatisation of ex-combatants. A large-scale survey found that 41% of Colombians look at ex-combatants with fear, and 81% distrust them more than other members of the community (Nussio, 2018). They are viewed with suspicion, as inconvenient newcomers who disturb the rhythm of society. This additional scrutiny, as if they were "strange insects", is evident to the ex-combatants themselves (Theidon, 2007, p. 80) who, in many cases, have decided to keep their former identities hidden to gain a certain level of acceptance (Rodríguez López et al., 2015). Furthermore, hiding their previous affiliation to a rebel group is also a safety measure, since there is a danger of retaliation by victims, members of other armed organisations or fighters who did not demobilise. The current DDR process serves as evidence of the risks that demobilised guerrillas face in a hostile environment. So far, the inability of the state to guarantee the physical security of the demobilised FARC members at a time of constant threats against them has led to 173 ex-combatants being killed since the signature of the peace agreement (UNVMC, 2019b).

Despite the numerous obstacles, the majority of FARC ex-combatants continue to trust the peace agreement. Recidivism estimates indicate that around 2,000 demobilised guerrillas returned to arms while more than 10,000 are still moving forward in their transit to civilian life (Avila, 2019). Of those ex-combatants who persist in the process, some have decided to leave the ETCRs and continue their reintegration individually. Others have decided to stay close to their comrades, either by staying in an ETCR or moving together to what they call New Regrouping Points (NPRs – *Nuevos Puntos de Reagrupamiento*), which are collective settlements of ex-combatants without legal recognition by the government. It is estimated that 3,100 ex-combatants are still living in an ETCR, and around 1,800 are congregated in more than 30 NPRs (Garzón et al., 2019; UNVMC, 2019a).

Those ex-combatants who decided to live collectively in an ETCR or NPR, have used their previous and recently acquired skills to develop businesses that could provide them with a stable income. They have created multi-purpose cooperatives that offer an array of goods and services, and are now producing and commercialising goods like coffee, beer, honey, handicrafts, clothing and shoes (Forero, 2019; Sierra, 2019). Among these proposals, a rather unconventional one that has increasingly drawn the attention of society and institutions has been tourism, which is the focus of the next sections.

METHODS

The social reintegration of ex-combatants cannot be fully understood by quantitative indicators. While the economic reintegration can be assessed by, for example, the number of ex-fighters who have a stable income, the social dimension requires approaches flexible enough to capture the different shades of grey that comes with the process of being accepted in society. Hence, we have opted for qualitative research methods, which were also best suited to build rapport with the research participants (crucial for the interaction with ex-combatants) and give them the freedom to express their thoughts in their own words.

The primary data for this contribution were gathered through fieldwork in Colombia in November and December 2019, which were complemented by content analysis of secondary sources such as policy and programme documents, resolutions and online news. Applied methods included non-participant observation and semi-structured interviews with key actors, including FARC ex-combatants who are leading tourism initiatives (5), civilians working together with the ex-combatants in these projects (3), tourism scholars (3), and officials from national government entities and international institutions who are supporting the reintegration process, including tourism initiatives (4).

Before going to the field, we conducted a press review to identify tourism initiatives led by ex-combatants that were getting increasing media attention. We decided to visit four of these projects. However, during the fieldwork, two events prevented us from making the visits we had foreseen. First, a national strike that was planned for one day resulted in several weeks of protests and restrictions in mobility throughout the country (from November 21, 2019, until January 21, 2020). Second, the death of a former combatant within an ETCR by a military put on hold the individual visits in various ETCR, since the ex-combatants feared for their safety. Despite the drawbacks, we managed to interview in Bogotá ex-combatants who are leading tourism projects in different parts of the country, and one ex-combatant accepted to be interviewed by phone. After the national strike, it was possible to visit an ETCR for two days and interview the ex-combatant who is leading the tourism offer on the site. This visit allowed us as well to have informal conversations with several demobilised guerrillas living in the area and thus complementing the formal interviews.

Respondents were reassured at the beginning of the interview about the confidentiality of the information and their identities; therefore, all participants cited in this chapter were assigned pseudonyms to ensure anonymity. The national and international officials were chosen based on their direct or indirect support to tourism projects led by ex-fighters. The interviews with the tourism scholars and civilians involved in the projects were obtained through snowball sampling. Due to the geographical distance to some participants and the mobility constraints, four interviews were conducted by phone. The average duration of interviews was 45 minutes for phone interviews, 2 hours for face-to-face interviews with ex-combatants and ca. 1.5 hours for personal interviews with the other participants. The quotes used have been translated from Spanish by the main author.

Our priority was to ensure that the tone of the interviews remained conversational so that the interviewees acknowledged that our purpose was not to interrogate them but to learn from them. The interviews started by presenting the purpose of the research, the reason why the authors reached the participant, and by addressing the participant's inquiries. We opted for open-ended questions, starting from general ones based on their role and participation in the organisation/project to more specific ones focused on points of views. No personal questions (e.g. age, level of education, political or religious beliefs) were asked to any of the participants.

We paid particular attention to the question order in the formal and informal interviews with ex-combatants. To the extent that the dynamic of the interview allowed, questions were asked so that the ex-fighters could talk first about the present, then reflect on the future and, finally, refer to the past. Questions on the present aimed at understanding the context and the current perceptions of ex-combatants. We started by asking them to describe their own experience with the reintegration process and the tourism projects, together with their motivations to engage in those projects. We also asked them to give a more normative stance on what they consider a good reintegration process. Questions on the future revolved around expectations about the tourism projects and coming plans, their fears and hopes about their social reintegration, and the contribution they want to make as a part of the civilian society. Finally, if the interviewer considered it appropriate, we continued with questions about the ex-combatants' pasts as active fighters, the major changes that they have experienced, the things that they do or do not miss from the times in war, and how easy or difficult has been to carry the identity of a demobilised. The interview that was conducted by phone did not address questions about the past.

Questions for non-ex-combatants were more focused on factual knowledge on the reintegration process, perceptions of the demobilised fighters, the motivations to support tourism projects to advance reintegration, and their assessment of the present situation and future scenarios. The formal and informal interviews were fully transcribed and processed using thematic analysis based on inductive and deductive coding.

DISCUSSION

The Tourism Offer of FARC Ex-combatants

The idea of engaging in tourism resonated in some FARC ex-combatants who saw that various groups were interested in visiting them, mainly to learn about their past and present situation. They had already gained certain experience receiving people in the demobilisation sites, such as government delegates, representatives from national and international entities, leaders from the private sector, and people from the surrounding communities. Little by little, ex-fighters developed options for accommodation, the provision of food, and a series of activities that they could offer to visitors.

According to 'José', a representative from the national cooperative of the FARC, there are currently 17 tourism projects led by ex-combatants around the country. One of them has received resources from the multilateral fund linked to the peace agreement, while the others have been financed by the ex-combatants with their cash allowances and with resources from international cooperation coming, for instance, from the United Nations Development Programme or the Norwegian government (PNUD Colombia, 2019a). 'Nora', a government representative, confirms that ten of these projects are receiving technical support from the *Inter-institutional Tourism Board for the Reintegration, Reconciliation and Territorial Development*, a panel of advisors from various entities, which aims to assist the development of tourism in areas where former FARC guerrillas are collectively advancing their reintegration.

The tourism initiatives have several elements in common. Van Broeck et al. (2019) indicate that most of the initiatives offer flora and fauna sighting tours, trekking in surrounding regions, or visits to the different projects that are being developed in parallel with tourism. Some of the projects focus on specialised services such as rafting or bird watching. Although the offer varies from site to site, they all include elements referring to the times of war, like replicas of camps where ex-fighters used to sleep, pictures of them before and after demobilisation, or adapted cooking spaces (e.g. how to produce minimum steam so as to not be detected by enemies). Ex-combatants also share their personal experiences during the conflict and their perceptions of the peace agreement and reintegration process.

The appeal of discovering unexplored places with great natural potential that were previously off-limits because of violence, added to the opportunity of meeting with demobilised rebels, has especially attracted people from urban contexts who had probably never interacted with a guerrilla before. As mentioned informally by 'Rodrigo', an ex-combatant living in an ETCR close to the capital: "The peasants know the guerrilla for a lifetime. We are not that exotic to them, as for the people from the city". Opening the doors of those reintegration spaces has also attracted national and international students and researchers interested in learning about contemporary peace and reintegration processes. 'José', who is also helping the FARC tourism entrepreneurs to consolidate their projects, reported that university groups had become one of the main target segments and, although not strictly considered as tourists, ex-combatants do treat them as such and believe that they can be a gateway to attract other types of visitors.

The progress reached so far has fuelled the enthusiasm of ex-combatants, who increasingly seem to believe that tourism can become a reliable option to generate income for them and the surrounding communities. Several initiatives have achieved key milestones, like the acquisition of equipment for their activities, the building of hosting sites, the formalisation of their company or the participation in tourism fairs. They are receiving more attention and along with it, more visitors. Aware of the potential that tourism has for them, FARC tourism entrepreneurs from around the country gathered in their first national tourism encounter in October 2019. According to 'José', this occasion allowed them to help each other to better define their tourism offer, framing them in the solidarity economy in order not to reduce tourism to economic gains for themselves or third parties only. In the words of 'Antonio', an ex-combatant involved in a tourism project in an ETCR, they want a tourism "that includes the communities and that they are the ones who truly benefit. Not to fill even more the financial pockets of large multinationals or large agencies".

Although the projects are still in their early stages and tourism plays for now only a limited role in providing a stable income for ex-combatants and surrounding communities, the potential of tourism to generate economic benefits is clear. The combination of natural assets with historical memory has proven attractive to various segments of society. Additionally, the willingness of ex-combatants and support entities to move these projects forward seems to be genuine. But ex-combatants have found in tourism not only an option for income generation but also a mechanism to empower themselves, to foster communication and collaboration with civilians and therefore to contribute to their social reintegration.

The Potential of Tourism for Social Reintegration

Promoting Empowerment, Visibility and Agency

Ex-combatants trained in a top-down manner have to go through a process of psychological adjustment, as they move from a predictable environment of following commands to a civilian life where they have to make their own decisions (Kingma, 1997). In the case of the FARC, whose members mostly followed a communist model during wartime, the need to adapt to a different societal model makes their adjustment even more challenging. 'Rodrigo', an ex-combatant living in an ETCR, admitted his troubles in adapting by mentioning that "the worst of all has been to leave a very hard but very beautiful reality in terms of community. To be back to a selfish society. To meet again with that reality, so complex". As a tourism scholar explained:

They realise that they enter into a completely different economic model and opposite to what they had learned all their lives, and now they have to face the capitalist model. There is nothing more to do. If they want to be reintegrated, they are going to be reintegrated into that system. Thus, they have to sell, to buy, to do business. That is shocking to them.

However, the overall experience of undertaking own businesses have strengthened ex-combatants' capacities for self-decision-making and adaptation to new contexts, as evidenced by tourism entrepreneurs. ''Marco', an ex-combatant supporting various tourism projects in ETCRs and NPRs, described that they were not prepared to face the formalities required to deliver tourism services, like acquiring certificates to be a tourism guide, even if "no one knows better some territories of Colombia than our people", but that they are progressively closing that gap. 'Nora' also recognised that ex-combatants have gone through a steep learning curve to understand the complexity of the sector, and although at the beginning they were very insecure, "now their level of empowerment is impressive: how they speak, how they embrace their ideas". In informal conversations held during the fieldwork, various ex-fighters conceded that being able to interact with different types of visitors and tourists has helped them to gain confidence in themselves and not feel intimidated because they do not have an educational degree, for example.

In addition to improving the perception of themselves and supporting their empowerment, the ex-combatants realised that perhaps the greatest contribution that tourism offers is to serve as a platform for their voices to be heard. They saw in tourism an opportunity to fight stigmatisation, not by trying to erase their identity as former rebels, but by offering direct exposure to it. Instead of anonymization, they found a counterstrategy to embrace their past and try to fit in society without having to reinvent who they were. The aim is that visitors improve their perceptions by getting to know them better. In words of 'Camilo', an ex-fighter leading a tourism project in a northern ETCR, the idea is "that people know us a little bit more because there is a lot of misinformation about us. Yes, we made mistakes, and we did very unpopular actions, but our whole life was not dedicated to that thing [war]". His comrade 'Felipe', who is involved in another tourism project in the Northwest, added to this by stating that "what has been shown is not all. [Among the ex-combatants] there are artists, there are poets, there are painters, there is a proposal for a [better] country".

Therefore, ex-combatants are using tourism to open spaces of encounter where they can accentuate that their identities are not reduced to being former rebels and that their lives cannot be understood just from the perspective of aggressors. They want people to know that not all of them killed or kidnapped people or benefitted economically from the war, but that they are also parents, sons and daughters, human beings with feelings and abilities. Moreover, with their life stories, ex-combatants are seeking to go beyond the rigid dichotomy that is usually perceived between victims and perpetrators. 'Antonio', an ex-combatant involved in tourism, expressed this as follows: "Tourism gives us the chance to talk about our history too, to tell why we enrolled in the FARC, to tell that we were victims before entering the guerrilla". He added that, because they were victims who decided to fight, for them it is clear that to avoid a recurrence in violence, "the fundamental part is to reintegrate rural communities into the social, political and economic life of the country", and not just ex-combatants who emerged from these.

Indeed, the fact that FARC guerrillas committed wartime crimes does not imply that their claims to engage in an armed struggle were unfounded. The armed conflict in Colombia is a consequence of the great inequalities and the political and structural violence perpetrated by elites that have permeated the country for decades (Grupo de Memoria Histórica, 2013). Thus, ex-combatants have seen in tourism an opportunity to continue their political struggle for a more inclusive society, albeit in a more informal way. A shared idea expressed by most is that they abandoned their weapons, but not their commitment to fighting for social justice, reflected in the peace accord that they signed with the government which includes strategies to address the causes that led to the armed confrontation. Their approach to tourism –very much in line with the principles of community-based tourism– also demonstrates their interest to promote inclusive development in the territories where they are settled. Their life stories and plead for a fairer society serve as a reminder to visitors and tourists that the reality to which the demobilised rebels are reintegrating remains marked by social inequality, lack of access to education and investment in the countryside, endemic corruption, and general unrest. Ex-combatants highlight that peace in a wider sense can hence not be reduced to a signature of an agreement, but has to be a process that involves everyone.

The message of mutual responsibility in building peace has motivated some civilians to support the tourist initiatives of former combatants. Ex-combatants were already working with members of surrounding communities, either because they wished so, or because it was a requirement of funding agencies. But tourism has opened spaces of collaboration with people from remote places and has made unlikely encounters possible. 'Maria', a citizen from the capital involved in one of the projects with ex-combatants, commented that for her there was "nothing better than to *hacer patria* [ensure the welfare of the nation] through tourism". 'Cesar', another civilian supporting a tourism project mentioned that by working with ex-combatants he has also faced stigmatisation, but for him "it is what needs to be done, to break those schemes. Or are we going to marginalise them forever?".

Hence, tourism is particularly appealing because it allows to venture beyond the surrounding communities, and open spaces for the sensitisation of people who might have never interacted with guerrillas and, consequently, might be more prone to prejudices. 'Felipe' related an experience he had in receiving people who were "haters" of the FARC, but that "at the end did not want to leave". He went on by stating that tourism is an activity which allows "looking into each other's eyes" and that it has been "one of the most beautiful exercises" he has had yet.

Another important contribution that ex-combatants are making by including their life-stories in the tourism narratives is to raise awareness about the cruelty of war. As the ex-rebel 'Camilo' expressed: "If we lay down our arms and leave the armed confrontation, it is because we want to send the message that the solution for the Colombian society cannot be violence". The ex-combatant 'Felipe', eloquently elaborated on their motivation to engage in conflict prevention:

We do not praise war. It is a tragedy that we do not want to repeat. We want to let the new generations know what happened in Colombia, what the tragedy that we lived was, how we killed each other between brothers, so that this does not happen again in the future [...] Those who call us traitors because we leave our weapons do not know the reality of war. I tell them, 'if you agree with the armed struggle, go ahead, leave your family, leave university, leave your job and go for it'. They will not do it. It is easy to encourage the armed struggle from ignorance. The romanticism of armed struggle. It is not like that. Go to the mountain a week to fight the Colombian army and you will understand.

Ex-combatants can testify against the use of weapons to attain political and social aims as they are experienced voices and direct witnesses of war. Their identity as former fighters is then a valuable asset that strengthens their message of non-violence. By making them visible and protagonists in the process, tourism suggests that ex-combatants can be part of the solution and not just the problem. Moreover, it proves that it is possible to advance the social reintegration of demobilised fighters in scenarios where they choose not to return to any community, as detailed in the next section.

Supporting Alternative Spaces for Social Reintegration

The decision of several FARC ex-combatants to stay together after demobilisation questions the traditional approach to reintegration in which the demobilised fighter is the one moving to a receiving community in order to be socially included. Since the peace negotiations, the FARC leaders have clarified that –although they would dismantle their armed structures– they would remain as a group and promote a collective reintegration into the society and into the political system (Zambrano Quintero, 2019). At the individual level, the FARC ex-combatants also have reasons to remain united. Many of them entered the group from their childhood or youth, and after decades of conflict, they have stronger ties with their war comrades than with their families, as the former combatant 'Antonio' explained. According to him, several people have nowhere to return to and fear facing a city life that is unfamiliar by themselves. This coincides with figures showing that 66% of the demobilised population are of rural origin and 77% did not have a place to live after putting down their arms (Universidad Nacional de Colombia, 2017). Another decisive factor mentioned by the ex-combatant 'Rodrigo' is the fear of retaliation. Without their weapons and trust in the state to preserve their physical security, many ex-fighters see in their comrades their only option for protection.

The concern that has been raised with this approach to reintegration is for ex-combatants to remain isolated, without much interaction with non-ex-combatants and therefore, not being truly accepted and included in the wider society (Faltas, 2004). The issue is not the place where ex-combatants decide to settle, but the possibilities of interaction with other groups that can happen in such places. Therefore, encouraging civilians to visit the areas where the ex-combatants decided to live can be a solution to stimulate inter-group interaction.

Findings on the Colombian context show that tourism has the potential to support the collective settlement of ex-combatants and still promote their social reintegration because the established logic of movement in the reintegration process is reversed. It is no longer just the ex-combatant moving to a community; it is about opening the possibility for members of civilian communities to visit the ex-combatants' settlement as well. It can be conceived as a 'reverse' approach to social reintegration. Of course, this does not preclude the need of ex-combatants to interact with civilians in other spaces and beyond a tourism experience, but it offers an additional way to create meaningful encounters that could lead to mutual empathy and recognition.

Moreover, as the settlements where the tourism projects are set up are located in areas of historical presence of the FARC, supporting the social reintegration of ex-fighters in territories they know and felt as their own can facilitate their transition to civilian life. According to 'Felipe', being able to identify anchor points in these territories can help modify his guerrilla thinking. He explains:

We slept here today, and tomorrow on another mountain and the day after tomorrow on another mountain [...] That is why the issue of housing [settling down in the territory] is so important, because the guerrilla goes from having a mobile and military mindset to having a civil mindset.

This rooting in the territory can be supported by tourism in the way that demobilised guerrillas can take pride in the place they inhabit. Previously, the ex-combatants exerted control over these territories by using fear to exclude people within their borders (Sack, 1986); now, they are conceiving a territoriality based on inclusion instead. They, as hosts, still exert power over the space where tourism encounters take place, but this power is rather channelled into sharing their knowledge of the areas they know extremely well and that Colombians avoided for years. Certainly, the change in settings in which reintegration can occur could also help address the stigmatisation of areas marked by violence. Tourism, as proposed by the FARC ex-combatants, could be a means to transform their image from sites of danger and domination to spaces where peace and reconciliation can take place.

In addition, ex-combatants share in their narratives that they want to promote the conservation of these territories that were off-limits not only to civilians but also to third parties with dubious intentions. Although a controversial statement, ex-combatants affirm that they were protectors of the environment and would like to continue with this role as tourism entrepreneurs. The ex-fighter 'Marco', explained his vision of tourism as a new way to promote environmental conservation without intimidation:

Before, with weapons, it was very easy to preserve nature, because we had military control. Nobody cut down trees because we did not allow it. We had at that moment the real possibility of doing so. Now we do not, but we do want to make visible through tourism that we have worked for conservation and that we want to continue contributing to the preservation of those areas […] No one lived as well in harmony with nature as we did. It was out of necessity as well because we lived on the monte [referring to natural setting like forests, mountains, jungles, etc.] and the monte guaranteed our lives, but it was a very harmonious relationship in the sense that we also contributed to the conservation of the natural environment.

As a way of example, one of the ETCR that is developing a tourism offer based on conservation got the support of scientists who spent fifteen days with the ex-combatants in a forest that the FARC controlled. They discovered fourteen new endemic species and produced a general inventory of the different types of fauna and flora, which could eventually help ex-combatants to become specialised eco-tourism guides and claim guardianship of certain natural resources (PNUD, 2019b).

The findings suggest that tourism has the potential to become a tool for social reintegration in Colombia, but its actual impact is yet to be seen. The tourism projects are just taking off, and for far, there has been more enthusiasm than there have been results. Much more remains to be done to turn these initiatives in sustainable sources of income and wider platforms to confront stigmas. An essential element to help these projects move forward is to guarantee security in neighbouring territories and reassure tourists and visitors that their life is not at risk. Although there might be interest in doing tourism with ex-combatants, the fear of visiting certain areas that are still mentally demarcated as dangerous might be greater, which would overshadow any effort done by ex-fighters in improving their tourism offers.

Furthermore, it is important to stress that the root causes of the conflict are still present and cannot possibly be solved through the development of tourism alone. Therefore, tourism is not as much a solution to achieve lasting peace, as it is a laboratory to foster an integrated approach of sustainable development, difficult encounters, non-violent exchange of ideas and building new (group) identities and (place) meanings.

CONCLUSION

The social reintegration of ex-combatants is a major challenge in societies that have gone through protracted conflicts. It constitutes a key component of peacebuilding as failing to reintegrate demobilised fighters could lead to recidivism and a new outburst of violence. This chapter intended to discuss the role of tourism in facilitating such reintegration by presenting the experience of FARC members who undertake tourism initiatives themselves after laying down their weapons.

In this novel attempt to link the study of post-conflict tourism with the scholarly discussion on DDR, we found potential for interdisciplinary research and argue that tourism offers possibilities to support unconventional approaches to reintegration. In the case of the FARC, it has served as a way to make ex-combatants visible and protagonists in their reintegration process, without having to renounce their identity as former fighters. In fact, such an identity gives them a privileged position to become active agents in conflict prevention. Instead of dealing with stigmatisation by hiding their past, Colombian ex-fighters are using tourism to share their life-stories in war and create encounters to demystify the coexistence with them. This research confirms that former combatants can take lead roles and that they are "capable of being productive and socially active agents after war instead of unstable and unworthy beneficiaries of assistance' (Mcmullin, 2013, p. 387 b). Ultimately, tourism can be a mechanism to promote reconciliation at the micro-level, parallel to macro-processes like transitional justice and reparation of victims. It can help pave the way for social cohesion and create support networks for ex-combatants.

We also highlight the fact that social reintegration involves a rapport with space. It is not just a matter on how to reintegrate (collectively, individually, in anonymity or not), but also a matter of where to reintegrate. The traditional way of understanding reintegration has been a return of ex-fighters where they unilaterally –and mainly individually– choose a community to settle. But many FARC ex-guerrillas have chosen to form their own communities and not 'come back' anywhere, which opens up the concern of whether social reintegration can truly be achieved when demobilised soldiers remain isolated from civilian communities. However, tourism offers a solution by making it attractive for people to visit the settlement places of the ex-combatants. Hence, as long as encounters are promoted with civilians, staying together with comrades and not returning to pre-war communities does not necessarily impede social reintegration.

Hopefully, the present study can unveil the potential of this line of inquiry and contribute to the study of the tourism-peace nexus through the lens of ex-combatant reintegration. The Colombian case is one of the multiple cases where tourism has been used in reintegration processes. The call is open to investigate what has been the impact of tourism in less recent cases of DDR. Furthermore, researchers interested in the process with the FARC ex-combatants could also assess the role of tourism in social reintegration from the perspective of the tourists and visitors, whose voices were not part of the chapter. From the narrations of research participants, we assume that encounters with ex-combatants normally lead to an improvement of perceptions towards them, but whether this is the actual case –and if so, to what extent– is still unknown.

Addressing an analysis at the group level is important to discover some mainstream developments, but this cannot prevent us from bearing the complexities of individual situations. Hence, to further the study of post-conflict tourism and the social reintegration of ex-combatants, researchers could explore how the involvement in tourism activities influences the perceived image that ex-combatants have of themselves and if there is any association with levels of self-esteem. In addition, it would be relevant to examine changes in their portrayed image. During wartime, combatants inevitably build a military identity and learned how to displayed an aggressive appearance (Theidon, 2009). Now, they have to reposition themselves as story-tellers and tourism guides. The risk lies on ex-combatants not being aware of this military-like and intimidating behaviour while interacting with visitors, which could instead reinforce stereotypes. On the other hand, it could also be possible that the tourism activity indirectly imposes requirements on the ex-combatants to perpetuate their image as fighters as to be attractive to visitors. Researchers should be attentive to how tourism development takes place and warn if it becomes economically viable at the expense of exploiting these violent pasts.

Beyond academic implications, this chapter proves timely for the Colombian context, where it is becoming increasingly difficult to maintain the progress made so far in terms of peace. With the pronunciation of few FARC leaders in taking on arms again and the emergence of new criminal organisation, the need to support the reintegration process of ex-combatants is of the utmost relevance to preventing men and women going back to war.

ACKNOWLEDGMENT

The authors would like to thank the anonymous reviewers for their careful reading of the manuscript and insightful recommendations.

REFERENCES

Agencia para la Reincorporación y la Normalización. (2019). *La Reincorporación en cifras* (Publication 10 Dec 2019). Retrieved from http://www.reincorporacion.gov.co/es/reincorporacion/La%20Reincorporacion%20en%20cifras/Reincorporación%20en%20cifras%20-%20corte%2030112019.pdf

Alluri, R. M. (2009). *The role of tourism in post-conflict peacebuilding in Rwanda*. Swisspeace.

Anson, C. (1999). Planning for Peace: The Role of Tourism in the Aftermath of Violence. *Journal of Travel Research, 38*(1), 57–61. doi:10.1177/004728759903800112

Autesserre, S. (2014). Going Micro: Emerging and Future Peacekeeping Research. *International Peacekeeping, 21*(4), 492–500. doi:10.1080/13533312.2014.950884

Àvila, A. (2019, Jul 17). Los dos posibles caminos de las disidencias de las FARC. *Semana*. Retrieved from https://www.semana.com/opinion/articulo/los-dos-posibles-caminos-de-las-disidencias-de-las-farc-columna-de-ariel-avila/623873

Barakat, S., & Chard, M. (2002). Theories, rhetoric and practice: Recovering the capacities of war-torn societies. *Third World Quarterly, 23*(5), 817–835. doi:10.1080/0143659022000028639

Berdal, M. R. (1996). *Disarmament and demobilisation after civil wars: arms, soldiers and the termination of armed conflicts. Adelphi Paper 303*. Oxford University Press for the International Institute for Strategic Studies.

Björkdahl, A., & Buckley-Zistel, S. (Eds.). (2016). *Spatialising Peace and Conflict: Mapping the Production of Places, Sites and Scales of Violence*. Palgrave Macmillan.

Bowd, R., & Özerdem, A. (2013). How to Assess Social Reintegration of Ex-Combatants. *Journal of Intervention and Statebuilding, 7*(4), 453–475. doi:10.1080/17502977.2012.727537

Causevic, S., & Lynch, P. (2013). Political (in)stability and its influence on tourism development. *Tourism Management, 34*, 145–157. doi:10.1016/j.tourman.2012.04.006

Clubb, G. (2014). "From Terrorists to Peacekeepers": The IRA's Disengagement and the Role of Community Networks. *Studies in Conflict and Terrorism, 37*(10), 842–861. doi:10.1080/1057610X.2014.941434

Colombia, P. N. U. D. (2019a, Feb 14). *El príncipe heredero Haakon de Noruega, embajador de buena voluntad del PNUD visita Mesetas para inaugurar Paraísos Ocultos*. Retrieved from https://www.co.undp.org/content/colombia/es/home/presscenter/pressreleases/2019/02/el-principe-heredero-haakon-de-noruega--embajador-de-buena-volun.html

Colombia, P. N. U. D. (2019b, Mar 25). *La casa de la vida. Recorrido turístico de naturaleza y reconciliación en el norte y nordeste de Antioquia*. Retrieved from https://pnudcolombia.exposure.co/la-casa-de-la-vida

De Vries, H., & Wiegink, N. (2011). Breaking up and Going Home? Contesting Two Assumptions in the Demobilization and Reintegration of Former Combatants. *International Peacekeeping*, *18*(1), 38–51. doi:10.1080/13533312.2011.527506

Donais, T. (2009). Empowerment or Imposition? Dilemmas of Local Ownership in Post-Conflict Peacebuilding Processes. *Peace and Change*, *34*(1), 3–26. doi:10.1111/j.1468-0130.2009.00531.x

Dwyer, C. D. (2012). Expanding DDR: The Transformative Role of Former Prisoners in Community-Based Reintegration in Northern Ireland. *The International Journal of Transitional Justice*, *6*(2), 274–295. doi:10.1093/ijtj/ijs014

Echandía, C. (2006). *Dos décadas de escalamiento del conflicto armado en Colombia 1986-2006*. Universidad Externado de Colombia.

Faltas, S. (2004). *DDR without Camps*. 15.

Forero, S. (2019, Apr 30). Los tres productos más taquilleros de los excombatientes de las FARC. *El Espectador*. Retrieved from https://www.elespectador.com/colombia2020/pais/los-tres-productos-mas-taquilleros-de-los-excombatientes-de-las-farc-articulo-857922

Fundación Ideas para la Paz. (2017). *La dejación de armas de las FARC y otros procesos de desarme en el mundo*. Retrieved from http://cdn.ideaspaz.org/media/website/document/598ca97c62763.pdf

Garzón, J. C., Prada, T., Silva, A., & Zárate, L. (2019). *Las trayectorias de reincorporación y seguridad de los excombatientes de las FARC*. Notas estratégicas 11. Fundación Ideas para la Paz. Retrieved from http://ideaspaz.org/media/website/FIP_NE_TrayectoriasFarc_Final_V02.pdf

Goffman, E. (1963). Stigma and social identity. *Understanding Deviance: Connecting Classical and Contemporary Perspectives*, 256-265.

Grupo de Memoria Histórica. (Ed.). (2013). ¡Basta ya! Colombia, memorias de guerra y dignidad (Segunda edición corregida). Imprenta Nacional.

Herrador-Valencia, D., & Riera, E. M. (2016). The Creation of a New Identity. *Defending the Forest in Cinquera in El Salvador.*, *19*, 25.

Humphreys, M., & Weinstein, J. M. (2007). Demobilization and Reintegration. *The Journal of Conflict Resolution*, *51*(4), 531–567. doi:10.1177/0022002707302790

Janzen, R. (2014). *Guatemalan Ex-Combatant Perspectives on Reintegration: A Grounded Theory*. 27.

Jennings, K. M. (2007). The Struggle to Satisfy: DDR Through the Eyes of Ex-combatants in Liberia. *International Peacekeeping*, *14*(2), 204–218. doi:10.1080/13533310601150800

Kingma, K. (1997). *Demobilization of combatants after civil wars in Africa and their reintegration into civilian life*. 15.

Kingma, K. (2002). Demobilization, Reintegration and Peacebuilding in Africa. *International Peacekeeping*, *9*(2), 181–201. doi:10.1080/113533312.2002.11978332

Lederach, J. P. (1997). *Sustainable reconciliation in divided societies.* USIP.

Lundin. (1998). *Mechanisms of Community Reception of Demobilised Soldiers in Mozambique.* 16.

Mac Ginty, R., & Firchow, P. (2016). Top-down and bottom-up narratives of peace and conflict. *Politics*, *36*(3), 308–323. doi:10.1177/0263395715622967

Martínez, H., & Lefebvre, I. (2019). *La reincorporación económica de los excombatientes de las FARC. Retos y riesgos a futuro.* Notas estratégicas 09. Fundación Ideas para la Paz. Retrieved from http://ideas-paz.org/media/website/FIP_NE_Reincorporacion.pdf

Mashike, L. (2004). "You are a time bomb…" Ex-combatants in post-conflict South Africa. *Society in Transition*, *35*(1), 87–104. doi:10.1080/21528586.2004.10419108

McEvoy, K., & Shirlow, P. (2009). Re-imagining DDR: Ex-combatants, leadership and moral agency in conflict transformation. *Theoretical Criminology*, *13*(1), 31–59. doi:10.1177/1362480608100172

McMullin, J. R. (2013a). *Ex-Combatants and the Post-Conflict State.* doi:10.1057/9781137312938

Mcmullin, J. R. (2013b). Integration or separation? The stigmatisation of ex-combatants after war. *Review of International Studies*, *39*(2), 385–414. doi:10.1017/S0260210512000228

Metsola, L. (2006). 'Reintegration' of Ex-combatants and Former Fighters: A lens into state formation and citizenship in Namibia. *Third World Quarterly*, *27*(6), 1119–1135. doi:10.1080/01436590600842407

Muggah, R. (2005). No Magic Bullet: A Critical Perspective on Disarmament, Demobilization and Reintegration (DDR) and Weapons Reduction in Post-conflict Contexts. *The Round Table*, *94*(379), 239–252. doi:10.1080/00358530500082684

Muggah, R. (Ed.). (2009). *Security and post-conflict reconstruction: dealing with fighters in the aftermath of war.* Routledge.

Nussio, E. (2018). Ex-combatants and violence in Colombia: Are yesterday's villains today's principal threat? *Third World Thematics: A TWQ Journal, 3*(1), 135–152.

Özerdem, A. (2012). A re-conceptualisation of ex-combatant reintegration: 'Social reintegration' approach. *Conflict Security and Development*, *12*(1), 51–73. doi:10.1080/14678802.2012.667661

Pachankis, J. E. (2007). The psychological implications of concealing a stigma: A cognitive-affective-behavioral model. *Psychological Bulletin*, *133*(2), 328–345. doi:10.1037/0033-2909.133.2.328 PMID:17338603

Paffenholz, T. (2015). Unpacking the local turn in peacebuilding: A critical assessment towards an agenda for future research. *Third World Quarterly*, *36*(5), 857–874. doi:10.1080/01436597.2015.1029908

Parkin, J. (2018, Dec 10). Colombian artist melts guns into tiles to make 'anti-monument' to conflict. *The Guardian: Americas Edition.* Retrieved from https://www.theguardian.com/world/2018/dec/10/doris-salcedo-colombian-artist-melts-guns-tiles-peace-monument

Pécaut, D. (2008). *Las FARC: fuentes de su longevidad y de la conservación de su cohesión*. 29.

Porto, J. G., & Alden, C. (2016). *From soldiers to citizens: Demilitarization of conflict and society*. Routledge. doi:10.4324/9781315583525

Pugel, J. (2006). *A Survey of Ex-combatants in Liberiai*. 74.

Richmond, O. P., & Mitchell, A. (2011). *Hybrid forms of peace: from everyday agency to post-liberalism*. Springer.

Rodríguez López, M., Andreouli, E., & Howarth, C. (2015). From Ex-Combatants to Citizens: Connecting Everyday Citizenship and Social Reintegration in Colombia. *Journal of Social and Political Psychology*, *3*(2), 171–191. doi:10.5964/jspp.v3i2.388

Rolston, B. (2007). Demobilization and Reintegration of Ex-Combatants: The Irish Case in International Perspective. *Social & Legal Studies*, *16*(2), 259–280. doi:10.1177/0964663907076534

Sack, R. D. (1986). *Human territoriality: its theory and history* (Vol. 7). CUP Archive.

Sierra, L. (2019, Jul 17). Estos son los productos que ofrecen los exguerrileros en Agroexpo. *Pacifista*. Retrieved from https://pacifista.tv/notas/farc-agroexpo-reincorporacion-productos-etcr/

Simone-Charteris, M. T., & Boyd, S. W. (2010). *Northern Ireland re-emerges from the ashes: The contribution of political tourism towards a more visited and peaceful environment. In Tourism, progress and peace*. CABI.

Sonmez, S. F., & Apostolopoulos, Y. (2000). Conflict Resolution Through Tourism Cooperation? The Case of the Partitioned Island-State of Cyprus. *Journal of Travel & Tourism Marketing*, *9*(3), 35–48. doi:10.1300/J073v09n03_03

Stedman, S. J. (1997). Spoiler problems in peace processes. *International Security*, *22*(2), 5–53. doi:10.1162/isec.22.2.5

Theidon, K. (2007). Transitional Subjects: The Disarmament, Demobilization and Reintegration of Former Combatants in Colombia. *The International Journal of Transitional Justice*, *1*(1), 66–90. doi:10.1093/ijtj/ijm011

United Nations, Environment Programme, United Nations, & Development Programme. (2013). *The role of natural resources in disarmament, demobilization and reintegration: Addressing risks and seizing opportunities*. United Nations Environment Programme.

United Nations. (2014, May 9). *Integrated Disarmament, Demobilization and Reintegration Standards*. Retrieved from https://www.unddr.org/iddrs-framework.aspx

United Nations Verification Mission in Colombia - UNVMC. (2019a, Oct 1). *Report of the Secretary-General*. Retrieved from https://colombia.unmissions.org/sites/default/files/en_-_n1928697.pdf

United Nations Verification Mission in Colombia - UNVMC. (2019b, Dec 26). *Report of the Secretary-General*. Retrieved from https://colombia.unmissions.org/sites/default/files/n1942147.pdf

Universidad Nacional de Colombia. (2017, Jul 6). *Caracterización comunidad FARC-EP. Censo socio-económico UN-CNR 2017*. Retrieved from http://pensamiento.unal.edu.co/fileadmin/recursos/focos/piensa-paz/docs/presentacion_censo_farc.pdf

Van Broeck, A. M., Guasca, M., & Vanneste, D. (2019). Tourism Initiatives as a Lever for Reintegration of Former FARC Combatants in Colombia. *Via. Tourism Review*, 15.

Willibald, S. (2006). Does money work? Cash transfers to ex-combatants in disarmament, demobilisation and reintegration processes: Does Money Work? *Disasters*, *30*(3), 316–339. doi:10.1111/j.0361-3666.2005.00323.x PMID:16911431

Zambrano Quintero, L. (2019). La reincorporación colectiva de las FARC-EP: Una apuesta estratégica en un entorno adverso. *Revista CIDOB d'Afers Internacionals*, *121*(121), 45–66. doi:10.24241/rcai.2019.121.1.45

ADDITIONAL READING

Bouvier, V. M. (2009). Colombia: Building peace in a time of war. US Institute of Peace Press.

Farmaki, A. (2017). The tourism and peace nexus. *Tourism Management*, *59*, 528–540. doi:10.1016/j.tourman.2016.09.012

Galtung, J. (1969). Violence, peace, and peace research. *Journal of Peace Research*, *6*(3), 167–191. doi:10.1177/002234336900600301

Higgins-Desbiolles, F. (2006). More than an "industry": The forgotten power of tourism as a social force. *Tourism Management*, *27*(6), 1192–1208. doi:10.1016/j.tourman.2005.05.020

Justino, P., Brück, T., & Verwimp, P. (Eds.). (2013). *A micro-level perspective on the dynamics of conflict, violence, and development*. Oxford University Press. doi:10.1093/acprof:oso/9780199664597.001.0001

Moufakkir, O., & Kelly, I. (Eds.). (2010). *Tourism, peace and progress*. CABI Publishing. doi:10.1079/9781845936778.0000

Nussio, E. (2018). Ex-combatants and violence in Colombia: are yesterday's villains today's principal threat?. *Third World Thematics: A TWQ Journal, 3*(1), 135-152.

Richter, L. K. (1983). Tourism politics and political science: A case of not so benign neglect. *Annals of Tourism Research*, *10*(3), 313–335. doi:10.1016/0160-7383(83)90060-9

Valentine, G. (2008). Living with difference: Reflections on geographies of encounter. *Progress in Human Geography*, *32*(3), 323–337. doi:10.1177/0309133308089372

Villarraga Sarmiento, Á. (2013). Experiencias históricas recientes de reintegración de excombatientes en Colombia. *Colombia internacional*, (77), 107-140.

KEY TERMS AND DEFINITIONS

DDR: Abbreviation for disarmament, demobilisation and reintegration. It is an approach widely used in post-conflict contexts to set the transition of combatants to civilian life.

Ex-Combatant: A person who participated in the fighting in a war and has decided not to take part any more in the armed confrontation.

FARC: Spanish acronym for Revolutionary Armed Forces of Colombia. It was a guerrilla movement that signed a peace agreement with the Colombian government in 2016.

Peace: A state referring to the absence of direct violence (e.g. killing or kidnapping) and the presence of social justice (e.g. equality, inclusive development).

Post-Conflict: It refers to a stage after the end of an open and large-scale violent confrontation between opposing parties. It does not imply a total absence of violence, nor can it be equated with peace.

Post-Conflict Tourism: Tourism development that follows an increase in security in regions that are trying to recover from violent conflict.

Social Reintegration: A process whereby a person who has acquired the status of a combatant during a war returns to civilian life and is recognised by others as a member of the society.

Tourist: A person whose primary motivation to visit a place is leisure.

Visitor: A person who visits a place for multiple reasons and not exclusively leisure.

Chapter 6
The Contribution of Tourism to Peacebuilding Processes:
The Case of African Peace Parks

Ana Margarida Teixeira
https://orcid.org/0000-0003-4776-7017
University of Coimbra, Portugal

Paula Duarte Lopes
University of Coimbra, Portugal

ABSTRACT

The concern for environmental protection within an increasingly globalised international system has led to the planning of Transfrontier Conservation Areas, among which Peace Parks aim to achieve or maintain peace across borders. Based on the growing importance that tourism practice has for the sustainability of Peace Parks, this chapter reflects on how the tourism activity developed within Peace Parks can contribute to peacebuilding processes. This dynamic is addressed by analysing a case study, focused on the Great Limpopo Transfrontier Park. It shows that tourism has the potential to positively contribute to sustainability, creating better living conditions for the local population and, consequently, promoting the achievement of peace. However, it also shows that tourism may have the opposite effect if national interests dominate, if there is insufficient consultation of local communities' interests or if sensitive border and land-use issues are not overcome.

INTRODUCTION

This chapter focuses on the impact tourism activity has on peacebuilding processes, in the context of African Peace Parks. The objective is to understand how tourism activities developed within Peace Parks can contribute to peacebuilding. The underlying research contributes to analyses of conflict transformation, focusing on concrete economic activities, which articulated with environmental protection objectives, promote the achievement of a positive peace. It moves the focus of analysis away from violence and

DOI: 10.4018/978-1-7998-5053-3.ch006

explores the potential for transforming a conflict into non-violent dynamics. By focusing on tourism's potential for promoting and consolidating peace, this perspective not only makes a theoretical contribution to Peace Studies and Tourism Studies, but also provides the basis for improving peace promotion policies.

The concept of peacebuilding adopted in this chapter is environmental peacebuilding. Besides the focus on the impacts that a conflict has at the socio-economic and political levels, this process addresses environmental issues as a means to promote long-term sustainable development (Bronkhorst & Bob, 2010). Environmental peacebuilding processes have already proven to have a positive impact on inter-state conflict resolution through the establishment of Transfrontier Conservation Areas (Ide, 2018) and are a strong instrument for conflict transformation (United Nations Environment Programme [UNEP], 2018). The concept of Peace Park is related not only to the concept of environmental peacebuilding, but also to the concept of conflict transformation. A Peace Park can be considered a practical example of a conflict transformation process, sustained by an environmental peacebuilding perspective. The economic activities that emerge in and around Peace Parks are driven by this initiative and, at the same time, they prove to be a crucial tool for the sustainable management and maintenance of those parks (Peace Parks Foundation, 2019c). Working along with other instruments, these activities enable Peace Parks to remain active and thus continue to be drivers for sustainable and positive peace (Hsiao, 2007). By 2018, the community development programmes implemented in some African Peace Parks had already proven to have a positive impact on the lives of over 20,000 people living in and around these parks (Peace Parks Foundation, 2018).

Tourism is the main economic activity analysed in this chapter, as it constitutes the main financial contributor to this type of conservation areas (Peace Parks Foundation, 2019c) and stimulates economic growth and job creation in rural and marginalised regions within and around the parks, thus contributing to poverty reduction (Southern African Development Community [SADC], 2012). In South Africa, 1 in 20 people depends directly on tourism for a job (Peace Parks Foundation, 2019c). With the growing importance of tourism for Peace Parks' sustainability, it becomes relevant to study how best to frame and plan this kind of initiatives, which have an intrinsic peacebuilding effect in the areas where they are located. Thus, the main research question addressed in this chapter is as follows: *how do tourism activities developed within Peace Parks contribute to peacebuilding processes?*

In order to answer this question, the analysis is framed within discussions on peace and violence, environmental peacebuilding and violent conflict transformation. The methodological approach adopted includes a literature review and document analysis, using a case study to better substantiate the argument that guides this chapter. Information sources include public data provided by Peace Parks Foundation, the main organisation responsible for African Peace Parks' management and maintenance, and other institutions that study and manage these parks, such as the Southern African Development Community, the International Union for the Conservation of Nature, the World Commission on Protected Areas, the World Wildlife Fund for Nature and the United Nations University for Peace. The chapter includes a detailed analysis of the Great Limpopo Transfrontier Park, established in 2002, which links the Limpopo National Park in Mozambique, the Kruger National Park in South Africa and the Gonarezhou National Park in Zimbabwe. Among the 18 existing African Peace Parks, the Great Limpopo Transfrontier Park was one of the first formally established in Africa and it is one of the largest in the world. Only African Peace Parks were considered, since this continent has the highest number of Peace Parks in the world.

The chapter is organised into two parts. The first section addresses the theoretical framework used to analyse Peace Parks, including peacebuilding concepts, setting the stage for discussing their articulation with tourism activities. The second section focuses on the analysis of Peace Parks as peacebuilders and includes an empirical examination of the Great Limpopo Transfrontier Park. The findings suggest tourism activities have contributed to peacebuilding processes in the context of this particular Peace Park, allowing for the identification of enabling and thwarting factors of such a process.

BACKGROUND

In general terms, Peace Studies overall objective is to better understand peacebuilding processes. For this effect, it is fundamental to distinguish positive peace from negative peace and direct violence from structural violence. Positive peace is defined as a process that goes beyond the absence of physical violence and is characterised as sustainable peace, achieved through the improvement of social, economic, political and security structures (Galtung, 1985). Although negative peace – the absence of physical violence – is important, positive peace addresses directly the dynamics that may lead to physical violence. Structural violence is identified as the result of structure and as a synonym of social injustice, as it creates unequal opportunities in society that structurally limit human potential and reproduces itself (Galtung, 1985). Again, direct violence is obviously important, but structural violence is what perpetuates the underlying factors that may enable direct violence. These concepts contribute to a clearer understanding of the role tourism may have in enabling the promotion and/or consolidation of (positive) peace and, concomitantly, in creating a framework to thwart structural violence.

The concept of environmental peacebuilding is essential for this analysis, since it focuses on creating resilience in communities affected by conflict and, consequently, on achieving sustainable peace, through environmental cooperation and integration of natural resources' management in conflict prevention and resolution processes (Environmental Peacebuilding, 2019). Environmental peacebuilding promotes the sustainability of ecosystems based on tailored cooperative policies (Carius, 2006). In other words, peacebuilding processes assume that sustainability is only achieved through cooperation and the sustainable management of natural resources, as well as an equitable distribution of those resources. Environmental cooperation processes respond to some of the criticisms raised to liberal peacebuilding, through a strong action at the local level and the recognition that community participation is fundamental to achieve sustainable peace (UNEP, 2004). Environmental peacebuilding presupposes that the ultimate goal of the process is to achieve a positive peace and, in its implementation methodology, seeks to address the causes of conflict. That is, to create mechanisms to respond to the structural problems that trigger violent conflict, in order to prevent the emergence of other conflicts and more violence. Environmental peacebuilding is also articulated with conflict transformation, since it works with the structural causes of violence and promotes positive peace.

Authors such as Lederach (2003), Frazer (2013) and Ghettas (2013) explore the above-mentioned concepts, addressing conflict transformation as an effective way of conflict resolution, in the sense that the conflict itself is not resolved, but it is transformed into non-violent dynamics. Conflict transformation can be defined as a process of change in the relationship between the parties to the conflict, which addresses the structural causes that gave rise to it (Lederach, 2003). Conflict transformation aims to build sustainable societies that choose to address their differences by non-violent means (Frazer & Ghettas, 2013). Peacebuilding is associated with conflict transformation, since both aim at promoting peace, sustainable development and participatory governance. In other words, the relation between peace and conflict, focusing on its structural causes based on development issues and participatory (or non-participatory) governance mechanisms, allows articulating peacebuilding with conflict transformation.

This process presents itself as an alternative approach to mere conflict management and aims to ensure sustainable peace – sustainable positive peace. To promote positive peace is to create an environment that goes beyond the absence of direct physical violence, characterised by social justice, equal opportunities, equitable distribution of power and resources, and the right to protection and security (Galtung, 1969). In post-violent conflict settings, the construction of social justice results from peace consolidation processes that are carried out. In contrast to initial peace processes internationally promoted, these required that international actors focus on supporting local actors in conflict transformation processes (Paffenholz, 2009, p. 5). They recognise the need for sensitivity in implementing a peace process, given that these processes involve a heterogeneous population, engage with very specific cultural values and are characterised by a broad intervention time horizon (Paffenholz, 2009, p. 5). The most important contribution conflict transformation offers is the shift in focus from international to local actors, giving greater emphasis to civil society and thus distinguishing itself from traditional liberal peacebuilding (Paffenholz, 2009, p. 5).

A Peace Park constitutes a concrete example in which a conflict transformation approach can be applied linked to an environmental peacebuilding framework. A Peace Park is a restricted category of Transfrontier Conservation Areas, which crosses territorial and/or legal boundaries, and declares a focus on peace and cooperation above the basic objectives of a Protected Area (Hsiao, 2007). The United Nations University for Peace defines Peace Park as a Protected Area with a significant history of conflict and where the establishment of a Peace Park brings a new meaning to the territory (Hsiao, 2007). In this chapter, the link between tourism and peacebuilding is independent of a history of violent conflict in the territory covered by the Peace Park. Environmental peacebuilding can be promoted in a territory that has not witness a history of violent conflict, but which, at any moment, due to structural causes of inequality, discrimination, forced dislocation, non-participation in decision-making, among other aspects, can escalate into hostile and even violent dynamics.

One of the main goals of a Peace Park is to generate sustainable economic development through tourism (Mearns, 2012), which has been the fastest growing industry in the world. Consequently, one of its underlying goals is to end poverty, in which many communities inhabiting in or around the territories under these parks live. The opportunities and benefits created by tourism are especially important in areas where communities living on the borders of Protected Areas suffer from high levels of poverty, illiteracy, unemployment and distorted patterns of resource use (Mearns, 2012). One result of creating a Peace Park is laying the foundations for a sustainable economic livelihood for the local community (Peace Parks Foundation, 2019b), as tourism activities provide a way to use the existing natural resources for the benefit of the population without depleting them, and stimulate economic growth and job creation (Mearns, 2012). At the same time, tourism has been the largest and most global sector contributing to the direct funding of Peace Parks (Peace Parks Foundation, 2019c), so in order to promote a virtuous cycle, it is crucial to develop tourism offers that can generate visitor returns. The World Wide Fund for Nature advocates that Protected Areas have the capacity to develop sustainable tourism strategies that bear the costs of conservation management, while creating employment and entrepreneurship opportunities for impoverished population in developing countries (Spenceley, 2005). Authors such as Mearns (2012) and Chiutsi & Saarinen (2017) argue that the local population living in or around Transfrontier Conservation Areas benefits from the economic opportunities they create, especially through tourism development. This helps these parks gain social legitimacy and follow the framework of global conservation priorities, focusing on the relationship between the community and the parks (Hottola, 2009). The World Conservation Strategy emphasizes the importance of relating the conservation area management to the economic activities of local communities (International Union for Conservation of Nature and Natural Resources et al., 1980). Within these dynamics, tourism plays an essential role, namely through ecotourism and community-based tourism, providing a direct link between tourism activities and the sustainability of communities living in and around Peace Parks. The concepts that emphasize the social component of tourism practice and its impact on the local community were born with Robert Glucksmann (1935), when he defined tourism as the set of relationships that are established between people who are passing through a certain region and those who live there (Andrade, 2006). Ecotourism is a crucial segment to be taken into account in this research, since Peace Parks base all their activity, including tourism, on natural resources and aim at their sustainable use to ensure their conservation (The International Ecotourism Society, 2019). Alongside ecotourism, community-based tourism is an approach to tourism practice that is crucial for the sustainable development of Peace Parks. In this type of tourism, the local population, often impoverished and marginalised, actively participates as land users, entrepreneurs, decision-makers and conservation agents (Mearns, 2012). The overall goal is for local community members to have power over decisions related to tourism development in their area of residence and to work together with other stakeholders to develop employment opportunities, create businesses and improve living conditions in their communities (SADC, 1999). This type of tourism implies that community members have a high level of control over the activities carried out and that they receive a significant proportion of the associated benefits (Scheyvens, 1999; Wood, 2002). The segments of ecotourism and community-based tourism allow for a dynamic relationship between them and the promotion of cooperation and peace among the countries involved, since they preserve the natural resources necessary for the sustainability of cross-border areas and ensure that local communities continue to be the main beneficiaries, through socio-economic development driven by tourism (Mearns, 2012).

The positive impacts that tourism has on the well-being of populations contribute to the achievement of the United Nations Sustainable Development Goals and its Agenda 2030 and, in particular, those which aim to eradicate poverty, protect the planet and ensure prosperity for all (World Tourism Organisation, 2019). Based on this universal, integrated and transformative vision, the World Tourism Organisation (WTO) strives to contribute with its technical assistance and capacity building, focusing on Goals 8 (decent work and economic growth), 12 (responsible consumption and production) and 14 (life below water), in which the role of tourism is highlighted.

TOURISM AND PEACE IN THE GREAT LIMPOPO TRANSFRONTIER PARK

Tourism activities associated with ecotourism and community-based tourism segments have already proved, in theory, the potential to have a positive impact on the sustainability of a Peace Park and, consequently, on achieving positive peace. Through the case study, concrete examples of these activities and their impact in the Great Limpopo Transfrontier Park region are discussed. Home to over 850 animal and 2,000 plant species, the Great Limpopo Transfrontier Park was one of the first Peace Parks formally established in southern Africa (Peace Parks Foundation, 2019a). This 35,000 km^2 park links the Limpopo National Park in Mozambique, the Kruger National Park in South Africa and the Gonarezhou National Park in Zimbabwe (Figure 1). It also links the communal land of Sengwe in Zimbabwe, and the Makuleke region of South Africa, which lies between the Kruger and Gonarezhou Parks (Peace Parks Foundation, 2019a). The Transfrontier Park is located within the Great Limpopo Transfrontier Conservation Area (GL-TFCA), covering almost 100,000 km^2, and which also includes the National Parks of Banhine and

Zinave, Massingir and Corumana, interconnecting regions in Mozambique, as well as several private and state-owned Protected Areas. In 2017, Greater Lubombos Conservancy in Mozambique became the first private area to be included in the GL-TFCA, contributing with 2,400 additional km^2 (Peace Parks Foundation, 2019a). The Great Limpopo Transfrontier Park is one of the few regions of southern Africa that still contains underdeveloped natural areas with little human influence, thus constituting a major historical resource (Great Limpopo Transfrontier Park [GLTP], 2019b).

The international treaty that created the Great Limpopo Transfrontier Park identifies six strategic goals (GLTP, 2019a). The first is to promote transnational cooperation among the countries involved in order to achieve effective ecosystem management (GLTP, 2019a). The second goal complements cooperation efforts by promoting alliances regarding biological natural resources management and other partnerships among all parties involved, including the private sector, local communities and non-governmental organisations (NGOs) (GLTP, 2019a). Cooperation among countries is a means to achieve sustainable peace, especially when local communities are involved in the joint management process. The second goal also explicitly states that natural resources management must go hand-in-hand with socio-economic development (GLTP, 2019a), creating benefits to local communities, namely through job creation, investment in infrastructure and in education and training of local people. These dynamics and associated results enable poverty reduction, which constitutes a requirement for achieving positive peace. The third goal is to strengthen ecosystem integrity, as well as ecological natural processes, by harmonising environmental management processes at international borders and removing artificial barriers that prevent wildlife's natural movement (GLTP, 2019a). The fourth and fifth goals relate directly to the promotion of sustainable regional socio-economic development, as they seek to develop appropriate frameworks, strategies and job creation plans for local communities through, for example, cross-border ecotourism activities (GLTP, 2019a). Through the fifth goal, a direct link is created between tourism and the sustainable socio-economic development of the Park and the communities within and around it. Consequently, a link is created between tourism and the promotion of positive peace. Finally, the sixth goal is to establish mechanisms that facilitate the exchange of technical, scientific and legal information for more effective joint ecosystem management (GLTP, 2019a). The goals of the Great Limpopo Transfrontier Park are thus focused on two broad areas. Firstly, transboundary cooperation and joint ecosystem management of the Park and, secondly, the promotion of activities that foster sustainable regional socio-economic development. Although there is no explicit goal to promote positive and sustainable peace, they all relate to the concept and help, directly or indirectly, to achieve this peace.

Figure 1.
Source: South African National Parks, 2019.

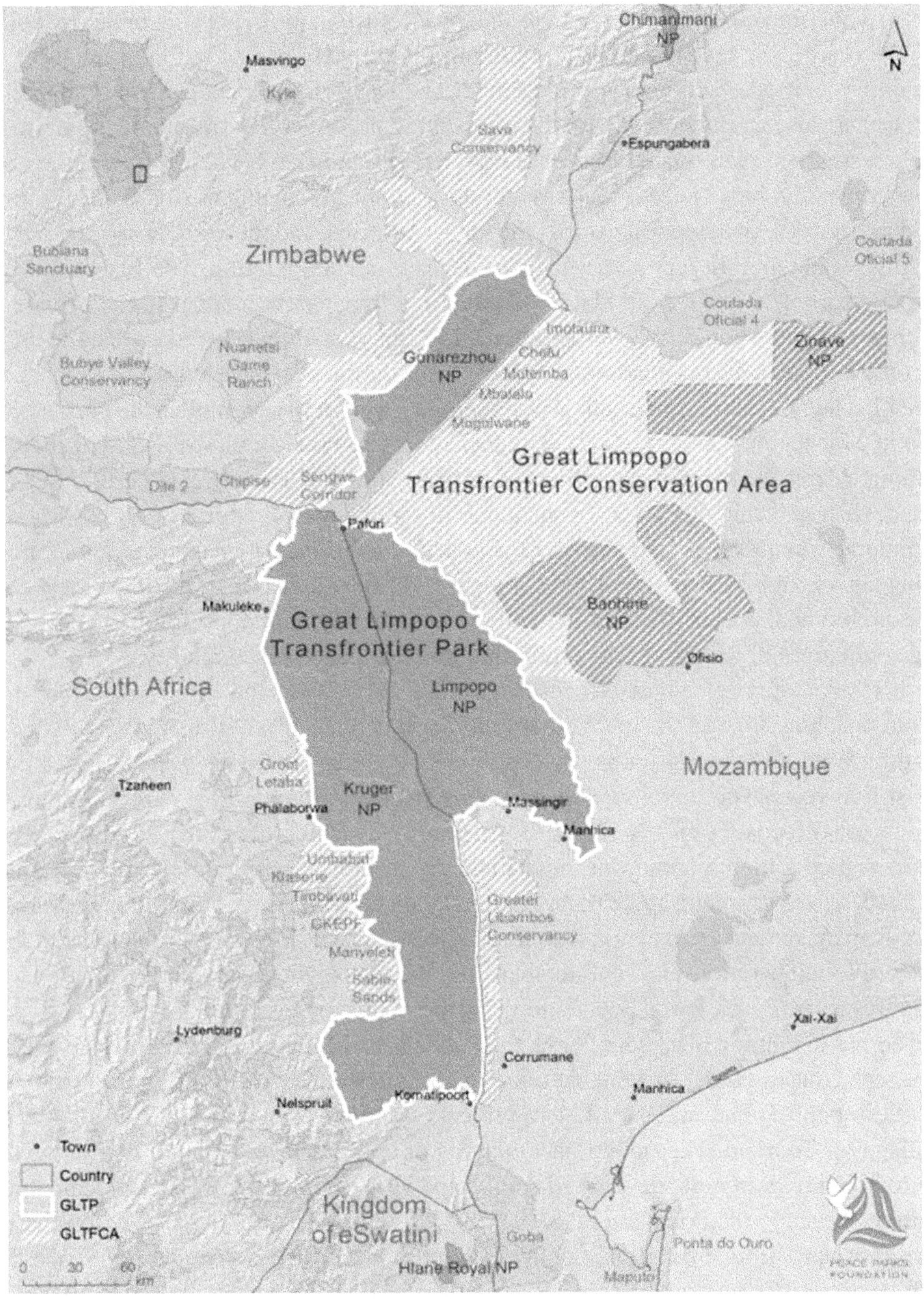

The Great Limpopo Transfrontier Park is committed to the development of tourism activities from its inception, as it recognises the benefits they bring both to the Park and to the local communities around and within it. Tourism activities contribute to the Park's sustainability, by financing its daily management and governance (Spenceley, 2005), promoting biodiversity conservation, crucial for supporting tourism, and regional socio-economic development. The Park's Joint Management Plan includes incentives promoting a close work relation with the tourism industry within a sustainable tourism framework and presents a strategic plan related to the tourism market and free access to the borders of the three countries to implement cross-border tourism activities (Joint Management Board, 2002). International cooperation promotes people mobility across the three countries, and these cross-country tourism activities, involving people mobility, promote further international cooperation. The Plan also specifies that tourism products implemented should be transferred in a transparent manner to the private sector, and that local entrepreneurs should be given preference in this transference process. The local population should not only be considered as potential employers, but should also have a stake in tourism companies' capital. The Plan lays out aspects to be taken into account, in order to accomplish the Park's goals through tourism (Joint Management Board, 2002). The local communities should be aware of the real impacts of establishing a Transfrontier Conservation Area, since some restrictions and certain rules need to be implemented, such as, achieving levels of sustainability in the use of natural resources. Concomitantly, local communities should be able to participate in the design of Park management and policies, as well as in training and personal/professional development activities. This is crucial, since local communities should have preferential access to education, culture and employment opportunities in the development and management areas of the Park, thus promoting positive peace sustainability.

The Park's Joint Management also focuses on the collaboration and implementation of policies to combat rhino poaching (one of the main conservation objectives of the Transfrontier Park), promoting the Combatting Wildlife Crime programme (Peace Parks Foundation, 2019a), that has been implemented over the past five years. The initiative aims to intensify efforts to control the illegal hunting of rhinos and elephants, also focusing on the social dimension of poaching. In April 2014, Mozambique and South Africa signed a memorandum of understanding on biodiversity conservation and management and, since 2015, a joint Limpopo and Kruger National Parks management committee is conducting, with the GL-TFCA management framework, a strategy to combat wildlife crime. This aims to facilitate the coordination of joint operations, law enforcement and the management and protection of national parks and adjacent reserves (Peace Parks Foundation, 2019a).

When it comes to training local population, the Park's Joint Management Plan reflects the assumption that it is through education that sustainable tourism activities are successfully promoted. In the context of the Great Limpopo Transfrontier Park, two projects stand out: the Hlawula Vutomi programme and the SA College for Tourism. The first project consists of a youth development programme that aims to create environmental awareness, develop life skills and help young people make sustainable decisions for their future (Peace Parks Foundation, 2019a). SA College for Tourism (SACT), on the other hand, is a non-profit organisation, created in 2001 by Dr. Anton Rupert, then President of Peace Parks Foundation, and developed in partnership with Rupert Family Foundations, as well as other international agencies and public donors (Peace Parks Foundation, 2019c). The School offers professional training to unemployed men and women from impoverished rural backgrounds and aims at preparing them for qualified positions in the ecotourism industry (Peace Parks Foundation, 2019c). After completing their training, the students are able to return to their communities as entrepreneurs in small tourism enterprises or to work in the local ecotourism industry that exists in or around the Transfrontier Conservation Area.

In essence, tourism is one of the main fields of action of the Great Limpopo Transfrontier Park, because the activity promotes regional socio-economic development, through local communities' participation in the tourism initiatives implemented and through job creation and promotion of education and training geared towards the local population. Also, tourism ensures local communities' livelihoods, as most of them base their subsistence and economic activity on natural resources exploitation (Convention on Biological Diversity, 2008). It is important to note that the tourism activities implemented in the Transfrontier Park can only be developed thanks to the natural diversity of Southern Africa, which is mostly covered and preserved by Protected Areas and attracts many visitors (GLTP, 2019b).

Consequently, the Great Limpopo Transfrontier Park tourism market is based on the ecotourism segment. The three countries hosting the Park have shown a strong demand for this type of tourism. In 2000, 78.9% of tourist arrivals in South Africa and 76% of arrivals in Zimbabwe were motivated by ecotourism (Spenceley, 2006). In that year, the revenue generated by this segment in the three countries was around USD 2.45 billion, and ecotourism contributed with 9% of the total Gross Domestic Product (GDP) of the Southern African Development Community region (Spenceley, 2006). It was also in 2000 that the South African National Parks (SANParks) – the organisation that developed and managed all tourism infrastructure in South Africa's Kruger National Park – initiated a process of power transfer, by granting concession rights to parastatal companies (Spenceley, 2006), in order to promote the economic development of the most impoverished local populations and to create business opportunities for local entrepreneurs (South African National Parks, 2000). Between 2002 and 2003, there was a decrease in international arrivals in the three countries, from 714 million, in 2002, to 694 million, in 2003 – a 2.8% drop, motivated by fear associated with terrorism, the emergence of the Coronavirus of Severe Acute Respiratory Syndrome (SARS-CoV) and little acceleration in the global economy (Spenceley, 2005). The number of visitors to the African continent continued to decline during the first decade of 2000 due to the growing tourist attractiveness of European and Asian continents (Roser, 2020). However, in 2014, the tourism sector represented 7%, 9.4% and 10.5% of the GDP of Mozambique, South Africa and Zimbabwe, respectively (Peace Parks Foundation, 2015). Until 2019, the overall trend dictated an increase in tourism demand, translated into growing values in tourism revenues and, in particular, ecotourism revenues, which benefited the development of Peace Parks. Nowhere else in the world was tourism growing faster than in southern Africa, where the number of hotels has doubled in just four years (Space for Giants & UNEP, 2019). By 2030, the number of international tourists in Africa was expected to rise from 62 million to 134 million (Space for Giants & UNEP, 2019). This growth rate had the power to transform the African continent and consequently the development of Peace Parks and their tourism activities, given that, already in 2019, tourism contributed with 8.5% of the continent's global economy, generating 24 million jobs (including 40% more full-time jobs than agriculture would generate with the same investment) and offering significantly more job opportunities to women, compared to other sectors (Space for Giants & UNEP, 2019).

According to the World Tourism Organisation, 80% of annual trips to Africa have been motivated by wildlife observation and ecotourism (Space for Giants & UNEP, 2019). At the same time, wildlife and natural resources will continue to be the main driver for tourism growth in Africa, which is something very positive in the context of the Great Limpopo Transfrontier Park, considering that the tourism activities implemented are ecotourism and community-based tourism. In addition, many of them are characterised as cross-border, i.e. covering two or more countries. The ecotourism segment promotes the conservation of natural and cultural resources, provides benefits for local communities by creating employment opportunities and generating economic flows to the region, and promotes environmental education (The

International Ecotourism Society, 2019). While tourism activity, in the context of Peace Parks, is based on natural resources, it is also based on communities. Community-based tourism promotes the active participation of the local population in the decision-making processes related to tourism development, so that the associated economic benefits are directed to the region itself (Mearns, 2012). Both concepts presented are drivers of socio-economic development in rural communities near the Transfrontier Park and relate to sustainable tourism, as they promote the sustainability of both the region and the tourism activity itself (UNEP, 2005). Thus, they also relate to positive peace promotion, by improving local populations' living standards, enlarging their development opportunities and establishing participative decision-making mechanisms, which empower local communities regarding the Park's management and development.

Cross-border Tourism Activities

In practice, several successful cross-border tourism activities have been established, including hiking trails, a series of adventure experiences, wilderness routes and trails (Peace Parks Foundation, 2019a), as well as safaris and tourist accommodations, located within national parks, which promote sustainability and conservation programmes. This is the case of the Chilo Gorge Safari Lodge (Figure 2), located in Gonarezhou National Park, Zimbabwe.

Figure 2.
Source: Chilo Gorge Safari Lodge, 2019.

The natural landscape of southeast Zimbabwe creates opportunities to develop unique tourism experiences. While the southern region of the National Park allows visitors to visit the common areas and panoramic views of the Governor's old house and enjoy the tranquillity of nature, the northern region focuses on the richness of the Chilojo Cliffs, which lead to Chilo Gorge and the community built 5-star hotel (GLTP, 2019c). From the hotel lookout, it is possible to observe many animal species, such as elephants, hippos and crocodiles (Chilo Gorge Safari Lodge, 2020). The Gonarezhou National Park is one of the last wild places in the African continent that is still intact, and where the animals are in their most natural state (Chilo Gorge Safari Lodge, 2020). The Chilo Gorge Safari Lodge also has a mission of sustainability and conservation, within which the participation of the local community is deemed essential.

The hotel and safari management promotes the **CAMPFIRE project** (Communal Areas Management Programme for Indigenous Resources), now replicated throughout the African continent and the rest of the world (Chilo Gorge Safari Lodge, 2020). This programme results from tensions and conflict that emerged between the local Shangaan tribe and those responsible for managing the Park (Chilo Gorge Safari Lodge, 2020). Initially, the community resorted to poaching to prevent wildlife from destroying its territory, and the Park's management framework, at the time, showed no interest in the issue and did not outline strategies for the community to benefit from their own land without resorting to poaching. On the contrary, it condemned the tribe for its actions (Chilo Gorge Safari Lodge, 2020). In this context, the CAMPFIRE project was born with the intention of creating a bridge between the local community and the Park management, and generating income for the socio-economic development of the region through tourism activity. In practice, the project has made it possible to build the community's first school, a clinic, a police station and a factory, from the profits made by tourism (Chilo Gorge Safari Lodge, 2020).

Figure 3.
Source: Covane Community Lodge, 2020a.

The project also made it possible to reduce poaching activities and the use of weapons by the local community. CAMPFIRE turned conflict into cooperation, thus, promoting non-violent means of addressing conflictual dynamics, and allowed the local community, wildlife and sustainable tourism to benefit themselves and each other. As a result, CAMPFIRE sustains local community livelihoods, promotes wildlife conservation and increases socio-economic development through tourism activities, and, thus, by addressing the root causes of the conflict between the local community and the Park management, enables a sustainable structural peace. Continuing the positive local impact created by the programme, Chilo Gorge Safari Lodge created the Mahenye Charitable Trust (Chilo Gorge Safari Lodge, 2020), to which guests of the lodge and visitors, in general, can donate money that will be invested in projects that directly benefit the local community. These projects are planned together with the population, and their implementation is overseen by the Chilo Gorge Safari Lodge management team (Chilo Gorge Safari Lodge, 2020). The continuous involvement of the local population in these projects' planning and development further contributes to sustaining the peaceful framework created, since it empowers local communities to participate in the definition of their own living conditions and future opportunities.

Another example of a tourism activity that has a high local impact is the **Covane Community Lodge** (Figure 3), a community-based tourism project located near the Massingir Dam in Mozambique.

The lodge is located about 7km from Massingir and opened in May 2004 (Spenceley, 2006). Besides the lodge, the project offers visitors the opportunity to learn about traditional dances and food, explore hiking trails, visit the villages, take boat trips and buy local handicrafts. The concept was initially presented to the local communities Canhane and Kubo, who, along with the Swiss NGO Helvetas, which was the driving force behind the project, studied the land law, for the delimitation of potential locations for accommodation (Spenceley, 2006). Helvetas eventually facilitated the delimitation of an area of 7,024 hectares by transferring full control over the land to the local community (Spenceley, 2006). A steering committee of ten local volunteers, chosen in a community meeting, was then organised and a team of twenty local volunteers was formed to work on the construction of the accommodation without assistance from the private sector. Already in 2004, the lodge employed nine community members and generated indirect employment for over 40 people (even the primary school students produced and sold handicrafts within the lodge for their own benefit) (Spenceley, 2006). Helvetas outlined the kind of qualifications the team should have and the community sought these qualifications within the local population. At the same time, the NGO promoted training courses in tourism and hospitality to all team members and organised visits to community-based tourism enterprises in Swaziland and Xai-Xai, Mozambique (Spenceley, 2006). The community decides every year at a community meeting how to invest the money generated by the accommodation in the local population. In 2004, the agreement stated that 50% of the profits should be invested in community infrastructure and the other 50% in improvements to the lodge (Spenceley, 2006). After one year of operation, the tourist resort had already allowed two wells to be rehabilitated and a classroom built. In the meantime, the management of the accommodation was taken over, together with the population, by the Mozambican NGO Lupa (Covane Community Lodge, 2020b). Always within a civil society framework, the Covane Community Lodge was planned, developed and is managed by local communities. The direct link between the tourist facility and the local community's benefits shows the potential for promoting a positive peace, in which socio-economic development, personal and professional enhancement skills and decision-making participation constitute crucial elements in this process. Furthermore, in this specific case, the land rights' transfer structurally contributed to the local community's empowerment, providing a legal stand for their endeavours and further improving their livelihoods' sustainability.

Figure 4.
Source: The Outpost Lodge, 2020.

Finally, the third example presented is also a tourist accommodation, where the focus is on creating better socio-economic conditions for the Makuleke community. Until 1969, this community lived on an area of 24,000 hectares, located in the north of the Kruger National Park, between the Luvuvhu and Limpopo Rivers, in South Africa. That year, the community was forced to move to another region in order for its territory to be included in the National Park (Spenceley, 2006). This displacement is a result of the interpretation of the 'national park' concept, which for all purposes consists of an area where 'human barriers' are removed. In 1998, the local community was compensated for its relocation, with the restitution of its land and the creation of a contractual park (Spenceley, 2006). However, although the local population was able to reclaim their land, conditions were created that limited their forms of exploitation: the area could only be used for conservation purposes. The community was not entirely free to decide how it wanted to use its land, but it was free to decide how to economically exploit the conservation values, eventually obtaining the right to grant hunting concessions (in the traditional non-hunting area of the Kruger National Park). A 25-year agreement between the Makuleke and SANParks was thus established, in which power and title over the territory were returned to the local community (Spenceley, 2006), empowering the once displaced, marginalised and excluded local population. The contract governing the incorporation of the Makuleke region into the Kruger National Park allows the population to sustainably use the natural resources and to develop low environmental impact tourism activities, such as **The Outpost** (Figure 4), a 24-bed luxury tourist accommodation located at the confluence of the Luvhuvhu and Mutale Rivers.

During the years before the lodge was built, the local population resorted to poaching activities to support themselves (Spenceley, 2006). As the lodge began to attract tourists and, consequently, income to the region, the management framework of the lodge wanted to gradually eliminate the practice, given the clear goal of biodiversity conservation contemplated in the contract. The development of the accommodation began in 2002 and resulted in a luxury development designed to have a low environmental impact (Spenceley, 2006). The accommodation planning was devised together with the population, and it was only initiated after a contract was signed between the accommodation management and the local community. The lodge is currently managed by locals (Spenceley, 2006). The Outpost allowed the development of local infrastructures and activities that directly benefit the local community, improving their living conditions and creating solid structures for the promotion of a sustainable peace. While it was being built, a fee was paid to the population in order to guarantee an alternative initial means of subsistence, as well as a deposit, to assure the community that the project was going to be implemented and would be in operation during the current contract period (Spenceley, 2006). It established that 8% of the gross income from the accommodation should be invested directly in the community, 2% should be put aside in a social development fund for the education of the Makuleke children (Spenceley, 2006) and, moreover, the accommodation management should pay a fee for each tourism activity carried out within the local territory that damaged it (for example, the lodge paid a monthly fee for each vehicle crossing the territory). As soon as the lodge operated at 60%, it had to pay an annual concession fee of USD 75,000 to the Makuleke community and about USD 150,000 in salaries to the 30 local employees, which constituted a significant amount of money in comparison to the average annual salary of the region (Spenceley, 2006). Although there is no information after 2006 on whether the targets set out in the agreement have been met, both the tourist resort and the contracted park continue to operate, contributing to a positive and structural peace.

From Tourism to Peace

After analysing in practice how tourism activities influence the region and the communities where they are developed, one can state that tourism activities in the Great Limpopo Transfrontier Park have fostered local socio-economic development and, therefore, promoted a sustainable positive peace in the region. These tourism activities contribute to the consolidation of regional peace and provide people with a sense of trust and well-being amongst themselves and in their relationship with the Park's promoters and/or managers and its natural resources. This has been achieved by the adoption and promotion of mutually reinforcing dimensions, which create a conducive environment based on tourism activities for promoting peace. The promotion of biodiversity conservation constitutes the overall framework for the creation of the Great Limpopo Transfrontier Park, with tourism activities developed in the ecotourism sector, and anchoring these in local environmental dynamics. Both the Chilo Gorge Safari Lodge and The Outpost, through the creation of an alternative livelihood for the local communities, have resulted in a decline of poaching activities in their regions. The tourism activities created became an alternative for poaching, thus, increasing the biodiversity conservation level of the regions covered, which then sustains the ecotourism activities. Improvement of living conditions constitutes another important dimension of positive peace. All the tourism activities developed in the Great Limpopo Transfrontier Park invest a significant proportion of their income in local community infrastructure, which directly improves those communities' overall living conditions. In order to promote local appropriation, it has been crucial to promote an active participatory dynamic. In this case, all tourism activities promote projects involving local communities in their planning and execution phases, and those communities benefit directly from these projects. These appropriation dynamics are also reinforced when tourism activities promote local values and culture. In the case of Covane Community Lodge, for instance, the visitors are given the opportunity to learn about traditional dances and food, and buy local handicrafts. The sustainability of these dimensions is reinforced by the creation of jobs for local community members and by the investment in local population's education and training. The former provides local community members direct participation in the socio-economic development of the region, either by working directly in the tourist sector or by working in tourist supporting sectors, which in turn allows them to escape poverty and/or an aid-dependency situation. The latter provides local communities with skills that can be used within the Great Limpopo Transfrontier Park or elsewhere, contributing to their empowerment as qualified workers. For instance, the Covane Community Lodge promotes training courses in tourism and hospitality to all team members, and The Outpost finances a social development fund for the education of Makuleke children. Both contribute to an enlargement of local population's opportunities.

As stated, these different dimensions are mutually reinforcing and constitute the basis for the sustainable transformation of conflict into cooperation and stability, i.e. peace. Tourism constitutes in these cases the linchpin that mediates between the necessary improvement of local communities living conditions and the creation of Peace Parks. The Chilo Gorge Safari Lodge was able to alleviate the existing conflict between the Shangaan community and the Park management, transforming their conflict into a fruitful cooperative project. The Great Limpopo Transfrontier Park thus presents a unique opportunity to use tourism development as an agent for regional ecosystem conservation, for socio-economic development (GLTP, 2019b) and for the achievement of the Sustainable Development Goals – tourism activities implemented in the Park contribute positively to the role of tourism in the sustainable management framework through Goal 8.9, which aims to "by 2030, design and implement policies to promote sustainable tourism and that create jobs and promote local culture and products" (World Tourism Organisation, 2019), overall contributing to a structure that sustains positive peace.

SOLUTIONS AND RECOMMENDATIONS

Although the different cases analysed in this chapter regarding the dynamics between peacebuilding and tourism in the Great Limpopo Transfrontier Park, until 2019, suggest a mutually reinforcing dynamics, it is important to address challenges identified in the process of this research that might work against peacebuilding or its sustainability. The first challenge is political. Peace Parks have been created based on the negotiation of international treaties and these, as history shows us, are as strong and reliable as the relations among the states that signed them. Consequently, if each or any state decides to address the tourism activities in effect, or the Peace Park itself, from a strict national interest perspective, with a protectionist and unilateral approach, the benefits achieved so far can easily be undermined and reversed. Also, another aspect of this political challenge resides with local-national dynamics, in which national authorities may overlook local necessities or demands in order to respond to national and/or international interests. The case of the Makuleke community renders this type of challenge visible, where the interests underlying the creation of the Park implied the forced dislocation of that community. The sustainability of the positive dynamics between tourism and peacebuilding reside in local appropriation, otherwise there is always the possibility for contestation, disruption and instability, that would directly affect the tourism activities and possibly increase levels of oppression and control over local communities. This aspect is also directly linked to land property rights that too often end up stripping local communities from their ancestral territories. Recognising and involving local communities as stakeholders of the Peace Parks and of the tourism activities is crucial to promote a sustainable and positive peace.

The second challenge is obviously socio-economic. The tourism activities analysed exist within a globalised neoliberal economic model, which, by definition, is more focused on profit maximisation than on local communitarian needs or specificities. Additionally, the division of benefits from the tourism activities needs to be just and proportionate regarding the local community, private actors involved, if any, and the political authorities. A disproportionate division of benefits, overlooking any stakeholder's expectations, including local communities, may lead to an unsustainable outcome. Consequently, it is crucial to promote a sustainable equilibrium between economic interests and local communitarian expectations and needs. These constitute essential factors to stimulate a regional socio-development that promotes and sustains positive peace.

The third challenge is environmental. Peace Parks have been created within an environmental conservation approach. Thus, tourism activities within these Peace Parks are focused on ecotourism. There is a continuous tension between conserving biodiversity in those territories and increasing the number of visitors to raise more funds to further sustain conservation activities and increase local communities' standards of living. All stakeholders involved need to recognise this continuous tension and manage it in a sustainable manner, keeping in mind that without biodiversity, ecotourism may decline, undermining local communities' human development and, consequently, undermining peacebuilding dynamics.

FUTURE RESEARCH DIRECTIONS

There are at least two factors that may thwart this potentially virtuous dynamic between tourism and peacebuilding in the context of Peace Parks and that requires further research. On the one hand, the impact of climate change and the occurrence of extreme weather conditions may destroy the tourism infrastructure created, as well as disrupt the natural habitats that constitute the Park's basis for existence. It is essential to research adequate climate change adaptation and mitigation measures to better prepare Peace Parks to face the coming challenges.

On the other hand, any occurrence that disrupts the international flow of people may undermine the Parks' sustainability. The analysis provided in this chapter was developed until the end of 2019, before the COVID-19 pandemic. The 2020 pandemic has structurally disrupted the circulation of people worldwide and consequently inflicted an enormous blow in the tourism sector (in fact in the economic sector, in general). Peace Parks' major reliability on tourism denotes a structural fragility when it comes to events curtailing people's movements. This type of pandemic may also constitute a risk to the natural ecosystems in these Parks, since there may be human-animal transmission of diseases that may wipe out the animal population of these territories. It seems this type of epidemics/pandemics may become recurrent, consequently, it is essential to research the width and depth of the impact of this type of events in ecotourism and in Peace Parks, in order to outline coping strategies.

Future research needs to take into account the challenges mentioned above – political, social and environmental – within climate change and pandemic frameworks. The peace promoted by ecotourism in Peace Parks may reveal to be too fragile in face of structural dynamics as those identified. Still, at the same time, the dynamics fostered by Peace Parks and ecotourism activities constitute an important basis for promoting confidence building and trust relations amongst all those involved. These constitute the basis for a long-lasting sustainable peace and, consequently, these dynamics will certainly outlast any interruption or disruption of tourism activities. Living conditions, participation in decision-making and local expectations as well as recognising local knowledge remain crucial for any adaptation, mitigation or coping strategy.

CONCLUSION

Peace Parks render visible the contribution tourism can make towards peacebuilding processes. These processes are much more than the absence of physical violence, since they are based on a notion of positive peace. A community characterised by a high human development, by social justice, by inclusive and participatory governance systems, by human rights, by sustainable livelihoods is a community living in a positive peace context. The tourism activities within the Great Limpopo Transfrontier Park showed how tourism activities can contribute to promoting and supporting positive peace in different communities. The initial direct link made is between tourism activities' income and local communities standards of living, but the examples analysed go well beyond. The material conditions are important, but the participation in the decisions regarding the activities developed, how they are developed and how the income is distributed is key to promote mutual understanding, confidence building and trust among communities and between local communities and national and Park authorities. Moreover, the training and education local communities benefit from provide them skills that can be used in the Park's activities but also elsewhere, and, therefore, have an impact in those local communities' possibilities beyond the actual Park. The Park itself is based on a conservation approach, grounded on natural habitats and ecosystems. Consequently, the tourism activities are developed in a structural articulation with local biodiversity.

The analysis suggests a win-win situation regarding tourism, biodiversity and peace (positive peace). However, these dynamics face constant challenges, which need to be acknowledged and addressed. Political turf wars among national authorities, between Park authorities and local communities and between national authorities and local communities should be avoided in order to strengthen the confidence building dynamics ecotourism brings to these territories. Also, economic considerations need to be balanced with local communities' needs and expectations. This is to say that maximising profit cannot hinder the consolidation of peace. Environmental concerns should also be clearly taken into account from the first moment. After all, ecotourism can only subsist if there is biodiversity for this effect. Even if more visitors in economic terms would allow for a higher income, it may also, in environmental terms, undermine the sustainability of the ecosystems.

Finally, and in face of the current world situation, future research needs to include an assessment of climate change impact in these Parks and identify adaptation and mitigation measures, as well as a scenario assessment regarding epidemics/pandemics. The 2020 pandemic revealed the fragility of the tourist sector with a temporary worldwide ban on the circulation of people. Still, there is evidence to suggest that the confidence building and the trust gained with the tourism activities within the Great Limpopo Transfrontier Park may resist the negative impacts from climate change or the 2020 pandemic.

REFERENCES

Bronkhorst, S. van J., & Bob, U. (2010). Environmental conflicts: Key issues and management implications. *African Journal on Conflict Resolution*, (2). Retrieved March 20, 2019, from https://www.accord.org.za/ajcr-issues/%EF%BF%BCenvironmental-conflicts/

Carius, A. (2006). *Environmental Peacebuilding: Conditions for Success* (12th ed.). Woodrow Wilson International Center for Scholars.

Chilo Gorge Safari Lodge. (2019). *The Lodge*. Retrieved April 7, 2020 from the Chilo Gorge Image Gallery.

Chilo Gorge Safari Lodge. (2020). *About*. Retrieved January 5, 2020, from http://chilogorge.com/about/

Chiutsi, S., & Saarinen, J. (2017). Local participation in transfrontier tourism: Case of Sengwe community in Great Limpopo Transfrontier Conservation Area, Zimbabwe. *Development Southern Africa, 34*(3), 260–275. doi:10.1080/0376835X.2016.1259987

Convention on Biological Diversity. (2008). Protected areas in today's world: Their values and benefits for the welfare of the planet (CBD Technical Series No. 36). Montreal: Secretariat of the Convention on Biological Diversity.

Covane Community Lodge. (2020a). *Galleries*. Retrieved January 5, 2020, from http://www.covanecommunitylodge.com/gallery.html

Covane Community Lodge. (2020b). *About Covane Community Lodge*. Retrieved January 7, 2020, from https://www.covanecommunitylodge.com/about.html

de Andrade, J. V. (2006). *Turismo: Fundamentos e Dimensões*. Ática.

Environmental Peacebuilding. (2019). *About Environmental Peacebuilding*. Retrieved December 3, 2019, from https://environmentalpeacebuilding.org/about/

Frazer, O., & Ghettas, L. (Eds.). (2013). *Conflict Transformation in Practice: Approaches to Conflict Transformation*. Retrieved from https://css.ethz.ch/content/dam/ethz/special-interest/gess/cis/center-for-securities-studies/pdfs/Conflict_Transformation_in_Practice_2013.pdf

Galtung, J. (1969). Violence, Peace, and Peace Research. *Journal of Peace Research, 6*(3), 167–191. doi:10.1177/002234336900600301

Galtung, J. (1985). Twenty-Five Years of Peace Research: Ten Challenges and Some Responses. *Journal of Peace Research, 22*(2), 141–158. doi:10.1177/002234338502200205

Great Limpopo Transfrontier Park. (2019a). *About GLTP*. Retrieved August 7, 2019, from https://www.greatlimpopo.org/about-gltp/

Great Limpopo Transfrontier Park. (2019b). *Tourism*. Retrieved August 8, 2019, from https://www.greatlimpopo.org/tourism/

Great Limpopo Transfrontier Park. (2019c). *Tourism Opportunities*. Retrieved August 2, 2019, from https://www.greatlimpopo.org/tourism/tourism-opportunities/

Hottola, P. (Ed.). (2009). *Tourism strategies and local responses in southern Africa*. CABI. doi:10.1079/9781845935085.0000

Hsiao, E. (2007). *Peace Parks: A Natural Alternative*. Retrieved February 13, 2019, from http://www.monitor.upeace.org/archive.cfm?id_article=456

Ide, T. (2018). Does environmental peacemaking between states work? Insights on cooperative environmental agreements and reconciliation in international rivalries. *Journal of Peace Research*, *55*(3), 351–365. doi:10.1177/0022343317750216

International Union for Conservation of Nature and Natural Resources, United Nations Environment Programme, World Wildlife Fund, Food and Agriculture Organisation of the United Nations, & UNESCO (Eds.). (1980). World conservation strategy: Living resource conservation for sustainable development. Gland, Switzerland: IUCN.

Joint Management Board. (2002). *Joint Policy and Management Guidelines for the Great Limpopo Transfrontier Park*. Retrieved from https://pdf.usaid.gov/pdf_docs/PNADD330.pdf

Lederach, J. P. (2003). *Little Book of Conflict Transformation: Clear Articulation of the Guiding Principles by a Pioneer in the Field*. Good Books.

Mearns, K. (2012). Community-based tourism and Peace Parks benefit local communities through conservation in Southern Africa. *Acta Academica*, *44*, 70–87.

Paffenholz, T. (2009). Understanding Peacebuilding Theory: Management, Resolution and Transformation. *New Routes*, *14*(2), 3–6.

Parks Foundation. (2018). *Peace Parks Foundation Annual Review 2018*. Retrieved from https://www.peaceparks.org/wp-content/uploads/2019/07/PPF_Annual-Review-2018_Draft-15_web_smaller2.pdf

Peace Parks Foundation. (2015). *Great Limpopo Transfrontier Conservation Area Integrated Livelihood Diversification Strategy*. Retrieved from https://maps.ppf.org.za/arcgis/apps/MapJournal/index.html?appid=6a4ecb43005247b1a8adb88d7b79a917# Peace

Peace Parks Foundation. (2019a). *Great Limpopo*. Retrieved August 1, 2019, from https://www.peaceparks.org/tfcas/great-limpopo

Peace Parks Foundation. (2019b). *Our Journey*. Retrieved February 19, 2019, from https://www.peaceparks.org/about/our-journey/

Peace Parks Foundation. (2019c). *SA College for Tourism*. Retrieved May 4, 2019, from https://www.peaceparks.org/sact/

Roser, M. (2020). *Tourism*. Retrieved January 13, 2020, from https://ourworldindata.org/tourism

Scheyvens, R. (1999). Ecotourism and the empowerment of local communities. *Tourism Management*, *20*(2), 245–249. doi:10.1016/S0261-5177(98)00069-7

South African National Parks. (2000). *Bidding memorandum for the tender of concession sites*. South African National Parks.

South African National Parks. (2019). *Great Limpopo Transfrontier Park*. Retrieved August 9, 2019, from https://www.sanparks.org/conservation/transfrontier/great_limpopo.php

Southern African Development Community. (1999). What governments can do for community tourism (Leaflet No. 2). Harare: Africa Resources Trust.

Southern African Development Community. (2012). *Transfrontier Conservation Areas*. Retrieved October 15, 2018, from https://www.sadc.int/themes/natural-resources/transfrontier-conservation-areas/

Space for Giants, & United Nations Environment Program. (2019). *Developing Nature-Based Tourism in African State Protected Areas*. (Working Paper No 1). Space for Giants. Retrieved from https://spaceforgiants.org/wp-content/uploads/2019/06/Building-Africas-Wildlife-Economy-Space-for-Giants-Working-Paper-1.pdf

Spenceley, A. (2005). *Tourism investment in the Great Limpopo Transfrontier Conservation Area: Scoping report*. doi:10.13140/rg.2.1.3315.4407

Spenceley, A. (2006). Tourism in the Great Limpopo Transfrontier Park. *Development Southern Africa*, *23*(5), 649–667. doi:10.1080/03768350601021897

The International Ecotourism Society. (2019). *What Is Ecotourism?* Retrieved July 17, 2019, from https://ecotourism.org/what-is-ecotourism/

The Outpost Lodge. (2020). *The Outpost Lodge*. Facebook Homepage. Gallery. Retrieved on January 7, 2020, from https://www.facebook.com/pg/theoutpostlodge/photos/?ref=page_internal

United Nations Environment Programme. (2004). *Environmental Management and Community Participation: Enhancing Local Programmes*. UNEP-IETC Urban Environment Series.

United Nations Environment Programme. (Ed.). (2005). *Making tourism more sustainable: A guide for policy makers*. United Nations Environment Programme & World Tourism Organisation.

United Nations Environment Programme. (2018). *Environmental Cooperation for Peacebuilding*. Retrieved January 11, 2019, from https://web.unep.org/regions/rolac/environmental-cooperation-peacebuilding

Wood, M. E. (2002). Ecotourism: Principles, practices and policies for sustainability. Paris: United Nations Environment Programme.

World Tourism Organisation. (2019). *Tourism and the SDGs*. Retrieved July 17, 2019, from http://icr.unwto.org/content/tourism-and-sdgs

KEY TERMS AND DEFINITIONS

Community-Based Tourism: Tourism practice where the local community has a high level of control of the design and implementation of tourism activities, and the profits generated by these activities are directed to the members of the local community.

Conflict Transformation: Peacebuilding process, which addresses the structural causes behind a conflict, in order to promote a sustainable peace.

Ecotourism: Tourism practice that bases its activity on natural resources. It promotes the conservation of these resources and the generation of benefits for the local communities.

Environmental Peacebuilding: Peacebuilding process that addresses environmental issues that are related to a conflict, in order to promote a long-term vision and sustainable development.

Peace Park: Transfrontier Conservation Area that, in addition to having clear objectives of biodiversity conservation, aims to promote peace and stability.

Peacebuilding: Process that aims to build a positive peace in regions who live in conflict.

Positive Peace: Sustainable, lasting, preventive peace. It is characterised by the absence of structural violence.

Structural Violence: Violence as a result of the structure. An inequality of opportunity that limits the potential of the human being.

Chapter 7
Post-War Tourism in the Reconciliation Process of New Zealand Vietnam War Veterans

Maria Dorsey
https://orcid.org/0000-0002-9096-0798
University of Otago, USA

ABSTRACT

Tourism has the potential to act as a positive force in reconciliation efforts between countries. The basis of tourism in facilitating reconciliation is premised on people coming into contact with one another in non-adversarial settings, which support a higher probability that positive effects can result from this contact. The investigation on post-war tourism and its role in moving the reconciliation process forward has been limited. Since the Vietnam War ended, there has been a growing phenomenon of Vietnam War veterans returning to visit Vietnam. This chapter examines the impact of New Zealand Vietnam veterans' visits to post-war Vietnam on the reconciliation process with the Vietnamese and with self.

INTRODUCTION

The emotional and physical trauma caused by acts of perceived wrong-doing and traumatic events such as war serves as an impetus for people to seek reconciliation with self and others. The complexity of reconciling relationships within a post-war context centres on the psychological as well as sociological elements that can either support or block the reconciliation process. Tourism as a facilitator of reconciliation is premised on the idea that people coming into contact with one another in non-adversarial settings supports a higher probability that positive effects can result from this contact (Moufakkir & Kelly, 2010). The positive effect of contact in a tourism setting can serve as a catalyst for the development of strategies that addresses past conflicts (Hewstone, Cairns, Voci, Hamberger, & Niens, 2006). In addition, the growth of trust, as well as improved intergroup relations and attitudes from these tourism contacts, may facilitate the reconciliation process among the affected intergroups at the individual level (Hayes & McAllister, 2001).

DOI: 10.4018/978-1-7998-5053-3.ch007

Intergroup contact is an essential element for reconciliation initiatives, but there are challenges for this contact to occur in a post-war context (Freeman, 2012). There has been a lack of research which investigates tourism and its role in moving the post-war reconciliation process forward. Building upon the conceptualization of reconciliation and previous research on contact theory as well as the collected data on the New Zealand Vietnam War veterans, this chapter provides a clearer understanding of tourism's impact on the post-war reconciliation process.

BACKGROUND

Reconciliation

People and institutions are motivated to reconcile for a variety of reasons, including the desire to continue a relationship, the need to reach psychological peace and/or moral reasons (Dwyer, 2003). Bar-Siman-Tov (2004) also raises the need for reconciliation when societies are involved in a conflict, in which widely shared attitudes and beliefs support adherence to the conflictive goals, maintain the conflict, and delegitimize the opponent. He believes that without reconciliation, there is less possibility for peaceful resolution and development of peaceful relations.

The advancement of reconciliation involves a "long-term process that consists of a multitude of initiatives and stages, where progress is an accumulation of small steps addressed at the individual–interpersonal, local, societal, and state levels" (Kosic & Tauber, 2010, pp. 83-84). A multitude of barriers may impede the post-conflict reconciliation process between groups and individuals. Elder (1998) deemed that for reconciliation to progress between those engaged in war, they must first begin by acknowledging the harm they inflicted upon their enemy. Taking responsibility for causing harm towards their enemy is in itself problematic. If both sides believe their cause is righteous, this will likely serve as justification to the harmful acts they carried out to further their agenda thus stalling the process of forgiveness and reconciliation (Branscombe & Miron, 2004). Even in cases when group members accept responsibility for harm-doing, they may still reason that committing this harmful action was necessary (Branscombe & Miron, 2004).

When feelings of collective anxiety arise from group threat, these feelings can motivate in-group members to become more protective of their own actions that in turn further validates their own negative behaviour towards the out-group (Wohl & Branscombe, 2008). Therefore, the in-group's negative perceptions of the out-group might lead to a greater willingness for them to forgive their own in-group for its aggressive actions against the out-group (Wohl & Branscombe, 2008). Large-group enmity can raise feelings of group belongingness that causes group identity to become more important than the identity of the individual. In addition, large-group enmity may increase the need to provide a clearer distinction between one's own group and the enemy group (Volkan, 1999a). Specific to war, practices may be enacted to help maintain a separate identity from the enemy and to insure that a psychological border is kept between the groups regardless of cost. When these practices fail to separate the identity of groups in conflict, mass anxiety may occur that increases intergroup hostility (Volkan, 1999b). Furthermore, the presence of emotions such as hate, contempt and disgust can lead to intergroup avoidance making it more challenging for positive intergroup contact to occur (Feddes, Mann, & Doosje, 2012).

Post-war reconciliation is hindered by the dehumanization that occurs between war adversaries. The enemy becomes dehumanized as it makes it easier to engage in violence against those who are not like self and viewed as not human (Aiken, 2008; Northrup, 1989). Moreover, the dehumanization process involves the rejection of valid and invalid information "about the enemy and relying on the psychological construction of the other" (Northrup, 1989, p. 74). Halpern and Weinstein (2004) believe that for reconciliation to progress there needs to be a "reversal of this dehumanization and a return of humanity to those from whom categorization has removed all individual attributes" (p 567). Furthermore, the dehumanization that occurs during war can hinder the post-war reconciliatory process by reducing support for adopting actions in order to repair the harm that was committed against the victim out-group (Zebel, Zimmermann, Tendayi, & Doosje, 2008).

Intergroup Contact via Tourism

Since the 1940s, academic researchers have explored the differing effects of intergroup contact on prejudice. Previous research on prejudice and its effects on individuals, groups and society have predominately studied it as a negative attitude needing to be eliminated (Brewer, 1999). Additionally, prejudice based on group membership is believed to be a contributing factor in the development and continuation of conflict in society. Both Williams Jr. (1947) and Allport's (1954) research findings became the foundation of contact theory which is principled on the idea that in-group members by learning new information about the out-group through direct contact will be less prejudiced towards members of the out-group[1]. Allport designated the optimal conditions supportive of contact's potential to reduce intergroup prejudice and promote positive intergroup outcomes (Pettigrew & Tropp, 2000). These optimal conditions are as follows; intimate and voluntary actions between individuals with equal status who share common goals in a supportive environment whose intergroup attitudes are not extremely negative (Allport, 1954, 1979).

In today's society in which countries and communities may have residents from a wide array of ethnicities, groups can still remain spatially separated (Dixon, 2001). Indeed, even when intergroup contact does occur in the everyday setting, individuals must "cross real or imagined barriers that continue to separate communities" (Dixon, 2001, p. 589). This statement holds true as well in the tourism setting. Tourists have the power to select the destination(s) and make decisions that impact their interactions with their hosts. This section examines the potential barriers to intergroup contact between hosts and tourists. This discussion is pertinent in providing the influential factors that affect the ability of returning war veterans to engage in intergroup contact with their Vietnamese hosts in post-war Vietnam.

A predominant obstruction for intergroup contact to occur between tourists and hosts is the existence of tourist bubbles. Tourist bubbles denote the "tendency of tourists to stay among themselves and to be physically in a foreign place but socially outside its culture" (V. L. Smith, 1989, p. 6). The term tourist bubble encompasses the physicality of place as well as psychological aspects of the tourist, such as the tourists' "eigenwelt"[2] of motivations, attitudes, and belief systems" (Jaakson, 2004, p. 45). Tourist bubbles or mobile enclaves enable the tourist to travel whilst maintaining a safe distance from other people (Edensor, 1998). Cohen (1972) describes how tourists do not abandon their everyday environment when visiting destinations, but instead travel within protective bubbles that resemble their daily life.

Often the modern tourist is not so much abandoning his accustomed environment for a new one as he is being transposed to foreign soil in an "environmental bubble" of his native culture. To a certain extent he views the people, places, and culture of that society through the protective walls of his familiar 'environment bubble, within which he functions and interacts in much the same way as he does in his own habitat (Cohen, 1972, pp. 166-167).

Psychosocial factors may impede tourists' contact with their hosts. For example, tourists motivated to travel as a form of interpersonal escape in which they desire to evade annoying people, to break away from a stressful social environment, and to circumvent social interactions with others may actively isolate themselves from participating in intergroup contact with their host (Snepenger, King, Marshall, & Uysal, 2006). In addition, tourists may spend the majority of their time in areas with high concentrations of tourist accommodations, attractions and facilities thereby limiting the amount and intensity of the intergroup contact with their host (Berno & Ward, 2005; Mathieson & Wall, 1982). As a result of this restrictive contact, there is less likelihood that intergroup contact will meet the conditions thought to reduce prejudice and promote positive attitudinal change.

Mass tourism is a barrier to intergroup contact producing a positive attitudinal change in tourists. Mass tourism[3] "commoditizes culture and human relations in which the limited contact between hosts and tourists dehumanizes both groups" (Seaton et al., 2001, p. 383). The result of mass tourism is an "increasingly dehumanizing assembly line production of tour experience, strictly controlled by the tour operators in terms of length of stay, choice of lodging, what to see, where to eat and shop, and with whom to interact" (Feng, 2012, p. 392). In particular, tours packaged for mass consumption have created "highly mediated experiences between host and guest that intensify the sense of outsiderness felt by each group, which in turn creates stereotypes" (McKercher, 2008, p. 345). As traditional hospitality cannot be maintained in the "face of the constant flow of tourists, tourists become objectified, faceless customers and locals are merely curious objects" (Seaton et al., 2001, p. 383; Feng, 2012). Furthermore, the relationships that form between the tourists and hosts are frail (Franklin, 2003). This frailty stems from feelings of uncertainty in maintaining strong, long-lasting relationships due to the transient nature of the contact (Franklin, 2003).

Tourism: Fostering Reconciliation

Scholars have argued whether tourism can foster reconciliation between nations in conflict and within a divided society (Braithwaite & Lee, 2006; Higgins-Desbiolles, 2003; Kim, Timothy, & Han, 2007; Sonmez & Apostolopoulos, 2000). Tourism can contribute to the development of positive images between long-term enemies, leading to more peaceful relationships on an individual level (Simone-Charteris & Boyd, 2010). When more conventional methods to reconciliation fail to reduce hostilities, tourism offers opportunities to improve relationships between adversaries as it increases intergroup contact and cooperation due to the need to work together in order to promote and make profitable their tourist goods and services (Simone-Charteris & Boyd, 2010; Sönmez & Apostolopoulos; 2000).

Tourists are drawn to visit war-related sites such as war memorials, battlefields, and war museums. War memorials and battlefield sites, seen as sacred, may serve to increase feelings of patriotism that may discourage a more critical view of the part that both sides played in the event (Kelly & Nkabahona, 2010). Rather than increasing peaceful attitudes amongst the visiting tourists, these sites can actively perpetuate hostility and hatred towards the (former) enemy, especially places that rely on shock factor to attract tourists (Kelly & Nkabahona, 2010). Conversely, these sites can help visitors become more knowledgeable about the past, which may help them realize the need to maintain a peaceful coexistence with others.

New Zealand's Involvement in the Vietnam War

New Zealand's military involvement in the Vietnam War began when they entered into an ANZUS alliance with Australia and the United States on September 1, 1951 (Munro, 1953). New Zealand did so to ensure that they had military support from the United States and Australia in case of a Japanese military resurgence (Heritage, 2008; Munro, 1953). Furthermore, this alliance was made with the mutual understanding that New Zealand and her fellow ANZUS allies would jointly counteract the spread of communism in the Asian- Pacific region (Heritage, 2008). After the French government vacated Vietnam in 1954, New Zealand provided political support to the citizens of the struggling Republic of Vietnam by offering them the opportunity to study in New Zealand (McGibbon, 2010). It was not until 1961 that New Zealand began to feel pressured by the United States to provide military support to the Republic of Vietnam to combat the North Vietnamese forces (McGibbon, 2010).

In 1963, medical assistance was provided by the New Zealand government to the Republic of Vietnam that eventually grew to include combat support in 1965 (Duggan, 2011; McGibbon, 2010). During the war, 3,500 New Zealand servicemen were deployed to Vietnam in whom the vast majority were drawn from the Army, and lesser numbers from the Royal New Zealand Air Force (61 airmen) and the Royal New Zealand Navy (28 navy men) (Challinor, 2009). The Vietnam War culminated in thirty-seven deaths and 187 wounded of New Zealand servicemen (Heritage, 2008). New Zealand withdrew all their forces by 1972 (Heritage, 2013).

As it became evident that the Vietnam War would not be an easy victory for the Free World Forces, an anti-war sentiment developed in New Zealand (Lavelle, 2006; McAdam & Su, 2002; McGibbon, 2010). Public demonstrations were held against the war, and returning servicemen from Vietnam were negatively targeted by war protesters (Challinor, 2009). The propaganda perpetuated by the New Zealand government and media during the conflict was complex in nature due to the biases of those reporting information on the war. Dependent on the source, there could be a call for support of New Zealand's participation in the war or a call to end New Zealand's military presence in Vietnam (Ed, personal communication, 2012) believed the news media was grossly unfair in its negative portrayal of the servicemen's actions in Vietnam and that the media backlash was not strongly counteracted by the New Zealand government.

Personal Motivations to Serve in Vietnam

Regardless of whether the New Zealand servicemen entered the military as conscripts or volunteers, they were given the option to decline deployment to Vietnam (McGibbon & Goldstone, 2000). Although Challinor (2009) argues it is a public myth that these servicemen volunteered to serve in the Vietnam War. The reality was once the men joined the military, there were clear expectations by themselves and the military that when called to serve would do so willingly despite personal reservations. In addition, the servicemen viewed serving in Vietnam as an opportunity to further their military career. This resulted in an environment that was not conducive for servicemen to decline deployment to Vietnam (Challinor, 2009). This sentiment is supported by the following veteran's account.

I didn't have to go to Vietnam, but on the other hand I was still a regular soldier. People often would say that oh you guys were all volunteers and you went to Vietnam because you were volunteers. The answer to that is quite simple- you were volunteers only once and that was when you decided to join the army. Once you joined the army you did what you were told to do. (Brian, personal communication, 2012)

The servicemen viewed Vietnam as an opportunity to utilize their military training, as well as the means to enhance and test their training and skills. Some servicemen believed Vietnam was a just war and that there was legitimacy to the domino theory and communist threat. Many servicemen saw going to Vietnam as a chance to have an adventure and escape the boredom of home service (Challinor, 2009).

War Propaganda and Military Training

From a broad perspective, propaganda was utilized throughout the Vietnam War by the Free World Forces to demonize the enemy, win the hearts and minds of the Vietnamese and garner public support for their political and military actions taken during the war (Campbell, Pickett, & Yoash-Gantz, 2010). In regards to the servicemen who were trained and deployed to Vietnam, the process of dehumanization enabled them to kill their enemy by helping them view these people as not human (Kelman & Hamilton, 1989). In general, war propaganda campaigns assign the enemy labels based on their ethnicity, nationality, religion, or politics that makes it easier to regard their demise as a means to the end (Kelman & Hamilton, 1989). In the case of the New Zealand servicemen, the war (Campbell et al., 2010) propaganda did affect their attitude towards the war and the Vietnamese people. Major Benfell (2006) remembered watching films such as the Night of the Dragon that caste the Viet Cong as murderers and pillagers.

My attitude towards the Vietnamese prior to going to Vietnam was influenced by the Time magazine articles. For instance there was a very graphic sequence of photos in the magazine of the Tet offensive attack around Saigon. These images glorified my concept of war especially as an 18 year old youth. I really wanted to go to Vietnam to personally experience combat but had no idea of what that really meant. (Duke, personal communication, 2012)

There was a variance amongst the servicemen in the level of information received about the Vietnamese people and the conflict prior to their deployment to Vietnam. Duggan (2011) was unaware of the reason that New Zealand was going to join the Vietnam War and some including himself was unsure where Vietnam was located. For Louis Brown, he was not taught the Vietnamese culture and language prior to his deployment to Vietnam (L. Brown, 2008). Brown believed that this type of training was not necessary as he would have had limited contact with the Vietnamese because the ARVN was charged in confronting the locals (Brown, 2008).

Servicemen were exposed to varying degrees to war propaganda, political indoctrination and training that dehumanized their enemy (Dennis, personal communication, 2012) voiced that being a young man at the time; he was more prone to believe in what he was being taught during military training about the war.

If you would like to call it indoctrination then you can. I bought into the domino theory. It was easy to point out on a map, how it was very possible that the domino theory could come to fruition. The domino effect was even more likely because we had just come out of the McCarthy era in which there were reds under every bed. (Dennis, personal communication, 2012)

The military training readied the servicemen to automatically react and kill their enemy. Tamatea (2007) asserted that "At that time, when you done the training you become a killer. You can kill anybody any time, yep". The process of dehumanization of the enemy began in training in which the recruits were taught to commit actions without having to make a conscious effort to think about them. Sisson, an infantryman, spoke about how the training impacted his feelings on the Vietnamese

It was very negative; the training was to turn us into machines which you sort of have to do to get people to be able to kill other people. We really regarded the Vietnamese enemy as non-people. This was the only way we could really function during the war. (Sisson, personal communication, 2012)

Not all servicemen felt that they had succumbed to the indoctrination and dehumanization process of the enemy during their training as exemplified by the following:

I always saw them as other human beings with the same emotions and so on as myself, and I never believed the propaganda, the American propaganda, that 8…] they were totally duped by the communists and therefore out of their brains. I believed that they were obviously highly motivated and believed in the cause they were fighting for, and I had a great respect for them, both for that and their military ability. (Challinor, 1998, p. 89)

Vietnam War Veterans Return to Former War Sites

War-related tourism has a long history dating back to the seventeenth century when people travelled to watch battles in progress (Butler & Suntikul, 2013). Indeed war-related tourism has grown to include travel to war memorials, cemeteries, battlefield sites, and war museums. The period that post-war tourism starts is at the end of the war that is marked by tourists' numbers beginning to rebound towards the pre-war period as there is a growing number of adventure tourists, family members of missing servicemen, and opportunists infiltrating the war zone (Weaver, 2000). Previous research has examined the conditions that increase the probability of war veterans in returning to locations they served in during a war (Leopold, 2003; Weaver, 2000). The passage of time is an influential factor in motivating war veteran in returning to the war sites. Weaver ascertained that older former soldiers are more likely to travel overseas to the battlegrounds 20 years after the conclusion of the war. Their willingness to travel is premised on the fact that it is during this period of time war veterans begin to enter into the family life cycle and occupation stage, in which these visits become more feasible and psychologically tolerable (Weaver, 2000). Leopold's (2003) findings support Weaver's results as her participants were returning to post-war Vietnam approximately 20 years after the Vietnam War ended. Leopold's survey indicated 29 New Zealand Vietnam veterans had returned to visit Vietnam predominately from 1996-2000 (Leopold, 2003). A major motivation of the New Zealand Vietnam veterans in returning to Southeast Asia was to see how the place has changed since they served in the Vietnam War (Leopold, 2003). This notion supports findings identified by Smith (1996), who acknowledged that veterans travel back to former war sites to "revisit and remember the days of their youth." A number of authors (Cameron & Gatewood, 2000; Cameron & Rutland, 2006; Garton, 2013; Smith, 1996; Timothy & Boyd, 2003) have pointed towards nostalgia representing a motivation for veterans.

METHODOLOGY

Qualitative research was conducted in which document analysis and in-depth interviews were drawn upon to explore post-war trips longitudinal effect on reconciliation between former enemies of war. Document analysis of archival research and in-depth interviews with New Zealand Vietnam War veterans allowed the researcher to track individual and social change and provided the historical background of the studied phenomenon that insured more accurate results (Bowen, 2009). By drawing on multiple sources of evidence and having the participants serve as co-researchers, the researcher avoided the project's results losing credibility because of the usage of a single source of information, single research method and single researcher's bias (Patton, 1989). Furthermore, the qualitative methods utilized in conducting the research generated data that provided a clearer understanding of the phenomenon of war veterans returning to post-war Vietnam and how these visits impact reconciliation between former foes.

This study included New Zealand Vietnam War veterans who served in differing capacities during the Vietnam War (1964-1973). Purposive sampling, which is commonly used in qualitative research (Tashakkori & Teddlie, 2003) was implemented in order to recruit participants. Purposive sampling was chosen because it empowered the researcher to make conscious decisions about choosing participants who would best answer the research questions (Crookes & Davies, 2004; Teddlie & Yu, 2007).

The war veterans who were interviewed (2011-2013) for this research were recruited by the following methods; electronic discussion boards and blogs posted by Vietnam War veterans, memoirs written by Vietnam War veterans, New Zealand's veterans' organizations, and through the snowball sampling method. Of the 12 participants that were interviewed 9 participants plan to or have made a return trip to post-war Vietnam and three participants had no intention to return to post-war Vietnam at the time of the interview. All of the participants were of European ethnicity and male.

The goal of the interviews used in this study was to explore the respondent's viewpoints, thoughts, and perceptions of the Vietnamese prior to, during, and post-Vietnam War. The in-depth interviews with the war veterans gave the researcher the opportunity to co-create meaning with interviewees by reconstructing perceptions of events and experiences related to their experience of serving in the Vietnam War and visiting post-war Vietnam (Crookes & Davies, 2004). The participants were co-researchers in that they were considered equals to the interviewer in the construction of the knowledge and meanings garnered from their own interview. This helped to ensure the accuracy of the project's results by reducing the effect of the researcher's biases and inexperience in fighting in the Vietnam War. Through encouraging a more involved relationship with the co-researchers that is fostered with mutual respect and trustworthiness, it was expected that this research method would help produce a more nuanced, relevant, and meaningful study. Specifically, the research was designed to involve the war veterans in the following aspects; to help determine the issues that should be addressed during the semi-structured interviews, to assist in analyzing the collected data, and to ensure the accuracy and relevancy of the results. In order to empower the interviewees as co-researchers, the interviewee had the options of conducting the interview in-person, via Skype, or by correspondence. The interviewees were encouraged to discuss issues that they believed relevant to the research. The duration of the in-person and Skype interviews was between one and two and a half hours. A summary of the interview was sent to the participants in order for them to make editorial changes that more accurately reflected their thoughts and feelings.

A triangulated approach to the data collection was utilized by drawing on secondary and primary sources. Secondary documents were primarily retrieved by utilizing library and museum resources, using internet and blog search engines, and soliciting documents from participants, historians and other academic researchers. A series of semi-structured interviews were conducted with New Zealand Vietnam War Veterans who have visited or plan to visit post-war Vietnam and those who have not returned to post-war Vietnam. Purposive and snowball sampling techniques were implemented in order to recruit participants for in-depth interviews. The interview phase of the study began in 2011 and was completed in 2013. Interviewees were given the option of answering the interview questions in-person, by telephone, via Skype or by correspondence either electronically or by mail. The following is a breakdown of the interviews; 11 face-to-face interviews, four e-mail interviews and 9 Skype interviews. The collected interviews included 22 male New Zealand Vietnam veterans in which 13 have returned versus nine who have not returned to post-war Vietnam. Interviewees consisted of 21 veterans of European ethnicity and one of Maori ethnicity. Moreover, there were two interviews with the children of New Zealand Vietnam veterans who accompanied their father on their post-war visit to Vietnam.

FINDINGS

War veterans wishing to return to where they served during the war may be obstructed by political violence, travel restrictions and psychosocial barriers. Tourism can be adversely impacted by political violence that occurs when governmental or anti-governmental groups are politically motivated to execute force through the commitment of violent acts (Neumayer, 2004). Political violence reduces tourism because people are pulled away from travelling to places in which recent conflict has taken place out of fear for their lives, anticipation of becoming involved in stressful situations, and the inability to visit sites due to travel restrictions by the host government (Neumayer, 2004). Although the Vietnam War ended in 1975, Vietnam continued to experience political violence intermittently with China and Cambodia until 1990 (Frost, 1991; Jencks, 1979). Throughout that time Vietnam enforced stringent travel restrictions for international visitors. As a result, it was unlikely that war veterans would be motivated to return to Vietnam until the conflicts were resolved and a more open travel policy was established by the Vietnamese government.

It was not until after the Vietnamese government adopted the open door or Doi Moi policy in 1986 that Vietnam began the process of building infrastructure and opening up their travel policies to attract international travellers (Elliott, 2012; Hobson, Heung, & Chon, 1994) . All of the interviewees who returned to Vietnam did so after the Doi Moi policy was put in place and peace agreements reached with China and Cambodia. Clarence (personal communication, 2012) travelled to Vietnam in 1993 and described his experience of being one of the first war veterans to go back to Vietnam after the war ended.

My travel agent told me to hide the fact that I was a Vietnam War veteran when I was in Vietnam. Back in those days, the Vietnamese government had passed a law that anyone that fought in the Vietnam War against them would be considered a war criminal and therefore subjected to being arrested. There was a tension filled moment when I was going through customs upon my arrival in Vietnam. I had planned to bury my old uniform and was terrified that they were going to search my bags. Fortunately they didn't. (Clarence, personal communication, 2012)

Clarence's experience signifies the anxiety and fear that the initial wave of war veterans experienced in travelling back to post-war Vietnam. Although fears for physical safety may no longer deter war veterans from travelling to Vietnam, there has been and still are travel restrictions that may impede some veterans from returning to Vietnam. For Hunt (2009) the inability to travel to certain sites has prevented his return to Vietnam.

Haven't been back to Vietnam although ...several opportunities to do so. I should and will at some stage. Part of the reason after speaking with friends that need permission from Vietnamese government to go to places on group tour. Difficult to have a chance to go where you want to go. Not interested in listening to the propaganda on Vietnam. In some ways don't regret that. I want to go at a time when I can go to some places that I want to go. I can't go back to old troop lines it is now a military area and there are no visitors allowed. (Hunt, 2009)

Even though travel restrictions may hinder some war veterans from returning to Vietnam, this constraint was neither a strong or prevalent factor in the 22 interviewees' decision to return or not to return to post-war Vietnam.

Financial constraints were identified by several interviewees as a deterrent to travelling back to post-war Vietnam. Amanda, a tourism operator that specializes in post-Vietnam trips for New Zealand Vietnam war veterans spoke that in planning group tours that there are veterans who would like to travel to Vietnam, but are unable to do so because they cannot afford the trip cost.

I am finding it more difficult to keep the costs down and trying hard to keep it under the NZ $4,000 mark. It is increasingly more difficult to get into good motels and maintain that price. I think that the government should pay for the whole lot for the vets. I know that some have had to say no because of the price and that makes me feel very disappointed that they can't do it. (Amanda, personal communication, 2012)

Financial constraints appeared to be a weaker rationale for veterans' decision to not return to post-war Vietnam as the majority of non-returnees cited other reasons. Furthermore, the New Zealand government provides travel allowances to assist war veterans who wish to attend commemorations and, or revisit battlefield sites. This weak reasoning was evident when Carl (personal communication, 2012) stated his inability to return to post-war Vietnam was based on a lack of monetary means. When I provided him information about the travel allowance he then revealed that a stronger reason for him not returning to Vietnam was the realization that he "had spent enough of my life living in the past and wanted to focus my life on living in the present" (Carl, personal communication, 2012).

Psychosocial issues stemming from war can serve as a major obstacle for veterans in returning to post-war Vietnam. Several interviewees acknowledged having a negative perspective of their time in Vietnam that has resulted in suppression of that part of their life history. Returning to post-war Vietnam would resurrect painful memories that the veterans want to remain in the past. For Vernon, he would have to face his feelings of guilt in leaving the Vietnamese people in grim circumstances. When asked if he was willing to return to post-war Vietnam, he responded with the following:

No. Simply put to many bad memories that I have no wish to revisit. Also I may meet some former South Vietnamese person I once knew and would have to explain my hurried exit in 1972. I was ordered out before my tour was to end because unknown to me we were running (Vernon, personal communication, 2012).

Additionally, veterans may not be psychologically ready to return to Vietnam. For example, Howard (personal communication, 2012) who has had many opportunities to go to Vietnam after the war recognized that he is not emotionally ready. Moreover, because he had a negative experience during the Vietnam War, it is apparent that he does not want to re-connect to that part of his past.

Although I have friends who have gone back and say how beautiful the country is now; this has not been the catalyst for me to return to Vietnam. I have been able to justify to myself that there are other places that I want to experience. During my time serving in the Vietnam War, I didn't have a marvellous or happy time. I have nothing to go back for. I don't have a need to go back there. My time in Vietnam was just a piece of my life (Howard, personal communication, 2012).

Furthermore, war veterans may not be motivated to go back to Vietnam because they feel shame and guilt at their actions during the Vietnam War. Specifically, they may feel their treatment of the Vietnamese people was unkind and unjust. For example, Daron does not want to go back because he views himself as an invader of Vietnam that has caused harm to the Vietnamese people (personal communication, 2012).

The people were so poor and while you were searching their house all you would find is an egg. The whole family would be hiding in a hole in the floor. It was like a bomb shelter. That is the only reason why I wouldn't want to go back because I am ashamed of having been there. It was a waste of time. We were really mean because we were invading their space. We would just enter their houses and look around. It was horrible to search people's houses. We would just bolt in their houses without any kind of greeting. It wasn't nice seeing people scared (Daron, personal communication, 2012).

Additionally, because Vietnam was eventually taken over by the communist forces despite their efforts, some war veterans may view the Vietnam War as a pointless military exercise that did not merit the loss of their fellow servicemen. Returning to Vietnam would bring up emotions that they are not ready or willing to experience. For example, Cory has decided not to return to Vietnam because "too many good people died for a cause that was eventually abandoned and so they died for nothing. So to revisit would stir up bad memories" (personal communication, 2012).

War veterans may be less inclined to take steps to reconcile with the Vietnamese as there is no need to engage with them on a daily basis. In contrast, there is a stronger need for the war veterans to reconcile with their fellow countrymen because of their shared coexistence on a day-to-day basis. This perception creates a barrier in the war veterans reconciling with the Vietnamese as their first priority is to reconcile with their own countrymen first. This sentiment was expressed by Ex-Vietnam Services Association president Chris Mullane.

There are some things to put right for our Vietnam veterans and families in New Zealand first and once they are put right then I think it would be good to establish some sort of reconciliation process with the Vietnamese from a veteran's perspective. (NZPA, 2007)

Subsequently, some veterans may choose not to travel back to Vietnam until reconciliation has first taken place with their own countrymen. For example, Howard (personal communication, 2012) felt that there was a need to reconcile with New Zealand before being able to engage in the reconciliation process with the Vietnamese. When asked if reconciliation with New Zealand was holding him back from returning to Vietnam. Howard responded "Yeah probably right in that question maybe I am not reconciled with the government, the war and my part in the war [...] that is getting a bit deep" (personal communication, 2012). As a result, Howard's inability to reconcile with the New Zealand government and his part in the war serve as barriers that have deterred him from travelling to post-war Vietnam and seeking reconciliation with the Vietnamese people.

Motivational Factors and Mitigating Barriers of Post-War Reconciliation

The primary motivation of the majority of war veterans in returning to post-war Vietnam was not based on wanting to reconcile with their former enemies. Rather, their motivation was influenced by the passage of time, nostalgia and curiosity, financial support by the government as well as positive word of mouth recommendations from fellow war veterans. Additionally, war veterans were motivated to travel back to Vietnam as they viewed the trip as an opportunity to address and resolve their psychosocial issues that stemmed from their war-related experiences. This section examines several mechanisms that enabled war veterans to mitigate the barrier of anxiety to enable them to be psychologically ready to return to Vietnam.

Time may be needed until the veterans move into a family-life cycle and occupation stage in which overseas travel becomes more feasible and psychologically tolerable; usually occurring approximately 20 years after the war has ended (Weaver, 2000). During the post-war process there comes a point in which war-related attractions and the return of the former combatants account for an increased growth in tourism (Weaver, 2000). This thesis supports Leopold's (2003) conclusions that the majority of New Zealand Vietnam veterans' who have returned to post-war Vietnam have chosen to do so after 1995. A few interviewees of this thesis, who had not returned, based their delay on being occupied with family and work obligations. Dennis explained that he had not returned to Vietnam yet because he has been busy working and caring for his family although he plans to in the future when he has the time (personal communication, 2012). Time also factors in pushing veterans to return before they reach the point that engaging in long distance travel is too difficult because of health and mobility issues related to old age.

The notion of nostalgia to "revisit and remember the days of their youth" is a major motivating factor for veterans to travel back to war sites (C. M. Cameron & Gatewood, 2000; Garton, 2013; V. L. Smith, 1996, p. 10; Timothy & Boyd, 2003). Serving in the war proved to be a stressful time, however war veterans still retain fond memories of the shared camaraderie and the more pleasant and humorous aspects of serving in Vietnam. Vietnam veterans are drawn to visit post-war Vietnam out of nostalgia and curiosity to see how the country has changed since the war ended. Jake (personal communication, 2012) was motivated to go back to Vietnam out of curiosity.

The reason why I felt motivated to go back to Vietnam was based on a need to see what had happened after they left. It is a bit inquisitive-this wanting to know. It is a bit like going back to a school reunion. You know that it is not going to be like it was back then (Jake, personal communication, 2012).

Additionally, these veterans in choosing to go back are actively trying to re-connect with their past. Joel (personal communication, 2012) also based his decision to return to Vietnam on nostalgia and curiosity.

My decision to go was influenced by the recommendations of some fellow veterans and some other friends who had holidayed there, plus of course my personal desire to go back to reminisce and remember and to see what changes had occurred in the many years since I had first been there (Joel, personal communication, 2012).

This nostalgia motivation is supported by Leopold's (2003) study which also reported that a major motivation for New Zealand Vietnam veterans returning to Southeast Asia was to see how the place had changed since they had served in the Vietnam War.

Financial support, the encouragement of others, and positive word of mouth have all served as motivation for veterans to travel back to Vietnam. The Minister of Veterans' Affairs Commemorations Discretionary Fund was established in 2001 to provide a one-off grant for eligible veterans who wish to attend commemorations and/or revisit the battlefield sites overseas where they served. The objective of this fund was to provide war veterans with an opportunity to travel back to where they served with the intention that they may achieve a sense of reconciliation with their experience ("VANZ News," 2011). James (personal communication, 2012) went back to Vietnam because he had the time available and was able to use government funds to finance the trip. "It was a big help and I don't know if I would have gone if I did not get the NZ$1800. I certainly would not have gone for as long as I did which was 6 weeks" (James, personal communication, 2012).

Positive word of mouth has been shown to increase the probable purchase of tourism products (Jalilvand & Samiei, 2012). Additionally, positive word of mouth by other veterans who travelled to Vietnam has pulled veterans to go back to Vietnam. The positive word of mouth recommendations may reduce the anxiety experienced by the veteran to a point that they are able to travel back to Vietnam. Since returning from a post-war trip to Vietnam, James has taken the initiative to convince his fellow veterans to travel to Vietnam (personal communication, 2012). By listening to his description of the beauty of the country and the friendliness of the Vietnamese people, four veterans decided to return to post-war Vietnam (James, personal communication, 2012). Brian (personal communication, 2012) felt an important factor in his decision to return to Vietnam was listening to his friends' positive experiences when they returned to visit Vietnam.

When war veterans emphasize the friendliness of the Vietnamese people this may reduce the anxiety level associated with worries of how they will be treated in post-war Vietnam. As a result more war veterans may be willing to return to post-war Vietnam. Furthermore, positive word of mouth recommendations by veterans who have travelled back to Vietnam has, according to Timothy, increased good will towards Vietnam which "I feel is evident by people travelling to Vietnam and conducting business with them" (personal communication, 2012). Conversely, James has encountered resistance from his fellow veterans when encouraging them to travel to Vietnam.

There are a lot of the guys that I have spoken with think that I am nuts for even going back for a visit. They couldn't be bothered. I had a huge argument with a guy the other week who said why the fuck would he want to go back there for-are you nuts! I tell them forget about the war-that is not why you go back. There is a country out there that is beautiful, economical, the people are friendly and the government is giving you NZ$1800 to go and visit it. What are you sitting here for? They are absolutely mad for not wanting to go (James, personal communication, 2012).

Pull factors such as financial support and positive word of mouth recommendations may not be enough to motivate some veterans to decide to return. The support of family members and fellow veterans also served as motivation for the veterans to return to post-war Vietnam. For example, Wayne made the decision to return to post-war Vietnam because of the encouragement he received from his daughter and agreement by his family to join him on the trip (personal communication, 2012). One theme that emerged from the data collection was that family members of the war veterans were motivated to travel with the war veterans to post-war Vietnam. Family members may be motivated to travel with the war veterans as they view the trip as an opportunity to learn more about them, wanting to serve as emotional support during the visit and feel that the trip is an opportunity for the war veterans to heal from the trauma of their war-related experiences. Iris (personal communication, 2012) a daughter of a Vietnam veterans explained her reasoning for accompanying her father to post-war Vietnam.

I wanted to support my dad. I thought the experience would be quite intense and emotional experience for him. I also wanted to go to gain a better understanding of my dad and what he had gone through in Vietnam. I felt that the trip might be healing for me as well but that mostly the trip was to be a healing experience for my dad. The trip would bring my family closer together because we would be facing something difficult together (Iris, personal communication, 2012).

It is important to investigate the role of anxiety in acting as a barrier to war veterans returning to post-war Vietnam and the ways in which it can be mitigated so war veterans are able to travel to post-war Vietnam. Many of the war veterans who have chosen to go back to post-war Vietnam saw the trip as an opportunity to address and work through their anxiety related to their war-related experiences. Anxiety has been identified as a barrier to intergroup contact reducing prejudice (R. Brown & Hewstone, 2005; Islam & Hewstone, 1993; Stephan & Stephan, 1985). Previous research supports the premise that travellers experience anxiety from perceived physical threat or maladaptive psychological issues associated with the risk of the mode of travel (Cheng, 2010; de Jongh, Holmshaw, Carswell, & van Wijk, 2011; McIntosh, Swanson, Power, Raeside, & Dempster, 1998; Reisinger & Mavondo, 2006). This thesis purports that war veterans too experience anxiety when contemplating visiting the country they once served at during the war. This anxiety stems from the traumatic experiences and deployment to a war zone. A recurrent theme of the interviews was war veterans being motivated to travel to post-war Vietnam to rid themselves of their "ghosts from the past" or feelings of fear and anxiety that stems from their war experience.

Often in research literature, there is a negative bias against tourist bubbles. This study discovered that returning war veterans to post-war Vietnam received positive benefits by travelling in a tourist bubble. In order to reduce psychosocial risks whilst travelling in post-war environment veterans may seek to travel within a group. Previous research has shown that tourists who are risk-averse are more likely to adopt a less risky strategy by choosing to travel in large groups as opposed to travelling alone (Money & Crotts, 2003). The accompaniment of supportive family and fellow war veterans may mitigate the psychosocial risk of experiencing anxiety and group threat associated with returning to post-war Vietnam.

The majority of interviewees who returned to Vietnam travelled within a group tour of fellow veterans, family members and in some cases both. Donald (personal communication, 2012) decided to return to Vietnam with a group of fellow war veterans in order to mitigate the anxiety and apprehension he felt in returning to Vietnam. "I needed to go with tight group of Vietnam veterans who would understand what it was like during the war" (Donald, personal communication, 2012). There is, to a great extent, the acknowledgement by the veterans that having the support of others was helpful in easing their anxiety in returning to Vietnam. Furthermore, by having conversations with their fellow veterans and family members served as part of the healing process. For example, Robert acknowledged that it was helpful to have his family members and veterans on the trip because "it enabled me to have a couple of moving conversations with my son and a few from Victor 5. We had the opportunity to draw on a bit as they say, and that was all part of the cleansing process" (personal communication, 2012).

Post-war Intergroup Contact: A Measure of Reconciliation

The occurrence of intergroup contact has been shown to facilitate the reconciliation process between individuals belonging to different groups (Ellison & Powers, 1994; Pettigrew, 1998; Pettigrew & Tropp, 2000; Sigelman, Bledsoe, Welch, & Combs, 1996; Sigelman & Welch, 1993). On the basis of contact theory, it is hypothesized that these post-war tourist exchanges will reduce hostility, increase understanding and move the reconciliation process forward between the war veterans and their Vietnamese hosts. This section examines intergroup contact via tourism that serves as a measure to determine the thickness of reconciliation achieved between New Zealand Vietnam veterans and the Vietnamese people. Moreover, political tolerance was utilized as a measure of reconciliation to determine the progress of reconciliation. Specifically, political tolerance was examined in relation to the post-war travel by veterans to Vietnam and its war-related sites.

The passage of time plays an important role during the reconciliation process between former war adversaries and allies. There is debate about whether the passage of time results in forgetfulness of the wrong that helps in the reconciliation process and whether the best approach to reconciliation is to do nothing in taking action to address the wrong (Gloppen, 2005). The counter-argument to this approach of letting time heal the wounds from the wrong-doing is that it is uncertain whether forgetting the wrong-doing results in reconciliation (Gloppen, 2005). Furthermore, there is a belief that there is a need to remember to prevent reoccurrences of the wrong-doing and that remembering is needed for long term reconciliation (Gloppen, 2005).

The passage of time may be a necessity in creating travel conditions for intergroup contact to occur between former adversaries within a post-conflict environment. This holds especially true for conflicts that involved military forces from foreign countries. Time may be required to re-build infrastructure damaged from the conflict to accommodate travellers and tourists. Additionally, time may be needed to convince tourists that there is a stable peace that will not risk their safety by visiting the location. Still the passage of time has shielded the New Zealand war veteran from witnessing first-hand the trials and tribulations that Vietnam endured to re-build their country after the war. By the time the veterans returned to visit Vietnam, the physical damage caused by the war was repaired and the Vietnamese people appeared to be doing well economically. As a result veterans may feel that there is no need for reconciliation with the Vietnamese based on the belief that the Vietnamese have recovered to the point that they no longer suffer from the adverse effects of the war.

The Vietnamese government has spun the political rhetoric that the Vietnamese do not seek to dwell on the past but instead want to move forward from the past. "This political rhetoric has been crafted strategically for the purpose of creating friendly ties with outsiders" (Kim, 2001, p. 627). This strategy appears to be working as a multitude of interviewees from this research expressed how it appeared that Vietnam has moved forward. For example, New Zealand Vietnam veteran (Michael, personal communication, 2012) expressed that there is no need to reconcile with Vietnam because Vietnam has moved on and is doing well.

In order for post-war travel to occur, governments must put in place policies that enable post-war travel to occur between former combatants. The growth of Vietnam's tourism can be attributed to thawing of post-war relations between Vietnam and the United States that occurred after the United States lifted the

Vietnam trade embargo in January 1994 (Jansen-Verbeke & Go, 1995). The lifting of the trade embargo opened the door for Vietnam to build infrastructure to better accommodate the influx of international tourists including returning Vietnam War veterans.

War veterans engaged in more intimate contact during the post-war visit with the Vietnamese people resulted in their gaining a greater understanding of post-war life for the Vietnamese people.

One of the things that you got to remember is when the North conquered the South, the people who had been fighting for South Vietnam during the war and/or abetting with the US, New Zealand and Australian coalition forces were treated very badly after the war. They were re-educated by the North Vietnamese and find it very hard to talk about it in public. If you have the opportunity to speak with them in private they will talk about it with you, but yes they were treated very, very badly. (Calvin, personal communication, 2012)

Reconciliation between former enemies was considered a goal with high priority within the context of psychosocial rehabilitation and social reconstruction (Sveaass & Castillo, 2000). There are indicators that demonstrate the thickness of achieved reconciliation based on tourism. War veterans who have chosen not to travel to post-war Vietnam signify a thin reconciliation. Although they are not engaged in violence, the lack of intergroup contact indicates that there is a broken or un-connected relationship. Limited contact between war veterans and locals is a barrier in the tourism environment in facilitating positive attitudinal change and reconciliation. The lack of intergroup contact is a result of the relatively brief time spent by war veterans in Vietnam and the limited contact with the local Vietnamese. Subsequently, there are diminished opportunities to engage in intergroup contact with the Vietnamese that would lead to greater understanding. For example, similar to other interviewee's experiences, Robert's (personal communication, 2012) contact with the locals was minimal as he was on a group tour. He did not engage in any significant conversation with the locals as the majority of his contact occurred through his interpreters and hotel staff. As a result, the limited contact with the Vietnamese offers less opportunity for the war veterans to gain a better understanding of the Vietnamese, thereby reducing the potential to reduce prejudice and produce thicker reconciliation.

Due to premise that reconciliation with other focuses on re-connecting a relationship between two or more individuals, it is important to point out the impact of the host and guest power dynamic on the reconciliation process. Within the travel environment there is the presence of an unequal relationship of power between hosts and guests that could affect the veterans' perspective of their host (Bruner, 1991; Tucker, 2001) . There is also uncertainty if the positive attitude displayed by the Vietnamese hosts towards the veterans is genuinely based on their willingness to reconcile with them or is based on their host role of pleasing their guests. For example, Michael stated that "I found that the people were so nice. There were kids in Saigon that we spoke with that were very, very helpful. They could see that we were tourists and wanting to spend money (personal communication, 2012). Moreover, the relatively superficial nature of host and guest relations may hinder the probability of engaging in more meaningful dialogue that would lead to a reduction of prejudice by promoting greater understanding of the other and facilitate the reconciliation process.

In addition, the message that the Vietnamese have forgotten the war was negated when Luke (personal communication, 2012) spoke with his tour guide

The tour guide had been a student and we had gotten on the subject on how people might feel and how we were received in Vietnam. The tour guide said that the Vietnamese look at it as they can forgive but they can never forget. (Luke, personal communication, 2012)

The significance of this conversation is that it indicates that even though it appears on the surface level that the Vietnamese have moved forward from the war does not necessarily mean that they have resolved their issues with foreign Vietnam War veterans. Moreover, the tour guide's statement indicates that forgiveness of the foreign servicemen who fought with the Free World Forces may be a part of the reconciliation process, at least from the Vietnamese perspective implying that they believe wrong-doing has been committed by the war veterans.

Several interviewees reported having negative encounters with their Vietnamese host. These encounters suggest that not all Vietnamese are accepting of the returning war veterans. Amanda (personal communication, 2012) shares her experience of being a tour operator who has been on multiple trips to Vietnam with New Zealand veterans.

There have been the odd times when the veterans have encountered going into some shops and not being served by the Vietnamese workers. I explain to them beforehand that hey look on the whole 99% of the time they will have an okay experience, but there may be one time that they walk into a shop and have problems. On my first trip to Vietnam that a woman wouldn't serve me at a shop and actually chased me out of the shop. Although the Vietnamese have never been physically aggressive towards the veterans I do warn them that it is possible they will encounter a Vietnamese who exhibits negative behaviour towards them. I advise them that they won't be aggressive towards them but to just walk away from the situation... walk two doors down to a different shop or whatever else they want to do. You might not be able to figure out why they don't want to serve you, but that they just need to move on. (Amanda, personal communication, 2012)

The weakness of travelling in a large tour group is that there may be fewer opportunities to interact with the Vietnamese. One reason for less contact is that the majority of time the veterans are travelling with their own group members visiting war-related sites rather than spending time actively engaged with the Vietnamese people. For example, Donald stated that "I did not have much contact and conversations with the locals because we were too busy touring" (personal communication, 2012). The barrier of limited contact with the Vietnamese may be mitigated when war veterans choose to travel individually or in small groups. For instance, Brian (personal communication, 2012) made the decision to travel with his partner and daughter rather than with a large tour group. As a result he was able to engage in more intimate contact with the Vietnamese because he visited and stayed in places catering to the local population rather than to foreign tourists.

The positive hospitality experienced by the returning veterans can reduce anxiety and serve as encouragement to engage in intergroup contact with the Vietnamese. The Vietnam veterans who have travelled to post-war Vietnam have experienced warmth and little resentment by the Vietnamese people ("Vietnam visit puts ghosts to rest," 1998). For instance, Jake, who has been to Vietnam multiple times, expressed how much he enjoys travelling to Vietnam because the Vietnamese people are so pleased to see him and his fellow war veterans come back (personal communication, 2012). Due to the curiosity of the Vietnamese people about the war, he has been able to speak to them about his experiences and perspective of the war (Jake, personal communication, 2012).

The willingness to engage and seek out intergroup contact with the other may be a measure of a thicker reconciliation. Thicker reconciliation can be measured by non-violent intergroup contact between former adversaries that results in a more intimate relationship. That is not to negate the potential of chance encounters between former adversaries in moving the reconciliation process forward. Rather, these chance encounters may have less likelihood to occur, thereby reducing the possibility of (re)building relationships damaged by conflict. The attainment of thicker reconciliation may be supported by the conditions of contact theory that increase the likelihood that positive attitudinal change will occur toward the other, thereby increasing the possibility that a stronger relationship will occur between those engaged in intergroup contact. The importance of these trips is that they are a means of engaging in intergroup contact that has the capability of improving intergroup relations through better understanding between the veterans and the Vietnamese thereby creating a thicker reconciliation. By recognizing the commonality of their war experience, former combatants may be able to recognize the humanity of one another that can be used to help build post-war relations with one another. For example, Ken explained that "the experience of meeting a former protagonist has been for me like meeting a long lost brother" (personal communication, 2012).

On a thick level, reconciliation implies there is a willingness and ability to co-exist peacefully with the former adversary. The number of visits to Vietnam by veterans' can serve as an indicator of the thickness of reconciliation achieved and the point reached in the reconciliation process. Veterans who are not willing to visit post-war Vietnam signify a thinner form of reconciliation than is the case with those who visit often. Even though these veterans are no longer engaged in physical violence towards the Vietnamese, they still harbour feelings of shame, guilt and hostility towards the Vietnamese that prevent them from taking actions to move the reconciliation process forward. Intergroup contact is a necessary component for any sort of working relationship to emerge, to re-connect and heal the relationship so that a thicker reconciliation can be reached and for the reconciliation process to progress. A thicker reconciliation occurs when veterans choose to return to Vietnam as it shows a willingness to re-connect with the Vietnamese people. There were several interviewees from this study who have engaged in numerous visits back to Vietnam including one veteran who recently re-located to Vietnam. After the Vietnam War, Timothy (personal communication, 2012) first travelled to Vietnam in 1994 for business purposes. At that time the Vietnamese government was just beginning to open up the country to outsiders and the United States had recently lifted their restrictions for their citizens to travel to Vietnam. Timothy initially felt anxiety when he travelled to Vietnam.

I felt there was an oppressive feeling in the air. The immigration officials at the airport were stiff and austere. I didn't feel very happy to be there because of this cold reception. My following business trips became easier and this anxiety receded over time as a result of becoming more familiar with the country. (Timothy, personal communication, 2012)

The reduction of anxiety and increased tolerance of the 'other' over time may be attributed to co-operation and the intergroup contact with the Vietnamese on a more intensive level in order to attain economic gain from the relationship. In 1999, Timothy travelled to South Vietnam to see how the country had changed since he had served there during the war. After that trip he has been to Vietnam numerous times on vacation and considers it a lovely place to go. Timothy felt that travelling back to Vietnam after the war can be seen as a step in the reconciliation process and feels that he is reconciled with them.

A better understanding of the other, an awareness of the commonality they share as servicemen and willingness to move the reconciliation forward may result from the intergroup contact between war veterans and their former Vietnamese enemy. For example, while Jake was having a frank discussion with his Vietnamese tour guide about Ho Chi Min, several war veterans from the Northern Vietnamese Army joined their conversation (personal communication, 2012). He felt that it was a friendly exchange about where they had all served during the war (Jake, personal communication, 2012). "It was no great leap of faith it was just that we all had been to the same war that so happened we were on different sides. That didn't matter. I could care less that they were from the other side. Frankly, I was delighted to meet them and have some discussion with them" (Jake, personal communication, 2012). Preston also had a positive interaction when he met with a former Viet Cong officer during his 2008 trip (personal communication, 2012).

In short during one of the NZ operations the enemy had been on one side of the wire and I on the other and neither one of us knew it at the time. This meeting was very cordial and we met as old enemies and parted as new friends with a new respect for each other. (Preston, personal communication, 2012)

Conversely, Robert's conversation with the former enemy indicates that there may be encounters that have hostile overtones (personal communication, 2012). By conversing with the man through an interpreter, he was able to discover that he had served with the D445 Company.

I then reached out to shake hands with the man who was a bit slow in shaking his hand. The man said that if we had crossed paths during the Vietnam War he would have cut my throat. (Robert, personal communication, 2012)

By having post-war contact with the Vietnamese there can be opportunities for the war veterans to witness the commonality that they as former combatants and allies share as groups who have both suffered from the adverse effects of the war. This increased sense of commonality can help re-humanize former adversaries and allies which can facilitate the reconciliation process. Robert viewed the unplanned encounter with the Vietnamese war veteran from the D445 Company as a cathartic experience and understood the man's hostile reaction to him.

I didn't feel any angst towards him after hearing that the man would have cut my throat during the war. It is understandable what the man said. He most likely would have had members of his family killed during that time. Possibly friends and soldier mates killed as well. I could have responded sod you and the way that you feel, but it is important to endeavour to make amends. I wasn't there to fight in Vietnam because I wanted to kill people; I was there for quite different reasons. If the conversation had continued I believe that part of it would have been to explain my reasons for being there. I felt that the handshake was the thing to do after a valiant fight or joust that at the end a handshake is quite appropriate because it is a signoff out of the battle and signals that it's time to move on. If I was able to in any way to help the man see that we were not the kind of people that he might have of thought of in Long Hai-that we were as human as him that would be good. (Robert, personal communication, 2012)

There are veterans who seek out contact with their former enemy during their post-war visits to Vietnam. For instance, Calvin sought a meeting with former Viet Cong soldiers to try to make the peace with his former enemy. During the meeting he felt respected by them and "a couple of them had a good old cry over why they ever did this-fought in the war" (Calvin, personal communication, 2012). As a result of this meeting, Calvin was able to understand that the Viet Cong soldiers respected the New Zealand soldiers for their fair treatment towards the Vietnamese during the war.

The war veterans' view of the Vietnamese has been impacted by their military training and the war setting in which first contact occurred between the two groups. For the majority of war veterans there was a lack of trust towards the Vietnamese people during their deployment in Vietnam. The return visit can provide experiences that support the re-building of trust of the war veterans towards the Vietnamese people. For example, during Calvin's visit he had made a purchase in which he did not receive the correct change. "The shop girl who served me ran about a quarter of a mile trying to find me to give me my money back. So that is the sort of people they are. The Vietnamese people now I would say I would trust them all" (Calvin, personal communication, 2012). This exchange between Calvin and the Vietnamese woman is significant because it exemplifies a marked change in how war veterans who returned to Vietnam now perceive the Vietnamese people. The change in attitude is such that during the war, they viewed the Vietnamese through a lens of mistrust and fear that has shifted to viewing the Vietnamese as trustworthy people. The interchanges between the war veterans and the Vietnamese during these post-war trips can build trust, an integral component in producing a thicker reconciliation.

CONCLUSION

This research has discovered that the barriers that hinder travel for some veterans, for others serves as their motivation to return to Vietnam. Veterans who have yet to or indicated an unwillingness to return to post-war Vietnam cited financial and time constraints as well as psychosocial issues in their choosing not to return to Vietnam. In these cases, positive word of mouth may not be enough to motivate some veterans to return to post-war Vietnam. Psychosocial issues may need to be resolved in order for some veterans to make the conscious decision to return. Moreover, the veterans who have returned to Vietnam indicate that there are coping mechanisms that they employ in order to mitigate the barriers hindering their return. The examination of New Zealand Vietnam War veterans who have gone back to Vietnam post- war brings to the forefront the complexities of providing a clearer understanding of the impact of these post-war visits on reconciliation between former enemies. New Zealand Vietnam War veterans may need to have the willingness to accept their former enemies' right to interpret the war as they see fit in order to achieve reconciliation with the Vietnamese people.

REFERENCES

Aiken, N. T. (2008). Post-conflict peacebuilding and the politics of identity: Insights for restoration and reconciliation in transitional justice. *Peace Research*, *40*(2), 9–38,111.

Allport, G. W. (1954). *The nature of prejudice*. Addison-Wesley.

Allport, G. W. (1979). *The nature of prejudice* (6th ed.). Addison-Wesley Pub. Co.

Bar-Siman-Tov, Y. (Ed.). (2004). *From conflict resolution to reconciliation*. Oxford University Press. doi:10.1093/acprof:oso/9780195166439.001.0001

Beaver, A. (2012). *A dictionary of travel and tourism*. Oxford University Press.

Benfell, G. (2006). *Interview by D. Challinor: New Zealand National Army Museum*. Waiouru.

Berno, T., & Ward, C. (2005). Innocence abroad: A pocket guide to psychological research on tourism. *The American Psychologist, 60*(6), 593–600. doi:10.1037/0003-066X.60.6.593 PMID:16173892

Bowen, G. A. (2009). Document analysis as a qualitative research method (Report). *Qualitative Research Journal, 9*(2), 27.

Braithwaite, D., & Lee, Y. L. (2006). Dark tourism, hate and reconciliation: The Sandakan experience. *IIPT Occasional Paper, 8*.

Branscombe, N. R., & Miron, A. M. (Eds.). (2004). *The social life of emotions*. Cambridge University Press.

Brewer, M. B. (1999). The psychology of prejudice: Ingroup love and outgroup hate? *The Journal of Social Issues, 55*(3), 429–444. doi:10.1111/0022-4537.00126

Brown, L. (2008). Interviewer: B. Beach. Vietnam War Oral History Project (Vol OHInt-0959-07), VOH, Wellington, NZ.

Brown, R., & Hewstone, M. (2005). *An integrative theory of intergroup contact* (Vol. 37). Advances in Experimental Social Psychology. Academic Press.

Bruner, E. M. (1991). Transformation of self in tourism. *Annals of Tourism Research, 18*(2), 238–250. doi:10.1016/0160-7383(91)90007-X

ButlerR.SuntikulW. (2013). *Tourism and war* Retrieved from http://otago.eblib.com.au/patron/FullRecord.aspx?p=1092763

Cameron, C. M., & Gatewood, J. B. (2000). Excursions into the un-remembered past: What people want from visits to historical sites. *The Public Historian, 22*(3), 107–127. doi:10.2307/3379582

Cameron, L., & Rutland, A. (2006). Extended contact through story reading in school: Reducing children's prejudice toward the disabled. *The Journal of Social Issues, 62*(3), 469–488. doi:10.1111/j.1540-4560.2006.00469.x

Campbell, T., Pickett, T., & Yoash-Gantz, R. (2010). Psychological rehabilitation for US Veterans. In E. Martz (Ed.), *Trauma rehabilitation after war and conflict* (pp. 159–176). Springer New York. doi:10.1007/978-1-4419-5722-1_8

Challinor, D. (1998). *Talking the talk: New Zealanders remember the Vietnam War*. University of Waikato.

Challinor, D. (2009). *Grey ghosts: New Zealand Vietnam vets talk about their war*. HarperCollins.

Cheng, Y.-H. (2010). Exploring passenger anxiety associated with train travel. *Transportation, 37*(6), 875–896. doi:10.100711116-010-9267-z

Cohen, E. (1972). Toward a sociology of international tourism. *Social Research*, *39*(1), 164–182. doi:10.2307/40970087

Colman, A. M. (2008). *A dictionary of psychology* (3rd ed.). Oxford University Press.

Crookes, P., & Davies, S. (2004). *Research into practice: essential skills for reading and applying research in nursing and health care*. Baillière Tindall.

de Jongh, A., Holmshaw, M., Carswell, W., & van Wijk, A. (2011). Usefulness of a trauma-focused treatment approach for travel phobia. *Clinical Psychology & Psychotherapy*, *18*(2), 124–137. doi:10.1002/cpp.680 PMID:20146201

Dixon, J. (2001). Contact and boundaries: 'locating' the social psychology of intergroup relations. *Theory & Psychology*, *11*(5), 587–608. doi:10.1177/0959354301115001

Duggan, P. (2011). *What have they done to the rain?* Kukupa Press.

Dwyer, S. (2003). Reconciliation for realists. In C. Prager & T. Govier (Eds.), *Dilemmas of reconciliation: Cases and concepts*. Wilfrid Laurier University Press.

Edensor, T. (1998). *Tourists at the Taj: performance and meaning at a symbolic site*. Routledge.

Elder, J. (1998). Expanding our options: The challenge of forgiveness. *Exploring Forgiveness*, 150-163.

Elliott, D. W. P. (2012). *Changing worlds: Vietnam's transition from Cold War to globalization*. Oxford University Press. doi:10.1093/acprof:oso/9780195383348.001.0001

Ellison, C. G., & Powers, D. A. (1994). The contact hypothesis and racial attitudes among Black Americans. *Social Science Quarterly*, *75*(2), 385–400.

Feddes, A. R., Mann, L., & Doosje, B. (2012). From extreme emotions to extreme actions: Explaining non-normative collective action and reconciliation. *Behavioral and Brain Sciences*, *35*(06), 432–433. doi:10.1017/S0140525X12001197 PMID:23164364

Feng, X. (2012). Who are the "hosts"? Village tours in Fenghuang County, China. *Human Organization*, *71*(4), 383–394. doi:10.17730/humo.71.4.f738385u0206j866

Franklin, A. (2003). The tourist syndrome: An Interview with Zygmunt Bauman. *Tourist Studies*, *3*(2), 205–217. doi:10.1177/1468797603041632

Freeman, C. M. L. (2012). The psychosocial need for intergroup contact:practical suggestions for reconciliation initiatives in Bosnia and Herzegovina and beyond. *Intervention (Amstelveen, Netherlands)*, *10*(1), 17–29. doi:10.1097/WTF.0b013e3283518de1

Frost, F. (1991). The Cambodia Conflict: The path towards peace. *Contemporary Southeast Asia*, *13*(2), 119–163. doi:10.1355/CS13-2A

Garton, S. (2013). Longing for war:nostalgia and Australian returned soldiers after World War I. In T. G. Ashplant, G. Dawson, & M. Roper (Eds.), *The politics of war memory and commemoration*. Taylor and Francis.

Gloppen, S. (2005). Roads to reconciliation: A conceptual framework. In S. G. E. Skaar & A. Suhrke (Eds.), *Roads to reconciliation*. Lexington Books.

Halpern, J., & Weinstein, H. M. (2004). Rehumanizing the other: Empathy and reconciliation. *Human Rights Quarterly*, *26*(3), 561–583. doi:10.1353/hrq.2004.0036

Hayes, B. C., & McAllister, I. (2001). Sowing Dragon's Teeth: Public support for political violence and paramilitarism in Northern Ireland. *Political Studies*, *49*(5), 901–922. doi:10.1111/1467-9248.00346

Heritage, M. C. a. (2008). *Fact Sheet 10: The Memorandum of Understanding and Tribute08*. Retrieved from http://www.nzhistory.net.nz/files/documents/vietnam-fact-sheet10.pdf

Heritage, M. C. a. (2013). *The Vietnam War*. Retrieved from http://www.nzhistory.net.nz/war/vietnam-war

Hewstone, M., Cairns, E., Voci, A., Hamberger, J., & Niens, U. (2006). Intergroup contact, forgiveness, and experience of "the troubles" in Northern Ireland. *The Journal of Social Issues*, *62*(1), 99–120. doi:10.1111/j.1540-4560.2006.00441.x

Higgins-Desbiolles, F. (2003). Reconciliation tourism: Tourism healing divided societies. *Tourism Recreation Research*, *28*(3), 33–44. doi:10.1080/02508281.2003.11081415

Hobson, J. S. P., Heung, V. C. S., & Chon, K.-S. (1994). Vietnam's tourism industry: Can it be kept afloat? *The Cornell Hotel and Restaurant Administration Quarterly*, *35*(5), 42–49. doi:10.1177/001088049403500514

Hunt, J. (2009). Wellington: VOH.

Islam, M. R., & Hewstone, M. (1993). Dimensions of contact as predictors of intergroup anxiety, perceived out-group variability, and out-group attitude: An integrative model. *Personality and Social Psychology Bulletin*, *19*(6), 700–710. doi:10.1177/0146167293196005

Jaakson, R. (2004). Beyond the tourist bubble?: Cruiseship passengers in port. *Annals of Tourism Research*, *31*(1), 44–60. doi:10.1016/j.annals.2003.08.003

Jalilvand, M. R., & Samiei, N. (2012). The effect of word of mouth on inbound tourists' decision for traveling to Islamic destinations (the case of Isfahan as a tourist destination in Iran). *Journal of Islamic Marketing*, *3*(1), 12–21. doi:10.1108/17590831211206554

Jansen-Verbeke, M., & Go, F. (1995). Tourism development in Vietnam. *Tourism Management*, *16*(4), 315–321. doi:10.1016/0261-5177(95)97356-U

Jencks, H. W. (1979). China's "punitive" war on Vietnam: A military assessment. *Asian Survey*, *19*(8), 801–815. doi:10.2307/2643723

Kelly, I., & Nkabahona, A. (2010). Tourism and reconciliation. In O. Moufakkir & I. Kelly (Eds.), *Tourism, progress and peace* (pp. 228–241). CAB International. doi:10.1079/9781845936778.0228

Kelman, H. C., & Hamilton, V. L. (1989). *Crimes of obedience: Toward a social psychology of authority and responsibility*. Yale University Press.

Kim, H. S. (2001). Korea's "Vietnam Question": War atrocities, national identity, and reconciliation in Asia. *Positions*, *9*(3), 621-635. doi:10.1215/10679847-9-3-621

Kim, S. S., Timothy, D. J., & Han, H.-C. (2007). Tourism and political ideologies: A case of tourism in North Korea. *Tourism Management*, *28*(4), 1031–1043. doi:10.1016/j.tourman.2006.08.005

Kosic, A., & Tauber, C. D. (2010). Promoting reconciliation through youth: Cross-community initiatives in Vukovar, Croatia. *Peace and Conflict*, *16*(1), 81–95. doi:10.1080/10781910903487050

Lavelle, A. (2006). Labor and Vietnam: A reappraisal. *Labour History*, 90.

Leopold, T. (2003). A provisional code of conduct for war heritage sites: Veterans' and managers' perceptions (Master of Tourism). University of Otago, Dunedin.

Mathieson, A., & Wall, G. (1982). *Tourism: Economic, physical, and social impacts*. Longman.

McAdam, D., & Su, Y. (2002). The war at home: Antiwar protests and congressional voting, 1965 to 1973. *American Sociological Review*, *67*(5), 696–721. doi:10.2307/3088914

McGibbon, I. (2010). *New Zealand's Vietnam War: A history of combat, commitment and controversy*. Exisle Publishing Limited.

McGibbon, I., & Goldstone, P. (2000). *The Oxford companion to New Zealand military history*. Oxford Oxford University Press.

McIntosh, I. B., Swanson, V., Power, K. G., Raeside, F., & Dempster, C. (1998). Anxiety and health problems related to air travel. *Journal of Travel Medicine*, *5*(4), 198–198. doi:10.1111/j.1708-8305.1998.tb00507.x PMID:9876195

McKercher, B. (2008). The roots of stereotypes about tourists. *Society*, *45*(4), 345–347. doi:10.100712115-008-9113-5

Money, R. B., & Crotts, J. C. (2003). The effect of uncertainty avoidance on information search, planning, and purchases of international travel vacations. *Tourism Management*, *24*(2), 191–202. doi:10.1016/S0261-5177(02)00057-2

Moufakkir, O., & Kelly, I. (Eds.). (2010). *Tourism, progress and peace*. CAB International. doi:10.1079/9781845936778.0000

Munro, L. K. (1953). New Zealand and the New Pacific. *Foreign Affairs*, *31*(4), 634–647. doi:10.2307/20030995

Neumayer, E. (2004). The impact of political violence on tourism: Dynamic cross-national estimation. *The Journal of Conflict Resolution*, *48*(2), 259–281. doi:10.1177/0022002703262358

Northrup, T. (1989). The dynamic of identity in personal and social conflict. In L. N. Kriesberg, A. Terrell, & S. J. Thorson (Eds.), *Intractable conflicts and their transformation*. Syracuse University Press.

NZPA. (2007). *NZ vets to get chance to visit Vietnam*. Retrieved from http://www.stuff.co.nz/national/48537/NZ-vets-to-get-chance-to-visit-Vietnam

Patton, M. Q. (1989). A context and boundaries for a theory-driven approach to validity. *Evaluation and Program Planning*, *12*(4), 375–377. doi:10.1016/0149-7189(89)90054-2

Pettigrew, T. F. (1998). Intergroup contact. *Annual Review of Psychology*, 49. PMID:15012467

Pettigrew, T. F., & Tropp, L. R. (Eds.). (2000). *Does intergroup contact reduce prejudice? Recent meta-analytic findings*. Lawrence Erlbaum Associates.

Reisinger, Y., & Mavondo, F. (2006). Cultural differences in travel risk perception. *Journal of Travel & Tourism Marketing*, *20*(1), 13–31. doi:10.1300/J073v20n01_02

Seaton, A. V., Smith, R. A., Inguanez, J., Pizam, A., Lebruto, S. M., Roper, A., & Boissevain, J. (2001). M. In J. Jafari (Ed.), *Encyclopedia of tourism* (pp. 366–403). Routledge.

Sigelman, L., Bledsoe, T., Welch, S., & Combs, M. W. (1996). Making contact? Black-white social interaction in an urban setting. *American Journal of Sociology*, *101*(5), 1306–1332. doi:10.1086/230824

Sigelman, L., & Welch, S. (1993). The contact hypothesis revisited: Black-White interaction and positive racial attitudes. *Social Forces*, *71*(3), 781–795. doi:10.2307/2579895

Simone-Charteris, M. T., & Boyd, S. W. (2010). Northern Ireland re-emerges from the ashes: The contribution of political tourism towards a more visited and peaceful environment. *Tourism, Progress, and Peace*, 179.

Smith, V. L. (Ed.). (1989). *Hosts and guests: The anthropology of tourism*. University of Pennsylvania Press. doi:10.9783/9780812208016

Smith, V. L. (1996). War and its tourist attractions. In A. Pizam & Y. Mansfield (Eds.), *Tourism, Crime, and International Security Issues* (pp. 247–264). Wiley.

Snepenger, D., King, J., Marshall, E., & Uysal, M. (2006). Modeling Iso-Ahola's motivation theory in the tourism context. *Journal of Travel Research*, *45*(2), 140–149. doi:10.1177/0047287506291592

Sonmez, S. F., & Apostolopoulos, Y. (2000). Conflict resolution through tourism cooperation? The case of the partitioned Island-State of Cyprus. *Journal of Travel & Tourism Marketing*, *9*(3), 35–48. doi:10.1300/J073v09n03_03

Stephan, W. G., & Stephan, C. W. (1985). Intergroup Anxiety. *The Journal of Social Issues*, *41*(3), 157–175. doi:10.1111/j.1540-4560.1985.tb01134.x

Stets, J. E., & Burke, P. J. (2000). Identity theory and social identity theory. *Social Psychology Quarterly*, *63*(3), 224–237. doi:10.2307/2695870

Sveaass, N., & Castillo, M. (2000). From war hero to cripple: An interview study on psychosocial intervention and social reconstruction in Nicaragua. *Peace and Conflict*, *6*(2), 113–133. doi:10.1207/S15327949PAC0602_2

Tamatea, P. P. (2007) Interview with Pepe Papa Tamatea/Interviewer: P. Diamond. Vietnam Oral History Project (Vol Reference Number OHDL-000513 digitial manuscript). Ministry of Culture and Heritage.

Tashakkori, A., & Teddlie, C. (2003). *Handbook of mixed methods in social & behavioral research*. SAGE Publications.

Teddlie, C., & Yu, F. (2007). Mixed methods sampling. *Journal of Mixed Methods Research*, *1*(1), 77–100. doi:10.1177/1558689806292430

Timothy, D. J., & Boyd, S. W. (2003). *Heritage tourism*. Pearson Education.

Tucker, H. (2001). Tourists and troglodytes: Negotiating for sustainability. *Annals of Tourism Research*, *28*(4), 868–891. doi:10.1016/S0160-7383(00)00084-0

VANZ News. (2011). Retrieved from https://www.veteransaffairs.mil.nz/about/whats-new/documents/vanznewsoct11.pdf

Vietnam visit puts ghosts to rest. (1998, July 20). *Waikato Times,* p. 10.

Volkan, V. (1999a). Psychoanalysis and diplomacy. Part 1: Individual and large group identity. *Journal of Applied Psychoanalytic Studies*, *1*(1), 29–55. doi:10.1023/A:1023026107157

Volkan, V. (1999b). Post-traumatic states: Beyond individual PTSD in societies ravaged by ethnic conflict. *Canadian Foreign Policy*, *7*(1), 27–38. doi:10.1080/11926422.1999.9673198

Weaver, D. B. (2000). The exploratory war-distorted destination life cycle. *International Journal of Tourism Research*, *2*(3), 151–161. doi:10.1002/(SICI)1522-1970(200005/06)2:3<151::AID-JTR196>3.0.CO;2-G

Williams, R. M. Jr. (1947). *The reduction of intergroup tensions*. Social Science Research Council.

Wohl, M. J. A., & Branscombe, N. R. (2008). Group threat, collective angst, and ingroup forgiveness for the war in Iraq. *Political Psychology*, *30*(2), 193–217. doi:10.1111/j.1467-9221.2008.00688.x

Zebel, S., Zimmermann, A., Tendayi Viki, G., & Doosje, B. (2008). Dehumanization and guilt as distinct but related predictors of support for reparation policies. *Political Psychology*, *29*(2), 193–219. doi:10.1111/j.1467-9221.2008.00623.x

KEY TERMS AND DEFINITIONS

Contact Theory: The idea that in-group members by learning new information about the out-group through direct contact will be less prejudiced towards members of the out-group.

Intergroup Contact: Interaction between in-group and out-group members.

Mass Tourism: Participation in tourism in large numbers.

Post-War Tourism: Starts is at the end of the war that is marked by tourists' numbers beginning to rebound towards the pre-war period as there is a growing number of adventure tourists, family members of missing servicemen, and opportunists infiltrating the war zone.

Prejudice: A negative attitude towards an individual, group or race based on their perceived characteristics.

Pull Factors: Attributes of a destination that encourage individuals to travel to that location.

Push Factors: Factors that encourage individuals to travel away from their home.

ENDNOTES

1 "Through a social comparison process, persons who are similar to the self are self-categorized with the self and are labelled the in-group; persons who differ from the self are categorized as the out-group" (Stets & Burke, 2000, p. 225).

2 Eigenwelt is "the personal world as it is experienced, including one's awareness of mind and body and the sense of who one is" (Colman, 2008).

3 Mass tourism refers to participation in tourism in large numbers (Beaver, 2012).

Section 2

International Relations and Tourism

Chapter 8
Tourism as an Agent of Peace and Reconciliation in Cross-Strait Relations

Jorge Tavares da Silva
https://orcid.org/0000-0002-2526-4745
University of Aveiro, Portugal

Zélia Breda
https://orcid.org/0000-0002-5882-063X
University of Aveiro, Portugal

ABSTRACT

There is a non-violent conflict over Taiwan's sovereignty, between the People's Republic of China (PRC) and the Republic of China (ROC). For PRC, this division cannot persist forever and does not exclude a possible military solution. While political divisions remain, the population on both sides of the strait interact, existing sociocultural and economic dynamics. These are usually interpreted as people-to-people dynamics, in which individuals act as peace agents or citizen diplomats. Tourism is a good example of this phenomenon, considering the increasing visitor flows between both sides. This dynamism sometimes pressures the political power to transform the conflict, but also acts as a throwing weapon in times of hostility. After 2016, the political landscape in Taiwan changed, and tourism became one of the sectors involved in political tensions. This chapter explores several dimensions of tourism in this conflict, particularly its role in peace and reconciliation between Mainland China and Taiwan, but also its vulnerabilities regarding high-level bilateral relations.

INTRODUCTION

Tourism emerged as a product of modernisations and has become one of the most popular industries in the contemporary world, providing opportunities for rapid economic development (Garg, 2012). Multiple sectors benefit from tourist flows, including transportation, hotels, and cultural activities. National finances

DOI: 10.4018/978-1-7998-5053-3.ch008

stabilise the accounts by collecting revenues, and are even a factor of resistance to unstable financial crises. About 1.4 billion people travelled the world in 2019, having increased by 6% in relation to the previous year (UNWTO, 2019). All this dynamism works as a factor of socialisation and a potential changer of conflicts, as defended by many scholars (Becken & Carmignani, 2016; D´Amore, 1988; Farmaki, 2017; Litvin, 1998; Wohlmuther & Wintersteiner, 2014). This can play a positive role in the reconciliation of countries in the aftermath of armed conflicts, and in cases when people are subjected to political barriers or even in contexts of not recognised governments or nations (Butler & Mao, 1996). Tourists could be a new kind of diplomats, may work as agents of peace, opening bridges and helping to reconcile peoples. Kim and Crompton (1990, pp. 353-366) considered that "tourism is a vehicle for implementing people-to-people diplomacy" and can significantly contribute to more integrated political processes, as they demonstrate with the case of the Korea conflict. This kind of interactions are considered as track two diplomacy, because they happen outside, but eventually link up, the formal government sphere (Davison & Montville, 1982). These new citizen diplomats have the potential to influence decision-makers as a pressure group, and facilitate discussion at the grassroots level (Azar, 2003). These initiatives are at the level of citizenship, both individual and collective, assenting in the maxim "peace is not made between leaders, but between peoples" (Garrison & Philips, 1989, p. xiii; Silva, 2012a, p. 118). Travel and flows of people can affect the thinking and action of government diplomacy by addressing root causes, beliefs and needs, thereby laying the base for formal negotiation processes or reframing policies. Comprised in a large group of informal diplomatic initiatives – including government contributions –, tourism belongs to a general system approach to peace named multi-track diplomacy in a kind of "pyramid of peace" (Diamond & McDonald, 1996; Lederach, 1999; Silva, 2012a). It goes beyond track two diplomacy, joining a formal and informal set of diplomatic dynamics, a systemic gear difficult to divide into parts.

We are aware that tourist flows do not have implications in only one sense. Many times, it deploys antagonism and fraction between people and countries. Even organised crime, including drug traffic and physical violence, may increase due to the growth of tourism. Local criminal groups develop their activities, taking into account the gradual presence of tourists in some areas. Another interesting point of view in tourism is the analysis of how previous war and violence scenarios can become attractive as spaces of memory (Ahmad & Hertzog, 2016). Other times, the development of 'slum tours' in South America – like *favelas* in Rio de Janeiro – are increasingly popular but lead us to question how ethical they are. In the media, tourism and other economic activities, as proactive forces in conflict resolution, usually do not receive the same attention and importance as its role on tensions and violence. In fact, it has two faces, in which one party does not invalidate the other. Our study fundamentally follows the line of the liberal theory of international relations, a legacy of the old idealist tradition centred on the peaceful Kantian cosmopolitism. In other words, we explore tourism as a potential builder of bridges, peace and reconciliation between people. Liberalism is the ideology of the industrialised world, basing its main idea in the notion of harmony amongst competing interests. Individuals or groups in defence of their own interests create a natural equilibrium. This chapter focuses on the "Taiwan conflict", which is frequently considered as one of the most complicated issues in global politics (Blanchard & Hickey, 2012; Silva, 2008). Like North and South Korea, China and Taiwan remain as a country politically divided, a nation divided into two states, if we consider the same identity[1]. Mainland China represents a frustrating paradox to islanders: it is the most relevant economic partner and an existential threat (Lin, 2016). It is a non-violent conflict – or crisis – over the sovereignty of Taiwan's territory, involving two ideologically opposed parties. According to the Conflict Barometer of the University of Heidelberg (2019), it is a level 2 conflict, on a 5-point scale. For Beijing, this division cannot remain in the *status quo* situation forever and does not exclude a possible military solution. While the political divisions remain, the population on both sides of the Strait are interacting in informal dynamics, including socio-cultural activities, sports, and business (Silva, 2012a; Silva, 2012b). Civil society actors interact with each other in people-to-people dynamics, acting as agents of peace or citizen diplomats. Tourism is one of the best examples of this phenomenon, with increasing flows of visitors on both sides of the Strait. These interactions work as a pressure factor on the political power for the transformation of the conflict, despite being also a throwing weapon in moments of greater hostility. For our analysis, we give more importance to the flow of tourists from Mainland China to Taiwan because this represents a significant change in the recent bilateral context. For over 40 years, Mainland tourists were banned from visiting the Republic of China (ROC). Many analysts believe that Chinese tourists visiting Taiwan contributes to peace across the Taiwan Strait. Those Chinese that visited Taiwan for the first time – particularly individuals – were less likely to support unification by force as a solution in cross-strait relations (Pan et al., 2018). Although tourism is a contributor to gradual socialisation between individuals on both sides of the Strait, it is dependent on political changes on the island.

This chapter intends to analyse changes in China-Taiwan relations after President Tsai Ing-wen came to power in Taiwan (2016). She belongs to the Democratic Progressive Party (DPP), a political party that reveals a strong propensity to the solidification of Taiwanese independent identity. In many ways, tourism is one of Beijing's weapons against the island's new government when it did not follow the "one China" position. This work begins by analysing the essence of the political problem that divides China and Taiwan. Then, it discusses how tourism has become an important element in the socialisation process between the population on both sides of the Strait. Finally, it examines bilateral tourism between China and Taiwan after Tsai Ing-wen came to power.

TAIWAN AND THE "ONE CHINA" DILEMMA

Between the People's Republic of China (PRC) and the ROC, there is a non-violent conflict over Taiwan's sovereignty that have last at about 70 years. The roots of the conflict come from the Chinese civil war in the first half of the 20th century, and the ideological context of the cold war that involved the United States in the region (Silva, 2012a). In 1949, communist military forces triumphed, and the nationalists moved to Taiwan, maintaining a resistant part of ROC. The two regimes maintained the mutual will to recover the "lost part". The tensions continued over time, existing no direct contact between islanders and mainlanders. The Taiwanese regime opened up in latter 1980s during the leadership of Chiang Ching-kuo, the son of the historical leader Chiang Kai-shek. In July 1987, the martial law was lifted and the military rule has ended. Press freedom improved, political parties were legalised, and new proposals for peaceful reunification were established, including the possibility of applying Deng Xiaoping's model of "one country, two systems". However, the democratisation of ROC brought new tension to the Taiwan Strait relations. The democratisation process in Taiwan, in 1995, also clashed with Beijing's interests, bringing DPP in power in several elections (2000, 2004, 2016 and 2019).

Currently, two parties dominate the political system: the Kuomintang (KMT) and the Democratic Progressive Party (DPP). The KMT ("the blues") protects the *status quo*, a model of strong bilateral cooperation and the "1992 Consensus", which is a verbal agreement between Taipei and Beijing. Both sides accepted the principle of "one China", despite different interpretations (Copper, 2010). The DPP ("the greens") is a sovereign party, which attracts the votes of people that have an anti-Chinese position.

On the one hand, there is a considerable power with an authoritarian political system that intends to recovery the domain of the "renegade province". On the other hand, Taiwan and a small group of islands live under a democratic political system. Beijing does not exclude the historical rights over ROC ("one China policy") and the possibility of a military solution to solve this issue.

In fact, prior to 1990, both parties accepted that there was only "one China" (Chow, 2008) and pretended to work together, finding a peaceful solution for unification. The problem is that each party has different interpretations of the "one China" principle. RPC considers that the island is part of its territory, in which its sovereignty is not negotiable. The ROC, commonly called Taiwan, is a country – *de facto* – but do not receive diplomatic recognition from most of the international community (it is not *de jure*). International isolation is one of the biggest challenges for the small island of 23 million inhabitants. At the same time, Beijing is trying to isolate the island internationally, putting pressure on international actors to refuse political recognition. The disparity in political regimes and systems precludes a short-term solution, conditioned by political changes in Taiwan.

The DPP is a party that since the early 1990s has had the clear objective of creating the Republic of Taiwan, with a strong propensity for the island's independence. Chen Shui-bian's leadership (2000-2008) was in this direction, leading to a breakdown in the relations between Beijing and Taipei. The DPP refused to affirm the existence of "one China", which would open the way for cooperation with a view to unification/reunification. Chen followed provocative speeches, defending controversial policies, such as the use of nuclear energy, the elimination of images of historical figures from the KMT party in public buildings, the abolition of the National Unification Plan and the attempt to carry out a referendum on Taiwan's possible access to the United Nations (UN). Chen Shui-bian's political career was affected by suspicions of corruption, in particular, the misuse of a secret presidential fund earmarked for international diplomatic missions. On May 20 2008, when Ma Ying-jeou took office as the newly elected president of the ROC, Chen lost his immunity and was put under surveillance by Taiwanese prosecutors (Silva, 2016). Ma Ying-jeou, from the traditional KMT party, introduced a new dynamics of openness and cooperation with the PRC. Ma created the conditions for reaching a set of joint agreements and initiatives with China. The opening of direct air and sea transport links between the island and mainland China is one of its policies. The unique historical rapprochement led to the acceptance of the principle of "1992 Consensus". Despite the tacit agreement, both parts had different interpretations of the specific meaning of the "one China" policy. Besides, Ma's openness cooperation had limits; he only admitted the situation "no unification, no independence, no use of force" (three noes) for Taiwan. This political guideline suggested he wanted to catch the median voters who wanted to continue the *status quo* (Lin, 2019). One of the most remarkable moments on this new climate of cooperation happened with the meeting between Ma Ying-jeou and Xi Jinping, at the Shangri-La Hotel in Singapore, in November 2015. The two leaders consolidated the "1992 Consensus", based on mutual acceptance of the existence of "one China". This informal agreement was established in 1992, in the sequence of track two semi-official dialogues between the Straits Exchange Foundation (SEF) in Taiwan and Association for Relations Across the Taiwan Strait (ARATS) in mainland China. These semi-formal organisations were created as a result of the establishment of secret channels by some unofficial organisations. The chairmen of both entities – Gu Zhengfu from SEF and Wang Daohan from ARATS – established contacts and gave hopeful signs of a gradual and constructive relationship between both parties (Wachman, 2007). The meeting took place during an election campaign in Taiwan; the initiative gained enormous political expression and was criticised by opposition parties. The intensification of cooperation between the KMT in Taiwan and the Chinese Communist Party (CCP) in the PRC was not to the liking of a substantial part of the younger population, affected by high levels of unemployment or precarious work situations. They complained that the opening of trade with Mainland China has damaged the national economy, particularly youth employment, and was a threat to the island's democracy and autonomy. A trade agreement in 2014, which allowed Chinese companies to invest in Taiwanese industries, triggered what became known as the "Sunflower Movement". A group of students occupied the Legislative Yuan (the Taiwanese Parliament) for 24 days, with the phenomenon having a significant impact on Taiwanese public opinion. Ma continued his policy of cooperation but gradually losing support from part of the electorate. Also, the Xi-Ma meeting in Singapore would not be enough for voting trends on the island, and Tsai Ing-Wen would eventually win the presidential elections in January 2016. The meeting did not produce the expected results, and the previous agreements did not reflect an improvement in living conditions, leading to a clear penalty in elections.

TOURISM AS A WINDOW OF OPPORTUNITY ON PEACEFUL CROSS-STRAIT RELATIONS

Political relations between China and Taiwan have improved at many times due to the dynamics of civil society (Silva, 2012). Business, trade, sports, education and travel have prospered in the Strait, particularly in the political era of Ma Ying-jeou (2008-2016). Ma assumed that Taiwan "should be more open, more pragmatic, and should not isolate itself" (Blanchard & Hickey, 2012). New policies gave priority to the so-called "mini-Three Links": postal service, transportation and trade between cities in Fujian (Mainland China) and the Taiwanese islands of Kinmen and Matsu (Lin, 2016). Some of these links started before Ma's presidency, although encouraged by him. Until then, Chinese tourists could not travel to Taiwan. After 1949, nationalist exiles who were in Taiwan – at about two million – were unable to visit their relatives, including children, brothers and women. Interactions between the two sides were almost non-existent; there was no post office, transport linkages, no telephone. At that time, travel to the mainland was a serious violation of national security laws (Clough, 1993).

The communist authorities had opened a campaign to bring Taiwan closer, opening the door to visits by KMT military and politicians who wanted to visit their family members. At this time, the PRC had its "anti-rightist campaign" underway, which contained a real risk of detention. Reforms and the opening-up process led by Deng Xiaoping made many individuals from Taiwan to visit the mainland clandestinely. The Taiwanese government was lenient in this phenomenon, which the PRC has long supported. The pressure of the Taiwanese population for the government to allow travel to Mainland China was growing. According to the Taiwan Compatriots Association of Fujian Province, at about 1,800 tourists from Taiwan visited the province in 1983 (Clough, 1993). Internal political discussion led Chiang-kuo to lift restrictions on visits to Mainland China. He feared that contacts between people from both sides would limit anti-communist perceptions in Taiwan. In July 1987, the Taiwanese government eased restrictions on travel to Hong Kong and, in October, to the mainland (except government employees and active military personal, for security reasons). Taiwanese tourists travelled to Mainland China via Hong Kong, as there were no direct connections. Applications for passports for such visits could be submitted to the Red Cross Society of China. In the absence of formal relations between the two parties, this institution served as an intermediary in the process. In the first week alone, 10,000 people submitted travel applications. After this, travel agencies took the opportunity to promote gift plans to family members (Clough, 1993). In 2020, KMT regained power, and new policies assumed a more rapprochement position. The two sides intensified cooperation and bilateral negotiations, including an extraordinary meeting between Xi Jinping and Ma Ying-jeou, the president of the PRC and the ROC, respectively.

Tourist flows had limitations on quotas and method of travel. Initially, Chinese tourists could only do it in groups and with a limited tour of seven days (Lin & Lee, 2018). Ma's government implemented several assertive tourism policies in order to develop the national economy. It included, for instance, the "Taiwan double arrivals plan", the project "Vanguard for Excellence in Tourism" and "Tourism 2020" (Lin & Lee, 2018). All these interactions would be considered informal diplomatic actions. As part of a system of multi-track diplomacy, tourism is as an activity that builds relationships and creates pathways for communication and joint actions (Diamond & McDonald, 1996). It enhances the economic condition of people from divided territories, thereby reducing poverty. In many ways, it is an entryway to pledge mutual trust, reduce tensions and propensity to provoke conflict, changing the perception of the "Other". Loosening up barriers leads China and Taiwan to become interdependent, creating a community of economic agents in both sides, linked to global business.

Breaking physical barriers between China and Taiwan was a decisive factor at an early stage, opening many doors of cooperation. Some considerer that the people-to-people tourism exchange may lead China to embrace Taiwan's democracy and liberal values. However, during Ma's rule, the high volume of cross-strait trade and people-to-people exchange provoked great concerns. The inbound immigration from the mainland was in the centre of this controversy, particularly a wave of students, tourists, investors and unskilled workers (Lin, 2016). Many Taiwanese believe that admitting an exponential number of mainlanders would erode national identity, similar to what is happening in Hong Kong. The civil society protest in 2014 – essentially formed by students – was mainly against the Cross-Strait Service Trade Agreement (CSSTA), signed with mainland China by the ruling party (KMT). The process of negotiation started secretly between SEF and ARATST, which provoked complains (Cole, 2015). The standard of living has worsened in Taiwan due to the growing presence of Chinese investors and tourists, including the rise of housing prices

TSAI ING-WEN ELECTION AND ITS IMPACT ON BILATERAL TOURISM FLOWS

On January 2016, the DPP regained power in Taiwan, bringing new political challenges to the relationship with the PRC. KMT had ruled Taiwan for eight years (2008-2016). President Ma Ying-jeou had managed a series of confidence-building measures, including new business and transport links with the mainland. This context culminated in the first meeting between the presidents of both parties in Singapore in 2015 (Brown, 2016). The acceleration in the agreements with the RPC created apprehensions in Taiwanese people, particularly amongst the youth. As mentioned before, in 2014, huge unrest surfaced in Taiwan, organised particularly by students ("Sunflower Movement"). Xi had served the party in Fujian province for 16 years, having embraced investment across the Strait and building up a good relationship with Taiwanese businesspeople (Brown, 2016). The economic results on Fujian province were one of the factors that fostered its promotion in the party structure. Despite this entire cooperative ambience, Ma refused the proposal of the "one country, two systems" principle to Taiwan. In the new electoral election (2015), Tsai Ing-wen defeated Eric Chu (31%) of the ruling KMT and James Soong of the People First Party (PFP). The new president – the first woman to be elected – gathered 56.1% of the popular vote, the highest for any DPP candidate in previous elections. The DPP was also able to win 68 out of the 113 seats in the Legislative Yuan election (Hsieh, 2019). Tsai remains evasive in the explicit identification of the territories to which the principle refers. One of Tai's greatest challenges is the recovery of the island's economy, strongly conditioned by the relations with Mainland China. Taiwan grew around 2% a year, a relatively modest growth compared to other Asian tigers. The unemployment reached 3.94% in January 2016, having been the cause of the discontent of the "Sunflower Movement". Low wages in Taiwan encouraged many Taiwanese to go to Mainland China (Beijing and Shanghai). In 2009, the previous KMT government sought to resolve this massive immigration problem by applying the 22K policy.

Tsai's speech was always in defence of Taiwan's official sovereignty and independence from China, although in the exercise of power, she revealed an adequate restraint. She reinforced the importance of the democratic system, the respect of the national identity and integrity, stressing that any form of interference will affect relations between the two sides of the Strait. The new political context confronts a more assertive foreign policy designed by Beijing, particularly in relation to the territories disputed in the South China Sea and East China Sea (Chen, 2017). PRC's president, Xi Jinping, transformed China's rule of a single political party into the rule of a single man. He tries the revitalisation of national pride, framed in the "China Dream" slogan (*Zhongguo meng*) (Xi, 2018). This context of national galvanisation developed over the hegemonic authority of the CCP.

The new political context brought a new direct influence on bilateral tourism. Since Tsai's election in 2016, the number of Chinese tour groups entering Taiwan has steadily declined. The total number of tourists visiting Taiwan over the 2000-2016 period grew from 2,62 million, in 2000, to 10.69, in 2016 (Figure 1), the fastest growth observed between advanced economies (Lin & Lee, 2018).

Figure 1. Number of Chinese tourists entering Taiwan (2000-2016)
Source: Lin and Lee, 2018

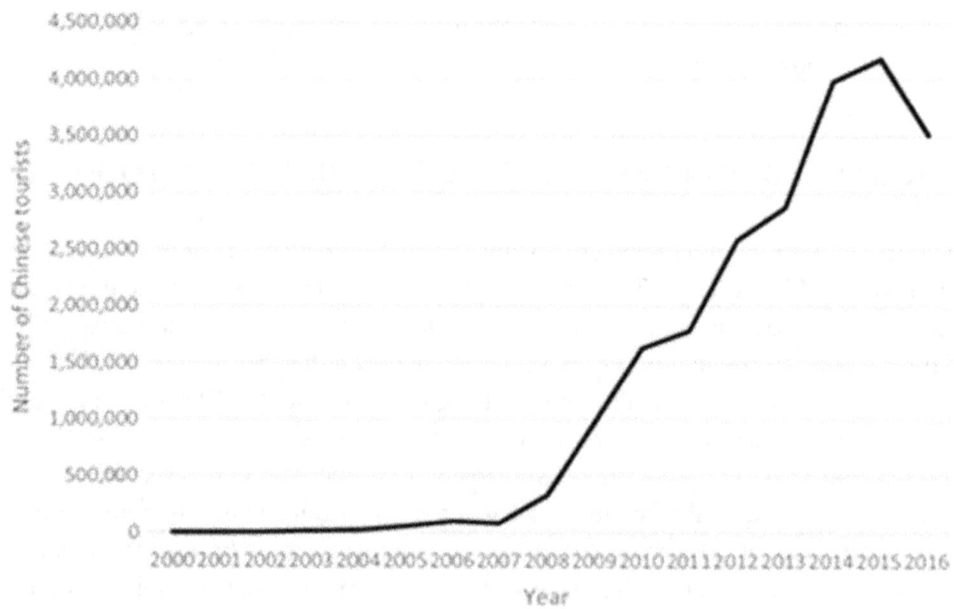

After 2016, a drop in tourism flows was registered, resulting from tension on cross-strait relations that worked as the most influential factor. China's Ministry of Culture and Tourism banned individual travel to Taiwan (implemented in 2011) as political retaliation. Tourist arrivals from China – particularly groups – promptly plunged 60% the following month. The program was directed to residents of Beijing, Shanghai and Xiamen in order to visit Taiwan. Taiwan president criticised Beijing for making tourism into a political issue: "Tourism should not be politicised" (The Diplomat, 2019). At the same time, the PRC conducted military exercises in the Taiwan Strait, carrying out an intimidation strategy on the island government. China looked at Tsai as a secessionist leader and implemented a set of coercive political measures to harm Taiwan's economy. At the same time, the PRC threatened to use force to solve the issue. According to Taiwan's Tourism Bureau, in 2009, the number of inbound visitors hit a record of 11.84 million (particularly from Japan and South Korea). There was a 7% increase from the previous year, which was the highest annual growth since 2014. Despite China's move to block its nationals from visiting Taiwan, visitors from China increased 0.5% from the previous year. In 2015, around 5 million Chinese tourists entered Taiwan and, in 2017, the number dropped to 3 million (Chinese group tours to Taiwan dropped by 30%), maintaining approximately this number per year until 2019 (Steger, 2020). The truth is that during the rule of Tsai Ing-wen, tourist flows from China to Taiwan have never achieved the pre-2016 figures. However, Taiwan tried to offset this decrease by attracting tourists from other countries. Taiwan government responded to the mainland's travel suspension by a new additional budget of US$115.8 million on tourism promotion (The Diplomat, 2019). Like in many other political conflicts that involve boycotts and economic sanctions, international players usually find other alternatives. The 2019 Taiwan elections were again a scene for new threats, also marked by the Hong Kong crisis. This context was very beneficial for Tsai, who wan the elections again. Consequently, the PRC used tourism as a throwing weapon against Taiwan, suspending travel outbound flows. It aimed to harm Taiwan's economy so that Tsai's supporters would reconsider their votes in the general election in 2019. International isolation and political and military pressure are increasing from Beijing, particularly in the era of Xi Jinping. Tourism, as a throwing weapon, is just one of the ways used by political power in a range of actions. In 2020, only 15 countries recognise Taiwan, a gradual process of political pressure from the PRC. In 2019, Taiwan was also locked out of the World Health Organization (WHO). At a time the world is struggling with Covid-19, this prevented Taiwan from helping the international community in fighting the pandemic. Beijing also banned Taiwanese films and filmmakers from attending the Golden Horse Awards that happened for the first time (South China Morning Post, 2019). It is impossible to predict the future, although the possibility of a more tragic scenario is not excluded. Until then, tourism will serve as an indicator of the political context, for the better or for the worse.

CONCLUSION

Tourism, as an agent of peace, is an area of increasing interest and importance for studies in international relations, admitting that in some cases provokes social conflicts. The attitude towards "Others" usually changes after they get to know their reality. It happened in the case of China-Taiwan conflict, when both parties eased physical barriers. After 1987, many islanders visited China for the first time and gradually improved the socioeconomic context. A major challenge resulted from the democratisation process in Taiwan (1995), immediately disapproved by the PRC. The two political parties that dominate the political system – KMT and DPP – cause mixed feelings in China. Tourism has revealed an enormous potential in this political struggle, in which the asymmetric differences between the two parties – economic, political and military capacity – are some of the mains factors that do not help to transform the conflict.

Taiwanese reject the strong presence of mainlanders in their territory; they fear to lose their identity and employment, damaging the economy. In this context, the growth of Chinese tourists in Taiwan – even though being an important source of revenue – is not helping to integrate the region. In other words, this does not follow the liberal perspective of international relations. In fact, tourism flows between China and Taiwan depend on political relations and cross-strait relations. The change of rule from Ma Ying-jeou to Tsai Ing-wen clearly shows the relationship between tourism and democratic alternations in Taiwan. During Ma Ying-jeou presidency, revenues generated by Chinese tourists in Taiwan were very relevant. Bilateral relations improved and many links were established. Tourism flows rose and, in 2015, approximately 5 million Chinese tourists visited the island. These flows contributed to changing the perception of mainlanders about Taiwan, as well as many Tawainese people changed their attitudes towards the Chinese. However, when DPP gained the elections – known for its independence ideas and refusal to accept the "1992 Consensus" – gave rise to more tense bilateral relations. The decrease of Chinese tourists after 2016, on Tsai Ing-wen's presidency, jeopardise the stability of cross-strait relations. Chinese propaganda initiated a set of hostilities towards its "renegade province" to demotivate interest in Taiwan. Chinese mainlanders who previously visited the island would be more resilient to this antagonistic information.

REFERENCES

Amad, R., & Hertzog, A. (2016). Tourism, memory and place in a globalising world. *Cultural & Tourism Studies*, *16*(3), 201–205.

Azar, E. (2003). Protracted social conflicts and second track diplomacy. In J. Davies & E. Kaufman (Eds.), *Second track/citizens's diplomacy: Concepts and techniques for conflict transformation* (pp. 15–30). Rowman & Littlefield.

Barometer, C. (2020). *The Heidelberg Institute for Conflict Research (HIIK) and Institute of Political Science at Heidelberg University*. Retrieved March 18, 2020, from: https://hiik.de/conflict-barometer/current-version/?lang=en

Becken, S., & Carmignani, F. (2016). Does tourism lead to peace? *Annals of Tourism Research*, *61*, 63–79. doi:10.1016/j.annals.2016.09.002

Blanchard, J. F., & Hickey, D. V. (2012). More than two 'sides' to every story: An introduction to new thinking about the Taiwan Issue. In J.-M. Blanchard & D. V. Hickey (Eds.), *New thinking about the Taiwan issue: Theoretical insights into its origins, dynamics, and prospects* (pp. 1–27). Routedge. doi:10.4324/9780203131497

Brown, K. (2016). *CEO, China: The rise of Xi Jinping*. London: I.B.Tauris.

Brown, K. (2017). *What does China want? China's World*. London: I.B.Tauris.

Butler, R. W., & Mao, B. (2007). Conceptual and theoretical implications of tourism between partitioned states. *Asia Pacific Journal of Tourism Research*, 25–34.

Chen, D. P. (2017). *US-China rivalry and Taiwan's mainland policy: Security, nationalism, and the 1992 consensus*. Palgrave Macmillan. doi:10.1007/978-3-319-47599-8

Chiang, M.-H. (2018). *China-Taiwan rapprochement: The political economy of cross-straits relations*. Routledge.

Chow, P. C. Y. (2008). An overview on the dilemma of 'One China' myth versus reality. In P. Chow (Ed.), *The 'One China' dilemma* (pp. 3–16). Palgrave Macmillan. doi:10.1057/9780230611931_1

Clough, N. R. (1993). *Reaching across the Taiwan Strait: People-to-people diplomacy*. Westview Press.

Cole, J. M. (2015). *Black Island: Two years of activism in Taiwan*. Scotts Valley: CreateSpace Independent Publishing Platform.

Copper, J. F. (2010). *The A to Z of Taiwan (Republic of China)*. Scarecrow Press.

D'Amore, L. J. (1988). Tourism: A vital force for peace. *Annals of Tourism Research*, *15*(2), 269–283. doi:10.1016/0160-7383(88)90087-4

Davison, W. D., & M., J. V. (1982). Foreign policy according to Freud. *Foreign Policy*, *45*, 145–157. PMID:11632029

Diamond, L., & McDonald, J. (1996). *Multi-track diplomacy: A system approach to peace*. Kumarian Press.

Farmaki, A. (2017). The tourism and peace nexus. *Tourism Management*, *59*, 528–540. doi:10.1016/j.tourman.2016.09.012

Garg, A. (2017). A study of tourist perception towards travel risk factors in tourist decision making. *Asian Journal of Tourism and Hospitality Research*, 47-57.

Garrison, J., & Philips, J. (1989). *The new diplomats: Citizens as ambassadors for peace*. Totnes: Green Books.

Hsieh, J. F. (2016). Taiwan's general elections of 2016. In W. Lee (Ed.), *Taiwan's political re-alignment and diplomatic challenges: Politics and development of contemporary China* (pp. 49–69). Palgrave Macmillan.

Kim, Y., & Crompton, J. L. (1990). Role of tourism in unifying the Two Koreas. *Annals of Tourism Research*, *17*(3), 353–366. doi:10.1016/0160-7383(90)90003-A

Lederach, J. P. (1999). *Building peace: Sustainable reconciliation in divided societies*. United States Institute of Peace Press.

Lin, C., & Lee, M. (2018). Taiwan's opening policy to Chinese tourists and Cross-Strait relations: The impacts on inbound tourism into Taiwan. *Tourism Economics*, 1–18.

Lin, G. (2019). *Taiwan's party politics and Cross-Strait Relations in evolution (2008–2018)*. Springer. doi:10.1007/978-981-13-5814-2

Lin, S. S. (2016). *Taiwan's China's dilemma: Contested identities and multiple interests in Taiwan's Cross Strait economic policy*. Stanford University Press.

Litvin, S. W. (1998). Tourism: The world's peace industry? *Journal of Travel Research*, *37*(1), 63–66. doi:10.1177/004728759803700108

MacDonald, J. (1993). Guidelines for newcomers to track two diplomacy. Occasional paper nº 2, *Institute of Multi-Track Diplomacy*, 1-10.

Mateus, A. (2017). O papel do turismo na reinvenção do crescimento económico em Portugal. Relatório encomendado pela Associação Portuguesa das Agências de Viagens e Turismo (APAVT), *Augusto Mateus e Associados*, 1-38.

Pan, H.-H. (2018). Does Cross-Strait tourism induce peace? Evidence from survey data on Chinese tourists and non-tourists. *International Relations of the Asia-Pacific, 0*, 1–33.

Silva, J. T. (2008). New trends in the Taiwan Strait conflict: The cooperative and quarrelsome framework. In L. Tomé (Ed.), *East Asia today* (pp. 321–340). Prefácio.

Silva, J. T. (2012a). Informal and nonofficial interactions in the new start of Cross Strait tRelations. In J. F. Blanchard & D. V. Hickey (Eds.), *New thinking about the Taiwan issue: Theoretical insights into its origins, dynamics, and prospects* (pp. 116–136). Routedge.

Silva, J. T. (2012b). The non-governmental channels of intermediation in the renewed context of Cross Strait relations: The Taishang's economic dynamics. In B. Guo & C. Teng (Eds.), Taiwan and rise of China (pp. 145-163). Lanham: Lexington.

Silva, J. T. (2016). Chen Shui–bian. In K. Brown (Ed.), *Dictionary of Chinese biography* (Vol. 4, pp. 62–67). Massachuttts.

South China Morning Post. (2019). *Mainland China to boycott Golden Horse awards in latest attempt to squeeze Taiwan*. Retrieved March 30, 2020, from: https://www.scmp.com/news/china/politics/article/3021731/mainland-china-boycott-asias-oscars-latest-attempt-squeeze

Steger, I. (2020). *China tried to threaten Taiwan by weaponing tourism, but it didn't work*. Retrieved March 30, 2020, from: https://qz.com/1780709/taiwan-tourism-hits-record-high-despite-china-boycott/

The Diplomat. (2019). *Taiwan slams China's decision to freeze individual travel to Taiwan*. Retrieved June 2, 2020, from: https://thediplomat.com/2019/08/taiwan-slams-chinas-decision-to-freeze-individual-travel-to-taiwan/

UNWTO. (2019). *World tourism barometer*. Retrieved March 30, 2020, from: https://www.e-unwto.org/doi/pdf/10.18111/wtobarometereng.2019.17.1.1

Wachman, A. M. (2007). *Why Taiwan? Geoestrategic rationales for China's territorial integrity*. Stanford University Press.

Webster, C., & Ivanov, S. H. (2014). Tourism as a force for political stability. In C. Wohlmuther & W. Werner (Eds.), International handbook on tourism and peace (pp. 167-180). Celovec: Drava.

Wohlmuther, C., & Wintersteiner, W. (Eds.). (2014). International handbook on tourism and peace. Celovec: Drava.

Xi, J. (2018). *Xi JinpingS: A grvernança da China*. Editora de Línguas Estrangeiras.

ENDNOTE

1 Many analysts consider that Taiwan has a particular national identity and should not be compared with Mainland China.

Chapter 9
Tourism as an Instrument of Cultural De-bordering and Economic Connectivity:
The Case of Macau

Francisco José Leandro
https://orcid.org/0000-0002-1443-5828
City University of Macau, China

ABSTRACT

Since the retrocession of Macau to the People's Republic of China (hereafter referred to as China) in 1999, the territory has strengthened interpersonal relations related to trade and commercial engagement between China and the Portuguese-speaking countries (PSC). Moreover, the tourism industry has served as a catalyst in promoting human social capital, technology transfer, and economic security. Therefore, the main research question stands as follows: to what extent has the Macau Special Administrative Region (MSAR) of the People's Republic of China delivered a sustainable contribution to the relations between China and PSC, using the tourism industry as apparatuses of cultural and economic de-bordering? The author argues in favour of a positive correlation between the role of MSAR and the tourism industry as an instrument of economic advancement. Moreover, China's Belt and Road initiative has put forward new opportunities to MSAR, based on the concept of people-to-people (P2P) exchange and the physical integration into the Greater Bay Area (GBA).

INTRODUCTION

Tourism and cultural industries are long-term promoters of human social capital and enhancers of human skills. They have been regarded as important contributors to development, employment, economic security, intercultural knowledge and social peace. Likewise, the policies associated with tourism industries are leading instruments of cultural de-bordering, as they hold the potential to contribute to multiculturalism, and as they constitute the first encouraging step towards interculturalism. In this line of reasoning, the

DOI: 10.4018/978-1-7998-5053-3.ch009

author sees education, development, economic security, and interculturalism as the foundation stones of peace transitions, democratic transitions, peacebuilding processes and conflict termination. Tourism development, as part of the Nation-building processes, can sometimes be a foundation stone of peace transitions as well.

The United Nations World Tourism Organization (UNWTO, 2020) asserts that in relation to the United Nations (UN) 2030 Agenda, tourism has the potential to contribute, directly or indirectly, to all goals. In particular, it has been included as targets in Sustainable Development Goals (SDG) 8 (decent work and economic growth), SDG 12 (responsible consumption and production), and SDG 14 (life below water). The first two goals apply directly to the case of Macau (China). "Tourism is one of the driving forces of global economic growth and currently provides for 1 in 11 jobs worldwide. By giving access to decent work opportunities in the tourism sector, society - particularly youth and women - can benefit from increased skills and professional development" (UNWTO, 2020). Indeed, in the Macau Special Administrative Region (MSAR) of the People's Republic of China (PRC), tourism, when associated with gambling, is a sector that generates more employment opportunities in the territory. In relation to responsible consumption and production, the tourism sector in MSAR has also adopted sustainability as an important reference point, in terms of planning and conducting operations. In addition, as recommended by the UN, MSAR believes that a well-trained and skilful workforce is crucial for tourism to prosper. "Tourism provides incentives to invest in education and vocational training and assist labour mobility through cross-border agreements on qualifications, standards and certifications" (UNWTO, 2020). Foremost, MSAR tourism-related policies are in line with SDG 16 and 17, as they believe that tourism promotes encounters between people of diverse cultural backgrounds, and therefore, the sector can foster multicultural and interfaith tolerance and understanding. Moreover, MSAR sees tourism as cross-sectorial, and that tourism has the ability to strengthen private/public partnerships and to engage multiple stakeholders. These two ideas (tourism as a promoter of multiculturalism and as a strengthener of business partnerships) are seen as anchors behind this chapter.

Since the retrocession of Macau to China in 1999, the territory has strengthened interpersonal relations related to trade and commercial engagement between China and the Portuguese-speaking Countries (PSC), using tourism and cultural industries as part of its strategic external relations policy. Moreover, tourism and cultural industries have catalytically promoted human social capital, technology transfer and economic security. Bearing in mind the extremely limited resources of the territory, MSAR stands as a unique case study. According to World Economic Forum, with a population density of around 20,800 per square kilometre, MSAR is the second most populated territory in the world (approximately 680,000 local inhabitants, in 2019, in an area of 32.9 sq km), as well as one of the top economies in the world in terms of Gross Domestic Product (GDP) per capita. For the United Nations World Population Prospects (2019) MSAR is placed placed among the most populated territories in the world. According to Trading Economics (2019a), MSAR received 36,231,160 tourists, which represents more than 50 times the number of local inhabitants. Official data from the World Bank and the projections from Trading Economics (2019b), mentioned that GDP in Macau was worth $58.50 billion in 2019, which represents 0.05% of the world economy, which means that the GDP per Capita in MSAR is equivalent to 460 percent of the world's average (Trade Economics, 2020). Likewise, according to the International Monetary Fund (2019) and World Bank (2019a; 2019b), MSAR ranked second in the world in terms of GDP purchasing power parity (PPP) per capita, just after Qatar. The 2020 Index of Economic Freedom, Macau score is 70.3, making its economy the 35th freest in the 2020 Index. Its overall score has decreased by 0.7 point due to a decline in the monetary freedom score. Macau is ranked 9th among

42 countries in the Asia–Pacific region, and its overall score is well above the regional and world averages. MSAR is the largest casino centre in the world and gaming tourism drives its growth. While more moderate than in the past, gaming and tourism revenue picked up as the economy returned to expansion since mid-2016 (IMF, Country Report No. 19/123). Those are substantial reasons to put forward the following main research question: to what extent has the MSAR delivered a sustainable contribution to the relations between China and Portuguese-speaking countries (PSC), using tourism as apparatuses of cultural and economic de-bordering?

MSAR AS A TOURISM PLATFORM FOR CULTURAL DE-BORDERING

Since the handover in 1999, especially after the 2001-2002 controlled-liberalisation of the gaming industry, MSAR engaged in positive cultural constructivism, as it has been exploring mutual-reinforcing paths to de-border culture, defined as "the collective programming of the mind" by (Hofstede, Hofstede, & Minkov, 2010). Since 2012, MSAR's government has been branding the political narrative of "Macau as a world-class centre of tourism and leisure", as a UNESCO Creative City of Gastronomy and, at the same time, as "a commercial platform between China and the PSC". Bearing these three components in mind, we must acknowledge that the economic security of MSAR is based on three intertwined pillars: (i) tourism, including gaming, hospitality, entertainment, and cultural industries; (ii) MSAR as a platform between China and the PSC; and (iii) the synergies within the Greater Bay Area (GBA) in terms of connectivity, services, and economic synergies.

Tourism in MSAR has two harmonising readings. On the one hand, it stands as a political initiative to attract mainland Chinese citizens to gamble, to enjoy the recreational facilities, and to participate in cultural-sports events organised by the different tourism promoters. This perspective aims to reinforce the national cultural identity, establish meaningful contact with a multicultural world, and at the same time, promote national unity through positive experiences and self-gratification, with the gaming industry at the core of the activities. Chinese tourism in MSAR stands as a decisive activity, from the economic, social-cultural, and political point of view. Indeed, it promotes Chinese national unity, performs an individual gratification role, strengthens Macau's sense of belonging or purpose, and advances domestic interculturalism.

On the other hand, tourism education, promotion planning, and activities are designed to lure key individuals and business groups from mainland China and PSC to meetings, incentives, conferences, and exhibitions (MICE). This second perspective is the driving force behind this study, which sees tourism as an instrument to bridge and bond cultural relations between China and the PSC, having in mind economic, trade, and development cooperation. These two perspectives are mutual reinforcing, and both deliver a contribution to cultural and economic intangible de-bordering. In this vein of reasoning, we understand de-bordering as referring to concurrent, voluntary, involuntary, multilevel, material, and immaterial processes of fading, weakening, trespassing, or removing the physical, cultural, legal, and economic barriers, which are preventing or impeding direct interactions between agents of two contiguous sovereign states, without compromising the exercise of sovereignty (Leandro, 2019a, p. 125). Tourism and cultural industries, when associated with MICE, generate contact opportunities, create a de-bordering mentality, curb attitudes of intolerance, shape the approach towards others, and introduce elasticity, which helps to accommodate different interests (Gelézeau, De Ceuster, & Delissen, 2015).

In MSAR, tourism industries and culture are seen as an instrument to foster intercultural sensitivity. According to the definition adopted during the 22nd Session of the General Assembly of the UNWTO, held in Chengdu, China, between 11 and 16 September 2017, cultural tourism is

A type of tourism activity in which the visitor's essential motivation is to learn, discover, experience and consume the tangible and intangible cultural attractions/products in a tourism destination. These attractions/products relate to a set of distinctive material, intellectual, spiritual and emotional features of a society that encompasses arts and architecture, historical and cultural heritage, culinary heritage, literature, music, creative industries and the living cultures with their lifestyles, value systems, beliefs and traditions.

The idea behind cultural tourism is precisely a personal and inner motivation to learn and discover tangible and intangible cultural differences. Discovering differences is the first step to develop intercultural sensitivity. Chen and Starosta (1997, p. 5) argued that intercultural sensitivity must be understood as "an individual's ability to develop emotion towards understanding and appreciating cultural differences that promotes appropriate and effective behaviour in intercultural communication". That is precisely the key argument in relation to establishing (bridging) and bonding (deepening) the commercial, trade, and development relations between China and the PSC; the ability to exercise and improve intercultural communication as an instrument to endorse multi-domain connectivity with consent.

Intercultural communication encompasses three important dimensions: cognitive, affective, and behavioural abilities. Combining the different readings (Chen, 2009; Chen & Starosta, 1996, 2000, 2003, 2005; Hammer, 1989) of the intercultural communication concept, the cognitive ability is represented by intercultural awareness (mutual knowledge), the affective ability by intercultural sensitivity (understanding and sharing feelings), and the behavioural ability by intercultural effectiveness or adroitness (practical skills with purpose). In the same line of reasoning, we must refer to the Model of Intercultural Sensitivity (DMIS) (Bennett, 1986, 1993, 2004, 2013) as a modelling-structure to describe how people experience and engage cultural differences. Bennett (1984, 1986) asserts that intercultural sensitivity is a developmental six-step process in which individuals are able to transform themselves from the "ethnocentric stage to ethnorelative stage." Indeed, it is precisely the "ethnorelativism" that nurtures empathy, which means the ability to understand another person's perspective (the cognitive concern identifies where you sense how the other person thinks about the world, which means you can put what you have to say in terms they will understand); the ability to sense what another person needs from you (the compassionate empathic concern identifies where you express the ways you care about the person by helping with what you sense they need); and the ability to feel what someone else feels (the emotional concern identifies where you instantly resonate with how the person feels) (Archer & Finger, 2018).

The World Economic Forum (2018, p. 29) identified essential skills: (i) concern for others (sensitivity, needs, and feelings), (ii) cooperation (natural attitude to contribute), (iii) social orientation (working with others), and (iv) social perceptiveness (understanding the reactions of others). Tourism and cultural industries foster "ethnorelativism", which is an outstanding contribution to concern for others, cooperation, and social perceptiveness. Ethnorelativism (and tourism and cultural industries) benefits from the fact that in MSAR there is an ancient historical-cultural heritage, which is the result of centuries of contact between Asia and the West, and there is also, as currently practised, a sort of "trilinguism" with Chinese (Mandarin and Cantonese), Portuguese, and English being spoken at private and official levels. Media, official press, business, attorneys, casinos, tour operators, small retail, transnational corporations, public services, hospitals and welfare services, public malls, and education services are available in three languages, or at least in Chinese and English.

In MSAR, tourism and cultural industries are seen as an instrument of soft power. Tourism has been developed as an entry door to seduction and enjoyment, taking advantage of immediate perceptions of a territory capable of offering innumerous immediate gratification opportunities. Studying cultural models, Hofstede, Hofstede and Minkov (2010, p. 240) argued that "indulgence stands for a tendency to allow relatively free gratification of basic and natural human desires related to enjoying life and having fun". The same authors recognise that Brazil, Portugal, and African cultures display a higher level of interest for indulgence, when compared with the restrained Chinese society. The ability of MSAR to emulate the intensified gratification appeal functions as an attraction, capable of facilitating the induction of social, cultural, political, and economic influence.

MSAR is not a sovereign state, and, therefore, cannot be perceived as such in terms of state soft power. However, the external policies of MSAR are in line with Chinese foreign affairs, especially as directed to PSC. MSAR performs a soft functional power as a platform between China and the PSC. In this line of thought, and recalling Nye (2008, p. 96), the soft power of a country "rests primarily on three resources: its culture (in places where it is attractive to others), its political values (Government policies at home and abroad are another potential source of soft power) […] and its foreign [external] policies (when they are seen as legitimate and having moral authority)". Indeed, tourism industries, when associated with meetings, incentives, conferences and exhibitions (MICE) and education opportunities, functions as a compelling window over multiculturalism, over the local population well-being, and over the political narrative of cooperation for development and mutual benefit. Moreover, MSAR displays a natural ability to attract through cultural and ideological appeal (Nye, 1999) and to contribute to Chinese economic security objectives by reaffirming Weber's (1947, p. 152) understanding of soft power as "the probability that one actor within a social relationship will be in a position to carry out his own will despite resistance".

Tourism industries, when associated with MICE and education opportunities, amplify the appeal of ideas that generate attraction, encourage seduction, promote reputation, and facilitate the establishment of personal bridges, capable of generating practical trade, business, and innovation results. McClory (2011) expanded Nye's three pillars of soft power to five and added education and business innovation. In fact, MSAR university and vocational education is tourism in reference to Asia, with particular reference to the City University of Macau (Faculty of Hospitality and Tourism Management) and the Macau Institute for Tourism Studies (IFTM). In relation to the later, the Shanghai Ranking Consultancy Global Ranking of Academic Subjects (2019) has rated IFTM among the best institutions of higher education in Asia for hospitality and tourism management studies. The institute was ranked the 5th-best in those fields in Asia and the 30th-best in the world (Macau Institute for Tourism Studies, 2019).

Cultural de-bordering is a long-standing process of developing cultural sensitivity, creating multicultural tolerance, and searching for nice intercultural perspectives. Cultural de-bordering aims at bridging and bonding interpersonal relations to promote human social capital for mutual benefit. In MSAR, tourism and cultural industries are seen as de-bordering instruments, aiming at the advancement of people-to-people (P2P) relations, supported by tourism education and MICE-business. Consequently, MSAR's role has been designed to reinforce high-level bilateral P2P (especially in light of its role as a platform between China and the PSC), which ultimate goal is to invest on promoting intercultural sensibility, and capitalises on the power of attraction as an instrument of connectivity with consent.

In MSAR, tourism industries and culture are seen as an instrument to develop P2P exchange. The idea is to deepening the existing mutual and consistent trust for pragmatic cooperation, between China and all the PSC, having in mind the potential of the GBA and the Belt and Road Initiative (BRI). As clearly stated by the National Development and Reform Commission (NDRC), P2P bonds are vital to an initiative with such scale as the BRI. Thus, the creation of P2P bonds begins by taking steps to overcoming barriers preventing the development of mutual interest, and fostering other levels of shared knowledge and understanding. P2P de-borders the cultural barriers of minds as it creates opportunities for dialogue, strengthens communication skills, and unveils cooperation projects.

Similarly, President Xi Jinping mentioned, in 2015, during the BOAO Forum for Asia that it is necessary to "promote inter-civilisational exchanges to build bridges of friendship". The P2P de-borders the cultural barriers of minds and encourages new long-lasting positive and functional bonds. Section 1 (cooperation mechanisms) of the 13th five-year plan for economic and social development of the Peoples Republic of China (2016-2020) stressed the importance of P2P inside and outside China. P2P concept is understood as the efforts made to promote exchanges and dialogues between different cultures, to form the basis for the advancement of regional cooperation. P2P economy (P2PE) aims at developing human social capital through the following ideas: (i) encouraging the establishment of human interaction at different levels; (ii) fostering new and reciprocal relationships to promote wider dialogue mechanisms; and (iii) promoting mutual benefits. Thus, P2PE complements state-to-state cooperation (S2SC) and fosters corporations-to-corporations exchange (C2CE) (Leandro, 2019b).

Initially, Putnam defined social capital (1993, p. 167) as "features of social organisation, such as trust, norms, and networks, that can improve the efficiency of society by facilitating coordinated actions". However, later the same author (2000, p. 19) developed it into the "… connections among individuals". Furthermore, Putnam (2000) distinguishes two types of human social capital: bonding (in the sense of connecting homogeneous individuals) and bridging (connections linking heterogeneous persons). Likewise, Putnam (2000, p. 19) states that the "core idea of social capital theory is that social networks have value". Bonding social capital is good for "enhancing within-group reciprocity and solidarity", and has the aptitude to implicitly use emotional intelligence[1] to establish and develop positive and meaningful relationships. Bonding social capital is, if not an instrument of multiculturalism, at least is a sort of path to multicultural coexistence, advancing new frameworks of connectivity. In this vein of reasoning, MSAR is positioning itself as the natural P2P exchange mechanism between China and the PSC, acting as a contributor and as a facilitator of Chinese national policy, and supplementing sovereign and non-sovereign vertical bilateral relations. P2P relations are the vital blood to energise long term-relationships and the decisive motivation for the very first step of a thousand miles' journey. The MSAR push for intensifying tourism and MICE activities is boosting the operation of informal mechanisms of proximity, and, therefore, giving a closer sense of hardheaded interplay among new networks of Sino-Lusophone private economic agents. MSAR has been building a Sino-Lusophone grid of exchange for mutual advancement, based on three equally harmonising P2P dimensions: (i) the frames of exchange, (ii) the instruments of exchange, and (iii) the characteristics of exchange. Tourism industries are part of all of them:

(1) Firstly, MSAR has identified a set of P2P frames of exchange to which tourism-MICE is contributing. The frames of exchange are multi-sectorial and inter-sectorial and often depart from the empiric shared interests of China and the PSC. Among these frames stand the professional and entrepreneurial sector; the education-academic and knowledge-skills transfer sector; the oppositions of overlapping social identities and emotional intelligence of key leadership; the horizontal

and vertical dimensions of the governmental and non-governmental relations sector; and the clear vision of the purpose of exchange. All of this can share a common language, which Ikenberry and Kupchan (1990) consider as an essential part of the socialisation process. Particular emphasis is placed on continuing to invest in the P2P education-academic sector, fostering another level of interplay with international standards.

(2) Secondly, MSAR has shaped P2P instruments of exchange, functioning to generate purposeful networks for permanent dialogue and interplay, which can be perceived to be of two types: institutions and policies.
   (a) Institutions: Office of the Secretary for Economy and Finance; Office of the Secretary for Social Affairs and Culture; Macau Government Tourism Office; Forum for Economic and Trade Cooperation between China and Portuguese-speaking Countries; Macau Trade and Investment Promotion Institute (IPIM); Macau Foundation; MSAR external commercial delegations; University of Macau; Polytechnic Institute of Macau; Institute for Research on Portuguese Speaking Countries of the City University of Macau; Macau Institute for Tourism Studies; with the functional collaboration of the Instituto Português do Oriente (IPOR) and the Portuguese School of Macau.
   (b) Policies: The concept of Macau as "one platform" and "three centres"; MSAR as a center for yuan settlement; the functional planning of the GBA; the establishment of the Macau Closer Economic Partnership Arrangement (CEPA); the Forum Macau cooperation fund between China and the PSC; the dynamism of the industries of MICE and rejuvenation of local creative industries; the creation of the Macau International Film Festival; the recognition of academic degrees between MSAR and Portugal; the full support of bilingual and multilingual development of local human social capital; the long-standing support of communication and media, in which the online "Lusophone Brief," the bilingual newspaper "plataforma" (澳門平台), and the TDM broadcasting and production agreements stand among the best examples; the consistent endeavor for benchmarking for financial, gaming, and tourism sectors; the promotion of sports to an international level - the "FIVB Volleyball Women's Nations League,", "Grand Prix Macau", "Macau International Dragon Boat Races" and the "Macau Marathon"; and the high number of events explicitly conceived to support China-PSC cultural relationship, namely: Procession of our Lady of Fátima, Script Road Festival, Lusophony Festival, International Parade, International Fireworks Display Contest, Macau Food Festival, Macau International Light Festival, Macau International Food Festival, Macau International Music Festival, International Youth Dance Festival, Macau International Travel Industry EXPO, Macau Lotus Flower Festival, International Trade & Investment Fair, Arts and Cultural Festival between China and the PSC, and PSC Products and Services Exhibition.

(3) Thirdly, bearing in mind the previous P2P dimensions, it is possible to identify the dominant characteristics of these processes of human exchange. Therefore, MSAR has:
   a) Clearly identified the main goal of P2P exchange as an instrument of mutual advancement, through pragmatic and nonpartisan cooperation for entrepreneurship, innovation, and growth;
   b) Promoted P2P to advance China-PSC cooperation in tourism, banking services, and education;
   c) Presented a protracted, consensual, inclusive, non-confrontational, and progressive P2P exchange involving ad hoc groups, with bilateral, trilateral and multilateral involvement;
   d) Built a P2P events-driven exchange directed to the enhancement of human social capital;

e) Promoted a transformational P2P exchange, based on non-politicized and secular collaboration, seeking tangible rewards and mutual gratification;
f) Implemented P2P practices, which raised the level of multiculturalism and inclusion, as the path for interculturalism.

MSAR AS A TOURISM PLATFORM FOR CONNECTIVITY AND ECONOMIC DE-BORDERING

When associated with entrepreneurship, tourism offers visitors a new perspective. The culture and economic perspective connect in a world of prospects, immediately available within the physical space of the GBA, where opportunities are looming: infrastructure, tourism, banking services, employment, digitalisation and technology, functional integration, and economies of scale.

The World Travel & Tourism Council (WTTC, 2020) indicates that, in 2019, the tourism sector worldwide directly contributed $2.9 trillion to the global economy, and $8.9 trillion to the global economy when taking wider indirect and induced impacts into account (10.3% of the world's GDP). The WTTC released, in 2019, a City Travel & Tourism Impact Report, which analysed and ranked 72 cities around the world, quantifying the economic contribution of tourism activities and related economic impact based upon several major indicators. MSAR ranked second regarding its tourism sector contribution to GDP growth and international visitor spending (WTTC, 2019a). The same report indicates that, in MSAR, one in five new jobs were created by travel and tourism over the previous five years. While the global economy grew by 3.2%, travel and tourism grew by 3.9%. In addition, the same report mentions that travel and tourism's contribution to GDP represented 72.2% of the total economy of MSAR (+12.7% since 2018). In the same line of thought, travel and tourism contributed to MSAR's employment with 194,400 jobs, which represents 50.3% of total employment and in 2029 is expected to contribute with 212,400 jobs.

In terms of infrastructural development, MSAR is located in an area equipped with state-of-the-art infrastructure: high-speed railways (Zhuhai and Hong Kong); seven airports (Zhuhai, Macau, Hong-Kong, Shenzhen, Huizhou, Guangzhou, and Foshan (planned); seven deepwater seaports (Jiangmen, Zhuhai, Hong-Kong, Shenzhen, Huizhou, Guangzhou, and Dongguan); and one pilot-free trade zone that consists of three areas (Hengqin, Nansha, and Qianhai). The two Special Administrative Regions are linked by the Hong Kong-Zhuhai-Macau Bridge, and they also benefit from the existence of the two Humen bridges, the Shenzhen-Zhongshan bridge, and the Guangzhou-Shenzhen-Hong Kong Express Rail Link. Macau Airport has non-stop passenger flights scheduled to 49 destinations in 11 countries, and 28 airlines are flying to and from Macau International airport (MIA). The land connections between MIA and Hong Kong International Airport (HKIA) operate almost 24 hours a day, by jetfoil or shuttle bus.

According to the Macau Government Tourism Office (2019a), the territory has 126 hotels, 41,148 rooms with an annual occupancy of 90.8%, 226 travel agencies, 1,972 registered tour guides, 478 restaurants, 152 bars, 77 spa, 39 karaoke establishments, one jockey club (MJC), six large entertainment facilities (with a special reference to the Cotai Arena, which is an ideal multi-purpose indoor area for any function), 26 swimming pools, 18 museums and memorials, 41 casinos, 6,739 gaming tables, and 17,009 slot machines.

In addition, the tourism office has organised 1,536 MICE activities involving 2,003,416 participants. Likewise, according to the Macau Government Tourism Office (2019b), the average length of stay of visitors was 1.2 days, and overnight visitors stayed 2.2 days. In 2019, visitor arrivals totalled 39.4 million, with a 10.1% growth comparing to 2018. Among those, overnight stay visitors increased by 0.8% to 18.63 million, while same-day visitors grew 20.0%, which led to the proportion of overnight stay visitors dropping by 4.4% from the previous year to 47.3%. In relation to visitors' arrivals by mode of transport, the majority of visitors arrived by land(29.3 million), while arrivals by sea fell by 39.5% (6.3 million), accounting for 15.9% of total visitors. Arrivals by air increased by 16.7% to 3.8 million, accounting for 9.8% of the total visitors. The major entry ports were Border Gate (53.1%), Hong Kong-Zhuhai-Macau Bridge (13.7%), and Macau International Airport (9.7%).

Figure 1. MSAR's contribution of travel and tourism to GDP (% of GDP)
Source: World Travel & Tourism Council

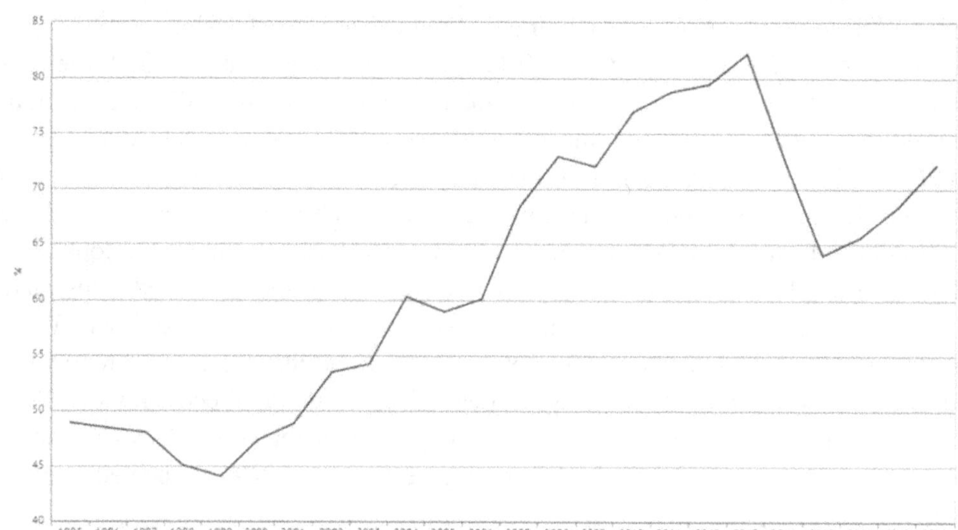

According to the Statistics and Census Service (DSEC) (2019), the MSAR unemployment rate was 1.7%. As of March 31, 2018, some 380,500 people were employed in the city, of which 280,100 were MSAR residents. According to Macau News (U, 2019), the number of workers engaged in activities related to gaming (83,000) and hotels, restaurants, and similar activities (56,000) increased by 35,000 and 1,800 quarter-to-quarter, respectively. According to the same source (2018), the majority of hotel revenue came from 5-star hotels (3.2 billion euros), accounting for 78.0% of total hotel revenue, 18.6% more than in 2017. Revenues from 4-star hotels were 4.39 billion MOP (492 million euros) and revenues from 3- and 2-star hotels were 3.78 billion MOP (424 million euros). In 2018, hotel expenses increased by 10% to 31.78 billion MOP (3.5 billion euros). According to data provided by DSEC, the 116 hotels and pensions (2018) employed 52,976 people, an increase of 3.7% over the previous year.

When we associate tourism and MICE for the purpose of MSAR as a platform between China and the PSC, then banking services also becomes relevant. In fact, since 2015, the banking sector has been advanced by the fact that Macau has turned out to be a renminbi clearing centre as well as a financial services platform. The BOC Macau branch has established agreements with four PSC (Portugal, Angola, Brazil, and Mozambique) and eight PSC banks to provide financial services in renminbi. Since 2018, the

Banco Comercial Português (BCP) and the *Caixa Agrícola* have offered financial services in renminbi via Macau. These financial services in *renminbi* (RMB) include transfer of money to mainland China. In May 2019, during the Lusophone Banking Association meeting held in Macau, which brought together Portugal, Cabo-Verde, and São Tomé and Principe representations, Portugal announced the very first emission of "Portuguese bonds" in RMB.

Most expat jobs in Macau are related to the banking or tourism industry. Therefore, according to DSEC (2019), the average salary in Macau as of 2018 was MOP 20,000 (approximately USD 2,437). The average monthly earnings of the total employed population in the MSAR – including temporary and permanent residents and working permit holders – is MOP16,000. In particular, those workers engaging in gaming and junket activities received an average salary of MOP 20,000, while those people employed in construction received an average salary of MOP 15,000. Nevertheless, Macau has a considerably lower average salary than Hong Kong, but it is still much higher than in mainland China. Tourism and banking sectors are expected to benefit from the current process of economic digitalisation. In 2019, the *Companhia de Telecomunicações de Macau* (CTM) announced the commercial 5G mobile network, which is expected to have the whole of Macau covered by the end of 2020. The digitalisation of the economy is expected to promote business connectivity, to explore the opportunities offered by e-commerce and m-commerce, and to facilitate e-payments within the territory and overseas. Digitalisation facilitates economies of scale and the integration of the supply chain, in the domestic and external contexts. Digitalisation, as an innovation modality triggers new dimensions of the production-client relationship, helps to control waste and inefficiencies, curbs production costs, increase market competitiveness, deals with customer fluctuations, and facilitates the new market-personalisation e-demands. In the same tourism technologic approach, Hong Kong PolyU and Companhia de Telecomunicações de Macau revealed plans to jointly conduct a spatial big data analytics study on the needs and preferences of tourists in Macau, leveraging the research expertise of Hong Kong PolyU's Laboratory for Smart City and Spatial Big Data Analytics (Hong Kong PolyU, 2019). The project aims to optimise the Wi-Fi facilities in Macau, and develop data-mining algorithm/tools to analyse the tourist activities and behavioural patterns in the city. These will not only help improve Macau's urban planning and tourism management as well as enhance tourists' experience, but also serve as a testbed for the development of smart cities in the Greater Bay Area.

Tourism and banking sectors are expected to benefit from the current process of regional integration into the GBA. During the GBA tourism economy summit held in 2019, Chen Yin, who is Chairman of Global Commercial Newspapers Union and President of Hong Kong Commercial Daily, stated that "with the support of the State, the GBA will further improve its cultural soft strength, connect with world-leading tourism cities and promote exchanges between Eastern and Western cultures [...] The GBA will strive to become a world tourism destination through cross-region tourism within the GBA". The GBA is focusing on tourism development with an integrated approach for well-being and with the purpose of creating synergies. The Establishment of the Association for the Promotion of Cultural and Creative Industries in GBA is helping to transform the GBA into a world-class tourism destination, based on cross-boundary protection of cultural heritage sites, multi-destination solutions, convenient transportation, and diversification of recreational options, namely: organised cruises, yacht piers, tailor-made tours, cultural sites, gastronomy, landscape, traditional ceremonies, religious heritage, international events, entertainment parks, music, opera, street shows, art galleries, international exhibitions, and the promotion of the Lingnan culture.

According to Talking Data, the GBA accounted for 13% of the total economy, with less than 5% of the total population nationwide (Deloitte China, 2019). "One Rail Link, One Bridge" has greatly promoted tourism in GBA; the Hong Kong section of Guangzhou-Shenzhen-Hong Kong Express Rail Link and the Hong Kong-Zhuhai-Macau Bridge have been put into service. The GBA has become one of China's top three airport clusters, with most cross-border connecting flights highly concentrated in Hong Kong International Airport. Airfreight traffic across the Greater Bay Area is higher than the combined figures of the bay area of San Francisco, New York, and Tokyo. Three of the world's top 10 container ports are also located there. In addition, online tourism service platforms become the main actors in cross-border tourism services: 9 entry points implement 144-hour visa facilitation and provide policy facilitation.

The GBA comprises Hong Kong, Macau, as well as the Chinese municipalities of Guangzhou, Shenzhen, Zhuhai, Foshan, Huizhou, Dongguan, Zhongshan, Jiangmen, and Zhaoqing in Guangdong Province. Nine cities in Guangdong province and the Special Administrative Regions of Hong Kong and Macau had a combined population of about 70 million in 2019, which makes up 5% of the total population in China inside the GBA. The three largest cities by population are Guangzhou (more than 14.5 million inhabitants); Shenzhen (approximately 12.5 million inhabitants); and Dongguan (about 8.3 million inhabitants).

According to the latest economic policy analysis (Chan & Rosario, 2017), Guangzhou is strengthening its functions as an international commerce and industry centre and integrated transport hub, and enhancing its role as a technological, educational, and cultural centre.

Shenzhen is transforming into a modern international city, striving to become a place of innovation and creativity with global influence. It is currently home to more than 3 million businesses, including such tech giants as Tencent, DJI, and Huawei. According to the Greater Bay Area Outline Development Plan, Shenzhen should continue its leading role as a capital of innovation and creativity.

Dongguan will develop as a hub for the service industry. It is a centre for manufacturing and home to about 9,000 enterprises backed by foreign investment, according to the Hong Kong Trade and Development Council. The city has five pillar industries, including electronic information, electrical machinery, textiles, food processing, and paper production.

Huizhou is a major producer of petrochemical, electrical appliances, smartphones, and lithium battery cells. The city is a popular tourist destination, with scenic mountains, long beaches, and around 100 islands and reefs that attract millions of visitors. The Luofu Mountain next to the Dongjiang River is famous for its Taoist culture.

Zhuhai is the regional leader in materials used by the printing industry and home to large manufacturers of yachts, clean energy vehicles, precision machinery, air conditioners, and electrical appliances. The city has become an educational and cultural centre with more than a dozen universities operating in the area. It also hosts the landmark Zhuhai Opera House, the Zhuhai Museum, the Urban Planning Exhibition Hall, and the Xiangzhou Cultural Centre.

Foshan is known for its electronics, information technology, new materials, and optics. It has developed the Nanhai high-tech zone and the Sanshui Leping Park, which has attracted major foreign companies, such as Volkswagen, Siemens, Osram, BASF, Fuji, Honda, and Toshiba. The city is a popular tourist destination due to the fact that the historic centre has a museum dedicated to Yip Man, the mentor of kung fu superstar Bruce Lee.

Zhongshan is foremost the hometown of Sun Yat-sen, the first president of PRC. The city has been developing as a world-class modern machinery manufacturing base, the home of a comprehensive transport network serving the western side of the Pearl River Delta, a centre for regional R&D for innovation, and an industrial base for lighting products, electrical appliances, textiles, electronics and communications, medical equipment, pharmaceuticals, steel, shipbuilding, information technology, LCD, and petrochemicals. The city is developing tourism in the following areas: thermal, well-being, and medical.

Jiangmen, or the gate river city, is developing a strong biotechnology sector, electronic information, medical equipment, computers, and office equipment in addition to such traditional sectors as furniture, garments, foodstuffs, motorcycles, and metallic products. Several international corporations are installed in the region such as Mitsubishi heavy industries, Panasonic, Hyundai, Swiss ABB, BP, Tesco, Emerson, and Mondelez. The famous "hanging-houses" were classified as a UNESCO heritage site in 2007.

Zhaoqing's economic sector is driven by electronics and information technology, machinery, oil refining, construction materials, papermaking, pharmaceuticals, and auto parts. It was in Zhaoqing where Portugal and China signed in 1582 the agreement (land lease) to set up a trading post in Macau, and it is home to the first nature reserve in China, which is a designated UNESCO natural biosphere reserve – the Dinghu Mountain National Nature Reserve. It was established in 1956 to protect evergreen monsoon rain forests of the south subtropical zone.

Hong Kong is consolidating its status as an international financial, transportation, and trade centre as well as an international aviation hub. It is engaged in a process to develop innovation and technology industries and nurture other emerging industries. Moreover, it has been established as the centre for international legal and dispute resolution services in the Asia-Pacific region. Macau is developing into a world-class tourism and leisure centre and a commerce and trade cooperation service platform between China and Lusophone countries, as well as into a base for exchange and cooperation where Chinese culture is the mainstream and diverse cultures coexist. Hong Kong and Macau are close neighbours, and both are international tourist destinations, and therefore there are synergies to explore. Hong Kong organises various types of mega-events to enrich the travel experiences of visitors and enhance the city's image as the Events Capital of Asia. Hong Kong is equipped with one of the best cruise terminals in the region (Kai Tak terminal), is the home of two major theme parks – Ocean Park and the Hong Kong Disneyland Resort – both attractive tourism destinations. In addition, Hong Kong boasts such major infrastructure projects as the Hong Kong-Zhuhai-Macau Bridge and the Guangzhou-Shenzhen-Hong Kong Express Rail Link, as well as the third runway of Hong Kong International Airport, which is currently under construction.

In 2017, China's trade with Lusophone countries was $117.6 billion, up 29.4% year to year. Brazil was the largest market, accounting for $87.5 billion of China's trade with the Lusophone world. It was followed by Angola, which made up $22.3 billion of trade with China, according to data released by the Chinese General Administration of Customs and compiled by the Monetary Authority of Macau (AMCM) (Sin, 2018). However, according to PortSEurope (2020), the value of trade between China and the Portuguese-speaking countries grew 1.55% year-to-year in 2019 to $149.639 billion, according to official Chinese figures released by Forum Macau (2020). In 2019, China sold goods to the eight Portuguese-speaking countries worth $44.064 billion (+5.3% on the year), and those same countries imported goods worth $105.574 billion.

CONCLUSION

Macau is perceived as a multicultural and business contact point, bridging human relations across continents. This is the reason why tourism and cultural industries is a vital economic sector to MSAR as a platform between China and the PSC. Tourism helps to reinforce the natural ability of MSAR as a platform for human relations, contributing to individual social capital by bonding and bridging. Since the retrocession of MSAR, the government has continuously invested in developing tourism-related areas, aiming to attract Chinese mainlanders and overseas tourism markets and to obtain synergies between tourism, education, banking services, and MICE. Tourism strongly contributes to prosperity, employment, technological advancement, social stability, economic security, and to a sense of "community bridging and bonding" between China and the PSC. Thus, to what extent has the Macau Special Administrative Region of the People's Republic of China delivered a sustainable contribution to the relations between China and PSC, using tourism industries as apparatuses of cultural and economic de-bordering?

The association between cultural tourism, MICE, and education activities has contributed to the instruments of human exchange, the frames of human exchange, and the formation of the dominant characteristics of the processes of human exchange. Cultural tourism, MICE, and education are strengthening the knowledge we need to turn concrete ideas into projects to promote mutual development. Consequently, it has strongly contributed to increasing the number of contact opportunities, to developing intercultural sensitivity, to enhancing mutual cultural understanding, and to establish a network of contacts between economic agents. Tourism has conveyed a positive image of MSAR, branding the territory as a "bridge" between China and the PSC as well as a "link" between East and West, therefore inducing social, cultural, economic, and political influence. MSAR's power of attraction is based on its perception as a territory capable of delivering responsible and sustainable gratification.

MSAR is creating synergies between tourism and cultural industries and banking services, technology, multilingualism, and education. MSAR has built an education system in the area of tourism, capable of providing a skilled labour force as well as exporting internationally recognised skilled workers. The existence of MSAR as a renminbi clearing centre and the Lusophone policy's association with tourism-education and MICE has helped Macau to differentiate itself from other Chinese economic hubs, namely Hong Kong. Indeed, the commercial relations between China and PSC are currently enjoying an excellent moment, with more and more opportunities looming. The MSAR integration into the GBA is creating synergies, which the tourism and cultural industries can easily explore, taking advantage of the high level of connectivity and regional integration as well of the infrastructure available for tourism, MICE, well-being services, and cultural events.

As Steve Orlins, the President of the National Committee on United States-China Relations (NCUSCR), asserted, "if the P2P relationship is weak, then there is no foundation that the G2G relations (government to government) can sit on". MSAR is investing decisively in the tourism and cultural industries as well as selected paired areas, and it has already delivered an excellent practical contribution to the national policy towards the relationship between China and the PSC. Tourism in Macau is seen as an economic sector, to produce economic benefits and promote cultural de-bordering, with particular attention to the PSC. Macau also sees cultural de-bordering (understood in Robert Putnam sense) as an economic instrument to promote connectivity and meaningful relationships. Tourism is and will continue to be a vital industry to bonding P2P relations, to contributing to cultural de-bordering, and to inducing growing flows of connectivity for cooperation and mutual benefit. Cultural de-bordering is the first step toward international cooperation, as tourism and cultural industries have been contributing and will contribute even more for this important purpose. Particularly, the future is promising for all tourism operators and the commercial relationship between China and the PSC. Without cultural de-bordering, the future will merely be a continuation of the past.

REFERENCES

Archer, D., & Finger, K. (2018). *Walking in another's virtual shoes: Do 360-degree video news stories generate empathy in viewers?* Retrieved on 26 March 2020, from: https://www.semanticscholar.org/paper/Walking-in-Another's-Virtual-Shoes:-Do-360-Degree-Archer-Finger/c67c231eecd6b476e8e4e3e3031d-ae24a9de0e05/figure/0/0

Bennett, M. J. (1984). *Towards ethnorelativism: A developmental model of intercultural sensitivity.* Paper presented at the Annual Conference of the Council on International Exchange, Minneapolis, MN.

Bennett, M. J. (1986). A developmental approach to training for intercultural sensitivity. *International Journal of Intercultural Relations, 10*(2), 179–195. doi:10.1016/0147-1767(86)90005-2

Bennett, M. J. (1993). Towards ethnorelativism: A developmental model of intercultural sensitivity. In M. Paige (Ed.), Education for the intercultural experience. Yarmouth: Intercultural Press.

Bennett, M. J. (2004). Becoming interculturally competent. In J. Wurzel (Ed.), *Toward multiculturalism: A reader in multicultural education* (2nd ed., pp. 62–77). Intercultural Resource.

Bennett, M. J. (2013). *Basic concepts of intercultural communication: Paradigms, principles, & practices.* Intercultural Press.

Chan, T., & Rosario, L. (2017). *Pearl River Delta: From the world factory to global innovator.* Instituto Internacional de Macau.

Chen, G. M. (2009). *The impact of intercultural sensitivity on ethnocentrism and intercultural communication apprehension.* Intercultural Communication Studies XIX, University of Rhode Island. Retrieved on March 25 2020, from: https://web.uri.edu/iaics/files/01Guo-MingChen.pdf

Chen, G. M., & Starosta, W. J. (1996). Intercultural communication competence: A synthesis. *Communication Yearbook, 19*, 353–384.

Chen, G. M., & Starosta, W. J. (1997). A review of the concept of intercultural sensitivity. *Human Communication, 1*, 1–16.

Chen, G. M., & Starosta, W. J. (2000). The development and validation of the intercultural sensitivity scale. *Human Communication, 3*, 1–15.

Chen, G. M., & Starosta, W. J. (2003). A review of the concept of intercultural awareness. In L. A. Samovar & R. E. Porter (Eds.), *Intercultural communication: A reader* (pp. 344–353). Wadsworth.

Chen, G. M., & Starosta, W. J. (2005). *Foundations of intercultural communication.* University Press of America.

China, D. (2019). *Decoding the outline development plan for Guangdong-Hong Kong-Macao Greater Bay Area (GBA).* Retrieved on March 27 2020, from: https://www2.deloitte.com/content/dam/Deloitte/cn/Documents/international-business-support/deloitte-cn-csg-decoding-greater-bay-area-plan-en-190415.pdf

Colman, A. (2008). *A dictionary of psychology* (3rd ed.). Oxford University Press.

Economics, T. (2019a). *Macau tourist arrivals*. Retrieved on March 23 2020, from https://tradingeconomics.com/macau/tourist-arrivals

Economics, T. (2019b). *Macau GDP*. Retrieved on March 23 2020, from https://tradingeconomics.com/macau/gdp

Gelézeau, V., De Ceuster, K., & Delissen, A. (2015). *De-bordering Korea: Tangible and intangible legacies of the sunshine policy*. Routledge.

Hammer, M. R. (1989). Intercultural communication competence. In M. K. Asante & W. B. Gudykunst (Eds.), *Handbook of international and intercultural communication* (pp. 247–260). Sage.

Hofstede, G. J., Hofstede, G., & Minkov, M. (2010). *Cultures and organisations: The software of the mind* (3rd ed.). McGraw-Hill.

Hong Kong Poly, U. (2019). *Turning Macau into a smart city for tourism*. Retrieved on March 24 2020, from: https://www.polyu.edu.hk/cpa/excel/en/201905/connections/c2/index.html

Ikenberry, G., & Kupchan, C. (1990). Socialisation of hegemonic power. *International Organization, 44*(3), 283–315. doi:10.1017/S002081830003530X

International Monetary Fund. (2019). *World Economic Outlook: GDP per capita*. Retrieved on March 24 2020, from: https://www.imf.org/external/datamapper/

Leandro, F. J. (2019a). The mesmerising journey from Gyeongju1 to Lisbon: The BRI as a mechanism of de-bordering, re-bordering, and co-bordering. *Baltic Journal of European Studies, 9*(2).

Leandro, F. J. (2019b). Macau Forum for three-continental dialogue: Perspectives and challenges for cooperation between Asia, Latin America and Central-Southeastern Europe. In S. S. Pajović, M. Andrijević, & S. C. Prado (Eds.), Europa Sudoriental, América Latina: Una mirada hacia la viabilidad de la cooperación triangular (pp. 39-40). Belgrade: Universidad Megatrend and International Federation of Study of Latin America and Caribbean (FIEALC).

Macau Government Tourism Office. (2019a). *Hotel's occupancy rate and average room rate*. Retrieved on March 24 2020, from: https://dataplus.macaotourism.gov.mo

Macau Government Tourism Office. (2019b). *Global indicators*. Retrieved on March 24 2020, from: https://dataplus.macaotourism.gov.mo/Publication/Report?lang=E

Macau Institute for Tourism Studies. (2019). *IFT in top 5 for Asian tertiary institutions teaching hospitality*. Retrieved on March 26 2020, from: https://www2.ift.edu.mo/NewsPortal/ift-in-top-5-for-asian-tertiary-institutions-teaching-hospitality/

McClory, J. (2011). *The new persuaders II: A 2011 Global ranking of soft power*. Institute for Government.

Nye, J. (1999). Redefining the national interest. *Foreign Affairs, 78*(4), 22–35. doi:10.2307/20049361

Nye, J. Jr. (2008). Public diplomacy and soft power. *The Annals of the American Academy of Political and Social Science, 616*(1), 94–109. doi:10.1177/0002716207311699

PortSEurope. (2020). *Trade between China and Portuguese-speaking countries grew by 1.55% in 2019*. Retrieved on March 24 2020, from: https://www.portseurope.com/trade-between-china-and-portuguese-speaking-countries-grew-by-1-55-in-2019/

Putnam, R. (1993). *Making Democracy Work: Civic Traditions in Modern Italy*. Princeton University Press.

Putnam, R. (2000). *Bowling Alone: The Collapse and Revival of American Community*. Simon and Schuster.

Sin, N. (2018). *Macau turns to a hub*. Retrieved on March 24 2020, from: https://www.euromoney.com/article/b17fnrfk46ln5m/macau-turns-to-a-hub

Statistics and Census Service. (2019). *Statistics*. Retrieved on March 24 2020, from: https://www.dsec.gov.mo/

U, C. (2019). *Macau | Average salary of local employees at MOP20,000 in Q1 2018*. Retrieved on March 24 2020, from: https://www.macaubusiness.com/macau-average-salary-local-employees-mop20000-q1-2018/

United Nations World Tourism Organisation. (2020). *Tourism in the 2030 Agenda*. Retrieved on March 23 2020, from https://www.unwto.org/tourism-in-2030-agenda

Weber, M. (1947). *The theory of social and economic organisation*. Free Press.

World Bank. (2019a). *International comparison program database*. Retrieved on March 24 2020, from https://www.worldbank.org/en/programs/icp

World Bank. (2019b). *GDP per capita, PPP (current international $)*. Retrieved on March 24 2020, from https://data.worldbank.org/indicator/NY.GDP.PCAP.PP.CD?locations=MO&view=chart

World Economic Forum. (2018). *The future of jobs 2018*. Retrieved on March 24 2020, from http://reports.weforum.org/future-of-jobs-2018/

World Travel & Tourism Council. (2019a). *City Travel & Tourism Impact Report*. Retrieved on March 24 2020, from: https://wttc.org/Research/Economic-Impact/Cities/

World Travel & Tourism Council. (2019b). *Macau: Country data*. Retrieved on March 24 2020, from: https://www.wttc.org/economic-impact/country-analysis/country-data/

World Travel & Tourism Council. (2020). *Economic impact reports*. Retrieved on March 24 2020, from: https://wttc.org/Research/Economic-Impact

ENDNOTE

1 According to Colman (2008), emotional intelligence is the capability of individuals to recognize their own emotions and those of others, discern between different feelings and label them appropriately, use emotional information to guide thinking and behavior, and manage and/or adjust emotions to adapt to environments or achieve one's goal(s).

Chapter 10
EU Tourism Contributing to Peace and Unity?
A Policy Rather Left for the Market!

Dina Sebastião
https://orcid.org/0000-0001-8988-5893
University of Coimbra, Portugal

Alina Stoica
University of Oradea, Romania

ABSTRACT

This chapter analyses the EU tourism policy regarding tourism sensitive to peace. It relies on the medieval, modern, and contemporary enlightenment philosophy of building a lasting peace in Europe, which were foundational ideas of European integration and keep being a reflex in its current values. Although the EU has been witnessing the longest period of this territory with peace, it is not taken for granted, and Euroscepticism and nationalism have been growing in Europe. The chapter assesses the conception of tourism as an intervening policy for the EU to contain nationalism, intolerance, and state conflict in Europe, using the theoretical framework of tourism sensitive to peace. It is concluded that Europe lacks an immaterial vision for tourism, as it is confined to the market functionality.

INTRODUCTION

Despite the post-war ambitions to create a project of European unity laid down in education and culture, that could be the pillars of European political construction, European integration started by the market. The aspiration of such a project is rooted in the European political philosophy contributions for the idea of European unity. In the light of the historical condition of war in Europe, a peaceful international order in the continent would not be achieved through instruments of international law and the balance of powers, philosophers argued, but rather through a political unity of the European states, a political system which should be based on a constitutional initiative, implementing a code of common education,

DOI: 10.4018/978-1-7998-5053-3.ch010

values and rights for the nationals of the member states. Such a system could rely on a federal base. Thus, peace, meaning not only the absence of material war between states but also the share of a common political value between all citizens (Galtung, 1964), grounded on the Republican principles of dignity, liberty and equality, is at the heart of the idea of political unity in Europe (Gruner, 1998a, 1998b). Nevertheless, the European integration did not start by a federalising constitutional perspective, but by the market integration. Displaying features of supranationality, the European Economic Community (EEC) was built as a common shared sovereignty between member states which preserved a large parcel of their own sovereignty. This primordial organisation evolved to a deeper form of integration, culminating in the creation of the European Union (EU), which has been reaching other areas of competence, with almost all the public policies within the reach of Brussels, although under different categories of power. Tourism policy came to be recognised for the first time by the Treaty of Lisbon as a scope where the EU can have complementary competence, thus helping member states to shape their national policy. This opens a great potential for the EU to use an economic activity as a means of educational propose, and specifically to use economic importance of tourism in the single market as a vehicle for promoting the value of peace. This normative orientation for European economic policies regains pertinence in the last decade, when the EU has been crossing different crises (the financial and economic crisis, the security crisis with terrorist attacks, the refugee crisis, the political crisis with Brexit), which have been resulting in the growth of nationalism, xenophobia and Euroscepticism, evident in the consecutive growth of populist and far-right parties. Peace is not thus taken for granted since a sense of national belonging and political mindset remains in the European electorate.

So the general aim of this chapter is to assess if the EU has been taking advantage of the market integration to implement and develop a strategy of a tourism policy "sensitive to peace". This research adds novelty to the European integration studies, where the trend of policy analysis in general, and tourism in particular, has been mostly centred in tracking the evolution of competences between the EU and the member states, and the policy-making in the EU within conceptual frameworks of multilevel-governance and Europeanisation (Bache, George & Bulmer, 2011; Dionysopoulou, 2012; Ertamay &Akbaba, 2019; Estol, Camilleri & Font, 2018; Graziano & Vink, 2008; Ladrech, 2010; Wallace, Polack & Young, 2010). Considering studies in the Tourism and International Relations fields about Europe, this chapter also adds novelty. Since most studies have been conducted from a national perspective of the nexus tourism and peace (Cousevi, 2010; Webster, & Ivanov, 2013; Wohlmuther, Wintersteiner & Wagner, 2013), this one relies on foreseeing a common European strategy under a normative orientation for peace.

To operationalise a normative approach to the EU policy analysis, interdisciplinarity is central in the research. It is assumed that the promotion of peace is a primordial value of European integration and is also instrumental for its future, since it can help to build up a common identity, enhance European citizenship and thus strengthen the EU polity. Such normative orientation is the background for the research question of this chapter: does tourism policy of the EU incorporates a strategy for a "tourism sensitive to peace"? As said, answering this question demands an interdisciplinary study, ranging from the political philosophy, and the historical perspective of European integration, passing through theoretical reviews about the role of tourism in international relations, and culminating in policy analysis.

Hence, the rationale of the chapter is as follows: In the first part, the historical motivation of peace for political unity in Europe is problematised *vis-à-vis* the real nature of European integration. An overview of the political philosophy and thought in Europe, mainly focusing the contemporary age, is made, in comparison with the political institutionalisation of European integration. The existence of an asymmetry between values and real competences opens a way to the normative orientation accomplishment, since market integration can be instrumental to the promotion of values. That is where the second part of the chapter fits, dedicated to a theoretical review of the role of tourism to the promotion of peace, stressing the concept "tourism sensitive to peace". Approaching some studies in the international relations field, it is argued that tourism is not an automatic driver of peace, although it can foster it when it is embedded in a set of ethical, moral and political guidelines and values. This conclusion backs the argument of policy-planning for tourism, and thus the pertinence of analysis of the EU tourism policy in this regard. This assumption supports the empirical analysis of our research, detailed in the third part of the chapter. It is based on a set of primary sources, composed by European Commission's communications, complemented by decisions of the Council of the European Union (hereinafter Council) and European Parliament (EP) resolutions.

As the tourism policy is a complementary competence of the EU, it is mainly subject to the soft instruments of intergovernmental coordination, where the Commission and the Council play the main roles, launching guidelines and fostering cooperation with and between member states. As a complementary competence, being not subject to legally binding acts from the EU, EP plays a marginal role here, since it is not exerting any co-decision power, reducing its action to the soft influence of resolutions. So, the main sources of analysis comprise the main instruments of EU governance in tourism policy. The timeframe of data analysis dates from 1982 to 2014, the years of the first and latest Commission communications on tourism. Although tourism is only given autonomy as a single policy by the Treaty of Lisbon, it is subject to EU action from earlier, within the internal market dynamics. Qualitative content analysis intends to assess the main goals set for tourism in the EU, and it is specifically focused on searching for segments of text formulating an immaterial vision for tourism, reflecting specifically the promotion of values of democracy, freedom, equality, common European identity, and historical educational perspective of war in Europe, that could underpin a common code for European national governments and tourism economic players. The result of this data analysis will be problematised in regard to the normative orientation of this study, taking conclusions in a prospective perspective for EU integration.

EUROPEAN INTEGRATION TOWARDS THE HISTORICAL NEED FOR PEACE

Mainly after the implosion of the Ancient Roman Empire, the history of Europe has been a succession of internal wars. Treaties of Westphalia, culminating in a long wave of conflicts between European nations and dynasties, incremented by religious dissidences, inaugurated an international order based on the recognition of state sovereignties, with the proposed balance of power to guarantee a peaceful environment (Wilson & Dussen, 1993, p. 40-44). However, warfare kept being a normal feature in Europe, aggravated by the formation of the nation-state, of which extremes took the form of nationalisms in the 20th century, with the two world wars (Hobsbawm, 2011). Therefore, war became a major concern for the European intellectuals engaged in conceiving a European political solution leading to "perpetual peace". Although it was mainly a term coined by the enlightenment philosophy, political proposals committed to guaranteeing a lasting peace date from the Middle Ages, with authors like the French jurist Pierre Dubois (1255 / 1260-1321?) suggesting the creation of an assembly that would occasionally meet to discuss and solve state conflicts (Soulier, 1997, p. 233). Or like Georges Podiebrad, King of Bohemia (1458-1471), who drew up a draft treaty proposing other monarchs to create an international assembly based on a mutual non-aggression pact between states, a constant instrument of diplomacy and arbitration of conflicts (Kurze, 1998, pp. 39-45; Soulier, 1997, pp. 233-234). In the Modern period, Duque of Sully

(1560-1641?), advocating religious tolerance, claims for the creation of a confederal body, a General Council, with states' representatives to take common decisions on commerce and religious affairs, and creating a common army (Malettke, 1998, pp. 93-100).

Towards the lasting effects of permanent internal war in Europe, with the advent of Republicanism, the foundation of USA, the French revolution and the Napoleonic Wars, the enlightenment philosophy will offer an exhaustive rationalisation for a federation as a solution for peace. In parallel with the affirmation of the nation-state in Europe, a group of avant-garde intellectuals proposed to transcend the nation-state model of domination of the international order. Inspired by the Hobbesian dialects of contractualism, they advocated that the absolute sovereignty of the state in the European international system was the cause of permanent conflict.

In this context, pacifist political proposals proliferated in Europe, like the ones from the English Quakers William Pen and John Bellers, strong advocates of religious tolerance and anti-militarism, who drew on experiences of the creation of the States-General representing the United Provinces of the Netherlands, and the Swiss confederation to suggest an international order based on a supranational political organisation in Europe (Gruner, 1998a, 1998b).

Rejecting the balance of forces between the House of Austria and that of France as sufficient to impede the outbreak of wars, and with disbelief in international treaties to guarantee permanent peace, Abbé de Saint-Pierre (François-Irénée Castel, 1648-1743) argued that the problem of Europe was at the absolute sovereignty of the state in the international order, proposing an alternative supra-state model to achieve perpetual peace in Europe (Grune, 1998, pp. 151-155; Heater, 1992, pp. 66-75). His project to make perpetual peace in Europe was widely cited and reviewed by exponents of the enlightenment, like Jean Jacques Rousseau, advocating that the solution could be a federation, in the contemporary and republican conception. Without it, Europe would constantly be in a "civil war" (Gruner, 1998a, 1998b; Heater, 1992; Seth & von Kulessa, 2011). As thought Emmanuel Kant, although his project goes beyond Europe.

Achieving perpetual peace would be for Kant an imperative of practical reason. Historical war consequences would lead individuals to deny warlike solutions and find a new political order. In "For Perpetual Peace" (1795), he predicted the evolution of Europe towards a federation of states, a process that would be extended to the entire world, assuming perhaps the shape of one sole Republic. The undertaking would take place in three phases, with the first being the establishment of a civil society based on Republicanism, the second the constitution of a political and legal order among all states, a federation; and the third the institution of a cosmopolitan political order. Europe was the most prepared continent to experience it and would be the prototype for a progressive constitution of a Federation of Peoples, or a World Republic as the ideal[1]. The ultimate goal of the World Republic was to institutionalise a Cosmopolitan Constitution, which would guarantee to all human beings the cosmopolitan right of common ownership of the Earth's surface. (Soulier, 1997, pp. 236-237; Santos, 2001; Seth & von Kulessa, 2011, pp. 20-22).

The intensification of trade relations and the growing interdependence of States, but above all the insurmountable limits of the Earth [...] are indications that lead Kant to think that the idea of a cosmopolitan right is not a chimaera nor is it just an idea of reason, but it is something which realisation the new conditions of the world will end up imposing on States and on all humanity with increasing urgency: as a recognised right, as a constitution or legislation that is valid worldwide and even as the establishment of forms of world power that can make it effective. (Santos, 2001, p. 60).

The idea of political unity in Europe constantly adapted to historical circumstances. Saint-Simon (1760-1825) made a political proposal to avoid destruction like the Napoleonic Wars and simultaneously address the needs of the new industrial society (Heater, 1992, p. 97-107). Industrialism pointed to a political internationalism, pushing for European planning of common projects of transport, exchanging scientific knowledge, but also guaranteeing civil rights and education for all. This could only be achieved with a European political organisation, based on a European Parliament (Rocha, 2003).

The continuing war in Europe renewed intellectual pacifism in the 19th and 20th centuries, with Victor Hugo's lyrical impetuosity calling for the creation of a United States of Europe, as the first stage of a universalist utopia of a World Republic (Gruner, 1998a, 1998b). Or with Coudenhove-Kalergi, calling for a European federation in favour of permanent peace, with his Pan-Europe proposal in the inter-war period. Inspired by Kalergi, Aristide Briand, French Foreign Minister, even launched the idea to create a "kind of federal tie" [2] in the Society of Nations in 1929 (Heater, 1992, pp. 116-143; Soulier, 1997, p. 253). The Spanish philosopher Ortega Y Gasset called for historical political realism, by creating a system that could overcome the nation-state model (Diego, 2003). And the Second World War would definitely bring political accreditation of those ideas.

The Start of the Union by the Integration of the Market

Despite several federalist proposals in the aftermath of the war, European integration did not start with the institutionalisation of universal citizenship, a moral code and educational rights, a fully federal system, or with external aspirations capable of universalising Kant's cosmopolitan law. The European Economic Community (EEC) was the politically possible institutional reality, although its political narrative is much embedded in the European pacifist philosophy. Not only in the initial aspirations assumed by the founders, but also in the following Commission, EP and national leaderships. Even the development of European competences and scope of action has been overcoming the strict limits of the market. The institution of the universal voting for EP, of the European citizenship, in the Treaty of Maastricht, comprising four common European rights additional to the national citizenship, the proclamation of the Charter of Fundamental Rights of the EU, which gained legal value with the Treaty of Lisbon, or even the European citizens' initiative are some examples that go beyond the borderlines of economic integration. Furthermore, the internal objective of economic and social cohesion and the assumptions of the EU as global actor, using soft power to spread normative influence to balance and stabilise the international system are indirect inheritances of a pacifist philosophy, in reconfigured narratives.

As a process of regional integration, it can be said that the EU has actually been achieving a lasting peace among its member states, historically the longest period without war. Despite the fact that the project was not started by the federalists' approach (Bossuat, 2009), it has been able to evolve, to deepen its political system and broaden the range of supranational competences (Loth, 2014; Silva, 2010, Sidjanski, 1996) from the strict functionalist and materialist low politics to touch some high politics issues. However, more recently, the EU has been subject to increasing public contestation, witnessing the possibility of reversing the project, experiencing disintegration with Brexit, and the general growth of eurosceptic parties, many of them relying upon nationalist and xenophobic messages. If the first decades of integration were merely an elite process, surrounded by a permissive consensus, the last decades have entered a constraining dissensus, where the national identity factor has been raised as instrumental in the political choice regarding EU issues (Hooghe & Marks, 2009).

Facing several crises in the last decade (financial and economic crisis, refugees crisis, terrorism, Brexit), with the degradation of social and economic conditions, with some political parties exploring migration issues associated to insecurity, the EU has awakened a complex mix of feelings of insecurity and deprotection by the Europeans citizens, who tend to find the solution in the national protection, thus turning permeable to nationalistic and xenophobic political messages. The literature on the politicisation of the EU has been reporting the association of financial and refugee crises to the third wave of populism in Europe (Pallaver, Gehler, & Cau, 2018) and perceptions about migration and economic issues to an interrelation of national contexts with the EU (Otjes & Katsanidou, 2017; Kriesi, 2016, p. 40; Sebastião, 2019, p. 95-98, Prat, 2013). The rise of Euroscepticism is, above all, a sign of the re-emergence of the national state as the safe political guardian (Sebastião, 2019).

Hence, lasting peace in Europe should not be taken for granted. Although the EU has been acquiring competencies beyond the market, that can act on the building of a genuine sense of European citizenship, it has not been competent of policies like education and culture (national competences) that could promote peace by an educational perspective of the history of Europe, of the damage of warfare, and of the European legacy and interdependence of nationalities. And not less important, peace in Europe is not only about the internal continental context, but also about avoiding to adopt a Eurocentric perspective of European politics, excluding the *Other* and refusing to recognise the world interdependency and the ethical responsibility in global inequality and conflict prevention. This is where tourism can be framed in a normative orientation as a policy by the EU, to promote a *Kantian* cosmopolitan view of Europe and the world, thus accomplishing the educational and cultural feature that the European integration lacks from the beginning.

HOW TOURISM CONTRIBUTES TO PEACE

Tourism has been one of the major growth industries in the last decades at the global level. The pleasure of travelling is already mentioned in Ancient Greece and Rome, and in the Modern Age the aristocracy did their grand tours, but the enjoyment of tourism by the middle classes is above all a phenomenon of the last half of the 20th century (Wintersteiner & Wohlmuther, 2013, p. 38). Benefitting from the post-industrial evolution of societies and from the emerging middle class post-World War II, and from the normalisation of the institution of holiday periods and subsidies, tourism became a service for the masses. Globalisation has increased the process, with the "rich" north travelling to the global south and more recently, the population of Asia and Latin America also benefitting from leisure travel. From 1950 to 2012, international tourist arrivals grew from 22 million to 1.035 billion (Upadhayaya, 2013).

We can define tourism as the act of a person changing place through different means of transport, with the intention of having leisure time, either to get to know other cultures and places, to practice sports, to visit family and friends, to participate in conferences or business events (Wintersteiner & Wohlmuther, 2013). So, in the first instance, tourism does not relate to peace, since the primary goal of the traveller is not to contribute to a peaceful world, to improve the human rights conditions, or to increase equality and political stability.

But some literature (Becken & Fabrizio, 2016; Levi & Haekins, 2010; Pizam, Fleischer & Mansfield, 2002; Webster & Ivanov, 2013) agrees that indirectly tourism can contribute to peace, as in the process of getting to know other cultures and the historical contexts of communities, it may result in the breaking of stereotypes and social barriers, an opportunity for self-growing emotionally and intellectually.

There are well-documented practices of tourism promoting peace, conflict resolution and global justice (Salazar, 2016). Yet, it is not consensual that tourism contributes to peace, and some authors (Blanchard & Higgins-Desbiolles, 2013; Fischer, 2013; Pratt & Anyu, 2016) consider such conclusions as generalist and idealistic, as they rely on theoretical assumptions hard to prove by empirical studies. It is highly controversial that the contact with the *other* may, in fact, result in the erosion of stereotypes, as there are other studies concluding that some touristic experiences incremented the negative perceptions, such as tourists valuing only the quality of the hosting experience, the relation quality-price, and the living conditions of the local population (Litvin, 1998; Wintersteiner & Wohlmuther, 2013, p. 44). So, as a controversial issue, peace through tourism has received great attention from the academic community (Alluri, 2009; Higgins-Desbiolles & Blanchard 2012; Leong, 2008; Litvin, 1998; Kelly, 2012; Wohlmuthe & Wintersteiner, 2013).

That is why some authors rather speak of a possible positive relation of tourism with peace, but not an automatic one (Upadhayaya, 2013). Instead of assuming an effective role of tourism in achieving peace, they (Higgins-Desbiolles & Blanchard, 2012; Webste & Ivanov, 2013; Wintersteiner & Wohlmuther, 2013) stress that the exercise of tourism in a responsible way can positively impact peoples' lives and the planet, promote prosperity and stability. If we get to think about the historical way of intercultural meeting with the "other", this was mostly accomplished in times of war, or colonisation, when the perverted relation led to constructed ideas of tension and subordination. It is true that nowadays tourism is not subject to such dangerous and biased relations. However, it is not free of experiences of tension and dissatisfaction, possibly enhancing previous existing stereotypes.

This leads to the importance of a cultural dimension of tourism, which is not about cultural tourism, but rather about seeing tourism as a framework of values, orientations and goals which will have an impact on social attitudes. The cultural dimension of tourism is a base of values that give an overall sense to its practices (Wintersteiner & Wohlmuther, 2013). This way, tourism is not only composed by a tour programme, but rather by a set of principles that reflect a certain vision of the world. As gathering a series of words, expressions and narratives about places and people, tourism is a social practice that turns out places pleasant and interesting to visit. As such, it is an arena where several agents interact inherently showing their perceptions of the outside world (Salazar, 2006).

This feature has been perceived by the international stakeholders who carried on some concertation for the development of responsible tourism. The World Tourism Organization has been reaffirming since the 80s the vital force of tourism for peace and international understanding (Salazar, 2006, p. 324). And the idea has consolidated between industry corporations and political leaders in the international community, which has engaged in declarations of peace through tourism, creation of international organisations, holding of international summits and conferences on the subject, charters and protocols for guidelines (Upadhayaya, 2013). For example, the Cape Verde Declaration (2002) recommends economic, environmental and social responsibility guidelines which can promote eco-tourism and fight the negative externalities associated with mass tourism (Wintersteiner & Wohlmuther, 2013).

Hence, implicit in this international mobilisation is the fact that the contribution of tourism to peace is not an automatic one. Tourism is an escape from the daily routine, a temporary experience abroad. Contrasting with a migrant, the tourist does not leave his home permanently. It is a voluntary interruption which does not result instantaneously in the reduction of prejudices since the intercultural experience is very simple and immediate. Experience shows that, unless there is a predisposition for that and proper conditions in the travel experience, it is difficult that tourism directly promotes peace environments (Webster & Ivanov, 2013; Wintersteiner & Wohlmuther, 2013). This is not to say that it is not possible, as there are reports of tourism experiences bringing citizens to understand the nature, causes and consequences of conflicts (Wintersteiner & Wohlmuther, 2013, p. 50). However, this does not happen as an isolated approach to tourism, but rather with framing the sector within a set of values related to a vision for the world based on democracy, equality and peace (Higgins-Desbiolles & Blanchard, 2012).

Tourism Sensitive to Peace

The concept of "tourism sensitive to peace" (Wintersteiner & Wohlmuther, 2013) does not deny the potential of tourism contributing to peace, but does not assume such contribution as an intrinsic feature of tourism itself. Other authors use the term "tourism responsive to peace" (Upadhayaya, 2013), which defines tourism attentive to the development of communities and emanates from a conduct code for peace. This concept assumes as a basic guideline that tourism must embed in all its activities an approach to the world based on democracy, equality, fraternity and human rights in general. It has not only to do with a message to be communicated in the tourist narratives, but with the involvement of several kinds of targets and guidelines in the diverse intervenient agents of the tourism industry.

If tourism is an "abstraction" of the world, and, in this way, a "mediation", there stands "the importance of how tourism displays its narratives and services" (Wintersteiner & Wohlmuther, 2013, p. 36). Whereas those narratives are reconstructed speeches about historical, economic, political and social contexts (Salazar, 2006, p. 327), some studies argue that there is a lack of awareness of the private travel agents for their involvement in the promotion of peace in post-conflict situations, which is aggravated by the low organisational capacity of the sector (Alluri, 2009; Alluri, Leicher, Palme & Joras, 2013). Another aspect is the local impact of tourism as an economic activity. Consigning that tourism corporations should operate in a responsible way, promoting local development, causing the minimum impact for residents and the planet, tourism acts towards sustainable development which can mitigate socio-economic differences and reinforce conflict prevention.

It is safe to say that tourism is a complex phenomenon in which different identities, industry traditions, and world vision exist and are continuously being performed, confirmed and reproduced (Salazar, 2006). This not only requires to place emphasis on all the inherent processes of the tourism industry but also defining an ethical and moral delimitation which cannot be dissociated from a wider political order. In this regard, the political rhetoric of tourism for peace should be backed by a political regulation for tourism, with a pre-definition of guidelines and principles, to define it as a public policy at the service of peace and democracy. This concern has been expressed from the beginning of the rising of mass tourism by some international organisations and initiatives, such as the Conference on "Tourism as a Passport for Peace", in Ireland (1987), or the "Global Conference of Tourism – Vital Force for Peace" (1988) (Wintersteiner & Wohlmuther, 2013), or resolutions by international organisations (e.g. UN resolution A/RES/52/13, 1997, "Culture of Peace"), stating that the culture of peace has to be a way of living and the whole cultural system must be orientated to non-violence.

The success of those initiatives depends not only on the regulatory policies and charters of intentions but also on the involvement of all players in the tourism industry. Even in the societies with the highest democratic and civil rights patterns, myths and elements of a culture of war remain latent. Wintersteiner and Wohlmuther (2013) verified this idea by comparing travelling experiences that replicate the already existing privileged or de-privileged positions of social groups in the world. Tourism is the opposite of migration. Whereas the latter is motivated by need, the former is by pleasure. The rich northwestern tourists appreciate very much the tropical and indigenous hospitality of the south when they travel there.

On the opposite, when migrants of the global south come to the rich north they are not welcomed in the same way tourists were welcomed, and the once tourists and now hosts do not look at them with the same grateful and sympathetic attention as when in the role of tourists (Wintersteiner & Wohlmuther, 2013). Such feeling, as was above mentioned, can be fostered by the growth of Eurosceptic parties, with nationalist xenophobic messages, in Europe. Migration and tourism are travel experiences, but on a complete world apart, evidencing the inequality and segregation of the world, which is ready to be politically explored in moments of crisis.

Here we meet the Kantian and cosmopolitan idea of hospitality. The idea of hospitality and non-discrimination is not a cultural practice, but a legal universal value. In the *Kantian* philosophy, hospitality was basic to universal and perpetual peace, a principle that nowadays regains significance with globalisation and climate change. And if the EU advocates the role of promoter of values (democracy, human rights, internal and international economic and social cohesion), it must incorporate them in its policies:

If tourism has a mission for world peace, then it is to spread the idea of (and to provide stimulating images of) world citizenship: the beauty of the whole planet, the diversity of human cultures and civilisations, the necessity of dealing with our differences in a political, i.e. nonviolent way – thus acting as global citizens. Tourism effectively contributes to peace if and when it helps tourists, as well as hosts, to learn that together we can make the world a better place for all human beings. (Wintersteiner & Wohlmuther, 2013, p. 42).

So tourism with peace goals must be involved in a broader cultural and political system, which provides it with a subjacent educational strategy.

This is what we intend to analyse in the EU tourism policy, assuming the EU as a mediator in two playing fields concerning tourism: i) at the internal level, as a political system from where public policies emanate to the national level with repercussions in the European transnational political arena and the sector stakeholders; ii) and as an indirect player at the extra-EU level, since European tourism industry operates in the international market, reaching non-Europeans citizens. The objective of this research is evaluating tourism as an EU public policy, not excluding that it also has an external consequence. Assuming that tourism sensitive to peace comprehends all the parts involved in the service provision it requires political action to induce change and specifically a culture of peace in the sector (Wintersteiner & Wohlmuther, 2013). Such guidance and normative role can be directed to tour managers, educators for tourism, travel and tour guides, to set financial incentives under conditionality, and other measures framed in an overall strategy for the sector.

PRIORITY FOR THE MARKET OVER A PEACE RESPONSIVE VISION

Tourism has not been a priority area in EU integration. The Treaty of Rome does not mention it, and the focus of the launching of the EEC was the liberalisation of goods. Although the Treaty already envisioned services as a sector to be liberalised (Article 3º), services in general, and tourism in particular, were not considered, in the post-war context of material reconstruction and economic recovery, as a strategic domain in the European integration.

Tourism as a service is only approached in the Single European Act (SEA), 1987, which launched the goal to complete the internal market and adding three additional freedoms to the already existing free movement of goods: free movement of services, capitals and people. As a result of the implementation of the SEA goals, tourism was targeted by EEC action as a service included in the action plan to complete the free market until the end of 1992. However, it was not referred to as a single public policy in the EEC Treaty. In the Treaty of Maastricht, tourism was mentioned as one of the sectors in which the EC could enact measures (Article 3, EC Treaty).

This statute had no change until the Treaty of Lisbon, which established tourism as a policy (Title XXII TFEU). However, it is categorised as a complementary competence, meaning that the EU is mandated to act only to develop actions of support, coordination or completion of the member state actions (Articles 6 and 195 TFEU). The scope of the complementary action of the EU is above all aimed at "promoting competitivity of companies of the Union in this sector" and fostering benchmarking among member states (Article 195, TFEU). Being a complementary competence of the EU, the competence for regulating tourism belongs to member states, while actions of the EU are confined to the soft role of defining general principles and orientation, and of leading the institutionalisation of coordination among national governments (Estol, Camilleri & Font, 2018).

Nevertheless, enlargement and integration in other areas favoured tourism development in the European economy. First, because the EEC/EU evolved from six to 28[3] member states and so territorial, historical and cultural diversity increased, expanding the touristic offer and demand. Second, because the free movement of people and capitals, culminating with the creation of the single currency, facilitated tourism, enhanced rationalisation of resources and thus profitability. Third, because the effects of the free-market in liberalising other sectors, like aviation, allowing a rapid expansion of the low-cost companies, broadened access to travelling for the Europeans and to Europe (Timothy, 2019). So, the development of the integrated market had a huge positive impact on tourism, which raised in the share of the European economy. Today, tourism in the EU represent 2.3 million businesses, mainly small and medium-sized companies, employs over 12.3 million people (EP, 2020). In 2018, the "travel and tourism" sector represented 3.9% of EU GDP (EP, 2020).

Despite the complementary competencies of the EU in tourism, much can be done by supporting member states, establishing guidelines and exchanging good practices in the sector. The EU can have a consequent action in tourism, on the level of bolstering intergovernmental cooperation that may have an effect through the national regulatory arena. Hence the European Commission (hereinafter Commission) and the Council of the EU (hereinafter Council) can use their current mechanisms to enforce national cooperation to set guidelines and create frameworks for policy principles that can have an impact on the conduct of the tourism industry. Based on this possibility, it is possible to define a vision for tourism sensitive to peace. That is what we are going to assess now.

MARKET PARADIGM REMAINS OVER A GENUINE VISION OF TOURISM SENSITIVE TO PEACE

In accordance with the EU competences in tourism, rarely have legal acts been aimed at specifically regulating the sector. The EU legal action affecting tourism is related with the functioning of the internal market, aiming at harmonising competition rules related with consumer protection, dealing with package travel and linked travel arrangements, passenger rights, bathing water quality, or safety in infrastructures of accommodation (EP, 2020).

Tourism is for the first time exclusively approached by the Commission in 1982, with a communication (EC, 1982) recognising the importance of the sector to achieve the ultimate goals of the EEC, stressing the role on the development of poorest regions and on balancing the balance of payments. The economic focus is confirmed within two years by a resolution of the European Parliament and the Economic and Social Committee (cit. in Estol & Font, 2016, p. 233), as by the Council (Council, 1984), underlining the importance of tourism for the integrated market. At the time, the 1985 White Paper of the Internal Market (EC, 1985), a core document setting the plan of action completing the internal market, did not refer to tourism. Although, within the context of the approval of the SEA, the Council committed to launch a coordination process in the sector, created the Tourism Advisory Committee (TAC), still running, formed by national government representatives to enhance national coordination.

In 1992, following a Commission's communication, the Council approved an action plan for tourism, stressing as central concerns the competitiveness of the sector expressed in the quality of destinies and products and in the consumer protection (Council, 1992). After the publication of the "Green paper on the role of the Union in the field of tourism" (EC, 1995), EP approved a resolution (EP, 1996) concluding that the EU was not succeeding in promoting the role of tourism for the economy and sustainable development. In the context of the European "Employment Strategy" launched in 1997, the Commission disclosed the Communication "Enhancing tourism's potential for employment" (EC, 1999), acknowledging the impact of tourism in employment, growth and sustainable development, where a set of governance tools and procedures, as well as the use of European financial instruments, are recommended, and specific goals are defined. In an attempt to accomplish such goals, a European Conference of Public Authorities of Tourism was held, collecting information and good practices to disseminate and improve strategies for tourism management.

As a result, several working groups were created to study specific fields of action, and in 2001 the Commission issued another communication (EC, 2001), followed by another position from the Council, both reaffirming and resetting previous lines of action and targets. As a result of the Lisbon Strategy (2000), predicting objectives of development for ten years in the EU economy, governance in tourism was reinforced, with greater and more frequent involvement of all the sector stakeholders and the TAC. From 2003 on, the value of sustainability is strengthened in the following instruments of tourism governance. Communication of the Commission (EC, 2003) placed sustainability as a central idea to which any action should converge to achieve competitiveness in tourism. The Tourism Sustainability Group is created as an instrument to reinforce intergovernmental coordination. The 2006 Communication "a renewed EU tourism policy: towards a stronger partnership for European tourism" (EC, 2006) reaffirms the same strategy on placing sustainability at the centre of any action for the sector. Within this framework, new mechanisms to boost coordination were created under the responsibility of the TAC.

Following a summit of the Ministers of Tourism, the Commission released in 2010 another communication (EC, 2010), based on the same paradigm, promoting sustainability in tourism connected with specific goals of growth, development and competitiveness in the global market. In 2012, the communication "Implementation and development of the common visa policy to growth in the EU" aimed at positively impacting the economy through a facilitated visa policy, specified tourism as one critical sector. And finally, in 2014, another Commission communication (EC, 2014) released the "European strategy for more growth and jobs in coastal and maritime tourism", addressing, among other issues, concerns to conciliate sustainability with competitiveness in those tourist areas.

The above description presents a chronological sequence of the Commission's initiatives for tourism, setting the main theme of the communications. Does a perspective of tourism sensitive for peace have a place in those guidelines? Some contributions for tourism sensitive to peace can be found, although they are indirect and inherent ideas to the overall model of the EU internal market growth. They are not a reflection of a primary assumption of regarding tourism as a tool for pacification and conflict prevention.

The value of sustainability is associated to tourism, linking it to the effects of equitable development and consequently to conflict prevention, but as far as subsequent communications are analysed, it turns clear that the value of sustainability is mainly connected with a whole economic long-term strategy. Concerns with environmental protection in tourism appeared already in the 1982 and 1995 communications (EC, 1982, 1995), promoting sustainable development as not only good for environmental preservation but also for the competitiveness of the sector. Later, the concept of sustainability was broadened and gained a polysemic meaning, not only related to natural resources conservation, as in the first communications, but now with an integrated vision of social progress. Thus, in 1999, the Commission extended sustainability to the preservation of rural areas, respect for the social and cultural identity of local communities and cultural heritage preservations. This conceptualisation is further strengthened in the next decades in a manner that all the stakeholders and individuals directly and indirectly reached by tourism activity must be positively impacted. Thus, sustainability is not only about the natural and cultural heritage preservation, but also it must be a cause and consequence of economic development, job creation, quality of jobs, respect for the integrity of local communities, and above all, a complex interdependency that must guide the entire value-chain of the sector (EC, 2003, 2006, 2010).

This approach was specified lately in 2014 with a specific concern with tourism in the coastal maritime areas (EC, 2014). In fact, such guidelines meet some criteria for tourism responsive to peace, in the way that it promotes the contribution of the sector for equitable development, involving all the stakeholders, which is an impacting condition for democracy and conflict prevention. However, in the last decade, the sustainability norm has had a greater economic approach in the Commission communications. In 2010 (EC, 2010), there was a great focus on growth and jobs creation, as in 2012 (EC, 2012), underlining the need for increased economic profitability of tourism, and even in 2014 (EC, 2014), where despite the concern with environmental protection of coastal maritime areas, the goal is about a strategy to explore the touristic potential of the maritime zone of the EU, with a prominent orientation for growth and job creation. This trend is much related with the post-2008 financial crisis juncture in Europe, with the subsequent sovereign debt crisis in the Eurozone, when some member states were under bailouts and consequent austerity measures, severely affecting demands on tourism.

Although there is an indirect orientation for responsible tourism, this is inherent to the whole values and strategies of the economic development of the EU, closely linked to sustainability granting environmental protection and consumer and civil society interests. Nevertheless, such a vision of tourism is merged in a functional perspective of the integrated market, i.e. it derives from a material and economic utilitarian perspective of tourism.

If one looks for an abstract vision of tourism for peace that can guide the national policies, economic corporations and individuals playing in the sector, one does not find it. If tourism is not only made of economic and material operations and actions but also of an immaterial condition, of narratives about historical, cultural and social contexts; if those narratives display abstractions of the world; and if the travel and tour guides are mediators between social realities and consumers, an ethical and moral code for tourism is pertinent for a sector to be responsive to peace. This code is even more needed when some studies argue that there is a lack of awareness for the political impact of such narratives, and when in most developed societies elements of war and conflict prevail in society. When we look at the communications of the Commission, such a vision is not found, although in two of them the potential contribution to the peaceful environment is approached:

[T]ourism can be a means of reducing the differences between the multifarious realities which constitute the Union, and by a continuous process of these differences being perceived in a non-confrontational context, can facilitate communication among citizens, and thus improve the basis on which exchanges, both cultural and economic, are made between the member states. [...] the tourist, who is not only a beneficiary of the products of the tourism industry and a user of tourist areas and heritages but also a factor in the political and economic construction of the Union." (EC, 1995, p. 16).

In 2010 the Commission also stated that tourism is an important instrument for "projecting our values and promoting the attractions of the European model, which is the result of centuries of cultural exchanges, linguistic diversity and creativity." (EC, 2010, p. 10) But after those statements, no further development was provided, no guidance was given. The value for peace in tourism is regarded here as an acquired characteristic, an intrinsic effect of the activity, something that research has already denied. In the EU policy for tourism, there is not a conceptual definition for tourism that can be politically consequent to contribute to democracy and peace.

Furthermore, because of its nationally fragmented nature, since tourism policy is not a supranational competence of the EU and relies on soft policy instruments, it is even more dependent on national and subjective conceptions vis-à-vis an idea of unity and European citizenship. This is also a cause for the lack of European identity of tourism (Akerhielm, Dev & Noden, 2003, p. 91). The lack of an immaterial vision for tourism contrasts with its economic weight in the EU and is a clear demonstration of the educational potential that is being wasted, not only for the European citizens but also for the world tourists, frequently visiting Europe. Despite the EU efforts to enforce instruments of national coordination on tourism, policy orientation has been vague if not completely omissive in creating a political vision for tourism related with peace. This gains even more meaning when support for tourism activities is provided by the EU budget funds, some of them aiding projects of transnational tourism related with cultural and heritage-based products (EC, 2012a, 2017, 2018).

CONCLUSION

As was evident in the conception for tourism sensitive for peace, if this activity can impact positively on education for peace, it must depart from a defined policy that can enable and orientate travel agents for it. Tourism is very much related to natural, cultural and material heritage, which have the potential to be explored with messages for peace. However, depending on the experience the tourist has, it can or cannot have that result. A clash between different narratives may also happen. If we look at the history of Europe, it has been actually built of a clash of territorial and national misconceptions. European integration was born of the will to finish a lasting conflicting international political environment, but recently national tensions have arisen in the EU. This has been evident in the expansion of political narratives for controlling borders and returning political powers to national sovereignty, increasingly gaining public support. In result, nationalist, xenophobic and eurosceptic parties have raised political representativeness in European and national parliaments. On the other side, this is not only an internal conflicting perspective, but it is also expressed in a Eurocentric perspective, seeing the *Other*, the non-European, the immigrant, the refugee, as the threat to the European way of life. The philosophic leagy of the EU, politically based on a multicultural and cosmopolitan rhetoric, is in practice questioned in several dimensions, and this is one of them.

The huge touristic potential of Europe could be used as an instrument to act on a political education of Europeans and non-Europeans. Visiting heritage places, monuments, places of war or freedom, museums which are symbols of peace and war, can be either used in a nationalistic perspective, with clashes among different national narratives, or they can be interpreted as a national and European symbiosis, providing that the visit to those places is an experience of learning with history to act on a peaceful future. But if we want this to be a basic principle to follow, a strategic direction for national policies and tour operators is needed. The EU tourism policy needs first a theoretical development in that way. Even with the soft power the EU has in tourism policy, it is possible to put in place such a conceptual framework that can guide all the stakeholders, and have an impact, such as the EU has been doing with the concept of sustainability.

The EU seems to follow the international trend. In the 1980s, the idea of peace through tourism was seductive (Wintersteiner & Wohlmuther, 2013, p. 43), but since the 90s tourism has been more regarded as a tool for economic development, and job creation to address the economic crisis. The economic value of tourism was always present on the EU agenda, and it has been enhanced since the last decade, in order to face the global financial crisis effects. And this trend may be reinforced in the future. In the last years, from 2015, Europe has entered a consistent phase of economic recovery and growth, but it was interrupted. The current times, with the economic crisis derived from the Covid-19 Pandemic, with a great impact on the tourism industry, pre-announce the reinforcement of the classical economic paradigm for the EU tourism policy. However, the EU has the lessons of history, particularly the moments of the economic crisis that potentiated social conflicts. Thus, it is time for a conceptual definition of tourism which embraces peace objectives in the EU.

REFERENCES

Akerhielm, P., Dev, C. S., & Noden, A. A. (2003). Brand Europe: European integration and tourism development. Cornell Hotel and Restaurant Administration Quarterly, 88-93.

Alluri, R. M. (2009). *The role of tourism in post-conflict peacebuilding in Rwanda.* Working Paper, 2. Bern: Swisspeace.

Alluri, R. M., Leicher, M., Palme, K., & Joras, U. (2013). Understanding economic effects of violent conflicts on tourism: Empirical refections from Croatia, Rwanda and Sri Lanka. In C. Wohlmuther & W. Wintersteiner (Eds.), *International Handbook on Tourism and Peace* (pp. 101–119). Drava Verlag.

Bache, I., George, S., & Bulmer, S. (2011). *Politics in the European Union.* Oxford University Press.

Becken, S., & Fabrizio, C. (2016). Does tourism lead to peace? *Annals of Tourism Research, 61*, 63–79. doi:10.1016/j.annals.2016.09.002

Blanchard, L., & Higgins-Desbiolles, F. (2013). Introduction. Peace matters, tourism matters. In L. Blanchard & F. Higgins-Desbiolles (Eds.), *Peace through Tourism* (pp. 1–15). Routledge. doi:10.4324/9780203370384

Bossuat, G. (2009). *Histoire de l'Union Européenne: Fondations, élargissements, avenir.* Paris: Belin.

Causevi. S (2010). Tourism which erases borders: An introspection into Bosnia and Herzegovina. In O. Moufakkir & I. Kelly (Eds.), *Tourism, progress and peace* (pp. 48-64). Oxfordshire: Cab International.

Council (Council of the EC). (1984). *Council Resolution of 10 April 1984 on a Community Policy on tourism.* Retrieved from: https://eur-lex.europa.eu/legal-content/EN/TXT/?uri=CELEX%3A31984Y0430%2802%29

de Prat, C. R. (2013). *Euroscepticism, Europhobia and Eurocriticism: The radical parties of the right and left vis-à-vis the European Union.* Peter Lang. doi:10.3726/978-3-0352-6347-3

de Villiers, D. (2013). Cornerstones for a better world: Peace, tourism and sustainable development. In C. Wohlmuther & W. Wintersteiner (Eds.), *International handbook on tourism and peace* (pp. 78–86). Drava Verlag.

Diego, E. (2003). Una Percepción de la idea de Europa en España durante el periodo de entreguerras 1918-1939. *Cuadernos de Historia Contemporâne,* 311-324.

Dionysopoulou, P. (2012). European tourism policy. *Knowledge Management in Tourism: Policy and Governance Applications*, *4*, 283–300.

EC. (1982). *Initial Guidelines for a Community Policy on Tourism* (COM/82/385 final). Retrieved from: https://op.europa.eu/en/publication-detail/-/publication/2c46eb35-049b-4fba-a8ae-2529521aa7fd/language-en

EC. (1985). *COM (85) 310 final.* White paper from the Commission to the European Council. Retrieved from: https://op.europa.eu/en/publication-detail/-/publication/4ff490f3-dbb6-4331-a2ea-a3ca59f974a8/language-en

EC. (1995). *COM (95) 97 final. The role of the Union in the field of tourism.* Commission Green Paper. Retrieved from https://op.europa.eu/en/publication-detail/-/publication/ed46c83e-95fc-4c5c-9dbc-da288647d7c2/language-en

EC. (1999). *COM 178/ Enhancing tourism's potential for employment Follow-up to the conclusions and recommendations of the High Level Group on Tourism and Employment.* Communication from the Commission to the Council, the European Parliament, the Economic and Social Committee and the Committee of the Regions. Brussels. Retrieved from https://eur-lex.europa.eu/legal-content/EN/TXT/?uri=CELEX:31999Y0623(01)

EC. (2001). *COM (2001) 665 final. Working together for the future of European tourism.* Commission communication to the Council, the European Parliament, the Economic and Social Committee and the Committee of the Regions. Brussels. Retrieved from https://eur-lex.europa.eu/LexUriServ/LexUriServ.do?uri=COM:2001:0665:FIN:en:PDF

EC. (2003). *COM (2003) 716 final. Basic orientations for the sustainability of European tourism.* Communication from the Commission to the Council, the European Parliament, the European Economic and Social Committee and the Committee of the Regions. Brussels. Retrieved from https://eur-lex.europa.eu/LexUriServ/LexUriServ.do?uri=COM:2003:0716:FIN:en:PDF

EC. (2006). *COM (2006) 134 final. A renewed EU Tourism Policy: Towards a stronger partnership for European Tourism*. Communication from the Commission. Brussels. Retrieved from https://eur-lex.europa.eu/LexUriServ/LexUriServ.do?uri=COM:2006:0134:FIN:en:PDF

EC. (2010). *COM (2010) 352 final. Europe: the world's No 1 tourist destination – a new political framework for tourism in Europe*. Communication from the Commission to the European Parliament, the Council, the European Economic and Social Committee and the Committee of the Regions. Retrieved from https://eur-lex.europa.eu/LexUriServ/LexUriServ.do?uri=COM:2010:0352:FIN:EN:PDF

EC. (2012). *COM (2012) 649 final. Implementation and development of the common visa policy to spur growth in the EU*. Communication from the Commission to the Council and the European Parliament. Brussels. Retrieved from https://ec.europa.eu/home-affairs/sites/homeaffairs/files/what-is-new/news/pdf/communication_on_visa_policy_to_spur_growth_com_2012_649_en.pdf

EC. (2012a). *Docs room*. Retrieved from http://ec.europa.eu/DocsRoom/documents?locale=en&tags=projects_on_cultural_tourism_2012

EC. (2014). *COM (2014) 86 final. A European Strategy for more Growth and Jobs in Coastal and Maritime Tourism*. Communication from the Commission to the European Parliament, the Council, the European Economic and Social Committee and the Committee of the Regions. Brussels. Retrieved from https://eur-lex.europa.eu/LexUriServ/LexUriServ.do?uri=COM:2014:0086:FIN:EN:PDF

EC. (2017). *Cultural Tourism – projects supported in 2017*. Retrieved from https://ec.europa.eu/docsroom/documents/29223

EC. (2018). *Tourism and Cultural and Creative Industries (CCIs) – projects supported in 2018*. Retrieved from https://ec.europa.eu/docsroom/documents/35247

EP. (1996). *Resolution on the Commission report on the evaluation of the Community Action Plan to Assist Tourism 1993-1995 - Council Decision 92/421/EEC (COM(96)0166 - C4-0266/96)*. Retrieved from https://eur-lex.europa.eu/LexUriServ/LexUriServ.do?uri=CELEX:51996IP0299:EN:HTML

EP. (2020). *Tourism. Factsheets of the European Union*. Retrieved from www.europarl.europa.eu/factsheets/en

Ertamay, S. I., & Akbaba, A. (2019). European Union tourism policies. *International Eastern Mediterranean Tourism Symposium*.

Estol, J., Camilleri, M. A., & Font, X. (2018). European Union tourism policy: An institutional theory critical discourse analysis. *Tourism Review*, *73*(3), 421–431. doi:10.1108/TR-11-2017-0167

Estol, J., & Font, X. (2016). European tourism policy: Its evolution and structure. *Tourism Management*, *52*, 230–241. doi:10.1016/j.tourman.2015.06.007

Fischer, S. (2013). Religious tourism: Business for peace in the Holy Land? In C. Wohlmuther & W. Wintersteiner (Eds.), *International handbook on tourism and peace* (pp. 120–142). Drava Verlag.

Galtung, J. (1964). An editorial. *Journal of Peace Research*, *1*(1), 1–4. doi:10.1177/002234336400100101

Graziano, P., & Vink, M. P. (2008). *Europeanization: New Research Agendas*. Palgrave Macmilan. doi:10.1057/9780230584525

Gruner, W. D. (1998a). Les idées de l'Europe politique au XVIIIe siècle. In K. Malettke (Ed.), *Imaginer L'Europe* (pp. 145–162). Éditions Belin.

Gruner, W. D. (1998b). Perceptions et conceptions de L'Europe au XIX siècle. In K. Malettke (Ed.), *Imaginer L'Europe* (pp. 163–190). Éditions Belin.

Heater, D. (1992). *The idea of European Unity*. Leicester University Press.

Higgins-Desbiolles, F., & Blanchard, L-A. (2012). Challenging peace through tourism: Placing tourism in the context of human rights, justice and peace. In O. Moufakkir & I. Kelly (Eds.), *Tourism, progress and peace* (pp. 35-47). Wallingford: CAB International.

Hobsbawm, E. (2011). *A era dos extremos*. Editorial Presença.

Hooghe, L., & Marks, G. (2009). A postfuncionalist theory of European integration: From permissive consensus to constraining dissensus. *British Journal of Political Science*, *39*(1), 1–13. doi:10.1017/S0007123408000409

Kelly, I. (2012). Peace through tourism: An implementation guide. *The Journal of Tourism and Peace Research*, *2*, 32–49.

Kriesi, H. (2016). The politicisation of European integration. *Journal of Common Market Studies*, *54*, 32–47. doi:10.1111/jcms.12406

Kurze, D. (1998). La republica Christiana et l'Europe Médiévale. In K. Malettke (Ed.), *Imaginer L'Europe* (pp. 11–50). Éditions Belin.

Ladrech, R. (2010). *Europeanization and national politics*. Palgrave Macmilan. doi:10.1007/978-1-137-06814-9

Leong, A. M. W. (2008). International tourism and world peace. *Journal of Macau University of Science and Technology*, *2*, 38–44.

Levi, S., & Haekins, D. E. (2010). Peace through tourism – Commerce based principles and practices: A global view. *Journal of Business Ethics*, *89*(S4), 569–585. doi:10.100710551-010-0408-2

Litvin, S. (1998). Tourism: The world's peace industry? *Journal of Travel Research*, *37*(1), 63–66. doi:10.1177/004728759803700108

Loth, W. (2014). *Building Europe. A history of European unification*. De Gruyter.

Malettke, K. (1998). L'Europe dans le 'Grand Dessein' de Sully. In K. Malettke (Ed.), *Imaginer L'Europe* (pp. 93–105). Éditions Belin.

Otjes, S., & Katsanidou, A. (2017). Beyond Kriesiland: EU integration as a super issue after the Euro-crisis. *European Journal of Political Research*, *56*(2), 301–319. doi:10.1111/1475-6765.12177

Pallaver, G., Gehler, M., & Cau, M. (2018). *Populism, populists, and the crisis of political parties: A comparison of Italy, Austria, and Germany 1990*-2015. Società Editrice il Mulino.

Pizam, A., Fleischer, A., & Mansfield, Y. (2002). Tourism and social change: The case of Israeli eco-tourists visiting Jordan. *Journal of Travel Research*, *41*(2), 177–184. doi:10.1177/004728702237423

Pratt, S., & Anyu, L. (2015). Does tourism really lead to peace? A global view. *International Journal of Tourism Research*, *18*(1), 82–90. doi:10.1002/jtr.2035

Rocha, A. (2003). Filosofia da Europa: Questões sobre a Europa. *Diacrítica, Filosofia e Cultura*, *17*(2), 7-22.

Salazar, N. B. (2006). Building a culture of peace through tourism: Reflexive and analytical notes and queries. *Universitas Humanistica*, *2*, 319–333.

Santos, L. R. (2001). Republicanismo e cosmopolitismo. A contribuição de Kant para a formação da ideia moderna de federalismo. In E. C. Leal (Ed.), *O federalismo europeu: História, política e utopia* (pp. 49–69). Edições Colibri. Instituto de História Contemporânea da Universidade de Lisboa.

Sebastião, D. (2019). Gregos e Troianos: A política de asilo na União Europeia, entre a politização e os direitos humanos. In I. Camisão & A. P. Brandão (Eds.), *O estado da União Europeia: Da(s) crise(s) à mudança?* (pp. 81–102). Editora Petrony.

Sebastião, D. (2020). Political parties in the EU political system. In R. Moldovan (Ed.), *The European Union: Policies, perspectives and politics* (pp. 131–152). Nova Science Publishers.

Seth, C., & von Kulessa, R. (Eds.). (2011). *The idea of Europe: Enlightenment perspectives*. Open Book Publishers.

Sidjanski, D. (1996). *O futuro federalista da Europa. A Comunidade Europeia, das origens ao Tratado de Maastricht*. Gradiva.

Silva, A. M. (2010). *História da Unificação Europeia: A integração comunitária (1945-2010)*. Imprensa da Universidade de Coimbra. doi:10.14195/978-989-26-0421-3

Soulier, G. (1997). *A Europa*. Instituto Piaget.

Timothy, D. J. (2019). Supranationalism and tourism: Free trade, customs unions, and single markets in an era of geopolitical change. In D. J. Timothy (Ed.), *Handbook of globalisation and tourism* (pp. 100–114). Edward Elgar Publishing. doi:10.4337/9781786431295.00018

Upadhayaya, P. K. (2013). Peace sensitivity in tourism codes of conduct at destinations in conflict: A case study of partnership actions in Pokhara, Western Nepal. In C. Wohlmuther & W. Wintersteiner (Eds.), *International handbook on tourism and peace* (pp. 230–246). Drava Verlag.

Wallace, H., Pollack, M. A., & Young, A. R. (2010). *Policy-making in the European Union*. Oxford University Press.

Webster, C., & Ivanov, S. (2013). Tourism as a Force for political stability. In C. Wohlmuther & W. Wintersteiner (Eds.), *International handbook on tourism and peace* (pp. 167–180). Drava Verlag.

Wilson, K., & van der Dussen, J. (1993). *The history of the idea of Europe*. Routledge.

Wintersteiner, W., & Wohlmuther, C. (2013). Peace sensitive tourism: How tourism can contribute to peace. In C. Wohlmuther & W. Wintersteiner (Eds.), *International handbook on tourism and peace* (pp. 31–61). Drava Verlag.

Wohlmuther, C., Wintersteiner, W., & Wagner, M. (2013). Connected by the trails that used to divide us: Peace trails in the Alps-Adriatic Region. In C. Wohlmuther & W. Wintersteiner (Eds.), *International handbook on tourism and peace* (pp. 321–334). Drava Verlag.

Chapter 11
Contexts of Political (Un) Stability in Asia and the Chinese Instrumentalisation of Tourism

Jiawei Xing
https://orcid.org/0000-0002-6721-5562
University of Aveiro, Portugal

Zélia Breda
https://orcid.org/0000-0002-5882-063X
University of Aveiro, Portugal

Jorge Tavares da Silva
https://orcid.org/0000-0002-2526-4745
University of Aveiro, Portugal

ABSTRACT

Tourism is one of the most vulnerable sectors in the face of economic, political, environmental, and social changes. The study regarding the relationship between political phenomena and the dynamics of international relations and tourism is still at an infancy stage. The authors refer, in particular, to contexts of crisis, catastrophes, and conflicts in their relationship with tourism indicators. This is the purpose of this work, focusing on the Chinese case due to its particularities. In other words, they want to see how some regional political phenomena—the Asian context—are reflected in the development of Chinese tourism. They find that there is an immediate correspondence between the number of Chinese tourists and the contexts of political instability, but in which tourist flows are very dependent on government guidelines. Although subject to market dictates, they often obey geopolitical imperatives.

DOI: 10.4018/978-1-7998-5053-3.ch011

INTRODUCTION

Tourism has been recognised as a dynamic sector for the development of global and regional economies. According to the World Tourism Organization (UNWTO, 2018), this sector contributed directly and indirectly to 10% of the world's Gross Domestic Product (GDP), a tenth of employment and 30% of service exports. In addition, it is an important sector for world trade and a major contributor to economic growth. The UNWTO also pointed to sustained growth: globally, international tourist arrivals grew to 1,326 million in 2017, registering a 7% increase over the previous year, which is the most significant increase since the global economic crisis of 2009 (UNWTO, 2018). According to the pace of growth, international tourist arrivals worldwide are expected to exceed 1.4 billion by 2020 and reaching 1.8 billion by 2030 (OMT, 2011).

Taking into account the volume and the rapid growth of tourism, it appears to be increasingly interconnected to other industries and fields. Note, for example, the relationship established between environmental conditions and tourism, health risks or international security. Although less discussed than the economic dimension in the tourism literature, the political dimension is very relevant (Richter, 1989). Tourism is often involved in political problems between states. Tourism development can be affected by its patterns, processes and direction (Hall, 2010). The interrelationship between tourism, politics and international relations is essential to understand how the international system works: from the exchange of ideas and changes in the image and representation of the "other", to formal diplomacy between states. Using Taoism symbolism, tourism has a positive and negative direction, both in terms of consequences, and in the way governments use it in their foreign policy actions.

Edgell (1990) underlines, in this perspective, that sometimes a government can prohibit outbound travel to war zones or territories of hostile nations. This was the case of the United States, when the country applied an economic embargo to Cuba in 1962, limiting the right of American citizens to trade, invest and travel to the "island of Fidel" (Hinch, 2010, p. 97). The cold war scenarios were conducive to the creation of political barriers between people, causing tourist losses. In fact, there was a drop of 75% concerning travel from the United States to the Soviet Union (Edgell, 1990). This action taken by the United States later led to a kind of "revenge" manipulated by the Soviet Union. In 1984, under the encouragement of the Soviet Union, the Eastern Bloc (except Romania) also boycotted the Los Angeles Olympic Games. Other examples show the same trend. The crash of a Korean civilian plane, shot down by Soviet jets in 1983, led several Western countries to ban flights to the Soviet Union or Soviet planes to land in their countries. Another case occurred in 1980, after the invasion of Afghanistan by the Soviet Union. Following this offensive, the United States denied the landing of Aeroflot flights in American territory (Richter, 1989). The President of the United States, Jimmy Carter, also encouraged the Americans and the Olympic Committee to boycott the 1980 Olympic Games in Moscow (Richter, 1983).

Asia is also conducive to cases that interrelate politics with tourism. If in many scenarios in the Far East, tourism is very conditioned to the will of political actors, in China this dependence is much greater. China has a political system of central leadership, exercising strong control over outbound travel, often instrumentalising this sector in achieving strategic objectives. Many travel constraints between countries in Asia are the result of local incidents. Examples are the tensions on the island of Diaoyu / Senkaku, the recurring crises in the Korean peninsula, the tensions in the South China Sea, terrorism in Central Asia, border tensions with India, Sino-American relations or the Taiwan Strait crisis. In this paper, we intend to analyse some of the regional political constraints and their relations with the tourism sector, centred on the Chinese context. We highlight the way in which this country has controlled outbound tourism and instrumentalised this sector to obtain strategic benefits. We start with an approach on the interdependencies between the domains of tourism and politics, and then analyse how China has control over this sector, which is managed according to its external relations, and, finally, we will illustrate some regional cases that exemplify the previous assumptions.

THE INTERDEPENDENCIES BETWEEN TOURISM AND POLITICS

For the sustainable development of international tourism, it is important that there is a good climate of cooperation between governments and good social relations (Elliott, 2002). In other words, the establishment of peaceful scenarios is essential to the development of this sector. Timothy and Kim (2015) and Prideaux (2005) state that, in the scope of international affairs, tourism has its role in promoting friendship with other countries, facilitating the development of good bilateral relations. In the same sense, Kim, Prideaux and Prideaux (2007) propose tourism as a potential mechanism for the promotion of peace, adding that the construction of this type of peace must be done through the establishment of positive relations between people.

Tourism is supposed to increase international understanding, respect and mutual tolerance between countries (Johnson & Thomas, 1992). D'Amore (2010) states that the tourism industry can play a vital role in promoting international understanding and collaboration between nations, in order to enable the construction of peace between the countries of the world. In addition, Edgell (1990) explains that tourism has its diplomatic function, it is, for a country, a way of gaining attention on the international arena, offering the opportunity to increase mutual understanding, goodwill and even peace.

A good relationship in the tourism sector can generate mutual understanding and, at the same time, also requires a stable relationship between countries. Since, on the one hand, tourism can act as a kind of catalyst that facilitates understanding and trust, on the other hand, bilateral tourism can become durable based on trust between people and governments. Therefore, a good relationship between governments can also encourage the development of tourism and "provide room for it to operate" (Kim, Prideaux, & Timothy, 2016). Hall (1998) indicates that good political relations between countries can be manifested by encouraging tourist flows, which can be regarded as an indicator of bilateral relations between countries (Richter, 1989). He adds that exchanges in the field of tourism between two countries increase or decrease depending, to a certain degree, on political relations. In other words, tensions dangerously affect tourism cooperation (Prideaux, 2005; Timothy & Kim, 2015).

Policies aimed at foreign tourists are formulated and implemented based on the state of the countries' bilateral relations, as in the case of China (Richter, 1989). Good relations can offer a good international environment for travellers, simultaneously, relations between countries are conditioned by security, instability and risk. The relationship between political tensions and tourism, including contexts of crisis and social upheavals, was analysed in detail by Hall (2010), noting that it is an under-researched area. However, he concluded that risk and instability caused by political conflicts are a major concern for international visitors. The effects are often devastating, leading to the ruin of many local operators. In other cases, diverting visitor flows from unsafe scenarios to other locations can cause a new economic dynamism. In other words, tourism has a very close relationship with events taking place in the international arena. It is noteworthy the case of the fall of the Berlin wall in 1989; there was an increase in tourist movements from various parts of Europe.

In the opposite direction, we can find harmful effects. Pforr and Hosie (2008) clarify that the tourism industry is particularly susceptible to these negative events since they can often put the sector under threat, disrupting its regular development. International tourist arrivals can be influenced by economic crises, energy and political issues, health issues, environmental or natural disasters (Hall, 2010). Nowadays, information about natural tragedies, such as earthquakes or typhoons, or tragedies caused by people, such as terrorism, are quickly disseminated through the media on television and the Internet. The perception of risk is an important factor, although often out of step with reality, very inflamed by the media or covered up by political power. Such information influences the search for alternative destinations (Zhang, Cho, & Wang, 2019). The attack on the twin towers in 2001, the 2008 economic crisis and the Covid-19 pandemic are some examples of international events with the most significant implications for the evolution of international tourism. In the case of the referred pandemic, the search for alternatives was not even possible, constituting an unprecedented blockade of international tourism flows. The international crisis resulting from 9/11 and the "war on terrorism" has led to the closure of many hotels, leading to a significant decrease in the number of tourists and the closure of many hotels. Baker (2014) reveals that events like this can have substantial impacts on the tourism industry, increasing tourists' anxiety about safety on destinations. The 2004 Indian Ocean tsunami also devastated the coasts of Indonesia, Sri Lanka, southern India, Thailand and other countries, causing a loss of one billion dollars in the region's tourism industry. Political conflicts between nations are also a factor that influences the normal functioning of tourism, also serving as an instrument of pressure in the foreign policy of some countries. They are particularly managed in authoritarian countries, such as China, a power in this field, among many others.

OUTBOUND TOURISM WITH CHINESE CHARACTERISTICS

The Chinese communist regime is an authoritarian political system that mixes traditional Leninism with meritocracy and legalism. In many ways, it is a unique system that, although having received external influences, has its own characteristics. The centrality of decisions places all areas of the economy and society under the will and guidelines of the Chinese Communist Party (CCP). Tourism is one of these areas, being added to central planning and coordination, conditioned by the evolution of external relations.

The internal movement between the countryside and cities has been managed over the decades with extraordinary care, avoiding mass displacement that could create social instability. According to Tse and Hobson (2008), this will be one of the reasons behind the creation of the 1958 household registration system, known as *hukou*. Fears of manipulation by the masses, mobilisations for action, the assimilation of ideas contrary to the purposes of the CCP, cause fear and induce the careful management of this whole process. Incidents such as the ones presented earlier are detonators of blockades in the field of tourism, often used as a weapon of political thrust against other countries.

China created the Approved Destination Status (ADS). Since the 1990s, the Chinese government has established this agreement, being negotiated with more than 120 countries. The granting of ADS to a country goes through a process of negotiations with the CCP. Chinese travel agencies can organise group trips and obtain visas according to the ADS guidelines (Arita et al., 2011). Chinese tourists can travel in groups to countries and regions that have signed the agreement, only by licensed travel companies. For the granting of ADS, of course, there must be good diplomatic relations. Any anomaly to the normal bilateral relationship may be a reason to withdraw the status. For example, the 15 countries that have official diplomatic relations with Taiwan did not receive the ADS. The "one China" principle is very relevant in Chinese foreign policy. The political dimension overlaps the economic dimension in this matter. The granting of ADS status also has a visa cooperation policy. The restrictive American policy in this area, to which is added the context of tension between Xi Jinping and Donald Trump, leaves the United States out of this status. The good conditions in terms of infrastructure, the solutions and the reception of Chinese tourists are also important factors in the ADS assessment (Tse & Hobson, 2008).

For the above reasons, we find that the political dimension is very relevant in the development of Chinese outbound tourism. It justifies the CCP's monopoly in managing the mobility of Chinese citizens, both internally, by limiting migration from the countryside to cities or by the mandatory confinement due to a pandemic, and externally, through the control of outbound travel. Three foreign policy scenarios illustrate the implications for regional security and their immediate effects on Chinese tourism: the territorial dispute on the Diayu / Senkaku Islands, the Korean anti-missile defence system and the naval confrontation with the Philippines. These will be analysed in the following section.

SCENARIOS OF REGIONAL INSTABILITY

The Islands Diaoyu/Senkaku Dispute

China claims historical sovereignty, in the Southeast China Sea (SCS), of a set of islands, reefs, rocks and shoals. It includes the Paracel Islands, the Spratly Islands, and other smaller islands, such as the Zhongsha, Dongsha and Huangyan Islands, in an area of more than three million square kilometres. The point is that six countries dispute territorial possessions in the SCS, including Japan and the Philippines (Wu, 2013). China has been occupying some of these islands – without resistance from third countries –, making earthworks and building airstrips on them, which causes tensions with the remaining sovereignty claimants (Brown, 2017). In the case of Japan, there is an old territorial dispute over the Diaoyu Islands (Chinese name) / Senkaku (Japanese name), sensitive in the bilateral political relationship. Occasionally some incidents raise tension levels, fearing an armed conflict between the two nations. The years 2012 and 2013 were particularly tense. Signs of tensions started to appear in early 2012, when it was reported that Japan intended to name the uninhabited islands, including the islands attached to the Diaoyu / Senkaku Islands. In March, Japan published the names of 39 uninhabited islands, including four islands affiliated to the Diaoyu / Senkaku Islands. In the face of this, the spokesperson for the Chinese Foreign Ministry declared that China opposed those actions, considering a unilateral measure taken by the Japanese side regarding the Diaoyu / Senkaku Islands, and illegal and ineffective the naming of the affiliated islands (MREPRC, 2012).

In April, Shintaro Ishihara, governor of Tokyo, expressed the idea of buying, on behalf of the Tokyo Metropolitan Government, some islands in the Diaoyu / Senkaku Islands from Japanese private owners, the Kurihara family, in his speech during the visit to the United States. He later got millions of yen donated by Japanese people as a shopping fund. In response to Ishihara's declaration on the purchase, the Japanese government proposed a nationalisation plan for the islands, in July, and decided to buy the three islands from the Diaoyu / Senkaku Islands in September to "nationalise" them. This, in fact, led to an official large-scale debate between China and Japan, in which the two sides both declared their sovereignty over the Diaoyu / Senkaku Islands several times.

Following the Japanese "nationalisation" of the islands, Chinese patrol ships, naval warships and air force planes repeatedly approached the waters near the disputed islands, whose main objective was to challenge Japan's control over the islands and pressure it to admit the existence of a territorial dispute. At the end of 2012, with the change in the leadership in both counties, the military and political balance between the two has been intensified throughout 2013. According to the Diplomatic Bluebook 2014 (Ministry of Foreign Affairs of Japan, 2014), from January to December 2013, Chinese ships entered Japanese territorial waters 52 times (involving a total of 180 vessels), and on one occasion, in August, remained in Japanese territorial waters for more than 28 hours, which constitutes a record of the Chinese intrusion with the longest time to date. In addition to the increasing involvement of the armed forces on both sides, the two countries demonstrated their attitude towards the sovereignty issue in the Diaoyu / Senkaku Islands (Smith, 2018). China strengthened its military power and demonstrated a resolute attitude towards defending the integrity of its territory, while Japan strengthened its surveillance capabilities and ties with the United States.

This political tension has deteriorated Sino-Japanese relations, also provoked anti-Japanese feelings in the Chinese people. In more than 100 Chinese cities, anti-Japanese demonstrations occurred, involving material damage and personal injuries; in some cities, these demonstrations have evolved to boycott Japanese products (Dreyer, 2016; Nakano, 2016), especially in the automotive industry and tourism (Drifte, 2014). As for tourism, data regarding Chinese tourists to Japan show a negative annual growth rate for 11 consecutive months, from October 2012 to August 2013. According to Siriphon and Zhu (2018), the Chinese government intervened before the National Day vacation in October 2012 in order to order state-owned companies to abstain from travel promotions to Japan; CITS also stopped advertising trips to Japan. China also decided not to send 50,000 Chinese to Japan to celebrate the 40th anniversary of the normalisation of the Sino-Japanese relations (Tse, 2013). In this context, there were cancellations of travel reservations from China to Japan. Chinese public policies clearly intended to discourage outbound travel to Japan. According to Cheng and Wong (2014), information collected through interviews with the Chinese indicates that, during the period of territorial dispute in the Diaoyu / Senkaku Islands, some state-owned companies also tried to avoid their employees travelling to Japan.

Japan also confirmed, through a White Paper on Tourism, that in 2012 Japan was unable to achieve the goal of receiving 9 million foreign tourists due, in part, to the uproar caused by the nationalisation of the Diaoyu / Senkaku Islands (MLIT, 2012). Specifically, after the seizure of the islands in September, there was a notable decline in the number of Chinese tourists, particularly group tourists (MLIT, 2012). According to the monthly reports on foreign tourists from the Japanese National Tourism Organization (JNTO), as of September 2012, there were few new bookings and cancellation of reservations made by Chinese tourists, despite the resumption of travel promotions for the Japan in major Chinese cities in late 2012 and early 2013, even at the height of the Chinese New Year, the golden season for making outbound trips, the number of Chinese tourists to Japan was down by 1.3% in February 2013. Subsequently, with the continued cooling of Sino-Japanese relations, there was also a reduction in air routes and the suspension of some flights between the two countries (JNTO, n.d.). However, as of September 2013, there has been a recovery with a 28.4% growth in the number of Chinese tourists compared to 2012, due to the depreciation of Yen (MLIT, 2013) and increased effort on the part of Japan to make promotions and disseminate information about travel to Japan in China.

South Korea's THAAD Missile Defence Incident

In the regional geopolitical framework, of competition between China and the United States, the establishment of an American Terminal High Altitude Area Defense (THAAD) anti-missile defence system in South Korea, in 2017, brought a new political crisis between Beijing and Seoul. According to the Chinese government, this facility is a threat to national security. Also in this scenario, there has been a substantial reduction in Chinese tourists to South Korea (Jin, Qu, & Bao, 2019; Juan, Choi, An, Lee, & Lee, 2017). In fact, the THAAD incident started in September 2016, when the South Korean Ministry of Defense announced that the Lotte Group's Sungju Golf Course would be the site of the THAAD system. On February 27, 2017, the Lotte Group also confirmed its willingness to provide the field for the implementation of such a system. Since March of that year, the Ministry of Defense of South Korea has released news that some equipment of the THAAD system has been delivered to the United States Air Force Base in South Korea. Political tensions between China and South Korea have intensified. In the face of that, the Chinese government has taken measures, such as restricting business with South Korea, boycotting the Lotte group, among others. The situation started to improve from October 2017, when South Korea declared it would freeze the installation of the THAAD system in December. The Korean president made his official visit to China, which was considered a sign of improved bilateral relations.

CCTV 4 (2017) reported that the China National Tourism Administration (CNTA) had announced a travel alert to South Korea on March 3. According to CNTA, several Chinese tourists were having difficulties in entering the Jeju Island, thus issuing a warning on the risks of travelling to South Korea. This constituted an attempt to pressure the reduction of travellers through official media, having also advised travel agencies not to operate group and individual travel to South Korea (Huang, 2017). According to information gathered, by Jin, Qu and Bao (2019), in interviews with operators of the six main Guangzhou-based tourism companies, they unveiled that some companies received a notification from the government stating that travel agencies should not organise travel to South Korea. In addition, online tourist services, including travel information, travel programs, and accommodation in South Korea have been removed. According to South Korea's tourism indicators, it appears that government intervention has had considerable effects on travel flows. Thus, in March 2017, the number of Chinese visitor arrivals to South Korea was 360,782, having decreased from 590,790 in the previous month. From March 2017 to February 2018, the monthly number of Chinese arrivals showed a negative trend compared to the homologue period of the previous year. March and April recorded a -40% and -66.6% decrease.

Sino-Philippine Maritime Tensions

Another case of tension involving Chinese foreign relations is the maritime disputes that the PRC has with the Philippines within the framework of "historic rights". In April 2012, a direct confrontation between China and the Philippines on the Scarborough Reef (Huangyan Island) raised tensions between the two countries. The Chinese embassy in Manila was the scene of demonstrations by many Filipinos against the Chinese stance. In May, many Chinese travel agencies suspended travel to the Philippines (BBC News, 2012), as a result of an action managed by the government. China's top three licensed tourism agencies, in terms of revenue, are all state-owned and only 8% of their 25,000 licensed travel agencies are authorised to sell international trips. Foreign agencies are not allowed to provide travel services to Chinese citizens (Anwar, 2019). The CNTA issued a statement recommending Chinese tourists to suspend their travel to the Philippines. Subsequently, it declared that all Chinese tourists travelling in the Philippines should leave the country before May 16 (VOA Chinese, 2012). Due to the size of the market, the drop of Chinese tourist flows was an important instrument of retaliation, having also been used in retaliation against Taiwan in the Tsai Ing-wen era (Financial Times, 2019). After the CNTA announcement in May, Chinese tourists in the Philippines accelerated their return to China, while charter flights from Beijing, Shanghai, Hangzhou and other cities to the Boracay island in the Philippines were also suspended (GOV. CN, 2012). According to a report by CCTV News (2012), a travel agent from the company Ctrip, one of the largest agencies in the country, revealed that safety is the priority of travel. In the face of protests by the Filipinos, the company had to suspend all trips to that country until the situation was resolved.

According to data from the Philippines Tourism Department, the number of Chinese visitors arriving in the country dropped from 26,807 in April to 15,204 in May, representing a decrease of 43.3%. The number of tourists continued to decline, reaching 12.283 visitors in June. Until the end of 2012, the stress of bilateral relations was maintained. Between April and December, the monthly average number of Chinese visitors was 17,159, with a total number being 154,428 in this period, representing a decrease of 18.2% compared to the same period of the previous year. The "China Tourism Industry Statistical Bulletin" also revealed that, in the previous two years (2010 and 2011), the growth rates of the number of Chinese tourists to the Philippines were 26.8% and 26%, respectively. However, in 2012 the increase was only 0.4%.

CONCLUSION

Conflicts between nations are a major cause of the drastic reduction in tourist flows, resulting from multiple factors. Insecurity and risk are some of these factors. The sustainable development of international tourism depends on political relations; in the case of Chinese foreign policy, this is almost imperative. It is one of the areas immediately affected, centrally managed by the mobility of travellers, used as a mechanism to induce pressure in other countries. Tensions between China and the Philippines, for example, showed how immediately Chinese outbound tourist flows decreased due to maritime incidents. Still, as usual in international embargoes and retaliations, international actors are quickly looking for alternatives. In many ways, we can say that there is a relative effect on the adoption of these measures. Instead of achieving political goals, they work more as an internal appeasement factor, and in the end, from an economic point of view, everyone loses with this action (lose-lose).

REFERENCES

Anwar, A. (2019). *How China is using tourists to realise its geopolitical goals*. Retrieved June 4, 2020, from https://www.eastasiaforum.org/2019/09/19/how-china-is-using-tourists-to-realise-its-geopolitical-goals/

Arita, S., Edmonds, C., Croix, S. L., & Mak, J. (2011). Impact of Approved Destination Status on Chinese travel abroad: An econometric analysis. 17 (5). *Tourism Economics*, *17*(5), 983–996. doi:10.5367/te.2011.0076

Baker, D. (2014). The effects of terrorism on the travel and tourism industry. *International Journal of Religious Tourism and Pilgrimage*, *2*(1), 58–67. doi:10.21427/D7VX3D

BBC News. (2012). *China travel agencies suspend Philippine tours.* Retrieved from BBC News: https://www.bbc.com/news/world-asia-18015538

Brown, K. (2017). *China's world.* I.B. Tauris. doi:10.5040/9781350985704

CCTV 4. (2017). *CNTA publicou alerta sobre as viagens para a Coreia do Sul.* Retrieved from http://tv.cctv.com/2017/03/04/VIDERw2uBWJ4chDEBjFpxIva170304.shtml

CCTV News. (2012). *A China suspendeu as viagens para as Filipinas.* Retrieved from http://www.iqiyi.com/w_19rs0d7n5x.html#curid=3278338409_dc21c91e039390dfc405cce5d1a6fe2b

Cheng, M., & Wong, A. I. (2014). Tourism and Chinese popular nationalism. *Journal of Tourism and Cultural Change*, *12*(4), 307–319. doi:10.1080/14766825.2014.914948

Chinese, V. O. A. (2012). *A China vai retirar todos os turistas chineses nas Filipinas.* Retrieved from VOAChinese: https://www.voachinese.com/a/article-20120514-tourists-return-151385825/958490.html

D'Amore, L. (2010). Peace through tourism: The birthing of a new socio-economic order. *Journal of Business Ethics*, *89*(S4), 559–568. doi:10.100710551-010-0407-3

Dreyer, J. T. (2016). *Middle Kingdom & Empire of The Rising Sun: Sino- Japanese Relations, Past and Present.* Oxford University Press.

Drifte, R. (2014). The Japan-China confrontation over the Senkaku/Diaoyu Islands: Between "shelving" and "dispute escalation". *The Asia-Pacific Journal*, *12*(30 (3)), 1–61.

Edgell, D. L. (1990). *International tourism policy*. Van Nostrand Reinhold.

Elliott, J. (2002). *Tourism: Politics and public sector management*. Routledge. doi:10.4324/9780203416136

Financial Times. (2019). *China suspends individual tourist permits to Taiwan before election*. Retrieved June 4, 2020, from: https://www.ft.com/content/6ba14934-b35e-11e9-8cb2-799a3a8cf37b

GOV.CN. (2012). *CNTA: Os turistas chineses estão a acelerar o seu retorno para a China.* Retrieved from Portal do Governo da RPC: http://www.gov.cn/jrzg/2012-05/13/content_2136105.htm

Hall, C. M. (1998). *Tourism and politics: Policy, power and place*. Routledge.

Hall, C. M. (2010). Politics and tourism: Interdependency and implications in understanding change. In R. Butler & W. Suntikul (Eds.), *Tourism and political change* (pp. 8–18). Goodfellow Publishers.

Hall, C. M. (2010). Crisis events in tourism: Subjects of crisis in tourism. *Current Issues in Tourism*, *13*(5), 401–417. doi:10.1080/13683500.2010.491900

Hinch, T. (2010). Tourism as an instrument of foreign policy: The US trade embargo on Cuba. In R. Butler & W. Suntikul (Eds.), *Tourism and political change* (pp. 97–107). Goodfellow Publishers.

Huang, E. (2017). *China is trying to punish South Korea by keeping its tourists away.* Retrieved from Quartz: https://qz.com/923890/china-retaliates-against-thaad-antimissile-system-and-bans-tourism-to-south-korea/

Jiang, Y. (2013). *Disputa das Ilhas Diaoyu e perspectivas das relações sino-japonesas*. Retrieved from China Institute of International Studies: http://www.ciis.org.cn/chinese/2013-07/01/content_6074003.htm

Jin, X., Qu, M., & Bao, J. (2019). Impact of crisis events on Chinese outbound tourist flow: A framework for post-events growth. *Tourism Management*, *74*(April), 334–344. doi:10.1016/j.tourman.2019.04.011 PMID:32287753

JNTO. (n.d.). *Estatísticas sobre visitantes estrangeiros no Japão*. Retrieved 11 2018, from JNTO: https://www.jnto.go.jp/jpn/statistics/data_info_listing/index.html?tab=block7

Johnson, P., & Thomas, B. (1992). *Perspectives on tourism policy*. Mansell.

Juan, Y., Choi, Y., An, S., Lee, C. K., & Lee, S. J. (2017). The constraints of Chinese tourists to visit Korea caused by THAAD using Q-methodology. *Asia Pacific Journal of Tourism Research*, *22*(12), 1261–1273. doi:10.1080/10941665.2017.1391306

Kim, S., Prideaux, B., & Timothy, D. (2016). Factors affecting bilateral Chinese and Japanese travel. *Annals of Tourism Research*, *61*, 80–95. doi:10.1016/j.annals.2016.08.001

Kim, S. S., Prideaux, B., & Prideaux, J. (2007). Using tourism to promote peace on the Korean Peninsula. *Annals of Tourism Research*, *34*(2), 291–309. doi:10.1016/j.annals.2006.09.002

Lickorish, L., Jefferson, A., Bodlender, J., & Jenkins, C. L. (1991). Developing tourism destinations: policies and perspectives. Harlow: Longman.

Ministério das Relações Exteriores do Japão. (2014). *Diplomatic Bluebook 2014 (period covered 2013 calendar year)*. Retrieved from Portal do Ministério das Relações Exteriores do Japão: https://www.mofa.go.jp/files/000055801.pdf

MLIT. (2012). *White Paper on Tourism in Japan 2012*. Retrieved from Japan Tourism Agency (MLIT): https://www.mlit.go.jp/common/001018364.pdf

MLIT. (2013). *White papers on tourism in Japan 2013*. Retrieved from Japan Tourism Agency (MLIT): https://www.mlit.go.jp/common/001211777.pdf

MREPRC. (2012). *Declaração do Ministério das Relações Exteriores da República Popular da China*. Retrieved from Portal do Ministério das Relações Exteriores da República Popular da China: https://www.fmprc.gov.cn/diaoyudao/chn/xwdt/t967820.htm

MREPRC. (2012). *Ilhas Diaoyu*. Retrieved from Portal do Ministério das Relações Exteriores da República Popular da China: https://www.fmprc.gov.cn/diaoyudao/chn/xwdt/t910599.htm

MREPRC. (2012). *Ilhas Diaoyu - Notícias*. Retrieved from Portal do Ministério das Relações Exteriores da República Popular da China: https://www.fmprc.gov.cn/diaoyudao/chn/xwdt/default_5.htm

Nakano, R. (2016). The Sino-Japanese territorial dispute and threat perception in power transition. *The Pacific Review*, *29*(2), 165–186. doi:10.1080/09512748.2015.1013493

Okumus, F., Altinay, M., & Arasli, H. (2005). The impact of Turkey's economic crisis of February 2001 on the tourism industry in Northern Cyprus. *Tourism Management*, *26*(1), 95–104. doi:10.1016/j.tourman.2003.08.013

Pforr, C., & Hosie, P. J. (2008). Crisis management in tourism preparing for recovery. *Journal of Travel & Tourism Marketing*, *23*(2–4), 249–264. doi:10.1300/J073v23n02_19

Prideaux, B. (2005). Factors affecting bilateral tourism flows. *Annals of Tourism Research*, *32*(3), 780–801. Advance online publication. doi:10.1016/j.annals.2004.04.008

Prime Minister of Japan and His Cabinet. (2012). *Press Conference by Prime Minister Yoshihiko Noda*. Retrieved from Prime Minister of Japan and His Cabinet: https://japan.kantei.go.jp/noda/statement/201208/24kaiken_e.html

Richter, L. K. (1983). Tourism politics and political science. *Annals of Tourism Research*, *10*(3), 313–335. doi:10.1016/0160-7383(83)90060-9

Richter, L. K. (1989). The politics of tourism in Asia. Honolulu: University of Hawaii Press.

Siriphon, A., & Zhu, J. (2018). Outbound Tourists and policy reforms under the Chinese mobility regime. In China's Reforms in the New Era and Their Implications for East Asia (pp. 1–29). Academic Press.

Smith, S. A. (2018). Japan and China: A fresh start in a difficult era. *Asia-Pacific Review*, *25*(2), 75–88. doi:10.1080/13439006.2018.1551823

Timothy, D. J., & Kim, S. S. (2015). Understanding the tourism relationships between South Korea and China: A review of influential factors. *Current Issues in Tourism*, *18*(5), 413–432. doi:10.1080/13683500.2013.776019

Tse, T. S. (2013). Chinese outbound tourism as a form of diplomacy. *Tourism Planning & Development*, *10*(2), 149–158. doi:10.1080/21568316.2013.783738

Tse, T. S., & Hobson, J. S. P. (2008). The forces shaping China's outbound tourism. *Journal of China Tourism Research*, *4*(2), 136–155. doi:10.1080/19388160802279459

UNWTO. (2011). *Tourism towards 2030: Global overview*. World Tourism Organization.

UNWTO. (2018). *Tourism highlights*. World Tourism Organization.

Wu, S. (2013). *Solving disputes for regional cooperation and development in the South China Sea: A Chinese perspective*. Chandos Publishing. doi:10.1533/9781780633558

Zhang, H., Cho, T., & Wang, H. (2019). The impact of a terminal high altitude area defense incident on tourism risk perception and attitude change of Chinese tourists traveling to South Korea. *Sustainability 2020, 12*(7), 1–13.

Zhang, H. Q. (2005). Impact of the tsunami on Chinese outbound tourism. *International Journal of Contemporary Hospitality Management*, *17*(5), 433–435. doi:10.1108/09596110510604850

Zhu, F. (2014). *China's first ADIZ decision: One year later.* Retrieved from https://amti.csis.org/chinas-first-adiz-decision-one-year-later/

Section 3

Human Interactions and Sustainable Development

Chapter 12
Hostels in Hostile Territory:
Tourist Spaces of Transformative Dialogue in the Israeli–Palestinian Context

Jack Shepherd
Mid Sweden University, Sweden

Daniel Laven
Mid Sweden University, Sweden

ABSTRACT

This chapter looks at the role youth hostels play in challenging some of the destructive narratives of the Israeli-Palestinian context, narratives revolving around Othering, and demonisation of the Other. To understand the role of hostels in this regard, the authors use the concept of transformative dialogue to demonstrate how hostels assist in the formation of partnerships between previously hostile groups, and also how hostels provide a tourism forum for self-expression and exposure to divergent narratives of space. The chapter is a comparative case study using qualitative data gathered at 14 hostels over a period of six years. The findings of the chapter stress the remarkable achievement of youth hostels in the region in challenging hegemonic discourses of separation and abjection, despite the enormous difficulty in doing so.

INTRODUCTION

This chapter explores how youth hostels can challenge some of the injurious master narratives of the Israeli-Palestinian context. Our aim is to examine the ways that hostels curate tourism spaces in order to facilitate engagement with the Other. In doing so, we employ the concept of transformative dialogue (Gergen, McNamee, & Barrett, 2001), which involves the engagement of antagonistic groups in respectful dialogue. Enacting transformative dialogue is extremely challenging in this context as shall be explored, but the results of our study suggest more hope than despair. We begin with the geopolitical and theoretical background, which is followed by findings from Israel and Palestine. We conclude the chapter with reflections on the role of hostels in peacebuilding efforts in the region.

DOI: 10.4018/978-1-7998-5053-3.ch012

BACKGROUND

The conflict between Israelis and the Palestinians has entered yet another decade. During the previous two, there has been little political will to reach a mutually satisfactory end to the conflict. The Israeli government appears more determined than ever to swallow the West Bank whole, supported by the current American administration and abetted by increasingly coy Arab states. The Palestinians, for their part, remain divided and struggle under the straightjacket of the occupation. One of the main reasons why the conflict has remained so intractable is due to the presence of two seemingly hegemonic opposing narratives of the region. These narratives are the product of over 70 years of enforced social distancing and are explored in a wide variety of texts (e.g. Masalha, 2019; Pappe, 2015; Said, 1979; Shavit, 2013). On the one hand, there is the Zionist narrative of the region, which excludes the presence of the Arab population in the historic land of Palestine. This has resulted in Arabs living in Israel being seen as an inconvenient reality, estranged from mainstream society, their parties ostracised in parliament and their status as citizens in Israel all but denied by the recent 'Nation-State bill'[1]. In the West Bank, this same discourse leads to the continuation of the Nakba or 'catastrophe', which refers to the displacement of Palestinians in 1948 following the creation of the state of Israel. It also leads to increasing encroachment of Israeli settlements on Palestinian land and the establishment of a military occupation in the West Bank, which restricts Palestinian mobility and ignores basic human rights (B'tselem, 2019). On the other hand, there is the Palestinian narrative of the region, which clings onto their 'right of return' to the houses and lands from which they were expelled in 1948 when the state of Israel was established. Although a right of all refugees as stipulated by the United Nations, the inability of Palestinian activist groups such as the BDS (Boycott, Divestment and Sanctions movement) to explain how this resettlement would work in practice, coupled with the violent rhetoric of Palestinian militant groups such as Hamas and Islamic Jihad in Gaza, lead to serious doubts over Palestinian willingness to accept the presence of Jewish Israelis in the historic land of Palestine.

In this way, the Israeli-Palestinian context is dominated by two opposing visions of the region. Both sides remain stuck in monologue forms of communication (Frank, 2005), where there exists no dialogue, and instead, both sides reside in their own echo chambers. In this way, they forget the reality that surrounds them – that the Other is going nowhere. A major consequence of this failure of dialogue is that the unseen Other becomes an increasingly abject and simplified being (Gergen et al., 2001; Sibley, 1995). As this process gains speed, "there are fewer challenges to one's descriptions and explanations" and "with the continuing tendency to explain other's actions in a negative way, there is a movement towards extremity" (Gergen et al., 2001, p.681). In this way, the Palestinian is always a terrorist and the Israeli always a brutal oppressor. These world constructions, where the Other is portrayed as dangerous and unwanted, permeate all facets of civil society from the head of government right down to the schoolbooks that children read (Peled-Elhanan, 2012). And, when there is contact, the result is usually what has been termed thanatopic communication, whereby dialogue is killed through the continual finding of fault in the Other and the reaffirmation of one's superiority (Gergen et al., 2001). Therefore, instead of realising that solving the region's strife begins with the hearing and understanding of the narratives of the Other, there is a tendency to avoid, simplify and demonise the Other.

Sadly, these patterns of communication are often transferred to the tourism arena. Israelis cannot visit many Palestinian tourism centres, such as Bethlehem, Jericho and Ramallah, due to their location within Area A, an area that current Israeli policy forbids them from visiting. Most Palestinians cannot visit Israeli tourism centres across-the-board due to the prohibitive permit system that denies most Palestinians access to Israel. Within Israel, domestic tourists of Palestinian origin (otherwise called Israeli Arabs) often find themselves subject to disproportionate and degrading treatment at the hands of airport security services (Riba, 2019), and even when travelling within Israel, tourist sites and attractions lack sufficient Arabic signposting or interpretation (Gilad, 2019).

When it comes to the millions of international tourists who visit the Holy Land, the picture is not much different. Brin (2006) described how tourists are routinely seen as potential agents of one of the competing narratives of this deeply contested space. Not surprisingly, tour guides are often accused of portraying a purified narrative of space. In Jerusalem, for example, Israeli tour guides will remove the presence of Palestinians from the city through their discursive interpretation of the city's neighbourhoods (Brin & Noy, 2010). It is for this reason that Israel uses tourism as a way of promoting narratives of belonging and therefore immigration (or Aliyah) through its Birthright (Taglit) tourism programme (Sasson, Shain, Hecht, Wright & Saxe, 2014), whilst discouraging or denying tourists of Palestinian origin to visit Israel. And it is for this reason that Palestinian activists encourage tourists to visit sites relating to the Israeli occupation in order to convert them into messengers for the Palestinian cause (Kassis, Solomon and Higgins-Desbiolles, 2017). Tourists are largely discouraged from crossing borders in the Israeli-Palestinian context, with visiting Israel from Palestine seen as collusion with the occupation and visiting Palestine from Israel seen as putting your life at risk (Shepherd and Laven, 2019). It therefore seems that no matter what you do as a tourist in the Israeli-Palestinian space, you are considered as collaborating with the abject Other.

We thus refer to 'hostels in hostile territory', not just in the sense that the hostels we will discuss operate within a region of ongoing sectarian conflict, but also because the tourism geographies of the Israeli-Palestinian context reflect the contours of the broader conflict. The master-narratives of Otherness are as evident in the region's official tourism discourse as they are in all other aspects of Israeli and Palestinian society. Perhaps tourism maps of the region offer the most logical depiction of the contested and confusing nature of these tourism geographies (Collins-Krenier, Mansfeld, & Kliot, 2006). Figure 1 is the official Israel Touring Map, which is available throughout the country's tourism information points. For any uninitiated observer, this map gives no sense of the presence of an independent Palestine nor of Palestinian life. Moreover, through a duplicitous use of colouring, the map uses the differing areas of control of the Palestinian Territories, as stipulated by the Oslo Accords, to turn Palestinian areas into what seems as mere topographical features of the landscape. Finally, the map refers to Palestinian areas as Judea and Samaria, which is terminology often used to reinforce Israeli territorial claims of these areas.

Figure 2 is the Official Tourism Map of the Palestinian Authority. While this map does clearly define the border between Palestine and Israel, it excludes references to Jewish Israeli sites and communities, most notably Tel Aviv.

We do not claim that either of these maps is right or wrong. However, these maps do reflect a purification of space (Newman & Passi, 1998) which reinforce exclusionary narratives of the Israeli-Palestinian space. The fact that tourism reflects the dominant or official discourses of the Israeli-Palestinian conflict is hardly surprising. As one Palestinian respondent in our research said, demonstrating the inescapability of the conflict in tourism:

Honestly, we are competing on the same land, tourism wise [...] The issue is that, at its core, this is a dispute over land. Everything that's build on that will finally fall into that logic.

Figure 1. Official Israel Touring Map.
Source: Collected by Jack Shepherd at Ovda Airport, February 10, 2018

Figure 2. Official tourism map of the Palestinian Ministry of Tourism and Antiquities
Source: Retrieved from http://travelpalestine.ps/; accessed January 23, 2020

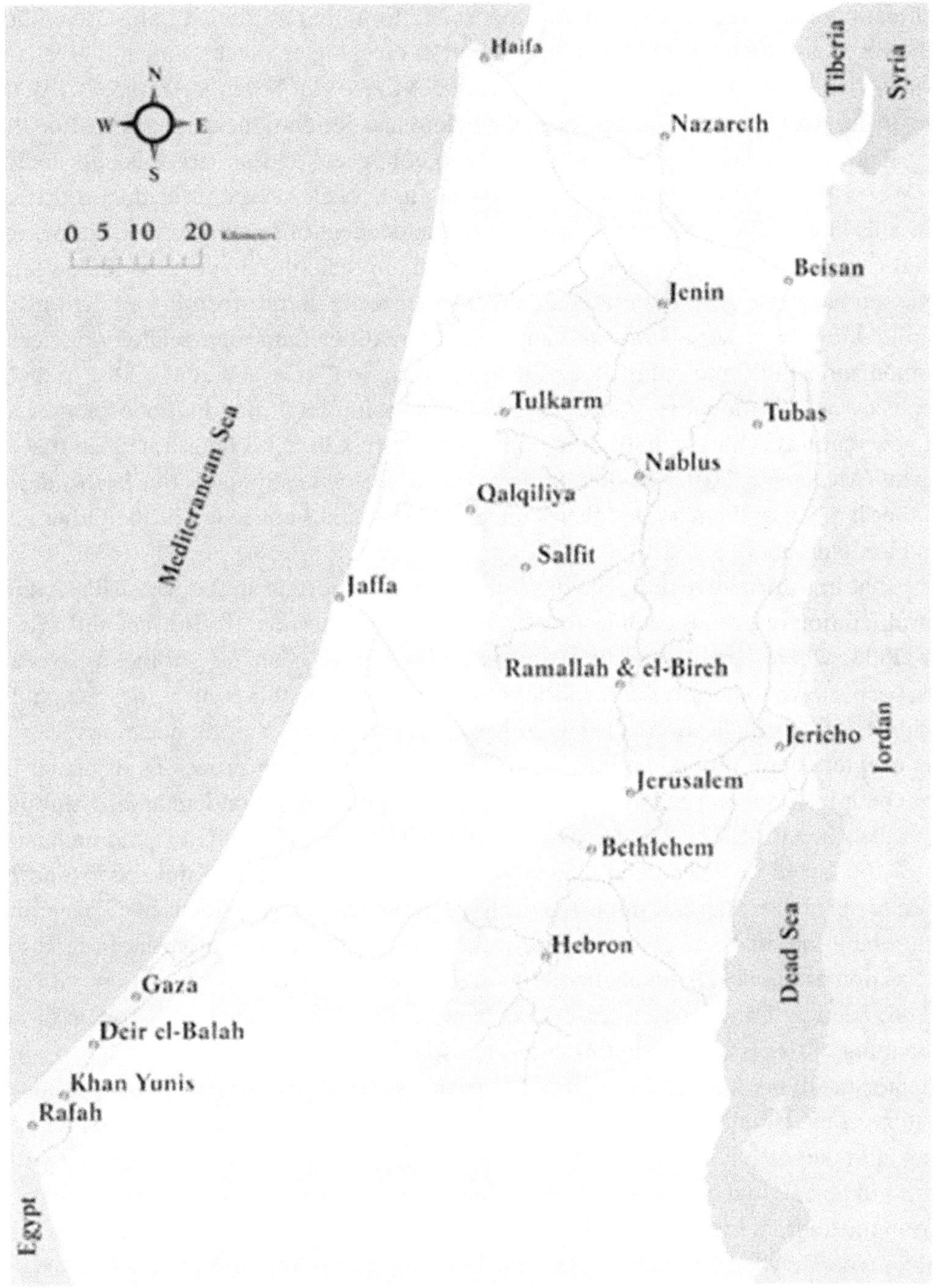

Despite these challenges, we feel there is still room for hope. Hope comes in a number of forms. First, tourism is the world's largest movement of people and offers tremendous potential for causing social change. With accelerating global communication networks and mobilities (Hannam, Sheller and Urry,

2006), there is hope that tourists will play a greater role in changing (or at least challenging) these kinds of destructive spatial dynamics. The second is that the Israeli and Palestinian tourism industries are, for better or worse, interdependent. It is nigh on impossible for a tourist to visit Palestine without visiting Israel. At the very least, they must cross an Israeli border crossing or checkpoint; and given the majority of tourists who visit Israel are Christian pilgrims (Israel Ministry of Foreign Affairs, 2014), the location of holy sites in cities such as Bethlehem, East Jerusalem and Jericho, means many visitors to Israel are also visiting Palestine. In addition, the growth in both countries' tourism arrivals appears to be highly correlated (World Tourism Organisation, 2018), largely as a result of complimentary tourism offerings.

It is from this hopeful perspective that we engage the concept of transformative dialogue. Emerging from a social constructionist philosophy of valuing and appreciating diverse worldviews, transformative dialogue represents "any form of interchange that succeeds in transforming a relationship between those committed to otherwise separate and antagonistic realities (and their related practices) to one in which common and solidifying realities are under construction" (Gergen et al., 2001, p.682). It places special emphasis on "relational responsibility, self-expression, affirmation, coordination and the co-creation of new realities" (Ibid., p.696). It is important to stress that this does not mean that there has to be a consensus (McNamee, 2010). Rather, transformative dialogue stipulates that both sides are willing to let the Other happen to them so that their "understanding and knowledge of the Other is enhanced" (Stewart and Zediker, 2000, p.232).

We know that transformative dialogue has been seen as important in the Israeli-Palestinian context given the proliferation of groups such as Roots (friendsofroots.net) or "To Reflect and Trust" (Bar-On & Kassem, 2004) where Israelis and Palestinians are invited to hear one another's narratives. These groups have been shown to improve perceptions of the Other within this context, despite public pressure against such acts of "normalisation" (Theissen & Darweish, 2018). Yet the questions we wish to deal with in this chapter centre around how tourism can be a part of such processes in which Israelis and Palestinians (both in Israel and in Palestine) can improve their mutual understanding through tourism. Moreover, at the same time, how can tourists, especially international tourists, gain understanding and knowledge of the Israeli or Palestinian Other, figures much-maligned and misunderstood by a global media whose focus on the region is almost entirely negative and conflict-focused? This point is crucial in a context where international actors appear to have as much influence on the progress (or not) of the conflict as domestic ones. A focus therefore on how tourists, through interaction with locals in the region, are provided with the space to co-create "new realities" (Gergen et al., 2001, p.682) is vital to our understanding of tourism's role in this contested space.

This chapter positions hostels as potential tourism spaces of transformative dialogue within the hostile territory of the Israeli-Palestinian context. We do this for two reasons. The first being the nature of hostels as intimate, convivial spaces where dialogue happens easily. The second being the apparent conduciveness of hostels in the region for challenging damaging hegemonic narratives. On the first point, it is clear from the tourism literature that hostels are seen as places of dialogue, both between backpackers themselves who are what Noy (2004, p.81) described as an "ad hoc community of storytellers", and between backpackers and the local community (Scheyvens, 2002). Backpackers are described as sociable beings, often packed in tight, intimate spaces such as dormitories and shared common rooms (Shepherd & Laven, 2019). Importantly, they are also in much closer contact with local people in a destination, not just because they depend on them for food and services that might otherwise have been provided by a hotel or tour company, but also because the hostel's common areas are spaces where the hostel staff work and relax (Moufakkir & Alnajem, 2017). Unlike in a hotel where the division of work and leisure is much stricter, and the environment sanitised for the tourist (Edensor, 2007), in a hostel the presence of staff and guests in close proximity, often sharing the work tasks of the hostel such as cleaning and bed-making, creates the possibility for meaningful engagement. Thus, the very act of staying in a hostel is an act of learning to accept and live with the Other…even if they snore!

On the second point, this chapter emerges from a series of works conducted in the Israeli-Palestinian context which appear to position hostels as spaces of dialogue (Gelbman and Laven, 2016; Hammami and Laven, 2017; Shepherd and Laven, 2019; Stenvall, Laven and Gelbman, 2017). These hostels were not chosen as case studies of cross-cultural dialogue due to their being as hostels, but because repeatedly, it appeared that hostels were sites of something extraordinary in the region; they were challenging the *status quo*, the master narratives, the monologue communication patterns we have described in the beginning of this chapter. Indeed, the intriguing role that hostels might play in this context was recently explored by a BBC News article entitled "Crossing Divides: The hostel promoting tourism to ease Israel's tensions" (Sherriff, 2020).

This chapter is an important extension of these previous works, and we draw on data from all of these prior studies to provide a comparative reflection, rather than focusing on another case study. As a result, in some instances, we offer insights directly from our previously published work, while in other instances we present previously unpublished data. We collected data at 14 hostels in the region between 2013 and 2019. During these visits, we conducted multiple semi-structured interviews with hostel managers, owners, staff and occasionally with guests. Our focus whilst interviewing was primarily on the "supply-side" as our interest so far has been on how the hostels provide the space for dialogue rather than the effect on the tourists. We also stayed in these hostels as guests, which amounts to several months of on-site participant observation. The hostels are located in both Israel (Jisr az-Zarqa, Nazareth, Tel Aviv / Jaffa, West Jerusalem, Eilat and Mitzpe Ramon) and Palestine (Bethlehem, Ramallah, Jericho). In some instances, we have provided the names of the hostels where it was appropriate to do so with the consent of the owners; in other instances, we have grouped together hostels or left out the hostel's name in order to ensure the confidentiality of the study participants. This is especially important when discussing Palestinian hostels. We first present the situation in Israel, and see how hostels there aim to promote transformative dialogue; and then we provide a view from Palestine on the same issues.

HOSTELS AS VENUES FOR TRANSFORMATIVE DIALOGUE

The View from Israel

In the spirit of transformative dialogue, our data demonstrated that numerous hostels in Israel represented spaces which promote "relational responsibility, self-expression, affirmation, coordination and the co-creation of new realities" (Gergen et al., 2001, p.696). They do this through providing tourism spaces for Israeli Arab self-expression, and for coordination, affirmation and dialogue between members of Israel's Arab and Jewish populations. They also do this by exposing the existence of multiple narratives of the Israeli-Palestinian space to tourists, whilst simultaneously challenging the monologue communication patterns of the region through opening lines of dialogue with Palestinian hostels. Each of these factors will be explored in this section.

Israeli Arabs (or Palestinians living in Israel) represent a sizable minority in Israel. Despite comprising almost a quarter of the country's population, Israeli Arabs are largely ostracised from mainstream Israeli society. Their contribution to Israeli political and cultural life is largely ignored. Within the tourism arena, the situation is no different. There is still today no museum in Israel that deals with an Arab history of Israel, nor one which looks at the trauma of the Palestinian Nakba. Israeli Arabs are, therefore, left in a situation whereby tourism offers them little room for self-expression and for sharing their narratives of life in Israel with tourists, both domestic and international (Hammami & Laven, 2017). There are, however, several Israeli hostels that address this issue. One such venture is the Fauzi Azar Inn, which was established in the Israeli Arab city of Nazareth in 2005 (Gelbman & Laven, 2016). The hostel is housed within a beautiful Ottoman-period property in the heart of Nazareth's Old City. The owners of the property, the Azar Family, opened their family home to tourism through a partnership with a Jewish social entrepreneur called Maoz Inon, who provided capital and a framework for running a hostel. The Azar family was able to use the guesthouse as a highly visible platform for sharing the experience of an Israeli Arab family of Palestinian origin. One member of the Azar family (the granddaughter of the inn's namesake, Fauzi Azar) and manager of the hostel at the time of Gelbman & Laven's (2016) first exploration of the site, described her experience in these terms (p.114):

I sit with guests – [and] many of them are Jews – and I share with them that [my] grandpa fought against the occupation. And [sometimes they] ask me, "Do you still call it occupation?" I say, "Excuse me, maybe for me it is still occupation." […] Sharing this story with guests is giving us [an opportunity] […] to tell [visitors] that here are Arab Christians. Ok, we have Israeli identity cards but it doesn't mean that we don't feel [a sense of] belonging to the Palestinians, or we are not Palestinians anymore. So these kinds of things are very challenging […] [But] It was very important for me what [our Israeli Jewish business partner] did […] this feeling that you don't have to hide anything, it's the truth.

This quote speaks to the power of the hostel to provide a platform for self-expression and exposing subaltern narratives that would never otherwise see the light of day through other official channels. The quotation also hints at the power of the hostel to promote transformative dialogue between the hostel's Arab staff and their Jewish guests. Israel's Arab and Jewish populations live in largely separate and separated realities – a direct consequence of the lingering tensions and hostilities left by the 1948 war that simultaneously caused the Palestinian Nakba and established the state of Israel, and by Israeli government policy, which has continually ostracised Israel's Arab population. Therefore, the creation of a space where Jewish Israelis can come and stay with an Arab family in the context of a backpacker's hostel is a valuable step towards creating room for transformative dialogue.

The same goes for the fact that this manager ran the hostel within a management structure that mixes both Israelis Jews and Israeli Arabs. As another Israeli Arab employee at the hostel affirmed, working at the hostel alongside Jews helped change her perceptions:

So he [our Israeli Jewish business partner] helped many people in this area, like me. He gives [me] a chance to have a job. So I tell [my Arab friends that question me for working here]… He's a nice guy, not like the other Jewish [people we have encountered]. Look, I did not like to live with Israeli Jews. So we have something in our minds that Israeli Jews are not good. Especially older Israeli Jews. It is like all Israeli Jews say all Arabs are not good. But when Israeli Jews meet Israeli Arabs and live and work together[…] this changes minds.

In 2019, this same study participant was promoted into a shared, senior management position at the hostel together with an Israeli Jewish counterpart. She described her career progression this way:

I started working at the Fauzi Azar Inn in 2008, as a cleaning lady. I cleaned the rooms. Then I worked at the reception. Then I opened a guesthouse in Cana [I am originally from the Arab-Palestinian village of Cana, which is now inside Israel and close to Nazareth], which I operated for one year. Then I came back to the Fauzi Azar Inn as the reception manager, and now I'm the manager of the entire hostel together with another lady. We now have two managers, and this is something, I am Arab and Muslim, and she is Jewish.

The experience described by this study participant directly reflects the notion of "new realities" (Gergen et al. 2001, p. 696) emerging from meeting and understanding the Other.

It is also worth considering the effect that this creation of a space for Israeli Arab self-expression has on the many international tourists who visit the hostel. Many visitors have no idea that Israel is a country with almost two million Arab citizens of Palestinian origin. The sensory experience of the Fauzi Azar Inn, with the sound of Arabic music playing in the reception, the waft of cardamom coffee from the kitchen and the Arabic spoken among its staff, tell a different story of Israel to that of the official discourse of the Israeli government and its tourism apparatus.

Another venture that is providing a similar space for self-expression and dialogue is Juha's Guesthouse in the Israeli Arab village of Jisr az-Zarqa (Stenvall et al., 2017). The village is often referred to as Israel's poorest village and suffers not only from extreme poverty (high unemployment, poor infrastructure, etc.), but also from isolation, having been effectively "walled off" from its neighbouring, much wealthier town of Caesarea, in what has been called "economic racism" (Galili, 2003). Domestic and international tourists dared not to visit Jisr az-Zarqa (TedxTalks, 2017), perceiving it as dangerous, despite being located on one of Israel's last strips of undeveloped coastline and its proximity to significant Roman archaeological sites. In response to this situation, a similar Arab-Jewish partnership led to the establishment of a hostel in the centre of the village. Once again, the hostel represents a space where Israeli Arabs can express themselves, and a new vision of Arab-Jewish relations can emerge through a tourism space. The discourse offered on the hostel's website is a good place to view this in action. For example, the first sentence visitors to the site encounter highlights the fact that Jisr az-Zarqa is the "only Arab village on the coast of the Mediterranean sea", while offering visitors the "opportunity to experience the authentic atmosphere of an ancient Arab fisherman's village". The site even promotes the village's proximity and easy access to Caesarea, which is especially interesting given the previous attempts to seal off Jisr az-Zarqa from its wealthy neighbour. Finally, the imagery of the site is highly symbolic and directly challenges the master narrative of racial and ethnic exclusion (Figure 3).

Figure 3. Screen capture from the Juha's Guesthouse website highlighting the Arab-Jewish partnership that established the hostel
Source: Retrieved from https://www.zarqabay.com/; accessed on January 23, 2020

Like the Fauzi Azar Inn, Juha's Guesthouse provides a platform for an Arab Israeli community to curate its own story, in its own community, on its own terms. The hostel highlights its Arab heritage as the primary attraction for tourists. Importantly, the hostel also emphasises its Arab-Jewish ownership in its promotional material, which is something not often seen in Israeli (or Palestinian) society. In these ways, the guesthouse offers another hopeful example of how hostels can represent spaces offering the possibility for transformative dialogue. From this perspective, the website's discourse and imagery are remarkable and stand in stark contrast with the master narratives that have endlessly weaponised the Israeli-Palestinian context.

Another Israeli hostel that offers the possibility for transformative dialogue between different ethnic groups, and also between tourists and Israelis more generally, is the Abraham Hostels in Tel Aviv and Jerusalem. To begin with, the outward face of the company, i.e. their logo, features both Hebrew and Arabic text in equal-sized font, which suggests a linguistic equality that has been denied by the recent Nation-State law (Figure 4).

Moreover, in both hostels, there is a clear mix of nationalities in the staff with many Israeli Jews and Israeli Arabs working side by side, alongside other staff and volunteers from some of Israel's many other ethnic groups. This, in turn, leads to the creation of a variety of different staff-led activities available to guests, which celebrate the different physiognomies of modern Israel. For example, guests can take part in Arabic cooking classes, Sudanese food tours and Shabbat dinners, all within the same hostel. Thus, the tourists (and the staff) who stay at the hostel are immersed in an Israel which shows itself to be more than just a "Jewish nation", but one which offers the possibility for a healthy dialogue between its various constituent ethnic groups.

These hostels, therefore, provide Israel's Arabs with the space to tell their own stories, to engage Israeli Jewish tourists in a dialogue about modern Israel through tourism, and the opportunity for international tourists to appreciate the multifaceted nature of Israel. Overall, they represent spaces where it is possible, as Stewart and Zediker (2000, p. 232) write, for "Letting The Other Happen to Me".

Some would argue that the real need for transformative dialogue lies between Israeli Jews and Palestinians living in the West Bank and Gaza, given the ongoing violence between these groups. The ability of our Israeli hostels to initiate dialogue between these two groups is severely restricted by, amongst other things, the mobility constraints imposed on Palestinians by the Israeli occupation of these areas, and by the restrictions on Israeli movements within Palestinian areas since the Oslo Accords of 1993. This means that there is not going to be any market for Palestinian tourists visiting Israel or vice versa anytime soon. Despite this situation, our research revealed that Israeli hostels are making serious challenges against the existing status quo, both in the establishment of links with Palestinian hostels, and in the provision of tours that offer tourists a multi-perspective approach to understanding the Israeli-Palestinian context.

In contrast to the official tourism discourse of the region, we have seen cases of Israeli hostels advertising Palestinian hostels. Figure 5 shows one such effort found in a hostel's reception in Jerusalem. It shows a leaflet advertising a hostel in Ramallah.

Figure 4. Abraham Hostels Logo
Source: Retrieved from www.abrahamhostels.com; accessed on March 8, 2020

Similar advertising was observed in a number of hostels in Tel Aviv, Mitzpe Ramon and Nazareth, where hostels in cities such as Ramallah and Bethlehem were advertised. In addition, when asked, the staff at numerous Israeli hostels could recommend a hostel in the West Bank to travellers. These leaflets and connections do not come from nowhere; and so they represent a reconstructive challenge against the non-communicative patterns of the Israeli-Palestinian context. In addition, the same could be said of the tours to Palestinian cities that are conducted from numerous Israeli hostels. From our experience on such tours with the Abraham Hostel chain in Israel, the actual guiding of the tours in cities, like Nablus, Ramallah and Jenin, is done by local Palestinians. Although we could (and perhaps should) critique such tours as just another example of Israel using its power to dominate the Palestinian tourism industry, the fact that Israeli hostels are providing employment possibilities to Palestinian tour guides and allowing Palestinians the forum to share their narratives of space is quite remarkable in a context that discourages all forms of fraternisation._

Figure 5. A leaflet advert for the Area D hostel in Ramallah at the Abraham Hostel, Jerusalem
Source: Collected by Daniel Laven in Jerusalem, January 9, 2016

A large part of transformative dialogue stems from a willingness to hear the perspectives of Others. As McNamee (2010) affirms, this does not mean one has to agree with these perspectives, but that through the process of dialogue, it is possible to understand how we relate to one another and how together we create meaning. Throughout our research, we have been struck by the multi-perspectivity that the tours run out of Israeli hostels offer. Such multi-perspectivity is epitomised by the promotion of dual-narrative tours associated with two of the hostels we visited. With titles such as "Jerusalem Dual Narrative Tour", whereby tourists are taken around contested cities and given both Israeli and Palestinian perspectives, it would appear such tours are the embodiment of transformative dialogue in the practice of tourism. Indeed, Schneider (2019) recently demonstrated how such tours increase empathy and understanding for both sides and help previously anti-Palestinian tourists reduce their Islamophobia and anti-Arab biases. Such tours are, of course, problematic as they seek to package an enormously complex situation into a one-day commercialised tourism experience and fail to address severe power imbalances (Strömbom, 2019). However, they can be seen as one potential positive avenue for "Letting The Other Happen To Me" (Stewart & Zediker, 2000, p.232) for tourists too scared to venture into the home of the Other, or into Palestine alone. Another example is the "Gaza Border Reality Tour". Unlike the majority of scholarship on political tours in the region which has stressed guides as simplifiers of the contested narratives of space (Brin & Noy, 2010; Gelbman & Maoz, 2012), here, tourists are invited to understand the complex history and daily reality of life along the Gaza border. On the tour, tourists meet Israelis affected by routine airstrikes on their schools and homes, as well as communities (*kibbutzim*) destroyed in the 1948 war. Yet crucially, the tour also stops at the ruins of a Palestinian village destroyed during the Nakba, and tries to provide multi-perspectivity in both the terminology the guides employ (i.e. referring to the events of 1948 as the War of Independence as well as the Nakba) and in the Skype call made with a Gaza resident during the tour. Through the provision of these multiple perspectives on the Israeli-Palestinian context, tourists are therefore encouraged to see the region through the prism of dialogue: that there are at least two sides in any dialogue, each with their own narrative that is worth comprehending.

The View From Palestine

In Israel, hostels can play a role in bringing together members of the country's Arab and Jewish populations in the hope of encouraging transformative dialogue. In Palestine, there is no such opportunity for direct Arab-Jewish dialogue. This is because the status of the Jewish minority in Palestine (and from here on, we mean the West Bank only) is very different from that of the Arab minority in Israel. Unlike the Arab minority in Israel who represent an indigenous people of the land, and who are accepted (albeit begrudgingly) as citizens of Israel, the Jewish Israelis who live in the West Bank are considered by most countries as living in illegally constructed settlements on stolen Palestinian land. It is also widely accepted that this Israeli presence in the West Bank is an impediment to peace between Israel and the Palestinians. Given the illegality of these Israeli settlements, there is no appetite amongst Palestinian hostels to be spaces of dialogue with Israelis living in the West Bank. Indeed, in all our research, we found only one example of a Palestinian hostel engaging with an Israeli tourism venture in the West Bank, whereby guests at a hostel in Jericho would get discounted entry to the Dead Sea resort of Kalya. Instead, what Palestinian hostels offer in the way of transformative dialogue are two things. The first is the provision of tourism spaces that demonstrate tentative relational responsibility and affirmation of Israel through cooperation with Israeli hostels. In other words, there is an acceptance of the presence of Israelis in the land of the former British mandate of Palestine as evidenced through certain behaviours in Palestinian hostels. The second is the provision of a tourism space that permits Palestinian self-expression and the corresponding possibility for international tourists, many of whom arrive with negative perceptions of Palestine and Palestinians, to experience the Palestinian Other within a safe space.

Relating to the first point, transformative dialogue includes the need to recognise one's relational responsibility. That is to say, a need to focus on what we create together, rather than a focus on individual responsibility (McNamee & Gergen, 1999). This is particularly important in a context where, as we described earlier, the Palestinian and Israeli tourism industries are in many ways co-dependent. All tourists to Palestine must come via Israeli entry points and the majority of tourists who visit Palestine also visit Israel. This situation is no different in Palestine's hostels. From extended observation in a variety of hostels in the region, it is evident that the vast majority of guests in Palestinian hostels also spend at least a couple of days in Israel, usually visiting destinations such as Tel Aviv, West Jerusalem or Ein Gedi. This situation, therefore, offers the possibility for both Israeli and Palestinian tourism actors to recognise their relational responsibility towards tourism in the region, and to take the opportunities that such interdependence creates such as sharing guests and collaborating on cross-border projects. Yet such cooperation is also deeply problematic and painful, especially for Palestinians, given the historical context and the daily violence of the Israeli occupation.

On the positive side and in the spirit of transformative dialogue, it is quite evident that the "separate and antagonistic realities" (Gergen at al., 2001, p.704) represented by the official tourism maps (Figures 1 and 2) are not shown to be the case with our Palestinian hostels. Compare, for example, Figures 1 and 2 with Figure 6.

Figure 6. A Palestinian hostel lobby tabletop
Source: Photo taken by Daniel Laven, November 30, 2018

Figure 6 is a picture of a table filled with leaflets in a Palestinian hostel. What is remarkable about this image is that alongside the leaflets advertising hostels in Hebron, Ramallah and Palestinian hiking routes, we see advertising for the Israeli Hostel Network, hostels in Tel Aviv and walking trails in Northern Israel. We saw similar things in other hostels in Palestine, with one hostel even advertising two Israeli hostels located in the disputed territory of the Golan Heights! If we see such tabletops in the light of the official tourism maps of the region, the comparison is rather striking. Such advertising is seen as a way of initiating some sort of dialogue with Israeli hostel operators, as one hostel owner attests:

I think that if collaboration actually ends up bringing more backpacker traffic into the West Bank then it's a good thing, and I have noticed that it does. People come and they come with a smile and they say that they were recommended by an Israeli hostel, and so having access to that market is kind of very interesting.

One hostel took their cooperation even further. The hostel was in early dialogue with an Israeli hostel based in Tel Aviv, getting advice on what kind of building to buy, what to expect in the first months and how to handle an ever-changing security situation. This hostel also became a member of the Israeli Hostels Network (ILH) to widen its marketing reach. A Palestinian member of the hostel's staff also had the opportunity to attend a professional development workshop within Israel itself. His experience there speaks directly to notions of transformative dialogue. He describes such interactions in the following way:

At the beginning, it was very hard for me to hear [what] the others [Israelis] have to say [...] but I could accept and see and understand why they say so. Like, the environment they grew up in, hearing the same things from their side which is very different than us; I could put myself in their position and understand why they are saying this, so that's how I accept other opinions.

For the most part, however, this kind of dialogue between hostel staff members on either side of the Green Line remains elusive. This is mostly due to the difficulty for Palestinians to get permits to enter Israel, but also due to personal and societal discomfort at working with Israelis. Even when there is a desire amongst Palestinian hostel staff to promote Israeli hostels, there is a deep fear that they might lose social capital amongst Palestinians who see such acts as treacherous. Similar fears were voiced when Lisa Strömbom (2019) looked at Palestinian views towards working as guides within the Dual Narrative Tours framework we discussed earlier. As one hostel owner in Ramallah lamented:

Unfortunately, what I cannot do is put up [...] brochures and posters related to hostels in Israel because we are under the watchful eye of our friends at the security services on the side of the Palestinian Authority.

Associated with these very real safety risks, there is the emotional turmoil for Palestinians when it comes to working with Israelis, as has also been noted in neighbouring Jordan (Buda, d'Hauteserre & Johnston, 2014). Given Palestinians are still coming to terms with the Nakba of 1948 and a military occupation that has been ongoing since 1967, it is very hard to think in purely business terms when it comes to cooperation with Israelis, as one hostel owner from Jericho puts it:

You have to always think that as a Palestinian, I think of Haifa as Palestinian, I think of Jaffa as Palestinian, and I think of Akko as Palestinian. So I cannot build my logic on a purely business [perspective] [...] I don't take your home and then sell you a ticket to come visit it.

Such feelings are compounded by the sense that even though there might be economic or social incentives to cooperate, there remains the question of inequality and injustice. Palestinian hostel owners feel that in any cooperation they are going to be the underdogs, given they operate within a tourism context where Israelis already dominate the tourism industry of the West Bank, with Israeli tour companies (some of which are owned by Israeli hostels) taking the lion share of tourists around West Bank cities such as Bethlehem. This naturally feels uncomfortable for Palestinian hostel owners who wonder why they should start a dialogue with Israeli hostels when Israel already dominates the Palestinian tourism market hostels work in. Yet due to the interdependence we spoke about earlier, Palestinian hostel owners are left having to recognise this fact despite all its injustice:

You know the numbers will always work out. Where there is demand, there is always a way to make money. And the demand is coming through Israel, and that's demand, but that's not because we have a choice about it. It's because we don't have an international airport or a border. (Hostel owner in Jericho)

Therefore, despite a tentative affirmation of Israel in Palestinian hostels, there exist very real issues relating to the nexus between dialogue and power that are left unresolved.

As noted earlier, self-expression is crucial for transformative dialogue (Gergen et al., 2001). If we do not give the Other the chance to express themselves, we are living in monologue. The chance for Palestinians to express themselves through tourism is extremely limited in the current context. With regards to expressing themselves to Israelis, this chance is all but impossible outside of the Palestinian-soldier duality for those living on the other side of Israel's Separation Wall that splits Israel from the West Bank (Isaac & Platenkamp, 2016). It is also limited by the fact that Israelis cannot travel to Palestinian areas (Area A) and Palestinians need hard-to-obtain permits to enter Israel. As for expressing themselves through tourism to international tourists, this ability is severely constrained by Israel's stranglehold on the Palestinian tourism industry, which means Israeli companies take the majority of tourism traffic to the West Bank, and by Israel controlling Palestine's entry points. For there to be dialogue, therefore between Palestinians and international tourists, there needs to be a space where Palestinians can express themselves on their own terms.

Before the arrival of hostels in the West Bank, backpackers and other low budget travellers did not have the same opportunities to engage with Palestinians. The opening of a string of hostels in Palestine in the last decade finally gave these travellers their perceived "safe space" in Palestine, where they knew other young, international tourists would congregate. This feeling of safety and comfort that hostels provide by being transnational spaces of youth culture is well-noted in the literature (Cohen, 1972; Ooi & Lang, 2010; Scheyvens, 2002). Our respondents also noted how relieved tourists look when they arrive at the hostel and see both local and international faces. Providing this space is a crucial first step in getting tourists to be brave enough to cross over from Israel into Palestine, and in the process, rectify some stereotypes or biases they may have inherited from a media obsessed with Palestinian violence or from Israelis, who often scare tourists from visiting Palestine (Shepehrd & Laven, 2019). It is worth also noting here that this "safe space" was also afforded to some Israeli tourists who also ventured into the West Bank to stay in Palestinian youth hostels. Across the West Bank, but particularly in Bethlehem, we heard from hostel staff how they were aware that Israelis would stay in their hostels, travelling under foreign passports. Some of them were quite happy with this happening, as they saw the transformative effect such sojourns had on these guests whose stereotypes were confounded by their stay. Of course, such fraternisation is done at great risk to the hostel, given that hosting an Israeli guest runs the risk of the hostel being accused of breaking the law, of being traitors to the Palestinian cause and even of kidnapping an Israeli.

By providing an identifiable space for international tourists to come and feel safe in Palestine, hostels, therefore, represent discernible spaces for "Letting The Other Happen To Me" (Stewart & Zediker, 2000, p. 232) in Palestine. Importantly, this is perceived as occurring within a tourism environment famed for its relaxed and intimate nature. This enables tourists to feel more comfortable broaching difficult subjects in a highly emotional context. The need for tourists to hear these stories is evident in the lack of knowledge most guests at Palestinian hostels have about "the situation", and by their timidity about raising the topic with locals, whom they feel might not want to be asked sensitive questions about life under the Israeli occupation. This is where Palestinian hostels can step in, as one owner of a hostel in Ramallah affirmed:

A lot of our guests, when we point out the refugee camp, they say "What do you mean a refugee camp? What kind of refugees?" So it's an opportunity to start a conversation with them by what we mean by that [...] You can't even talk about the issue without someone jumping down your throat in many contexts [...] So I wanted to remove that element, make a comfortable space for people where they could come and explore on their own.

Exploring on their own also involves encouraging backpackers to go out and have dialogue with the local community. As one staff member in Ramallah saw it, hostels are "*a middle man without the commission*". So instead of being bused around "safely" from one sanitised version of Palestine to another, tourists in Palestinian hostels are encouraged to eat locally, visit other hostels and take tours with locals. In this way, the tourists get to engage and create dialogue with people from a variety of the region's different backgrounds. As hostel staff are aware of, finding tourism information and booking tours online in Palestine is tricky, given that the quality of information provision is low. This situation is aggravated by the strength of Israel's marketing muscle; if one enters "West Bank Tour" into Google, the first page of results are all Israeli companies. This is why hostels can play a crucial role as spaces of dialogue as they form a network of places where tourists can be connected to Palestinian tour operators and therefore exposed to Palestinian narratives of Palestinian space. As one hostel worker in Ramallah put it:

I guess through the hostel here, you can see and learn how someone is living here. [Tourists] think it is a war zone, they think everyone is involved in politics which is not correct.

The provision of these tours and the actual space of the hostel give Palestinians a form of self-expression in a context that has largely robbed Palestinian's of choice and voice. In the process, the tourism space of the hostel provides the chance for transformative dialogue to take place and for tourists to think differently about the situation:

I care about sharing with them how a person could live here; what kind of life he can have here, and sometimes I come to open their eyes to things they thought differently about before [...] I need to show them that I have my reasons to stay here, to decide to stay in Palestine. (Hostel staff member in Ramallah).

Hostel staff also expressed such sentiments:

I changed a lot of my ideas, thoughts and beliefs [...] I got to know the different points of view about Palestine. Like at some point I never called Palestine and Israel a conflict, I don't think it is but I didn't know it was called this way. (Hostel staff in Ramallah, as reported in Shepherd & Laven, 2019, p. 18)

It is worth remembering that transformative dialogue is not about consensus. Instead, it is about recognising that "there are other salient systems in concurrence with ours" (Gergen et al., 2001, p.696). The intention, therefore, of the hostels is not to *force* tourists into supporting the Palestinian cause, but rather, opening up a dialogue whereby tourists can experience the Other, hear a Palestinian narrative which they are rarely exposed to and then make their own decisions::

We are not trying to force opinions or create opinions but we are some sort of connection between Palestinian society [and] the international community in such a small space. So yeah, inside the hostel, we tell what facts are through maps and statistics and readings and articles. Also we tell opinions and stories of outside the hostel through the staff or the people who stay here, who come here and work here (Hostel staff in Ramallah).

You could lecture people about how many refugees left in 1948 and how many left in 1967, those numbers are unlikely to stick unless you're going to use them in a lecture to somebody else and make a living out of it. But sharing stories and asking "Who is this guy you have a picture of on the wall?", "Oh yeah that's my cousin, he's in jail" "What for?" "He's a political prisoner; he's a hero of the family". This kinda stuff. And people say this is not because they want to promote [an agenda] but because it is their daily reality (Hostel owner in Jericho).

In sum, our data have stressed the importance of Palestinian hostels in providing spaces which are perceived as safe, sociable environments where Palestinians have the chance to tell their own stories and express themselves in ways not usually afforded to them. Tourists gain the chance to engage in transformative dialogue as they learn about aspects of Palestine often omitted from a global media representation of Palestine, such as hiking trails and beer festivals. One could, of course, trifle about the idea of tourists and Palestinians existing in "separate, antagonistic realities" (Gergen et al., 2001, p.682) and therefore be in need of transformative dialogue. Many tourists come to Palestine as keen supporters of the Palestinian cause. Yet what we believe to be undeniable is that the majority of tourists who visit the region do not visit the West Bank (excl. Jerusalem).[2]. The reasons for this vary, but there is a general perception of Palestine as dangerous. From a Palestinian perspective, Palestine is a perfectly safe place for a tourist to visit. Therefore, there is clearly a need for dialogue between these two groups. As to whether or not such exposure to Palestinian life through hostels changes the perceptions of the guests towards Palestinians is difficult to say. Similar exposure has been shown to improve tourist perceptions in other forums (Schneider, 2019) and our own observations largely support this assessment. However, it is important to remember that hostels cannot control this outcome, but they can control the spaces they curate. The spaces they curate for dialogue.

CONCLUSION

In conclusion, it is quite evident from our years of research on hostels in the hostile territory of the Israeli-Palestinian context that they represent distinctive tourism spaces where some of the master narratives of the Israel-Palestinian conflict can be challenged. Viewed through the lens of transformative dialogue (Gergen et al., 2001), we have seen how hostels are spaces of relational responsibility and coordination as Israelis and Palestinians (both in the West Bank and in Israel) are working together in hostel management structures, in marketing across the Green Line, and in trying to make the most of a situation which has created interdependency, for better or worse. We have seen how they are spaces of self-expression and affirmation of the stories of Others. Particularly so for the suppressed narratives of Israeli Arabs and Palestinians, long estranged by current tourism trends but now provided an opportunity through hostels and the tours hostels offer to tell their stories and shape foreign perceptions of this much-misunderstood region. These dynamics lead to the creation of new realities – Jewish-Arab ownerships, cross-border cooperation and tourists leaving the region with a greater understanding of the various narratives that constitute the Holy Land. Our hostels, therefore, epitomise the transformative power of "Letting The Other Happen To Me" within tourism spaces (Stewart & Zediker, 2000, p.232) and are a clear rebuttal to the purification of space presented in the official tourism maps of both Israel and Palestine (Figures 1 and 2). They embarrass those maps, sometimes overtly with the examples of cross-border marketing, and sometimes subtly, by providing a space with which they explore the "grey" areas of those sanitised tourism geographies.

Naturally, the ability of each hostel to promote transformative dialogue or create new realities through transformative dialogue varied depending on where they were located in this hostile territory. We saw how in Palestine, it is much harder to have cross-cultural dialogue and thus challenge the processes of Othering, mostly because Palestinians literally cannot escape such processes as they have become embodied in the form of an occupation with Apartheid characteristics (Weizman, 2017). Therefore, whereas signs of "peacebuilding" in Israel could be a Jewish-Arab hostel partnership, such a partnership in the West Bank would be unthinkable. It is for that reason that in Palestine, it is better to focus on the role tourists can play as potential agents for disrupting the status quo. Our work, therefore, supports a peace geographies approach to studying conflict, whereby peacebuilding looks different in different spaces and at different times (Williams & McConnell, 2011).

Despite all the success these hostels have had in challenging the master narratives of the region, they are of course situated in one of the world's most intractable conflicts; a conflict "that has always smouldered and sometimes blazed" (Bowen, 2020). It is safe to say that transforming the Israeli-Palestinian context wholesale through hostels, or even tourism generally, is pie in the sky stuff. Every positive act of cross-cultural contact and every tourist made aware of some of the intricacies of the situation is a small counter-strike stacked against the might of the discursive weight of the shadow of the Holocaust, the ongoing devastation of the Nakba, the geopolitical arm-wrestling of the Middle East and the competitive victimhood of refugee and diaspora narratives. What is more, tourism itself is also somewhat Janus faced, giving with one hand and taking with the other. In this chapter, we have described how tourism spaces such as hostels can provide a forum for transformative dialogue, which is liberating and empowering, but tourism can equally constrain and disempower. Even some of the examples we have given that permit self-expression or dialogue such as dual narrative tours come with their own problems relating to injustice, power imbalances and the role of profiting making in conflict-based tourism (Strömbom, 2019).

Based on our study, we encourage further research into the role of tourism spaces in facilitating dialogue, both between acrimonious groups, and between tourists and locals, in contested spaces such as Israel-Palestine. Yet at the same time, we also encourage studies that take a critical approach to understand the relationship between dialogical tourism spaces (such as the hostels we present) and issues of justice and power. As one of our Palestinian respondents pointed out, how is it just or peace enabling, when after working with Israelis to provide a dual narrative tour in the West Bank, at the end of the day, the tour guide sees that "*the reality after you get off the bus does not change, it only gets worse*". We consider such studies to be essential in order that scholars do not end up inadvertently promoting concepts or ventures which fail to tackle the burning questions of injustice.

ACKNOWLEDGMENT

Funding for this study was provided, in part, through the Swedish Foundation for International Cooperation in Research and Higher Education (STINT).

REFERENCES

B'tselem. (2019). Retrieved from https://www.btselem.org/

Bar-On, D., & Kassem, F. (2004). Storytelling as a way to work through intractable conflicts: The German-Jewish experience and its relevance to the Palestinian-Israeli context. *The Journal of Social Issues*, *60*(2), 289–306. doi:10.1111/j.0022-4537.2004.00112.x

Bowen, J. (2020, January 29). Trump's Middle East peace plan: 'Deal of the century' is huge gamble. *The BBC*. Retrieved from https://www.bbc.com/news/world-middle-east-51263815

Brin, E. (2006). Politically-orientated tourism in Jerusalem. *Tourist Studies*, *6*(3), 215–243. doi:10.1177/1468797607076672

Brin, E., & Noy, C. (2010). The said and the unsaid: Performative guiding in a Jerusalem neighbourhood. *Tourist Studies*, *10*(1), 19–33. doi:10.1177/1468797610390982

Buda, D. M., d'Hauteserre, A.-M., & Johnston, L. (2014). Feeling and tourism studies. *Annals of Tourism Research*, *46*, 102–114. doi:10.1016/j.annals.2014.03.005

Cohen, E. (1972). Toward a sociology of international tourism. *Social Research*, *39*(1), 164–182.

Collins-Kreiner, N., Mansfeld, Y., & Kliot, N. (2006). The reflection of a political conflict in mapping: The case of Israel's borders and frontiers. *Middle Eastern Studies*, *42*(3), 381–408. doi:10.1080/00263200500521230

Edensor, T. (2007). Mundane mobilities, performances and spaces of tourism. *Social & Cultural Geography*, *8*(2), 199–215. doi:10.1080/14649360701360089

Frank, A. (2005). What is dialogical research, and why should we do it? *Qualitative Health Research*, *15*(7), 964–974. doi:10.1177/1049732305279078 PMID:16093373

Galili, L. (2003, December 19). Long division. *Haaretz*. Retrieved from https://www.haaretz.com/1.4802471

Gelbman, A., & Laven, D. (2016). Re-envisioning community-based heritage tourism in the old city of Nazareth. *Journal of Heritage Tourism, 11*(2), 105–125. doi:10.1080/1743873X.2015.1044993

Gelbman, A., & Maoz, D. (2012). Island of peace or island of war: Tourist guiding. *Annals of Tourism Research, 39*(1), 108–133. doi:10.1016/j.annals.2011.05.008

Gergen, K. J., McNamee, S., & Barrett, F. (2001). Toward transformative dialogue. *International Journal of Public Administration, 24*(7), 679–707. doi:10.1081/PAD-100104770

Hammami, F., & Laven, D. (2017). Rethinking heritage from peace: Reflections from the Palestinian-Israeli conflict. In D. Walters, D. Laven, & P. Davis (Eds.), *heritage and peacebuilding* (pp. 137–148). The Boydell Press.

Hannam, K., Sheller, M., & Urry, J. (2006). Mobilities, immobilities and moorings [Editorial]. *Mobilities, 1*(1), 1–22. doi:10.1080/17450100500489189

Isaac, R., & Platenkamp, V. (2016). Concrete U(dys)topia in Bethlehem: A city of two tales. *Journal of Tourism and Cultural Change, 14*(2), 150–166. doi:10.1080/14766825.2015.1040413

Israeli Ministry of Foreign Affairs. (2014). *Christian tourism to Israel.* Retrieved from https://mfa.gov.il/MFA/AboutIsrael/Spotlight/Pages/Christian-tourism-to-Israel.aspx

Juha's Guesthouse. (2020). *Home*. Retrieved from https://www.zarqabay.com/

Kassis, R., Solomon, R., & Higgins-Desbiolles, F. (2017). Solidarity tourism in Palestine: The alternative tourism group of Palestine as a catalysing instrument of resistance. In R. Isaac, C. M. Hall, & F. Higgins Desbiolles (Eds.), *The politics and power of tourism in Palestine* (pp. 37–52). Routledge.

Knesset. (2018). *Basic Law: Israel: The Nation State of the Jewish People*. Retrieved from https://knesset.gov.il/laws/special/eng/BasicLawNationState.pdf

Masalha, N. (2019). *Palestine: A four thousand year history*. Zed Books.

McNamee, S. (2010). Research as social construction: Transformative inquiry. *Saúde & Transformação Social. Health and Social Change, 1*(1), 9–19.

McNamee, S., & Gergen, K. J. (1999). *Relational responsibility: Resources for sustainable dialogue.* SAGE.

Moufakkir, O., & Alnajem, M. (2017). Hostel front desk employees' perceptions of customers: A comparative analysis based on cultural background. *International Journal of Culture, Tourism and Hospitality Research, 11*(3), 355–371. doi:10.1108/IJCTHR-07-2016-0068

Newman, D., & Passi, A. (1998). Fences and neighbours in the postmodern world: Boundary narratives in political geography. *Progress in Human Geography, 22*(2), 186–207. doi:10.1191/030913298666039113

Noy, C. (2004). This trip really changed me: Backpackers' narratives of self-change. *Annals of Tourism Research, 31*(1), 78–102. doi:10.1016/j.annals.2003.08.004

Ooi, N., & Lang, J. (2010). Backpacker tourism: Sustainable and purposeful? Investigating the overlap between backpacker tourism and volunteer tourism motivations. *Journal of Sustainable Tourism*, *18*(2), 191–206. doi:10.1080/09669580903395030

Pappé, I., & Chomsky, N. (2015). *On Palestine*. Haymarket Books.

Peled-Elhanan, N. (2012). *Palestine in Israeli School Books: Ideology and Propaganda in Education*. I.B. Tauris. doi:10.5040/9780755608195

Riba, N. (2019, March 10). Asked to strip by Security Guard, Palestinian artist refuses to fly to Israel. *Haaretz*. Retrieved from https://www.haaretz.com/israel-news/.premium-asked-to-strip-by-security-guard-palestinian-artist-refuses-to-fly-to-israel-1.7038673

Said, E. (1979). *The Question of Palestine*. Vintage Books.

Sasson, T., Shain, M., Hecht, S., Wright, G., & Saxe, L. (2014). Does Taglit-Birthright Israel Foster Long-Distance Nationalism? *Nationalism & Ethnic Politics*, *20*(4), 438–454. doi:10.1080/13537113.2014.969149

Scheyvens, R. (2002). Backpacker tourism and Third World development. *Annals of Tourism Research*, *29*(1), 144–164. doi:10.1016/S0160-7383(01)00030-5

Schneider, E. M. (2019). Touring for peace: The role of dual-narrative tours in creating transnational activists. *International Journal of Tourism Cities*, *5*(2), 200–218. doi:10.1108/IJTC-12-2017-0092

Shavit, A. (2013). *My Promised Land*. Scribe.

Shepherd, J., & Laven, D. (2019). Providing counter-narratives: The positive role of hostels in the Israeli-Palestinian context. *Tourism Geographies*, 1–24. Advance online publication. doi:10.1080/14616688.2019.1669215

Sherriff, L. (2020). Crossing divides: The hostel promoting tourism to ease Israel's tensions. *The BBC*. Retrieved from https://www.bbc.com/news/world-middle-east-51446690

Sibley, D. (1995). *Geographies of exclusion*. Routledge.

Stenvall, A., Laven, D., & Gelbman, A. (2017). The influence of social entrepreneurship in tourism on an Arab village in Israel. In P. J. Sheldon & R. Daniele (Eds.), *Social entrepreneurship in tourism: Philosophy and practice* (pp. 279–293). Springer. doi:10.1007/978-3-319-46518-0_17

Stewart, J., & Zediker, K. (2000). Dialogue as tensional, ethical practice. *Southern Journal of Communication*, *65*(2-3), 224–242. doi:10.1080/10417940009373169

Strömbom, L. (2019). Exploring Prospects for Agonistic Encounters in Conflict Zones: Investigating Dual Narrative Tourism in Israel /Palestine. *Alternatives*, *44*(2-4), 75–93. doi:10.1177/0304375419857421

TedXTalks. (2017, September 17). *Me-Rational / Neta Hanien / TedxJaffa* [Video File]. Retrieved from https://www.youtube.com/watch?v=ZdNcOy_eAB0

Thiessen, C., & Darweish, M. (2018). Conflict resolution and asymmetric conflict: The contradictions of planned contact interventions in Israel and Palestine. *International Journal of Intercultural Relations*, *66*, 73–84. doi:10.1016/j.ijintrel.2018.06.006

Travel Palestine. (2019). Retrieved from http://travelpalestine.ps/

Weizman, E. (2017). *Hollow Land: Israel's architecture of occupation*. Verso Books.

World Tourism Organization. (2018). *UNWTO tourism highlights* (2018 ed.). UNWTO.

KEY TERMS AND DEFINITIONS

Area A: An area under the sole control of the Palestinian Authority (PA), as stipulated by the Oslo Accords of 1993. Area B is under joint Israel and Palestinian control and Area C is under Israeli control.

Dual-Narrative Tour: A tour around contested cities with two guides, one from each side in the conflict.

Green Line: The border that represents the boundaries of Israel pre-1967, before its occupation of the West Bank.

Hostels: Here, hostels are defined as cheap accommodation facilities for backpackers and other travelers, characterised by shared communal living spaces and dormitories.

Israeli Arab: The official Israeli government term for non-Jewish citizens of Arab/Palestinian origin.

Israeli Hostels Network (IHL): A network of independent youth hostels across Israel.

Nakba: The displacement of Palestinians in 1948 following the creation of the state of Israel.

Oslo Accords: An agreement reached between Israel and the Palestinians in 1993 that aimed for mutual recognition of each other's right for self-determination. It envisaged Israeli troop withdrawals from Palestinian areas. The Area system (A, B & C) are a product of these negotiations.

Palestinian Authority (PA): The representative government of Palestine established since the Oslo Accords in 1993.

Settlements: A term used to describe Israeli communities that are illegally constructed on Palestinian land in the West Bank.

ENDNOTES

1 The Nation-State bill, formally the Basic Law, passed in July 2018 stipulates that Israel is the nation state of the Jewish People, omitting any reference to any of Israel's numerous minority groups. Arabic was also downgraded from an official language of the state to one with 'special status'. Furthermore, the bill states that "The State views the development of Jewish settlement as a national value, and shall act to encourage and promote its establishment and strengthening" (Knesset, 2018).

2 From a survey conducted by Jack Shepherd in May 2019, only 16 out of 302 tourists surveyed (5%) in Tel Aviv visited the West Bank (excl. Jerusalem) on their trip. Results from this study are yet to be published.

Chapter 13
Can Events Facilitate Intercultural Understanding?
A Critical Investigation at Llangollen International Musical Eisteddfod

Karen Davies
Cardiff Metropolitan University, UK

Caroline Ritchie
Cardiff Metropolitan University, UK

ABSTRACT

The founding philosophy of many cultural events established after the Second World War was to enhance the dynamics of peace through supporting and developing multicultural understanding. Over 50 years after their establishment, this chapter investigates the potential of such iconic events to achieve this aim and contribute to the concept of peace through tourism, based on a longitudinal ethnographic case study of Llangollen International Musical Eisteddfod. The results show that this aim can be achieved by such events if they provide enough time and space for participants (performers and audiences) to interact. However, the study also identifies current cultural, political, and fiscal challenges in providing these temporal and physical spaces.

INTRODUCTION

Multicultural festivals are a strong and flourishing sector of the international tourism industry. The chapter presents the details and results of an investigation into the potential of these multicultural festivals to develop international understanding and contribute to cohesion, friendship and peace. This is in a world rocked by terrorism, where information on other cultures becomes infinitely more accessible through the world-wide-web and modern transportation and yet where true intercultural understanding is often very limited or shallow. The research investigates the processes and mechanisms which are used to encour-

DOI: 10.4018/978-1-7998-5053-3.ch013

age and promote intercultural understanding, global citizenship and peace at Llangollen International Musical Eisteddfod, North Wales. The study critically aims to answer the following research questions:

- What are the processes and mechanisms at Llangollen International Musical Eisteddfod that are used to encourage and promote intercultural understanding, global citizenship and peace?
- What different groups of people are involved in these processes and mechanisms?
- Have these processes and mechanisms changed over time, and what can be done to improve them in order for the event to continue to promote intercultural understanding and world peace?

Llangollen International Musical Eisteddfod was specifically set up to help achieve the ideals of reconciliation between nations and is one of the few current festivals which can state their overall objective as world peace. Klaic (2006, p. 55) commented that "after the end of the Cold War, festivals have [...] become the backbone of international cultural co-operation", and indeed, there were many other multicultural events set up in the period that Newbold, Jordan, Bianchini & Maughan term the "age of reconstruction" (2015, p. xi), as an opportunity to "re-affirm 'appropriate' cultural values, encourage social stability and boost morale" (Quinn, 2010, p. 268).

The annual six-day event has a traditional music and bardic competition structure, which is modelled on the more well-known National Eisteddfod of Wales, and takes place in Llangollen in north Wales in the first week of July. The first Llangollen Eisteddfod event took place in 1947 in order to provide a platform for international groups to compete in choral music competitions within a supportive and friendly environment. Now as then, amateur competitor groups from around the world and the UK attend the event to enter the competitions. In the early years around 60, 000 to 80,000 competitors would take part, but these days this has reduced to approximately 5,000 due to a number of factors, including the expense for groups to travel to the event, restrictions on visas, and changes in the funding schemes available. The competitions have been extended from the initial choral music to include dance. In line with other festivals, the event has faced many challenges and issues, not least the financial pressures of non-profit organisations that rely heavily on volunteers to ensure their longevity and sustainability and often need to "change business activities in ways that may compromise their mission in order to gain market shares" (Andersson & Getz, 2009, p. 252). At the Llangollen Eisteddfod this is reflected in the relatively recent addition of a series of evening concerts at which professional musicians play and which attract additional audiences to the event who do not necessarily attend the rest of the festival. Whilst the event has responded well to these and other external pressures, such as changes in legislation in relation to housing competitors, and competition from other similar events, there have inevitably had to be changes within the organisation which may have impacted on the overall initial intercultural friendship and peace objectives of the festival. Therefore, Llangollen International Musical Eisteddfod is particularly well suited to be a case study to investigate how far present-day festivals can promote intercultural communication and thus contribute to global citizenship and world peace.

BACKGROUND

Peace applied to the tourism context was pioneered by D'Amore (1988; 2009), who established the International Institution for Peace Through Tourism in the United Nations International Year of Peace, 1986. Since then, the United Nations World Tourism Organisation (UNWTO) has promoted the concept in line with their responsible tourism policies and codes of conduct. Galtung's (1996) and Haessly's (2010) conceptualisations of the notion of peace suggested that it should be defined not merely as an absence of conflict, war and violence with a focus on what people (specifically governments) should stop doing, but as a presence and process with a focus on what people should start doing. Peace through tourism is associated with Galtung's idea of cultural power where positive peace is enhanced via holistic, global partnerships and where equality, justice and life enhancement prevail to promote global citizenship. According to Edgell (2014), while global interest in the role of tourism in the promotion of peace is continually growing, research in this area remains limited. Tourism and attitude change has been the closest theoretical paradigm in tourism literature (see Gomez-Jacinto, Martin-Garcia, & Bertiche-Haud'Huyze, 1999; Pizam, Jafari, & Milman, 1991), but these early studies found more evidence of negative potential of tourism activities to change attitudes. In more recent studies, Steiner and Reisinger (2003) have looked in some depth at cultural behaviour and tourism, and Tomljenovic (2010) has explored the potential for tourism to encourage intercultural understanding. The latter study suggests that a number of issues need investigating to gain a full picture, namely: type of travel, levels of language familiarity, number of previous visits, motivations for travel, initial attitudes (including prejudices) and number and quality of contact opportunities.

Despite a wealth of literature surrounding cultural capital and social interaction at events, there is very little literature on the role of the events industry in the promotion of world peace. Moufakkir and Kelly (2013) did investigate the role of events within the world peace concept as a sector of the tourist industry rather than a discrete discipline. They concluded that the majority of events contribute nothing to the notion of peaceful understanding and, in fact, can be counterproductive due to their frequent ability to divide rather than unite communities. However, they also pointed out that whilst there are not many events which have peace as an overall objective, events by their very nature bring people together in circumstances conducive to interactions and shared experiences, one of the pre-requisites for the development of understanding and peace being that those involved (both hosts and guests) are open-minded and inclined toward goodwill. Moufakkir and Kelly (2013) produced a SWOT analysis of events for world peace, shown in figure 1. This was adapted to the case study event as part of the conclusions of the study, as can be seen at the end of the chapter.

This SWOT analysis was conducted on events in general and shows some areas of strengths that may or may not be applicable to all scenarios. It is reasonable to suggest that the growth in the number of events globally may have the potential to promote the concept of peace further, but this will very much depend on how the event is run, its specific objectives, and the nature and authenticity of its cultural elements. Festivals, arguably, are one type of event that has a higher potential to promote and facilitate intercultural understanding, global citizenship and the notion of peace. They can be seen as institutions that are structured in order to draw and reproduce distinctions in society (Sassatelli, 2011). Politically festivals mirror ideologies emerging within the public sphere (Habermas, 1962) – from the right the politicians are demanding increased accountability for public investment (in the arts) and from the left, communities believe that they should play a role in defining and developing the services that are available to them. Consequently, festivals have a role to play in the creation of a "post-national form of cultural citizenship" (Fabiani, 2011, p. 93). Literature on the role of festivals in society and their juxtaposition with culture, globalisation and citizenship is well developed (see for example Mouffakir & Pernecky, 2015; Newbold, Jordan, Bianchini, & Maughan, 2015; Quinn and Wilks, 2013; Richards, 2015; Richards, de Brito & Wilks, 2013; Spracklen & Lamond, 2016) and several authors (e.g. Baiett, 2015; Moufakkir & Pernecky, 2015; Newbold et al., 2015; Quinn, 2010; Szabo, 2015) have formed detailed historical accounts of festivals from early traditional community rituals through to the contemporary events which make up the global festival industry today. Baiett (2015, p. 25) highlights their changing nature from the purely organic, to the organised, commercial organism, demonstrating that the notion of 'organic festivity' has been severely diminished for a number of reasons including the "fast-paced world of unlimited cyber-entertainment options… extreme shopping alternatives, and an insatiable demand for hedonistic experiences". Described as festivalisation (Getz & Page 2016; Negrier, 2015; Quinn, 2013), the rationalisation of the festivals industry has meant that a standardised outdoor music festival model has become evident incorporating a significant similarity in programming content (Finkel, 2009). This lack of creativity resulting from event managers trying to emulate the success of established festivals puts enormous pressure on organic temporal and physical spaces thus potentially reducing opportunities for intercultural exchange and intercultural understanding (Matheson, 2008).

Figure 1. A peace through events SWOT analysis
Source: Moufakkir and Kelly, 2013, p. 136

STRENGTHS *Event attributes which bring people together in non-adversarial circumstances:* • Growth in numbers and variety • Cultural elements • Associated economic benefits • Locational and temporal flexibility • Community involvement • Perception of authenticity	OPPORTUNITES *External development which can contribute to the desired event outcomes:* • Media support / publicity • Globalisation • Multiculturalism • Advances in technology • Growing sophistication of tourists • Sustainability concerns • Tourism education
WEAKNESSES *Attributes which hinder or restrict events in achieving the desired outcomes* • Controversial themes • Poor visitor behaviour • Community disruption • Ethnocentrism / competition • Limited time span	THREATS *Developments likely to increase hostility or contribute to a decline in event effectiveness:* • Terrorism • Focus on war / atrocity themes • Tourist demands – volume and nature • Rejection of sustainability measures • Dismissal of peace objective

Despite this rationalisation of festivals, their role in encouraging social interaction is still widely accepted. As noted in Moffakir and Kelly's (2013) SWOT analysis, opportunities arising from the growing multicultural nature of global societies mean that festivals are increasingly recognised as "shared spaces where learning and knowledge exchange can happen" (Comunian, 2015, p. 54). For example, Chalcraft, Magauadda, Solaroli and Santoro's study of the Womad (World of Music and Dance) Festival showed that "music festivals are classic examples of participatory experiences that unite aesthetic and social ideals and imaginaries in discrete but temporary spaces" (2011, p. 27). Similarly, Morgan (2008) and

Nordvall, Pettersson, Svennson and Brown (2014) discuss how the design of the event space can impact on the number and types of interactions that take place as part of the overall event experience. Additional research into these different interactions considers the development of cultural capital through bridging and bonding relationships (Quinn & Wilks, 2013; Wilks, 2011), as well as the potential for festivals to create feelings of communitas amongst their audiences (Andrews & Leopold, 2013; Turner, 1969), leading to transformational experiences on the cognitive, conative and affective levels (Getz, 2008). However, while intercultural understanding and global citizenship do come into these discussions as a potential benefit of certain types of festivals, the processes and mechanisms that facilitate and encourage this outcome as a form of positive peace remain underexplored.

The proposal of this chapter is that multicultural events are tourist environments which allow people to act in a peaceful manner and can go some way to encouraging intercultural understanding through shared experiences and knowledge exchange. These events are "hybridised cultural spaces" (Sorrells, 2013, p. 76) with their own sets of rules, norms and behaviours which can be designed in such a way so as to allow for a high number of quality contact opportunities as suggested by Tomljenovic (2010). Lee, Arcodia and Lee (2012, p. 336) define the multicultural festival as a "themed celebration at which people of a range of ethnicities, including members of both the ethnic minorities and dominant population, have extraordinary and significant experiences deemed in some way beneficial". Lee et al. (2012) stated that there was a need for further research into these types of events as "useful instruments for promoting social harmony and integration". Their study used two multicultural festivals in South Korea, namely the 'Global Village Multicultural Festival' and the 'Colourful Multicultural Festival' to demonstrate what these events can do to contribute to the formation of a successful multicultural society in the age of globalisation. Findings concluded that there are two major benefits of attending this type of event. Firstly, cognitive benefits such as the development of cultural capital, or the potential to learn new things, were found to be high in this environment, a point that is supported by further research on festivals as learning environments (see for example Comunian, 2015; Deventer, 2015; Szabo, 2015). Second, Lee et al. (2012) point out the transformational benefits that can be gained from the multicultural festival environment, which are described as discarding old ways of thinking, incurring changes in attitude and perception. According to Walmsley and Franks (2011) and Richards (2015), it is often the immersive and co-creative nature of such event environments that encourage communias and lead to such transformations.

These events in particular often act as an intensified means of transmitting messages to their audiences beyond the traditional forms of verbal and non-verbal communication, for example, via performances (Andrews & Leopold, 2013); interpretation techniques, such as images, signs and symbols, including those produced in marketing materials, both hard copy and via social media (Getz & Page, 2016); and with the use of outreach projects, especially to those of younger age groups (Schulenkorf & Edwards, 2010). Whilst traditional intercultural communication literature provides a sound basis of theory from which to analyse this phenomenon within the event setting, it omits a number of forms of intercultural exchange present at festivals. The most surprising of these is the music and dance performances themselves (Andersen & Wang, 2009) which often transcend cultural barriers and are a powerful form of aesthetic cultural expression which initiates activities of communitas and transfer of knowledge through tacit learning (Comunian, 2015). This omission is partly attributed to the complex ongoing debate into authenticity in tourism and events in general and in particular into music and dance performances (see Andrews & Leopold, 2013; Chhabra, Healy, & Sills, 2003; Matheson, 2008; Wang, 1999). When applying this debate to the world music context, Haynes (2005, p. 355) argued that different musical styles are actually "racialised aesthetic constructions", which are "readily deployed to order and evaluate cultural forms" and that such cultural constructionism may simply serve to reproduce racialised aesthetic distinctions, therefore, acting to reinforce stereotypes as opposed to being a means to bring people together. It could be said, therefore, that unless the music or dance performances are truly authentic and in their original cultural form they are not communicating anything of value to the audiences other than through feelings of shared experiences (Matheson, 2008). This puts into question the potential for events based primarily on music performances to contribute to real intercultural understanding.

Further debate going against the idea that cultural events can facilitate intercultural understanding, lies in the notion that touristic activities in any form do not allow for depth of cultural exposure and, therefore, opportunities for intercultural communication are limited, due to the short-term nature of such activities (Bruner, 1991; Selstad, 2007). Hofstede (1997) also mentioned in the same vein, that contact between hosts and guests of different cultural backgrounds are one of the most superficial relationships as their perceptions are often based purely on symbols such as clothing or music or from the outcomes of commercial exchange. Steiner and Resinger (2003) further this proposal in highlighting the asymmetrical nature of host-guest interactions, which they attribute to a number of factors: different roles and goals, different situation status, different motivations and behaviour, different access to wealth and information, different commitment and responsibilities, and different socio-cultural position and cultural identities. Hosts and guests in the festival environment normally refer to the local community, festival organisers and performers (as hosts) and the visitors or attendees (as guests). However, in the multicultural festival environment, these distinctions can become more complex. This is especially true in the case of Llangollen International Eisteddfod, where the international competitors could be seen as both hosts (in that they are performers) and guests, in that they are also attendees that visit the event from international locations and experience many elements of the Welsh culture during their stay.

The question, therefore, is whether the festival environment does enough to overcome some of the asymmetrical relationships to which Steiner and Reisinger refer, in order to facilitate and encourage depth of intercultural understanding. The case study event was used to investigate the processes and mechanisms that are used to encourage and facilitate intercultural understanding, in order to highlight how these types event could do more to encourage global citizenship and add to the ideal of world peace.

METHODS

The investigation of the phenomenon of intercultural communication and exchange at the Llangollen Eisteddfod took place over a period of three years, involving attendance by the researcher at three successive editions of the annual event, for a total of 6 days per event (the whole duration). The researchers immersed themselves into the case study event as an attendee and used self-reflection to produce anecdotal evidence of their own experiences (Ellis, Adams, & Bochner, 2011), which were later used alongside other participants' experiences and opinions to build up a constructionist case study account. A number of methods of data collection were adopted with all the key stakeholder groups involved with the event; organisers, volunteers, competitors and attendees in order to gain a deep insight into the beliefs and understanding of the phenomenon from a variety of perspectives. The data collection design evolved throughout the three-year period reflecting the outcomes of each previous phase of the research, as per De Certeuau's (1986) model.

The first data collection phase developed an in-depth understanding of the organisational structure, aims and objectives, history, politics and development of Llangollen International Musical Eisteddfod over time as well as ascertaining the organisers' perspective of the main processes and mechanisms for intercultural communication and exchange during the festival. In-depth interviews were conducted with seven members of the organising committee at the Eisteddfod offices prior to the researchers' attendance at the first event. The researcher also gained access to archived materials held at the Eisteddfod's offices which provided an additional rich historical data set. This included press cuttings of stories from local and national newspapers dating back to the first-ever International Eisteddfod, as well as historical editions of the Friends of Llangollen Eisteddfod newsletter. Using a number of themes, this material helped formulate a historical narrative where the researcher was able to develop a richer understanding of how change had occurred in relation to the phenomenon being studied, not just limited to the three years of primary data collection. In addition, it enabled the significance of events, actions or beliefs which may have originally appeared to be insignificant to be reviewed, and vice versa, as per the advantages of longitudinal studies highlighted by Wond and Macauley (2005).

At all three editions of the event, the primary researcher collected photographic evidence of perceived incidences of intercultural communication and/or exchange. A total of 194 photographs were taken over the three years, and sorted for analysis into areas of the event and type and depth of exchange or communication (signs and symbols, verbal communication, non-verbal communication). The first visit focused purely on the photographic evidence collected by the researcher. The second visit to the event also included semi-structured interviews with four event competitors from a variety of cultural backgrounds and a questionnaire survey with event attendees. This survey aimed to provide details on the demographics of attendees, the number of times they had visited the event, their motivations to attend, and their opinions and attitudes towards intercultural exchange at the event, using a series of questions and statements with Likert scales. Questions were also asked on the areas and elements of the event they considered to be most conducive to intercultural exchange. The results were systematically analysed to allow for comparison with qualitative data in later phases.

In phase three, interviews were conducted with three event attendees (one of which was a volunteer) to ascertain their views on the meanings behind intercultural communication and exchange. These participants were then asked to take photographs (a maximum of 27 each using a disposable camera) of any aspects and incidences they felt captured the phenomenon when they were at the event. These images were elicited alongside the researchers' own photographs in post-festival interviews. The benefits of photo-elicitation interviews are plentiful, not least the fact that the participants can discuss their own images with a feeling of being in the driver's seat and taking part ownership of the research (Matteucci, 2013; Pink, 2008), thus producing very thick descriptions. A grid was produced (Figure 2) to aid the process of elicitation; the participants were asked to discuss why they took their photographs whilst placing them on the grid, considering what type of communication or exchange was taking place (i.e. verbal or non-verbal) and the perceived depth of the exchange. The interviews were video recorded to allow for analysis in a 'hybrid' format to include both text and images (Vince & Warren, 2012).

Figure 2. The photo-elicitation grid

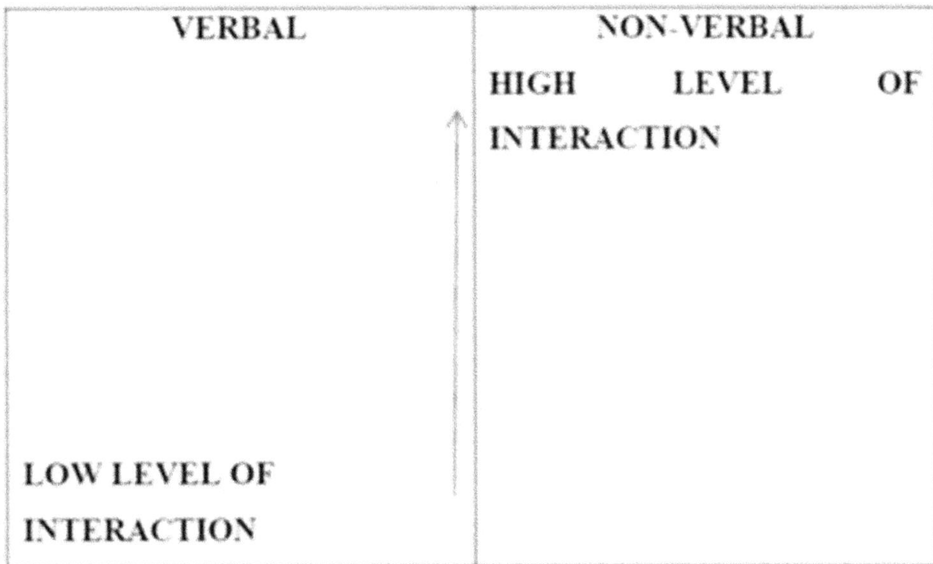

Following the final phase of data collection, the individual incidences of intercultural communication and exchange identified in all interviews and visual evidence were analysed using models of the symbolic exchange process of intercultural communication (Neulip, 2015; Ting-Toomey & Chung, 2012). This critical analysis identified a belief that there were spaces where different types of intercultural communication and exchange (a mix of verbal or non-verbal) occurred to varying degrees and perceived depth during the festival. This data was then triangulated with the results of the questionnaire, the interviews with organisers and the historical narrative of archived material to produce a detailed discussion of all sets of primary data collection that considered the nature, depth and processes and mechanisms of intercultural communication at the event and how these had changed over time.

ANALYSIS: THE PROCESSES AND MECHANISMS THAT ENCOURAGE INTERCULTURAL UNDERSTANDING AT LLANGOLLEN INTERNATIONAL MUSICAL EISTEDDFOD

The historical account formulated from the archived material and interviews with the organisers and long-term volunteers revealed that, in the early years, large numbers of competitors and their families attended the event and were hosted by families within the host town of Llangollen. This meant that the depth of interaction and intercultural understanding was high, and that there was physical and temporal space away from the competitive arena for in-depth intercultural interaction to occur. A good example of this process was described within the archived materials for the event, where during the late 1940s a family hosted both a German volunteer and a Russian folk instrumentalist. The German volunteer had been very wary of the Russian due to the treatment her grandfather had received when captured by the Russians during WWII, but by the end of the week they had become friends (Adams, 2005).

Since many of the organisers that were interviewed had been involved with the festival for a number of decades, they commented upon two specific issues which they believed had impacted upon the event, reducing the potential for intercultural exchange and impacting the processes and mechanisms available to the organisers with which to facilitate intercultural understanding and promote peace. The first was that these days competitors are accommodated separately in more expensive hotels and education centres which can be up to 30 miles from Llangollen. The second issue raised was that of competition from other events and visas, their expense and the difficulty of getting them. As a result of these issues, the cost of participating in these festivals had risen significantly, and most competitors needed to participate in several festivals/competitions during a single visit to the United Kingdom (UK) to justify this expense. Consequently, most competitors are no longer accompanied by family members, which has changed the nature of the audiences, reduced their cultural diversity, and reduced the amount of downtime available in which to form in-depth relationships with the community, other competitors and attendees. These challenges reflect Biaett's (2015) views on the difficulties festivals face in maintaining organic festivity and Andersson and Getz's (2008) research on the problems facing non-profit festivals.

The questionnaires revealed that most attendees did not have a primary motive of cultural exploration or meeting new people, potentially limiting the transformational element of the festival; an increasing number were staying for shorter periods, perhaps visiting just for one day with the main aim of seeing the professional evening concert rather than to experience an immersive multicultural festival. This might suggest that the event is not doing enough to create meaningful exchanges for those who attend the event only once, but it could also be argued that one day is not enough to fully immerse oneself into a learning environment. More positively, however, the results also supported Tomljenovic's (2010) findings, that the more often an attendee visited the festival and the longer they stayed the greater chance there was of them experiencing and wanting to experience deep and meaningful engagement opportunities which could lead to higher levels of intercultural understanding. The questionnaire results also revealed that there were certain areas and aspects of the event that were perceived to be more conducive to intercultural interaction, namely the general walkways, the trade and exhibition stands and the 'external' performance areas.

External Processes and Mechanisms

Media and public relations are the main external mechanism for intercultural communication at the Llangollen Eisteddfod. Moufakkir and Kelly (2013), in their SWOT analysis, suggested that the use of media before, during and after the event is vital in the promotion of the message of peace. However, the organisers of the event actually have very little control in relation to how and where the media represents the festival. They also questioned whether the external messages sent by the media effectively portray what happens during the event. For example, in 2014, while Russia was supporting the insurgency in Ukraine the fact that groups of Russian and Ukrainian dancers were participating in the same festival was portrayed on TV and social media as a positive news story of hope and harmony. However, there seemed to be little or no actual communication between individual competitors during their short time at the festival, from the visual evidence produced by the researcher and comments made within the elicitation interviews by research participants:

I watched a staged and filmed dance display by Ukrainian and Russian dancers, [...] and this was well-attended and strongly applauded, with the encouragement of the film crew. I felt very cynical about this, although many of those watching were obviously quite excited by it as a demonstration of peace and understanding! I tried to notice if the dancers actually 'communicated' among themselves, but couldn't. (Photo elicitation interview, participant 2, 2014)

The organisers also made mention the use of an online television programme 'Llangollen TV', with which they could retain more control of content, but this had not been very well developed. They also expressed a strong interest in developing outreach programmes and working with schools as a form of external intercultural communication in line with the effective interpretation techniques that events can use, as suggested by Getz and Page (2016).

Internal Processes and Mechanisms

Reflecting the work of Andrews and Leopold (2013) and Comunian (2015), the performances were seen to be the most important and effective process of non-verbal intercultural exchange in the way that they induced strong emotions from the attendees and volunteers enabling feelings of communitas to emerge in particular spaces. According to questionnaire data and participant observations, the types of performance that were most conducive to these experiences were usually non-competitive and staged in a more intimate setting where the proxemics of audience and performers were reduced and where, sometimes, audiences were encouraged to take part in the performance. It was obvious within the photo-elicitation setting that the research participants had taken very similar photographs to the researcher to capture such moments (Figure 3).

Figure 3. The external stages: Researchers' image and participant's image

Stories were recounted by organisers of moments where intercultural exchange had taken place within the Pavilion setting (the main performance space), for example, performers mingling with the general audience after they had finished their own performance to watch and applaud others. Co-performance, the sharing of the stage and audience space by groups whose nations are in serious conflict, often initiated a particularly emotive response from the audience. As Walmsley and Franks (2011) and Richards (2015) have shown, co-creative experiences can be effective in providing immersive cultural experiences which have great transformational potential. However, the organisers, some of whose families had provided the original home accommodation, also commented on the reduced amount of immersive co-creation which now occurred. From the archives, incidences were recounted of communitas taking place outside the festival arena, for example, where ad hoc performances would spring up out of nowhere (see figure 4 and accompanying quotation).

Figure 4. 'Street rehearsal by dancers from St. Malo'
Source: Liverpool Echo, 1953, p. 3

[…] never did audiences have better value for money than the audiences get at Llangollen, the streets of the town itself being free shows where one can meet with people of many countries in national costume, people ready to talk, dance or sing at the drop of a hat". (Jones, 1954, p. 2, archived material)

They suggested that exchanges like this rarely took place now because once the competitors had finished performing they usually had to rush back to their distant accommodation or to the next festival they were performing at and so attendee and competitor interaction was greatly reduced.

Non-verbal communication via costume and dress was also identified as a mechanism with which to stimulate interactions, the costumes acting as physical elements of culture. While the authenticity of the competitors' traditional costumes was queried by the organisers, the costumes frequently acted as an icebreaker for attendees, often leading to verbal communications of one form or another. The costuming usually occurred when the competitors were participating in particular activities; competing, walking around the site or town or actively seeking to promote their culture. There was more opportunity for verbal conversation when the competitors were relaxed and had been given time – between performances and within the performance areas which did not contain a competitive element (the external stages and the general walkways), as shown in figure 5.

Figure 5. General walkways and external stages, participant image and researcher image

However, the event organisers commented that the wearing of national costume away from the competition areas had significantly decreased in recent years. Articles within the archives of the Friends of Llangollen newsletter suggested, along with the organisers, that this was due to two factors; first, the reduced downtime competitors had to spend informally at the event and second, the cost and care of the costumes which were now required for multiple performances. The organisers suggested that this reduction in visual symbolism also contributed to the reduced amount of contact between attendees and competitors.

The social context of verbal intercultural communications is important, and perceptions and expectations from both sides can vary greatly depending on this factor. The results from the attendees and personal observations suggested that verbal interaction with the competitors did still occur at the event, although they may not have been often or deep. Reflecting Steiner and Reisinger's (2003) work, when deeper exchanges did happen they were in spaces which were conducive to interaction and enabled by the rules and norms of behaviour of the Eisteddfod, for example around the external stages, at the trade and exhibition stands and event walkways. The following exchange took place on one of the walkways on the festival site where the Indian ladies were between competitions and walking around looking for opportunities to engage with attendees, a temporal factor enabling to them to engage in such activities:

Figure 6. Trade and exhibition stands: Researcher image and participant image

[…] an occasion arose on day 4 when we were approached by a group of Indian ladies in their traditional costumes. The performers were handing out bracelets with the word 'friend' inscribed on them, and after giving some to my daughters we had a conversation about children and the differences in the ways they are brought up in our respective cultures. (Researchers' field note, visit 1, 2012)

The agendas of the transaction seemed to be equal from both sides – both were motivated to know about each other's culture in relation to a common area of interest (children). Therefore, there was a similarity in attitude which Steiner and Reisinger (2003) pinpoint as conducive to learning. The exchange developed into a conversation due to a number of factors – the breaking down of barriers to verbal communication through physical appearance, the offering of a material element of culture and a common point of interest, all of which led to quality interactions (Tomljenovic, 2010). The exchange was a good example of the act of bridging in terms of cultural capital (Quinn & Wilks, 2013; Wilks, 2011).

There was also quite a lot of verbal interaction between competitor groups and the people hosting exhibition stands. One South African competitor, at the event for the first time, had had conversations with at least four strangers. One of his conversations took place on the Rotary Club International exhibition stand and appeared to be relatively deep and challenging; relating to the charity's work in South Africa versus the UK.

[…] why go all the way to Africa, why not doing the good work here? I told him that he must remember that we are waiting for government change all over Africa and then we can start financially thinking of the future again, investing again." (South African competitor, 2013).

It could be said that the competitor felt relaxed enough to be able to engage in these types of conversations which suggests that the Llangollen Eisteddfod as a cultural space creates some opportunities for people from different cultures to communicate at depth.

It appeared that most verbal communication occurred where there was some sort of commercial transaction involved (e.g. trade stands and hospitality units), but the depth of the interactions was generally considered average in these situations, during the elicitation interviews. Steiner and Reisinger (2003) point out that commercial activities between hosts and guests are often imbalanced and so the nature of the interactions may not be wholly positive. However, it is evident that, in this particular festival, the nature of host and guest differed depending on the stall. For example, on the Around the World trade stand each competitor group was given the opportunity and time to promote their country and or group on a day to day basis; the stand hosts were competitors as well as hosts and the festival attendees, and other competitors were their guests. This stall was popular and provided material elements of culture through which to reduce barriers to intercultural communication and learning via symbols (flags) and verbal communication. A number of photographs of this stall were taken by both researcher and participants (see figure 6 for examples):

On other stands where there was a material element and an area of common interest depth of exchange and engagement was also noted.

I think they were Polish or something, or Eastern European anyway [pointing to the people browsing the stall]. [...] And they were really interested and they were chatting to the man for a long long while, so this was a high level of verbal communication. (Photo elicitation interview, participant 2, 2014)

As with the Indian ladies offering bangles for children, beer provided the material element which facilitated in-depth conversations. These interactions may go somewhat against the criticisms made of cross-cultural behaviour in tourism by Steiner and Reisinger (2003) that interactions of a commercial nature are imbalanced and superficial and, therefore, asymmetrical. However, it does support the view that multicultural events have the potential to alleviate some of the issues relating to the asymmetrical nature of host-guest interactions through shared motivations, behaviours and activities, common interests and goals.

On the whole, interactions between competitors and attendees varied from the desire to merely take photographs to in-depth discussions on politics and religion. Many attendees were happy to just take photographs, and, if they did enter into conversations, most appeared to be relatively superficial – a co-presence as opposed to focused interactions (Steiner & Reisinger, 2003). This seemed to be due to a number of factors, including the lack of open physical spaces available for interaction, lack of time within the competitors' programme, language barriers, lack of motivation and what was referred to by the organisers as the 'great British reserve'.

SOLUTIONS AND RECOMMENDATIONS

Mouffakir and Kelly's (2013) SWOT analysis of peace through events was adapted and applied to Llangollen International Musical Eisteddfod in order to visualise the conclusions from the study and provide recommendations. This can be seen in figure 7.

As the results also show, even where the aim of the festival is still retained over time, modern multicultural festivals face ever-increasing fiscal, political and temporal challenges in ensuring that they do not become simple commercial vehicles from which to display traditional cultural performances. The fiscal challenges include from the competitors' perspective the increased expense of travel, accommodation and visas; visa issues are also linked to the UK political situation where obtaining visas for competitors from countries deemed politically unfriendly becomes ever more difficult. Both the organisers and the competitors now seek external funding in order to run and attend the festivals, and should continue to work with organisations such as the British Council to try and improve these matters. While the organisers are required to show fiscal probity as well as a cultural legacy, raising funds can mean, as at the Llangollen Eisteddfod, that the authentic amateur competition offering has to be balanced with professional commercial evening concerts in order to attract sufficient funding for the core festival to take place. Paradoxically while the evening concerts may bring in much-needed funding and audiences these particular festival attendees have little or no interest in participating in intercultural experiences; nor do the professional acts they have come to see. For the competitors in order to maximise the benefit of the funding (and visas) they receive they try to dance at multiple competitions or exhibitions in one visit and seek out cheap accommodation at places which may be a long-distance from the festival site.

Figure 7. Adaptation of Moufakkir and Kelly's (2013) peace through events SWOT analysis

STRENGTHS *Event attributes which bring people together in non-adversarial circumstances:* • Cultural elements • Associated economic benefits • Location • Innovative programme of events open to change • Outreach programmes • Co-creative aspects through Friends of LIME • Community involvement • Sincerity in delivery of organisers and performers	OPPORTUNITES *External development which can contribute to the desired event outcomes:* • Media support / publicity • Globalisation • Multiculturalism • Advances in technology – online programmes, interactive exhibits and displays • Growing sophistication of tourists • Sustainability concerns • Tourism education • Redesign of event site and programme to allow more space and time for interaction
WEAKNESSES *Attributes which hinder or restrict events in achieving the desired outcomes* • A reliance on evening concert income • Limited age span of audience • Diminishing numbers of competitors and attendees • Lack of spaces for interaction to occur • Limited time span – no space in programme for 'down time' • Accommodation for competitors too far away	THREATS *Developments likely to increase hostility or contribute to a decline in event effectiveness:* • Terrorism • Commercialisation • Media attention on conflict themes • Changing tourist demands – volume and nature • Dismissal of peace objective • Lack of control over media • Problems with obtaining visas • Performers attending many events in one visit to the UK leads to potential for dilution of opportunities for ICE • Change of health and safety legislation

This has the effect of reducing the time that they have available to interact with other competitors, festival visitors and local residents (who may be organisers, festival visitors and or volunteers).

The temporal and fiscal restrictions on the competitors have also reduced the opportunities for the traditional festival attendees to participate in focused multicultural interactions. Essentially, the Eisteddfod's challenge is to keep its original ethos alive in an increasingly commercialised world of festival management, whilst at the same time recognising how society has changed, and finding new ways to promote the in-depth cross-cultural understanding which supports the message of intercultural understanding and peace. Since the results have shown, both historically and currently, that the most in-depth focused interactions occur when all those participating in the experience are not under pressure, spaces, both temporal and physical, need to be created allowing for more organic festivity and in-depth intercultural exchange experiences to occur. One example of how these spaces might be created would be to provide a competitor's village adjacent to the festival site which would enable more competitors to have more downtime, after their competitions, on the festival site.

FUTURE RESEARCH DIRECTIONS

There is potential for the methodological approach to be taken to other case study festivals in order to investigate the same phenomenon so that a growing body of knowledge can be produced on the effectiveness of multicultural events as an important element of the tourism industry in facilitating intercultural understanding and contributing to global citizenship and peace. Of particular note is the effectiveness of longitudinal studies at events that consider changes over time in line with external environment, as well as case studies that include a visual element. The study has shown that both these approaches allow for greater depth of the analysis of data leading to rich, credible and rounded case studies. The applicability of this approach internationally would help event managers and tourism organisations involved in multicultural events to become more aware of the intricacies of intercultural communication in order to be able to design their events effectively to meet specific wider intercultural objectives.

CONCLUSION

The question of whether multicultural festivals such as the Llangollen Eisteddfod can still contribute to the concept of intercultural understanding, friendship and peace is complex. Whilst the original objectives of these post-war festivals centred on reconciliation through the development of friendship, hope and peace, it can be argued that despite there now being much more knowledge about other cultures and perspectives this knowledge is frequently shallow and lacks depth of understanding. This would suggest that there remains a need for multi-day festivals to provide spaces in which perceived knowledge could be challenged and developed by in-depth interaction with other cultures, breaking down stereotypes, prejudices and ethnocentricity as part of international responsible tourism policies. A key factor in this is that organisers, as the most influential stakeholder group, continue to believe that they have a moral and global duty to promote peace, cohesion and humanism through the festivals which they produce.

Analysis of the various processes and mechanisms through which intercultural exchange has occurred and currently occurs at the Llangollen Eisteddfod has provided a good insight into both the environment needed and the current pressures which are challenging the existence of that environment. Levels and depths of intercultural exchange between competitor groups and attendees were variable depending on the spaces and times in which they took place. However, on the whole, the event environment went some way to diminishing these variables, breaking down barriers to communication through shared experiences.

REFERENCES

Adams, N. (2005). Just like it used to be. In J. Davies (Ed.), A world in harmony: A celebration of the Llangollen International Eisteddfod (pp. 18-19). Llangollen: Friends of the Llangollen International Musical Eisteddfod.

Andersen, P. A., & Wang, H. (2009). Beyond language: Non-verbal communication across cultures. In L. A. Samovar, R. E. Porter, & E. R. McDaniel (Eds.), *Intercultural communication: A reader* (12th ed., pp. 264–281). Wadsworth Cengage Learning.

Andersson, T. D., & Getz, D. (2008). Stakeholder management strategies of festivals. *Journal of Convention & Event Tourism*, *9*(3), 199–220. doi:10.1080/15470140802323801

Andrews, H., & Leopold, T. (2013). *Events and the social sciences*. Routledge. doi:10.4324/9780203070741

Biaett, V. (2015). Organic festivity: A missing element of community festival. In A. Jepson & A. Clark (Eds.), *Exploring community festivals and events* (pp. 7–30). Routledge.

Bruner, E. M. (1991). Transformation of self in tourism. *Annals of Tourism Research*, *18*(2), 238–250. doi:10.1016/0160-7383(91)90007-X

Chalcraft, J., Magauadda, P., Solaroli, M., & Santoro, M. (2011). Music festivals as cosmopolitan spaces. In European Commission (Ed.), European arts festivals: Strengthening cultural diversity (pp. 25-36). Luxembourg: Publications Office of the European Union.

Chhabra, D., Healy, R., & Sills, E. (2003). Staged authenticity and heritage tourism. *Annals of Tourism Research*, *30*(3), 702–719. doi:10.1016/S0160-7383(03)00044-6

Comunian, R. (2015). Festivals as communities of practice: Learning by doing and knowledge networks among artists. In C. Newbold, C. Maughan, J. Jordan, & F. Bianchini (Eds.), *Focus on festivals: Contemporary European case studies and perspectives* (pp. 53–65). Goodfellows Publishers. doi:10.23912/978-1-910158-15-9-2640

D'Amore, L. (2009). Peace through tourism: The birthing of a new socio-economic order. *Journal of Business Ethics*, *89*(S4), 559–568. doi:10.100710551-010-0407-3

D'Amore, L. J. (1988). Tourism: A vital force for peace. *Tourism Management*, *9*(2), 151–154. doi:10.1016/0261-5177(88)90025-8

De Certeau, M. (1986). *Heterologies: Discourses on the other*. Manchester University Press.

Deventer, K. (2015). The public festival: Inspiration and interconnectivity at the heart of festivals. In C. Newbold, C. Maughan J., Jordan, & F. Bianchini (Eds.), Focus on festivals: Contemporary European case studies and perspectives (pp. 255-264). Oxford: Goodfellows Publishers.

Edgell, D. L. (2014). Travel and tourism: The language of peace. *Journal of Hospitality and Tourism, 12*(2), 30–41.

Ellis, C., Adams, T. E., & Bochner, A. P. (2011). Autoethnography: An overview. *Forum. Qualitative Research, 12*(1).

Fabiani, J. (2011). Festivals local and global: Critical interventions and the cultural public sphere. In C. Newbold, C. Maughan J., Jordan, & F. Bianchini (Eds.), Focus on festivals: Contemporary European case studies and perspectives (pp. 92-107). Oxford: Goodfellows Publishers.

Finkel, R. (2009). A picture of the contemporary combined arts festival landscape. *Cultural Trends, 18*(1), 3–21. doi:10.1080/09548960802651195

Galtung, J. (1996). *Peace by peaceful means: Peace and conflict, development and civilisation*. International Peace Research Institute.

Getz, D. (2008). Event tourism: definition, evolution, and research, *Tourism Management, 29*(3), 403-428.

Getz, D., & Page, S. J. (2016). *Event studies: Theory, research and policy for planned events* (3rd ed.). Routledge. doi:10.4324/9781315708027

Gomez-Jacinto, L., Martin-Garcia, J. S., & Bertiche-Haud'Huyze, C. (1999). A model of tourism experience and attitude change. *Annals of Tourism Research, 26*(4), 1024–1027. doi:10.1016/S0160-7383(99)00063-8

Habermas, J. (1962). The structural transformation of the public sphere: An inquiry into a category of bourgeois society. Cambridge: Polity.

Haessly, J. (2010). Tourism and a culture of peace. In O. Moufakkir & I. Kelly (Eds.), *Tourism, progress and peace* (pp. 1–16). CABI. doi:10.1079/9781845936778.0001

Haynes, J. (2005). World music and the search for difference. *Ethnicities, 5*(3), 365–385. doi:10.1177/1468796805054961

Hofstede, G. (1997). *Cultures and organisations: Software of the mind*. McGraw-Hill.

Jones, J. (1954, July 7). Annual miracle of Llangollen. In 1954 International Eisteddfod Supplement. *Liverpool Daily Post*.

Klaic, D. (2006). "Festival", in Lexicon. *Performance Research, 4*(11), 54-55.

Lee, I., Arcodia, C., & Lee, T. J. (2012). Benefits of attending a multicultural festival: The case of South Korea. *Tourism Management, 33*(2), 334–340. doi:10.1016/j.tourman.2011.04.001

Matheson, C. M. (2008). Music, emotion and authenticity: A study of celtic music festival consumers. *Journal of Tourism and Cultural Change, 6*(1), 57–74. doi:10.1080/14766820802140448

Matteuccii, X. (2013). Photo elicitation: Exploring tourist experiences with researcher found images. *Tourism Management, 35*, 190–197. doi:10.1016/j.tourman.2012.07.002

Morgan, M. (2008). What makes a good festival? Understanding the event experience. *Event Management, 12*(2), 81–93. doi:10.3727/152599509787992562

Moufakkir, O., & Kelly, I. (2013). Peace through tourism: A sustainable development role for events. In T. Pernecky & M. Luck (Eds.), *Events, society and sustainability: Critical and contemporary approaches* (pp. 130–150). Routledge.

Mouffakir, O., & Pernecky, T. (Eds.). (2015). *Ideological, social and cultural aspects of events*. CABI. doi:10.1079/9781780643526.0000

Negrier, E. (2015). Festivalisation: Patterns and limits. In C. Newbold, C. Maughan, J. Jordan, & F. Bianchini (Eds.), *Focus on festivals: Contemporary European case studies and perspectives* (pp. 18–27). Goodfellows Publishers. doi:10.23912/978-1-910158-15-9-2668

Neulip, J. W. (2015). *Intercultural communication: A contextual approach* (6th ed.). Sage Publications.

Newbold, C., Jordan, J., Bianchini, F., & Maughan, C. (2015). Introduction: Focusing on festivals. In C. Newbold, J. Jordan, F. Bianchini, & C. Maughan (Eds.), *Festivals in focus* (pp. xv–xxvi). Goodfellows Publishers. doi:10.23912/978-1-910158-15-9-2599

Nordvall, A., Pettersson, R., Svennson, B., & Brown, S. (2014). Designing events for social interaction. *Event Management, 18*(2), 127–140. doi:10.3727/152599514X13947236947383

Pink, S. (2008). Analysing visual experience. In M. Pickering (Ed.), *Research methods for cultural studies* (pp. 125–149). Edinburgh University Press.

Pizam, A., Jafari, J., & Milman, A. (1991). Inñuence of tourism on attitudes: US students visiting USSR. *Tourism Management, 12*, 47–54. doi:10.1016/0261-5177(91)90028-R

Quinn, B. (2010). Arts festivals, urban tourism and cultural policy. *Journal of Policy Research in Tourism, Leisure & Events, 2*(3), 264–279. doi:10.1080/19407963.2010.512207

Quinn, B. (2013). *Key concepts in event management*. Sage Publications. doi:10.4135/9781473914742

Quinn, B., & Wilks, L. (2013). Festival connections: People, place and social capital. In G. Richards, M. P. de Brito, & L. Wilks (Eds.), *Exploring the social impacts of events* (pp. 15–30). Routledge.

Richards, G. (2015). Imagineering events as interaction ritual chains. In G. Richards, L. Marques, & K. Mein (Eds.), *Event design: Social perspectives and practices* (pp. 14–24). Routledge.

Richards, G., de Brito, M. P., & Wilks, L. (Eds.). (2013). *Exploring the social impacts of events*. Routledge. doi:10.4324/9780203108123

Sassatelli, M. (2011). Urban festivals and the cultural public sphere: Cosmopolitanism between ethics and aesthetics. In L. Giorgio, M. Sassatelli, & G. Delanty (Eds.), *Festivals and the cultural public sphere* (pp. 12–28). Routledge.

Schulenkorf, N., & Edwards, D. (2010). The role of sport events in peace tourism. In O. Moufakkir & I. Kelly (Eds.), *Tourism, progress and peace* (pp. 99–117). CABI. doi:10.1079/9781845936778.0099

Selstad, L. (2007). The social anthropology of the tourist experience, exploring the 'middle role'. *Scandinavian Journal of Hospitality and Tourism, 7*(1), 19–33. doi:10.1080/15022250701256771

Sorrells, K. (2013). *Intercultural communication, globalisation and social justice*. Sage Publications.

Spracklen, K., & Lamond, I. R. (2016). *Critical event studies*. Routledge. doi:10.4324/9781315690414

Steiner, C. J., & Reisinger, Y. (2003). *Cross-cultural behaviour in tourism: Concepts and analysis*. Butterworth-Heinemann.

Szabo, J. S. (2015). Festivals, conformity and socialisation. In C. Newbold, C. Maughan, J. Jordan, & F. Bianchini (Eds.), *Focus on festivals: Contemporary European case studies and perspectives* (pp. 40–52). Goodfellows Publishers. doi:10.23912/978-1-910158-15-9-2662

Ting-Toomey, S., & Chung, L. C. (2012). *Understanding intercultural communication* (2nd ed.). Oxford University Press.

Tomljenovic, R. (2010). Tourism and intercultural understanding or contact hypothesis revisited. In O. Moufakkir & I. Kelly (Eds.), *Tourism, progress and peace* (pp. 17–32). CABI. doi:10.1079/9781845936778.0017

Turner, V. (1969). *The ritual process*. Penguin.

Vince, R., & Warren, S. (2012). Participatory visual methods. In G. Symon & C. Cassell (Eds.), *Qualitative organisational research: Core methods and current challenges* (pp. 275–295). SAGE. doi:10.4135/9781526435620.n16

Walmsley, B., & Franks, A. (2011). The audience experience: Changing roles and relationships. In Walmsley (Ed.), Key issues in the arts and entertainment industry (pp. 1-16). Oxford: Goodfellow Publishers.

Wang, N. (1999). Rethinking authenticity in tourism experience. *Annals of Tourism Research*, *26*(2), 349–370. doi:10.1016/S0160-7383(98)00103-0

Wilks, L. (2011). Bridging and bonding: Social capital at music festivals. *Journal of Policy Research in Tourism, Leisure & Events*, *3*(3), 281–297. doi:10.1080/19407963.2011.576870

Wond, T., & Macaulay, M. (2011). Extending time- extended benefit. *Public Management Review*, *3*(2), 309–320. doi:10.1080/14719037.2010.536059

Chapter 14
Factors Affecting the Impact of Tourism on Attitude Change:
A Qualitative Research

Sedat Çelik
https://orcid.org/0000-0002-4541-6355
Şırnak University, Turkey

ABSTRACT

Tourism plays a vital role in getting to know and understand each other. In this regard, the main purpose of this research chapter is to understand the relationship between tourism and attitude change and the factors affecting this relationship. Firstly, the role of tourism in changing attitudes is discussed within the framework of Allport's Contact Hypothesis, and then qualitative research results are given. The phenomenology research design was used in the research, and the interview method was preferred in obtaining data. Six open-ended questions were asked to 12 tourists, who came to Şırnak, determined by the purposeful sampling method, with a semi-structured interview form. The answers given by the participants were determined by the descriptive analysis method. The research reveals that tourism is critical in eliminating problems between societies, opening social communication channels, reducing prejudices, and making existing attitudes more positive.

INTRODUCTION

The tourism sector, which is developing day by day, has become a very important phenomenon for both politicians and researchers due to its economic returns. However, the social-psychological effects of tourism are not taken into consideration. The main purpose of this research is to reveal the relationship between tourism and attitude change in terms of the social-psychological effects of tourism. As a matter of fact, the negative attitudes (prejudice, hostility, discrimination, etc.) that exist among societies in today's world constitute an important obstacle for societies in terms of understanding each other. The major difficulty in overcoming this obstacle is the inability to meet a common denominator and prejudices (Ruesch, 2011). Ellis (2006) emphasises that there must be a level of communication to reduce inter-

DOI: 10.4018/978-1-7998-5053-3.ch014

communal conflicts. On the other hand, Allport (1954) introduces the "Intergroup Contact Hypothesis" for reducing hostilities and prejudices among societies. According to Allport, when people interact, they break communication barriers and begin to get to know each other. Therefore, prejudices and hostilities can be diminished by this activity (Allport, 1954). From this point of view, "tourism" is one of the most important tools that can provide social interaction (Gamper, 1981).

Tourism allows people to see and recognise the similarities and differences of each other in different cultures. This can increase interaction and reduce prejudice. Tourism reveals an important and very strong ground where different societies and cultures can come together and contact with people of their destination. In this respect, most of the studies have been done by using an attitude scale with applying quantitative research (Milman et al., 1990; Pizam et al., 2002; Sirakaya-Turk et al., 2014). Also, many studies (Amir & Ben-Ari, 1985; Anastasopoulos, 1992; Pettigrew, 1979; Çelik, 2019b) demonstrates that tourism has important effects on changing attitudes as positive or negative ways. The question then arises: What are the factors causing this change? The number of applied studies that can answer this question is quite limited. To find appropriate answers to this question, the more qualitative research analysis is needed to be applied. However, the depth analysis with applying qualitative research method for the points of what factors effects this change has also remained limited. Therefore, this study aims to clarify these points by conducting a qualitative research method.

For this research, selected domestic tourists who came to the province of Sırnak which is the Southeastern Anatolia Region of Turkey, and applied a qualitative research method to analyse the effects of tourism on attitudes change by author. By using the phenomenology pattern, the tourist experiences were analysed in-depth, and it is expected to reveal the causes of tourism-attitude change between the lines. Besides, interviews were applied to local tourists who came to Şırnak province. In these interviews, the thoughts on the differences between the attitudes of the tourists before and after the visit, the reaction of their close circles (relatives, friends, etc.) when coming to Sirnak, the level of contact with the local people, and the satisfaction levels and the relationship of tourism attitude change were discussed. This study is expected to contribute to the literature since the relationship between attitude change in the study of tourism has remained limited in terms of a social psychological context. The results found in this study present some substantial data to private and public officials in the tourism sectors. It can also contribute to the awareness of decision-makers and tourism stakeholders on the social-psychological effects of tourism on society.

BACKGROUND

Attitude is one of the most basic and broad concepts of social psychology. Indeed, Allport, (1954) stated that attitude is the basic building block of social psychology. Many definitions have been made for the concept of attitude, and different components have been addressed in these definitions. Eagly and Chaiken (1993, p. 1) defined attitude as "a psychological tendency expressed by the evaluation of a certain being with some degree of goodness or dissatisfaction". Fishbein and Ajzen (1975, p. 216) defines attitude as "positive or negative emotions (evaluations) of an individual about an object related to target behaviour". "Attitude; is a term that includes topics such as preference, emotions, excitement, belief, expectation, perception, values, principles, thoughts and intentions" (Bagozzi 1994).

When attitudes are taken as positive and negative, negative attitudes refer to prejudice and positive attitudes refer to stereotype. Stereotypical is "generalising about a group of people by attributing the same characteristics to almost all members of a group, regardless of the differences between them" (Aronson et al., 2012, p. 752). Prejudice is "irrational negative attitudes towards some ethnic groups and their members." (Pettigrew, 1982, p. 2).

Prejudices against individuals and societies can lead people to discrimination, hostility, and even physical violence (Allport, 1954). The biggest obstacle in the formation of these attitudes is the lack of communication and interaction of people (Hewstone 1989). Although it is difficult to change these negative attitudes, it is not impossible. Many theories (learning theories, functional theories, cognitive coherence theory, mental contradiction theory, citation theory, etc.) related to the change of attitudes have been introduced (İnceoğlu 2010). One of these theories, the "Intergroup Contact Theory," which was introduced by Allport (1954) for attitude change, is quite remarkable. In Allport's theory, it is stated that people will increase their interaction when they communicate under certain conditions (groups having equal status, cooperation among groups, having common goals, support of the authorised person). In this way, he states that negative attitudes, such as prejudice, will decrease.

Change of attitudes can be achieved through communication and interaction. In this case, the tourism sector can be an important tool. Thanks to tourism, tourist groups and locals have the opportunity to get to know each other. This opportunity will help reduce prejudices and demolish stereotypes between communities. In this case, changes in the attitudes of workers in the tourism sector towards different groups are a good example (Bizjak et al., 2011). Some studies on tourism and attitude change (Çelik, 2019; Günlü et al., 2015; Milman, Reichel, & Pizam, 1990; Pearce, 1980; Pizam, Fleischer, & Mansfeld, 2002; Nyaupane, & Uysal, 2014) demonstrates that negative attitudes can turn into positive attitudes. Conversely, a few studies (Anastasopoulos, 1992; Sirakaya-Turk et al., 2014) reveal that attitudes change negatively in this respect. Tomljenovic (2010) proposed a model that demonstrates the effects of the contact hypothesis in local people-tourist interaction. This model reveals that the different factors of the variety and number of activities in the destination, destination familiarity, length of stay, travel motivation, personality traits, socio-cultural attitudes and quality of contact, among others, may affect tourist attitude change.

Sirakaya-Turk et al. (2014) showed that the attitudes of German tourists who came with a package tour were negatively changed towards Turkish people after the trip. The reasons for this have been identified as dissatisfaction with the tour guide and shopping experiences. Anastasopoulos's (1992) study, the attitudes of Greek tourists visiting Turkey were also changed negatively. The reasons for this change were the quality of life in Turkey and the effects of institutions and culture. On the other study, Ajanovic, Çizel, and Çizel (2016) dealt with whether or not the bias of Erasmus students has diminished within the framework of the theory of contact change. As a result of that study, there was a decrease in the prejudices of the students. Çelik (2019b) determined in his study that attitudes of tourists who visit the Southeastern Anatolia Region of Turkey were more positive than people who never visit this region. Also, the tourists coming to Gaziantep and Şanlıurfa determined that there were positive changes in their attitudes after their arrival. Fisher & Price (1991) found in their research that holiday satisfaction and intercultural interaction were directly effective in post-holiday attitude change, and travel motivation was indirect.

MATERIALS AND METHODS

The main purpose of this study is to understand the relationship between tourism and attitude change, along with the factors affecting this relationship. The sub-goals are subjected as the analysis of the attitudes of local tourists before coming to Şırnak, the view of the tourists' circle (relatives, friends) after coming to Şırnak, the level of communication of domestic tourists with local people, whether there is any change in the attitude after visiting Şırnak. The level of satisfaction of the tourists and travel to show how the attitude relationship is.

Data Collection and Interview Form

For the research, a qualitative research method was conducted. To obtain in-depth information for the experiences of tourists, a phenomenology pattern (Creswell 2014) in which individuals about a phenomenon depicted their experiences was used. Also, a semi-structured interview form consisting of open-ended questions was used. One month after the tourists ended the tour, phone calls were made to collect the data. Permission was obtained from the participants to record the interviews. During the interview, the phone calls recorded and noted the important points.

While preparing the research questions, the literature (Ajanovic et al., 2016; Anastasopoulos, 1992; Günlü et al., 2015; Milman et al., 1990) was examined, and the questions were finalised by taking the opinions of two academicians who are experts in their fields. Demographic questions and six open-ended questions were asked to participants.

Firstly, age, gender, educational status, occupation, and hometown information were obtained to determine the demographic status of the participants (Table 1). Then the following questions were asked:

1. What are your feelings and thoughts about Şırnak before you come to Şırnak?
2. How did they meet this situation when you tell your relatives (friends, family, etc.) that you will come to Şırnak?
3. How was your contact with the locals during the Şırnak trip?
4. Has there been a change between your point of view before coming to Şırnak and your point of view after coming and returning? If so, please state how positive or negative it happened.
5. If I asked you to evaluate your Şırnak tour in general, what topics would you like to talk about? Describe the elements you are satisfied with and not.
6. Do you think that travelling is effective in changing your attitudes/prejudices? Can you explain your previous experiences by thinking?

The more violent and intense attitude, the harder it is to change. In other words, negative attitudes change more difficult than positive attitudes (Basheer, 2019; Kağıtçıbaşı & Zeynep, 2014, p. 137). The first question was asked to determine whether the attitudes (positive, negative, neutral) of the tourists before coming to Şırnak, if any, what happened. The environment and reference groups (relatives, friends) have an important effect on tourist behaviour and attitudes (Harlak, 2000; Kozak & Decrop, 2009). Therefore, the second question was asked to reveal what their relatives' reactions were when they told them about Şırnak. Also, the level of tourist contact with the local community is an important factor affecting their attitude (Hewstone 1989). Therefore, the third question was asked to determine whether they contacted the local community and, if so, how they contacted (during shopping, staying at a hotel, eating, and drinking, etc.) to people. Many studies (Ajanovic et al., 2016; Günlü et al., 2015; Milman et al., 1990) show that tourism has changed attitudes, either positively or negatively. The fourth question was asked to reveal whether the attitudes against Şırnak have changed during the tour. Also, studies (Anastasopoulos, 1992; Sirakaya-Turk et al., 2014) determined that satisfaction is effective in attitude change. The fifth question was, thus, asked to determine how satisfaction affects the attitude. Lastly, many studies illustrated that travelling changes attitudes and tourism provides social peace (Çelik, 2019a; Jafari, 1989; Kim et al., 2007; Pizam et al., 2002). For this, the *sixth question* was asked to determine whether travelling and seeing different places were effective in changing attitudes, along with the opinions of the participants.

Sample Design

Homogeneous sampling, which is one of the purposeful sampling methods frequently used in qualitative research, was used in this study. The data were obtained from 12 volunteers (3 men and 9 women) from the local tourists who came for the Şırnak tour between May and September 2019. Most of the participants are female, retired, university graduates, and people from outside Şırnak Province. The sample was considered sufficient since saturation was achieved with the answers received (Charmaz 2006). Since most of the answers given to the questions were similar, it has been seen that saturation was achieved in the answers.

Table 1. Information about the participants

Participant	Age	Gender	Education	Vocation	Hometown	Duration
P1	52	Female	High School	Retired	Ankara	20:09
P2	58	Female	High School	Retired	Karadeniz/Pazar	46: 35
P3	59	Female	University	Retired	Ankara	17: 47
P4	62	Female	High School	Retired	İstanbul	13: 58
P5	45	Female	University	Retired	Ankara	13: 39
P6	62	Female	University	Retired	İstanbul	18: 40
P7	59	Female	University	Retired	Çorum	21: 29
P8	63	Female	University	Retired	İstanbul	27:23
P9	58	Female	High School	Tourist Guide	Giresun	23:25
P10	45	Male	University	The technique	Kayseri	18:36
P11	58	Male	University	Retired	İstanbul	21:25
P12	52	Male	Master	Retired	Sivas	18:40

Source: Created by the author

Reliability – Validity

Reliability-validity is very important in qualitative research. Research validity-reliability was provided based on studies of Christensen et al. (2015) and Creswell (2016). The demographic characteristics of the participants demonstrate descriptive validity. What the participants said was given as a direct quote. Confirmation of the participants regarding the results was obtained. Perceptions of the participants for the interpretation validity were discussed (general information about the thoughts of the tourists was taken from the tour guide). Also, results for theoretical validity and previous research results are discussed (Christensen et al. 2015).

Researcher Position

The role of the researcher should be stated in qualitative research (Creswell, 2016). The researcher is an academician in Şırnak province. Şırnak and its surroundings (Southeastern Anatolia Region) have created a negative perception in people's minds due to terrorist incidents, border security, and the media for many years. This situation leads to a social distance between the East-West society in Turkey. For this reason, tourist visitation was limited due to these conditions. In fact, this position of the city was the motivation to analyse the city as a research topic on attitudes. Besides, it is a duty to reveal the role of tourism that can increase awareness of people on the removal of communication barriers between communities. Also, if tourism is well managed, it is the key to world peace.

Research Site

This research was conducted in the Southeastern Anatolia Region of Turkey's Sırnak province. Sırnak, which borders Iraq and is located in southeastern Turkey near the Syrian is a province of Turkey. The reason why the research is done here is for the location of the Southeastern Anatolia Region, an important region where the history of humanity is written, hosting various peoples such as Turks, Kurds, Assyrians, and Arabs. Due to some conflicts, there has been much migration from the Southeast to the West (Saraçoğlu, 2009) and prejudices emerged between the East-West community in Turkey (Yılmaz et al., 2018). For this reason, this study was carried out in Şırnak province in this region.

Data Analysis

Firstly, the interviews were transcribed and transmitted to the words. Later codings, categories, and themes were created. The findings were examined by an interpretive-inductive approach. While doing this analysis, coding and theming were done according to the analysis approach of Straus & Corbin (1990). The codes were created, and the frequencies of the themes were shown by tabulation. This analysis was carried out with the help of the NVIVO 12 package program.

RESULTS

Attitudes Pre-Travel

There would be a bias in the tourists coming to Şırnak before starting the research. However, in the results, most of the participants stated that their attitude towards Şırnak was neutral (i.e. neither positive nor negative) (before coming). The answers to the question and the number of participants in the codes are shown in Table 2.

Table 2. Theme 1 "Attitudes before coming"

Theme	Category	Codes	F
Attitudes before coming (28)	Neutral (10)	Familiarity	1
		Loving, accepting, respecting all people	3
		I had no negative thoughts	3
		People like myself	1
		I was not afraid	2
	Positive (9)	Sincere people	1
		I had no worries about Şırnak people	1
		Family origin	1
		I had no prejudice	3
		I came with love	2
		Part of the homeland	1
	Negative (7)	Disclaimer	1
		Border security	2
		Terror	4

Source: Created by the author

I had nothing negative. I have no fear. I am a bit courageous. Nothing came to my mind, only I will go on a trip with his happiness and it was the happiness to move away from Istanbul for 6 days. I even looked at the travel list just like this. I did not think of any negative thoughts. I am happy to see a different region. I did not know that we were the first tourists. (P2)

I had no negative thoughts, I came with more love because I came from my roots, I was curious about nature. It made me happy to see people like myself a bit, I came very positively, I had no prejudices. (P3)

Just like everyone else, I only had my prejudice about terrorism. I don't know before I leave because I think we get nervous because there is bad news every day. (P9)

I have no negative thoughts before I come. I am familiar with that area because of my wife's duties. So I had no prejudice. (P4)

I had no negative thoughts, it was part of our country, I came with a lot of curiosity and love. I had no negative thoughts. (P12)

The Reaction of the Relatives

The second question refers to the reaction of the relatives regarding the decision to travel to Şırnak. As a result, there were intensive negative reactions to this question, only one person was neutral, and one person was received positively. The answers to the question and the number of participants in the codes are shown in Table 3.

Table 3. Theme 2 "The reaction of the relatives"

Theme	Category	Codes	F
The reaction of the relatives (family, friends, etc.) (25)	**Neutral (1)**	There was no reaction.	**1**
	Positive (1)	They supported my departure to Şırnak	**1**
	Negative (23)	You were wearing a vest	**2**
		Are you crazy?	**2**
		Security-related fear, anxiety, anxiety	**7**
		What are you doing there?	**5**
		Prejudice	**1**
		You can go to Şırnak only as a candle.	**2**
		Dangerous	**1**
		Terror	**3**

Source: Created by the author

The people around me are doing what you are doing, there is terrorism. There is terrorism killing only those who come from the West. If there is terror, you will die from a bomb in Istanbul. Since we will be in touch with the public, I have never looked at those thoughts. Whatever the forehead is, I am always a positive person. (P1)

Their attitude was negative, are you crazy? What was needed, it was said to go back as soon as you left. (P11)

When I said I would go to Şırnak, the reactions came were the concerns for me. Although the news about the "terrorist" incidents has been cut for the last one or two years, they still expressed their concern that "if something happens to you". In other words, our trip was not found safe by my circle. (P8)

No, I have travelled to many parts of the world, why would they react when they were going to Şırnak? My brother-in-law is Sason. I came to those places several times. (P5)

Contact Between Local People and Tourists

The third question was asked whether they had contact with local people. The contact of tourists with local people is an important factor in changing their attitudes. In all of the answers, it was revealed that they contacted the local people. It can be said that the contact states are sufficient when looking at the narratives. The answers to the question and the number of participants in the codes are shown in Table 4.

Table 4. Theme 3 "Contact between tourist and local people"

Theme	Category	Codes	F
Contact between tourist-local people (23)	Accommodation in the house of the local people (6)	We have seen great interest in homes	3
		We are still meeting	1
		Exchanging gifts	1
		They accompanied us to the bus	1
	While walking on the street (8)	We attended their wedding	1
		They invited us to their home	3
		We were very comfortable while travelling	2
		Friendly and warm communication	2
	During eating and drinking (5)	Eating and drinking	5
	During Shopping (4)	Shopping	4

Source: Created by the author

We had public communication, of course, we stayed at the house of people we never knew in Beytüşşebap (County of Şırnak), we gave each other gifts, we had no problems. We did not encounter any negative situations here. I met 3-4 people in the church, they called home and offered tea. The people I met were educated, conscious, and respectful, everyone was very good. (P1)

"We stayed in their homes, we stayed in the brother of the mayor in Beytüşşebap, we did shopping from there, the family was very welcoming. We went to Dengbej house, we were able to communicate with the people who played the music there. I thought about how those there met us. But they welcomed us very well, there was a common point. There were sincere communications. (P12)

People's attitudes were very good. When they passed the street, they accepted us at their wedding, there was no problem. (P10)

Table 5. Theme 4- pre-post attitudes for Şırnak

Theme	Category	Codes	F
Pre-post attitudes for Şırnak (17)	More positive than before travel (14)	My perspective has changed a lot	5
		I was not expecting so friendly	2
		I wasn't expecting geographically like this	2
		I became more conscious	2
		I left happier	3
	Neutral (As I expected) (3)	I was familiar before	3

Source: Created by the author

Pre-post Attitudes for Şırnak

From the results, we can see that the attitude of most of the participants changed positively. Just a few people stated that they were as they expected, so there was no change in their attitudes. The answers to the question and the number of participants in the codes are shown in Table 5.

After coming to Şırnak, I think more tours of Şırnak or its eastern region from now on. My thoughts turned even more positive, and prejudices diminished in those who came with me. (P9)

My perspective has changed. I said what a beautiful geography, what a beautiful place, people are very friendly, I was not expecting that much. They welcomed very lovingly. (P7)

There was not much change in my point of view, I encountered the things I had anticipated, I got the full value of my expectations. (P4)

General Evaluation about PackageTour

When the answers given about this question are evaluated, it is seen that every participant focuses on travel satisfaction. Only one participant (P3) stated that she was generally satisfied but complained about hygiene and cleanliness. The answers to the question and the number of participants in the codes are shown in Table 6.

Table 6. Theme 5 "General evaluation about package tour"

Theme	Category	Codes	F
General evaluation about Packet Tour (30)	**Satisfaction (29)**	Impressive and beautiful	**3**
		Accommodation is clean	**1**
		Hospitality	**13**
		Trust the guide	**5**
		Tourist group was good	**1**
		Very pleased	**6**
	Dissatisfaction (1)	Hygiene-Cleaning	**1**

Source: Created by the author

I am satisfied. There is a lack of hygiene and service quality in the hotels, they can improve themselves on these issues. There is no anxiety about the public, come without knowing them, be a guest, etc. good, but there must be progress on these issues. (P3)

I was very pleased with the tour, I knew the guide beforehand, it has done everything like a man, so to speak, I did not see any deficiency in any subject, there is no problem either in terms of humanity or behaviour, it is necessary to develop a little more for the person to continue his own culture. I am a person who loves to read anyway, I have been reading research books for years, it corresponds exactly with what I read. (P4)

I loved our tour, the team was very nice, we were satisfied with everything. (P10)

Tourism and Attitude Change

The perspectives of the participants, not only for the tour of Şırnak but also for tourism and attitude change, have been presented. According to the answers given, most of the participants thought that travelling changed attitudes and reduced prejudices. The two participants stated that this change would be influenced by the level of knowledge of the tourists about the destination, the personality structure (tolerant or intolerant) and the severity of the attitudes (prejudice, anxiety, etc.) before arriving. The answers to the question and the number of participants in the codes are shown in Table 7.

Table 7. Theme 6 "Evaluations of the tourism effect on attitude change"

Theme	Category	Codes	F
Tourism-attitude change relationship (18)	**Varies from person to person (5)**	It does not change the attitudes of the people who are searching where they are going.	2
		Anxiety level, the severity of prejudice	2
		Tolerance level	1
	Changes in the positive direction (13)	My prejudice is changing	2
		Recognising new people	5
		New places new perspectives	5
		West-east fusion	1

Source: Created by the author

I have not changed my attitude, but I have seen changes in the attitudes of my partner friends. I think that traveling necessarily changes people's ideas, I think that these tours should increase, that they should be made constantly and that people experience more by seeing them. if more people go, for example, we came to 15 people, each of them reached 15 people. For example, my daughter did not want to come. After my tour, my daughter says let's go. (P4)

It changes a lot from person to person. When you go very anxious in your mind, you cannot open yourself much. If you go with a more open mind and tolerance, there are different changes. Surely, travelling is also effective in changing attitudes. It is necessary to bring the West and the East together and combine more. (P12)

It changes people's attitudes, definitely. I do not think it changed my attitude anywhere. Because I'm doing research. I see that people who have not researched and believe what they hear are surprised. This was the case in 1992 in Van and Tunceli. On this tour, many people were concerned about coming, but were surprised and surprised that the local people were so hospitable after coming, and they were very happy with the trips to Central Anatolia. (P11)

DISCUSSION

In this research, what are the attitudes of tourists coming to Şırnak (prejudice, stereotypes, etc.) before the arrival of Şırnak, the reaction of the tourists reference groups on Şırnak tour, whether there has been a change in their attitude after arrival, general evaluation of tourist on tour (satisfaction and dissatisfaction), the relationship between tourism-attitude change and the local public contact situations affecting these changes, and tourism-attitude change in general, are handled from the perspective of domestic tourists.

Due to terror events and the distinction between the regions of East and West developing in the historical process in and around Şırnak, before the resource, tourists were to be biased, though. Newcomer tourists would have prejudices about the region and toward people who live around the region. But, when you look at the answers to the first question, it is wrong thought about bias. It was found that just a few participants had negative thoughts. The negative attitudes were determined to be related to local security problems and terrorism. Also, this study determines that there are two important reasons for the lack of bias of tourists; the first is that tourists have humanistic and open-to-experience personality structures (codes can be viewed); and the second is that their spouse, friends, relatives come to the Southeast and know more or less about the region. Similarly, Günlü et al. (2015) showed that none of the 12 people who went to Armenia had a negative attitude. Most of them had a neutral and positive opinion.

It was observed that the tourists coming to Şırnak received negative reactions from the family and friends' circles regarding their participation in the Şırnak tour. Looking at the coding, the most important reasons for the negative opinions of the reference groups are terrorist events and security problems in the region. This situation is in line with the negative attitudes of the participants that appeared in the first question. In general, reference groups, which include family and friends, affect people's attitudes (Harlak 2000). This can make people's thoughts more negative and more severe. As a result, it may be difficult to change a violent attitude. From a different point of view, tourists came despite the negative thoughts of their reference groups. However, the reasons for their arrival are, as seen in the codings; wondering, researching, and personality structures.

The contacts with the local people affect the attitude of both the tourist and the public (Tomljenovic 2010). In the answers given, it is seen that all tourists who participated in the interview could contact the local people. These contacts take place during shopping, during interacting with people while staying at local people's homes and during walking down the street. It turns out that these interactions are positive. As a matter of fact, according to Allport's contact hypothesis, positive or negative contact affects the attitude change. Similar to the findings of this study, the quantitative study of Joo et al. (2018) in Texas also demonstrates that the frequency of interaction and the interaction of local people with tourist activities have increased emotional solidarity and reduced social distance. In another study, Fan et al. (2017) found that the quality of social interaction has a negative effect on the perceived cultural distance, that is, the cultural distance decreases as the social interaction increases. In their study on the attitudes of local people, Carneiro et al. (2018) found that the intensity of social interaction and satisfaction with this interaction positively affected the attitudes of local people as the effects of tourism.

After the Şırnak tour, this study showed that there was a change in the attitudes of the tourists. There was a more positive change in the attitude of the majority. In this case, it is seen that the prejudices of the tourists are only related to security concerns, and there is not a very violent attitude in this, because the local people were friendly, and the confidence in the tour guide, the satisfaction of the tour and the personality structures of the tourists are also effective. Also, some participants saw Şırnak geographically better than they expected. Others, on the other hand, were found to be familiar with Şırnak because of their kinship and what they heard, and therefore, there was no change in their attitudes. Similarly, Nyaupane et al. (2016) investigated whether there was any change in the attitudes of students who went abroad and before and after the abroad program towards Austria, Netherlands, and Fiji. While the attitudes of the satisfied ones changed positively, the attitudes of the dissatisfied changed negatively. Similarly, Günlü et al. (2015), in their study with 12 people who went to Armenia, found that the participants became more positive than their previous attitudes. In the literature, it is seen that due to the nature of the package tour, the tourist-local people interaction cannot be achieved adequately (Noy & Cohen, 2005; Sirakaya-Turk et al., 2014). However, in this study, the disadvantage of the package tour was not illustrated, since the tourists coming with the package tour have had a high interaction with the local people.

As for the general evaluation, this study shows that they were satisfied with the tour. However, there are some deficiencies in service quality, hygiene, and cleaning point, but they do not care much about them. The reason for this is that the tolerance levels of incoming tourists are high. People with a high level of tolerance tend not to see minor errors. Çelik (2018) determined that people with higher tolerance levels were more satisfied with their personality traits. The participants emphasise that tourism positively changes attitudes. However, some participants state that this situation may vary from person to person. They state that bias severity with low tolerance level is high, and people who do not research before coming will not change their attitude. This is a very important result. Tomljenovic (2010) states that the best change is in soft attitudes. As a matter of fact, this situation also emerged in our fourth question in his thoughts towards Şırnak.

As seen in some studies, tourism has changed attitudes as the sample of this study. In the emergence of this change, the separation of tourists, contact with the local people, their pre-travel attitudes are not too neutral, and their personality structures are important factors. This result is very important in terms of contributing to the literature. Tomljenovic (2010) suggests similar results to the model of the contact hypothesis in tourist-host interactions. Considering these results will help future studies to understand tourism-attitude change.

CONCLUSION AND RECOMMENDATIONS

Research results show that tourism is a very important tool in intercultural interaction and in understanding people. It is an important phenomenon in ensuring world peace. Therefore, this aspect of tourism is a point that politicians, state officials, and non-governmental organisations should emphasise.

Today, there are important problems and communication barriers between communities around the world. At this point, it is important to consider the social psychological contribution of tourism, which contributes to the economy, development, cultural and environmental development. Nowadays, state officials talk more about the economic effects of tourism rather than the social-psychological effects. Even this can reveal that tourism is not yet socially and psychologically ignored.

Decision-makers should see the impact of tourism on peace and related development strategies. They can do this with various activities. For example, societies that do not like each other can encourage inter-country travel by signing travel protocols among themselves and ensure cross-cultural interaction. In this way, prejudices can be reduced by interacting societies that do not like each other. Moreover, educational institutions can make student attitudes more moderate among young people. It is easier to change the attitudes of young people than the attitudes of adults and older people (Krosnick & Alwin 1989).

Destination marketers can increase their advertising and promotion activities where tourism reveals a culture of unity and peace. While using many elements (sea, belief, culture, food and drink, rest, etc.), promotions are made for the needs of people. With togetherness, intercultural interaction, recognition of different cultures, and humanity themes, people can be persuaded to travel. Because living together and respecting people has become a need in this complex world. Indeed, some places use it. For example, Hatay Province uses the slogan "City of Peace, Culture, and Tolerance" using its multicultural, faithful, and linguistic structure. Similarly, Siirt Province uses the slogan "Trilingual City". Such slogans can encourage communities with large distances to interact by encouraging them to travel.

This study has some limitations. Firstly, conducted interviews with 12 local tourists in this study cannot be generalised. It was also done with tourists coming to Şırnak in May-September 2019. This situation may reveal a lack of measuring the perception of tourists temporally. Therefore, it is possible to work with tourist groups coming in different months in a longer period. Another shortcoming of the research was that foreign tourists were not selected for the interview. The reason for this is that foreign tourists do not come to Şırnak. When foreign tourists come, if we talk to them, the difference between local and foreign tourist perspectives can be observed. Also, researchers can examine the issue of tourism and attitude on larger samples by using mixed methods for future studies. In the studies to be carried out, more detailed modelling can be explored concerning tourism personality, satisfaction, the role of tourist guides, destination quality perception, tolerance level, duration of stay, culture, and the effects of other tourists.

ACKNOWLEDGMENT

I would like to thank the Özgür Bozan and Fırat Yılmaz for their support.

This study was presented at the 20th National Tourism Congress when nine interviews were held. In this study, the literature and methodology were developed, and the number of interviewers was increased to 12.

REFERENCES

Ajanovic, E., Çizel, B., & Çizel, R. (2016). Effectiveness of Erasmus programme in prejudice reduction: Contact theory perspective. *Turisticko Poslovanje*, *17*(17), 47–60. doi:10.5937/TurPos1617047A

Allport, G. W. (1954). *The nature of prejudice*. Addison-Wesley Publishing Company.

Amir, Y., & Ben-Ari, R. (1985). International tourism, ethnic contact, and attitude change. *The Journal of Social Issues*, *41*(3), 105–115. doi:10.1111/j.1540-4560.1985.tb01131.x

Anastasopoulos, P. G. (1992). Tourism and attitude change: Greek tourists visiting Turkey. *Annals of Tourism Research*, *19*(4), 629–642. doi:10.1016/0160-7383(92)90058-W

Aronson, E., Wilson, T. D., & Akert, R. M. (2012). *Sosyal psikoloji* [*Social psychology*] (O. Gündüz, Trans.). Kaknüs Yayınları.

Bagozzi, R. P. (1994). *Advanced methods of marketing research*. Blackwell Business.

Basheer, K. A. (2019). *Psychology for B.SC. Nursing* (2nd ed.). PHI Learning Pvt. Ltd.

Bizjak, B., Knežević, M., & Cvetrežnik, S. (2011). Attitude change towards guests with disabilities: Reflections from tourism students. *Annals of Tourism Research*, *38*(3), 842–857. doi:10.1016/j.annals.2010.11.017

Carneiro, M. J., Eusébio, C., & Caldeira, A. (2018). The influence of social contact in residents' perceptions of the tourism impact on their quality of life: A structural equation model. *Journal of Quality Assurance in Hospitality & Tourism*, *19*(1), 1–30. doi:10.1080/1528008X.2017.1314798

Çelik, S. (2018). Impacts of personality traits and tolerance levels of local tourists on perceived destination satisfaction. *Tourism Academic Journal*, *5*(1), 154–165.

Çelik, S. (2019a). Does tourism reduce social distance? A study on domestic tourists in Turkey. *Anatolia*, *30*(1), 115–126. Advance online publication. doi:10.1080/13032917.2018.1517267

Çelik, S. (2019b). Does tourism change tourist attitudes (prejudice and stereotype) towards local people? *Journal of Tourism and Services*, *10*(18), 35–46. doi:10.29036/jots.v10i18.89

Charmaz, K. (2006). Constructing grounded theory: A practical guide through qualitative analysis. *Sage (Atlanta, Ga.)*.

Christensen, L. B., Johnson, R. B., & Turner, L. A. (2015). *Araştırma yöntemleri-desen ve analiz* [Research methods-pattern and analysis] (A. Aypay & E. Gumus, Eds. & Trans.). Anı Yayıncılık.

Creswell, J. W. (2014). *Research design: qualitative, quantitative, and mixed methods approaches* (4th ed.). SAGE Publications.

Creswell, J. W. (2016). *30 Essential skills for the qualitative researcher*. SAGE Publications Ltd.

Eagly, A. H., & Chaiken, S. (1993). *The psychology of attitudes*. Harcourt Brace Jovanovich College Publishers.

Ellis, D. G. (2006). *Transforming conflict: Communication and ethnopolitical conflict*. Rowman & Littlefield Publishers.

Fan, D., Zhang, H. Q., Jenkins, C. L., & Lin, P. (2017). Does tourist–host social contact reduce perceived cultural distance? *Journal of Travel Research*, *56*(8), 998–1010. doi:10.1177/0047287517696979

Fishbein, M., & Ajzen, I. (1975). *Belief, attitude, intention, and behavior: An introduction to theory and research*. Addison-Wesley.

Fisher, R. J., & Price, L. L. (1991). International pleasure travel motivations and post-vacation cultural attitude change. *Journal of Leisure Research*, *23*(3), 193–208. doi:10.1080/00222216.1991.11969853

Gamper, J. A. (1981). Tourism in Austria a case study of the influence of tourism on ethnic relations. *Annals of Tourism Research*, *8*(3), 432–446. doi:10.1016/0160-7383(81)90007-4

Günlü, E., Özgen, H. K. Ş., Dilek, S. E., Kaygalak, S., Türksoy, S. S., & Lale, C. (2015). Turkish visitors in Armenia: Any changes in attitudes and perceptions? *Journal of Tourism and Hospitality Management, 3*(1), 29–43. doi:10.15640/jthm.v3n1a3

Harlak, H. (2000). *Önyargılar: Psikososyal bir inceleme* [Prejudices: A psychosocial investigation]. Sistem Yayıncılık.

Hewstone, M. (1989). Changing stereotypes with disconfirming information. In D. Bar-Tal, C. F. Graumann, A. W. Kruglanski, & W. Stroebe (Eds.), *Stereotyping and Prejudice: Changing Conceptions* (pp. 207–223). Springer. doi:10.1007/978-1-4612-3582-8_10

İnceoğlu, M. (2010). *Tutum algı iletişim* [Attitude perception communication] (5. Baskı). Beykent Üniversitesi Yayınları.

Jafari, J. (1989). Tourism and peace. *Annals of Tourism Research, 16*(3), 439–443. doi:10.1016/0160-7383(89)90059-5

Joo, D., Tasci, A. D. A., Woosnam, K. M., Maruyama, N. U., Hollas, C. R., & Aleshinloye, K. D. (2018). Residents' attitude towards domestic tourists explained by contact, emotional solidarity and social distance. *Tourism Management, 64*, 245–257. doi:10.1016/j.tourman.2017.08.012

Kağıtçıbaşı, Ç., & Zeynep, C. (2014). *Günümüzde insan ve insanlar: Sosyal psikolojiye giriş* [Today, human and humans: Introduction to social psychology] (16th ed.). Evrim Yayınları.

Kim, S. S., Prideaux, B., & Prideaux, J. (2007). Using tourism to promote peace on the Korean Peninsula. *Annals of Tourism Research, 34*(2), 291–309. doi:10.1016/j.annals.2006.09.002

Kozak, M., & Decrop, A. (2009). *Handbook of tourist behavior theory & practice*. Routledge. doi:10.4324/9780203881804

Krosnick, J. A., & Alwin, D. F. (1989). Aging and susceptibility to attitude change. *Journal of Personality and Social Psychology, 57*(3), 416–425. doi:10.1037/0022-3514.57.3.416 PMID:2778632

Milman, A., Reichel, A., & Pizam, A. (1990). The impact of tourism on ethnic attitudes: The Israeli-Egyptian case. *Journal of Travel Research, 29*(2), 45–49. doi:10.1177/004728759002900207

Noy, C., & Cohen, E. (2005). *Israeli backpackers: From tourism to rite of passage*. SUNY Press.

Nyaupane, G. P., Teye, V., & Paris, C. (2016). Tourism and attitude change: The case of study abroad students (No. 64). Tourism Travel and Research Association: Advancing Tourism Research Globally.

Pearce, P. L. (1980). A favorability-satisfaction model of tourists' evaluations. *Journal of Travel Research, 19*(1), 13–17. doi:10.1177/004728758001900103

Pettigrew, T. F. (1979). The ultimate attribution error: Extending Allport's cognitive analysis of prejudice. *Personality and Social Psychology Bulletin, 5*(4), 461–476. doi:10.1177/014616727900500407

Pettigrew, T. F. (1982). Prejudice. In T. F. Pettigrew, G. M. Fredrickson, D. T. Knobel, N. Glazer, & R. Ueda (Eds.), Prejudice-dimensions of ethnicity (pp. 1–30). Harvard University Press.

Pizam, A., Fleischer, A., & Mansfeld, Y. (2002). Tourism and social change: The Case of Israeli eco-tourists visiting Jordan. *Journal of Travel Research*, *41*(2), 177–184. doi:10.1177/004728702237423

Ruesch, M. (Ed.). (2011). *A peaceful Net? Intergroup contact and communicative conflict resolution of the Israel-Palestine conflict on Facebook*. Inter-Disciplinary Press.

Sirakaya-Turk, E., Nyaupane, G., & Uysal, M. (2014). Guests and hosts revisited: Prejudicial attitudes of guests toward the host population. *Journal of Travel Research*, *53*(3), 336–352. doi:10.1177/0047287513500580

Straus, A., & Corbin, J. (1990). *Basics of qualitative research: Grounded theory procedures and techniques*. Sage.

Tomljenovic, R. (2010). Tourism and intercultural understanding or contact hypothesis revisited. In O. Moufakkir & I. Kelly (Eds.), *Tourism, progress and peace*. CABI Pub. doi:10.1079/9781845936778.0017

Yılmaz, O., Cesur, S., & Bayad, A. (2018). Psychological correlates of negative attitudes of Turks and Kurds toward one another. *Turkish Psychological Articles*, *21*, 82–99.

Chapter 15
Tourism, Citizenship, and Education:
A Virtuous Triangle for Peacebuilding

Adalberto Dias de Carvalho
https://orcid.org/0000-0001-5826-7582
Higher Institute of Business and Tourism (ISCET), Portugal

ABSTRACT

Tourism is not necessarily an activity that contributes to the promotion of peace. Well-known testimonies and criticisms point out their adverse effects and consequent rejection, namely in terms of intensive occupation of territories and tampering with local cultures and identities. These harmful impacts undermine the sustainability of tourism, in its complexity (which imply a harmonious but fragile system of interdependent variables, considering the existence of open and universal hospitality) and multidimensionality (because its variables are of a different, but coherent, nature). In this context, sustainability, as necessary and vulnerable, can be easily threatened by immediate economic interests and by significant gaps in civic awareness and the exercise of citizenship by all or part of the tourism protagonists and responsible. Thus, the importance of an education that takes into account the perspective of universal solidarity, which privileges, among others, the contributions of John Dewey's "cultural criticism" and of interactive constructivism.

INTRODUCTION

Many studies focus on the importance of tourism for peace or, conversely, how peace can contribute to the development of tourism. These studies, supporting several and conflicting perspectives, in addition to their own density, confirm the controversial, therefore problematic, character of this subject. Not only because they end up referring to a debate that goes beyond the frontiers of scientific research, first of all, because the subjective representations of people and groups intervene in it, but also because they depend enormously on the more or less unstable realities that, directly or indirectly, serve as a reference. At the same time, basic, as well as polysemic, concepts, such as tourism and peace, are involved.

DOI: 10.4018/978-1-7998-5053-3.ch015

This chapter intends to present some of the main positions that identifying, in their confrontation or complementarity, non-coincident aspects and valuations about the relations between tourism and peace, justify their problematisation. First of all, it is impossible to ignore the meanings of representations that refuse in the name of certainties that become truths and values that, in the final analysis, have nothing more than to impose the monopoly and universality of a given conception. On the other hand, we will have to accept the inconclusive nature of a problem like this, given the ambivalence it contains.

Considering this hermeneutic assumption as unavoidable, it is argued that the formation of civic and ethical consciences of the tourist activity agents, through an interactive and constructivist education open to plurality provides the essential anthropological condition for a potential positive relationship between tourism and peace.

STATUS AND IMPACTS OF TOURISM ON CONTEMPORARY SOCIETIES

The cultural diversity of people who cross paths through tourism is significant. It indicates, for example, that, even with apparently similar tourist practices and as widespread as those linked to the enjoyment of the beaches, they are lived differently in the West and in the east. In reality, tourists seek different experiences according to their cultures and, in this case, according to the way they value the aesthetics of their body. There is also diversity in what is sought: for example, the nature in Bhutan, the historical heritage in Florence, the boldness of modernity in Dubai. In several places, such as Jerusalem or Auschwitz, it is the memories that have been stored there: in the first, with the Holocaust recorded in the Auschwitz museum and in the Birkenau camp; in the second, with the several sacred places that eternalise moments and practices of the Christian, Jewish and Muslim religions. With the growing urbanisation of our societies, there tends to be an aesthetic and cultural appreciation of rurality, of bucolicism and of the natural landscapes associated with it, while inhabitants of less economically developed areas preferentially seek the fascination of the urban paradigm. From complementarity to conflict between these perspectives often goes a small step whenever the meetings translate into misunderstandings due to reciprocal ignorance.

In any case, in a globalised society such as ours, there are new ways of contact between peoples, which may allow the development of relations of harmony when, in particular, the current means of communication, namely the internet, expand the possibilities of mutual familiarity. Nevertheless, it all depends largely on the attitudes of the actors involved – visitors, host communities, tourism professionals – and, first and foremost, on their awareness of the nature and importance of otherness as an ethos of the relationship. All this to the extent that only the recognition of the other as another person with its own identity and dignity allows a true interpersonal encounter and not a simple subject-object relationship that tends to assume itself as a power relationship. As Gérard Leclerc (2015, p. 270) states, "contact is the fortuitous result of physical proximity" while "the encounter is the product of an intentional approach".

In recent history, tourism has developed greatly, at least in the West, because of the peace achieved with the end of World War II, though with some local disruption when crises arising from social and political conflicts, as in the case of the Balkan War, or when terrorist acts occur, as in the case of Egypt. Once overcome, there is usually a re-establishment of tourist flows. Regions with more instability are also those with a smaller tourist movement, such as central Africa. That is, there is a direct relationship between the founding or restoration of peace and tourism flows, and it is important to bear in mind that their emergence is at the outset generating wealth and jobs, which may, in turn, favour the calm of tensions and increased wellbeing that support the existence of situations of social harmony. In addition to activities directly linked to tourism, such as hotels and restaurants, others are a result of tourism, such as agriculture, fishing and handicrafts. All of these can serve the wellbeing of the population and the tourists or, if their development is not harmonious and concerted, entails, for example, the disappearance of those who are most dependent on them and, in the end, unemployment, distortion of social organisation and the emergence of new pockets of poverty. These remain significant to the wellbeing of the local community as well as the tourists. Therefore, endeavouring to sustain them is critical for they support employment, prevent the distortion of social organisation and the emergence of new pockets of poverty.

Algarve, a tourist region par excellence of mainland Portugal, stands out as an example. This region, as can be read in the RIS3 - Algarve 2014-2020 of the Algarve Regional Coordination and Development Commission (2020, pp. 31, 62), experienced, during this period, difficulties in reconciling the explosion of tourist activity and other traditional sectors of the economy, "as was the case of agriculture and the sea, since tourism absorbed most of the attention of policymakers and 'captured' most of the investments and the mobilisation and entrepreneurship capacity of private actors". To overcome the social dysfunction resulting from this imbalance but taking advantage of the fact that tourism remains the anchor activity of the others, it is recognised that "the sea sector and the agro-food/agro-industrial sector seem to be particularly useful for job creation, especially for intermediate age groups and less qualified workers, while other sectors may be more important for job creation among younger age groups and/or those with higher levels of education".

Another example of the difficulties that sometimes arise in the relationship between tourism and the traditional economic activities that depend on it, we emphasise the considerations that Rogerson (2015) makes regarding the connections between agriculture and tourism in South Africa. The author highlights the tendency of large hotel units to use imported foods instead of those produced locally due to various constraints, among others those of availability, quality, transport and prices, alongside the tourists' own preferences. In addition, when local production units had attempted to meet demand requirements, they experienced major difficulties due to their size and weak management mechanisms, that is, an insufficient economy of scale.

With mass tourism, the possibility of its impacts being intensely and extensively negative is revealed at the outset, with changes in the balance of natural and human resources serving them as destinations. The risks of rejection by the host communities increase, and the tendency for visitors to look at the visitors as mere objects for the enjoyment and realisation of common places such as those of exoticism and authenticity grows. That is, the differences tend to be overstated for this purpose, without prejudice to the search for convenience imposing the organisation of support structures (hotel structures, assistance structures and others), similar to those of the countries of origin, usually economically or socially more developed.

Gentrification, identified in relation with overtourism, is also one of the consequences of tourism that is not the result of a balanced relationship between tourists and hosts, which thus favours the growth of tensions and even of tourismphobia. Thus, as stated by Milano (2019), "The renewed interest in the adverse impacts of tourism, as implied in the term overtourism, is related to a variety of well-established causes, such as the loss of sense of belonging, diminishment of sense of place, increased congestion and privatisation of public spaces, explosive growth of cruise tourism, the rapid growth in numbers of seasonal and day visitors".

Meantime, other types of tourism seek to be alternatives, namely the ecotourism and the voluntary tourism, which cultivate as their main motivations not leisure by itself, but the respect for the natural and human contexts of the destinations, as well as the solidarity with those who face greater social and economic difficulties.

At least in 20 of the world's 48 poorest countries, tourism becomes the first or second source of foreign exchange (cf. Sharma & Thapar, 2016, p. 164). The World Tourism Organization (2011) even has a program called "Tourism and Poverty Reduction". However, we also know that in several of these countries, still according to Sharma and Thapar (2016), the largest returns on tourism end up in the hands of multinationals and nations that have the economic and political power to set up and exploit their structures. In these geographical environments, tourism tends to look for a natural and human exoticism that is usually no more than the product of colonial processes in which the idea of an ahistorical primitivism is transformed into an ideal of humanity untouched by civilisation. In its essence, it is a representation of the "good savage" in a context that would approach a native world preserved by the backwardness of its inhabitants.

Although these are findings that emerged in the context of a reflection on the political and social problems of the arrival of thousands of refugees in Europe seeking decent living conditions, this seems to be the main theme for a reflection on the connections between tourism and peace. In fact, we can wonder if tourism, through the visits it provides favoured a change of mentality in the inhabitants of the tourist source countries or, on the contrary, motivated a repetition or even a reinforcement of some of the obstacles to the opening of the relationship between visitors and hosts, regardless of who the visitors and visitors are. In fact, the arrival of migrants, bringing out xenophobic feelings and acts of rejection, has often been associated by common sense with a threat of invasion. Thus, we can wonder if – and in this case to what extent – tourism opens doors for overcoming the anthropological impasses of peaceful coexistence, allowing, among other things, that thousands of people can escape war, misery and stigmatisation. In Germany, the situation is ambivalent because, as Moser et al. (2019, p. 6) states, there is "a widespread concern about the impact of large-scale immigration on Germany's society and economy", with "most Germans support only subsidiary protection, and want people to be sent back once conditions in their home country improve". However, it is also true that "in contrast there is strong support for refugee protection".

As can be seen in the Atlas of Migration (Dario et al., 2019) published by the European Union, tourist mobility often runs in the opposite direction to migration because of the escape from poverty, war and authoritarian regimes, which means that the geographical and social contexts of tourist destinations are thus opposed to those of territories sought by immigrants. The former tend to be idealised, ignoring the disinherited of the potential emigrants who make up a large part of the host communities. Apparently, there is no arrogant xenophobia here, no walls. However, we can claim its existence when one finds that, behind the admiration of the local populations, there are overpowering attitudes that want to impose cultural and social stagnation in the name of defending an alleged authenticity. In this context, the walls become more subtle and, instead of physical expression, they become psychological, sociological and ideological, thus ending up imposing strong regional asymmetries and, thus, a discriminatory territorialisation of poverty and war. Under these conditions, the transcultural potential of tourism does not exist, being replaced by ethnic segregation or acculturation.

That is, even if there is some economic growth and increased employment, the asymmetry between tourists and hosts does not allow, for the absence of social harmony, the effective construction of global peace scenarios.

REVIEW OF LITERATURE AND DEFINITION OF KEY TERMS

In order to understand the relationship between tourism and the problem of peace, we must introduce the distinction between positive peace and negative peace, that is, between the promotion of equity, social justice, harmony, cooperation among humans, and the mere absence of overt violence. Since the latter may be more apparent than real, only the former can guarantee a deep and lasting peace as it seeks authentic and humanly demanding coexistence and not just the perception and practice of acts of physical violence such as they are caused by war and terrorism.

In our opinion, it is particularly important to distinguish between negative and positive peace because only the second fits within the solidary horizons of tourist activity, while the first has an essentially political and even military tendency. In some way, we can say that positive peace is built and has a greater chance of being durable while negative peace is intended to be imposed by force, especially in situations where violence is already installed, and therefore has less chances of lasting.

Significantly, the World Travel & Tourism Council (WTTC), which represents the Travel & Tourism private sector globally, collaborated with the Institute of Economics and Peace (IEP) promoted a research about the links between tourism and peace. In its report summary "Tourism as a driver of peace" (WTTC, 2016, p. 1), we can read:

- An open and sustainable tourism sector means a higher level of positive peace: namely the attitudes, institutions, and structures that create and sustain peaceful societies.
- Countries with an open and sustainable tourism sector are more likely to enjoy higher levels of positive peace in the future.
- The more sustainable a country's tourism sector the lower the country's level of violence and conflict is likely to be.
- In non-conflict-affected countries, tourism is resilient to increases in violence and conflict.
- Tourism is resilient to increases in terrorism, even when terrorism is targeted at tourists.

To reach these conclusions, the study carried out measures based on three domains of peacefulness in what concerns a country's level of negative peace ("ongoing domestic and international conflict; societal safety and security; and militarisation") and through indicators such as "military expenditure as a percentage of GDP and nuclear weapons capabilities" and the "level of perceived criminality in society and homicide rate". In turn, the Positive Peace Index "is based on eight pillars that have been determined as essential to a peaceful society: a sound business environment; good relations with neighbours; high levels of human capital; acceptance of the rights of others; low levels of corruption; a well-functioning government; free flow of information; and the equitable distribution of resources".

Stephen Pratt & Anyu Liu (2015, p. 1) examining the relationship between peace and tourism across 111 different countries concluded that "tourism is the beneficiary of peace rather than grounds for peace". According to the authors, "the answer to whether tourism contributes to world peace, in general, is still yet to be confirmed", especially because "what holds for individual tourists may not hold for nations in the aggregate", as happens, for example, in Israel in relation to Jordan, where the most favourable opinions of tourists who visited this country did not have a significant impact on political relations between these. The study also showed "the effect of peace on tourism development to policymakers and government officials in different destinations", especially in those where the economic relevance of tourism is higher.

For S. Litvin (1998, p. 1), tourism research should abandon the discourses that glorify the virtualities of tourism as a promoter of peace in a way that is not compatible with the rigour that scientific research should demand. In this context the author asks "tourism create peace or is tourism, along with other many other industries, a fortunate beneficiary of peace?", to add that if "tourism is never successful in the absence of peace, it cannot, therefore, be a generator of peace". This author also notes the absence of empirical analyses that objectively show the causality between tourism and peace. Rather, it was found that the massive presence of tourists often motivates attacks that thus have great media impact. Litvin concludes that there is undoubtedly a "co-relation between successful tourism and absence of war", although it cannot be deduced from this that there is a causal relationship.

In contrast to this doubtful tone, it is important to take into account the positions that, particularly following the First Global Conference "Tourism: A Vital Force for Peace", organised by Louis D'Amore in 1988, proclaim the defence of the crucial role of tourism in the implantation and maintenance of peace, considered in terms of a universal harmony to be achieved by sharing and respecting different cultures and overcoming the voids generated by isolation. In fact, Litvin (1998, p. 2) makes the inventory of these papers that he considers as laudatory speeches, because finally they have little objective, until the date of publication of the article referenced above. He then quotes Knopf (1991), Holland (1991), Canneday (1991), Khamouna and Zeiger (1995) and McIntosh, Goeldner and Richtie (1995), who defend, respectively, ideas such as tourism is the:

"Fundamental key to world peace"

"The equivalent of atomic energy by being a positive force for wold peace", which promotes "the mutual understanding and respect"

"One of the great forces for peace and understanding in our time"

"If not a passport to peace, is at least a worthy effort toward building peace"

According to Wintersteiner and Wuohlmuther (2014, p. 43), "Moufakkir and Kelly state that the interest in the concept of tourism and peace can be presented from 'euphoric recognition in the 1980s to scepticism in the 1990s followed by an almost total rejection', and by a renewed interest in the 21st century through the investigation of critical situations in different concrete scenarios, such as Bosnia and Herzegovina, Israel-Palestine, Rwanda and other sensitive regions in terms of their stability. On the other hand, Kelly (2012, pp. 32-49) came to propose other areas of action in which the problem of the connections between tourism and peace should be considered. They are as follows:

•ntergroup contact

- Ethical concerns (respect for the environment and human rights)
- Positive impact of tourism on negative elements of globalisation (poverty eradication)
- Awareness raising among suppliers and consumers by codes of conduct, peace parks and tourism education.

In this context, Wintersteiner and Wuohlmuther (2014, pp. 43-44) highlight the cultural and social dimensions of tourist activity and the decrease of violence by peaceful means. The authors state, "the task of tourism is to provide travel experiences which encourage appropriate attitude change, for which a certain predisposition to learning about the other is essential", demanding the adoption of the concept of "peace-sensitive tourism" instead of "peace-through-tourism idea", being that one "a less ambitious expression", place "tourism in the context of other peace-promoting activities and designs" (Wintersteiner & Wuohlmuther, 2014, p. 31). This new perspective presupposes a broader conception of sustainability and, with it, of hospitality because, in addition to valuing the economic impacts of tourism, now also include the social, cultural and environmental impacts, alongside human rights and justice, in a broad conception of citizenship.

Van Den Dungen (2018, p. 1) talks about "peace tourism" clarifying that this expression "involves visits to places, at home and abroad, which are significant because of their association with such notions as peace-making, peaceful conflict resolution, prevention of war, resistance to war, protesting war, nonviolence and reconciliation". In our opinion, this notion will benefit if it is extended beyond cities, museums and other places that, in one way or another, witness to peace, as the author's scope seems to be. Without prejudice to considering this dimension, it should also serve to express tourist practices that carry a broad sense of hospitality, recognise and implement through them the solidarity and positive tolerance that can now build transcultural cores that aggregate diversity and detonate not conflicts – evident or hidden – but rather of social spaces of conviviality. This is what Ravi Kumar (2018, p. 1) means when he states: "Tourism provides an effective tool for promoting peace which in turn leads to cross-cultural interaction resulting in tolerance, national integration, compassion, goodwill, educational standards, justice and mutual respect for each other".

Several authors argue that instead of providing peace, tourism needs peace to exist. Hence, in a context of negative peace, tourist itineraries usually avoid spaces in which there are conflicts, abandoning places that experience violence despite having settled in them. The correlation between tourism and positive peace would arise, for example, when, for this purpose, we seek to highlight the meaning of cities or other places marked by peace-related events and not just places where war leaves its physical and social traces. These tourism paradigms impose demanding goals on education for human rights, tolerance and nonviolence and on politicians and tourism professionals as well, as they have to draw and provide itineraries for peace and references as monuments, names of streets and squares, etc. that enhance peace as the means and end of humanity.

IMPORTANCE OF SUSTAINABILITY IN THE COMPLEX RELATIONSHIP BETWEEN TOURISM AND PEACE

In our opinion, the connection between tourism and peace involves the operational intermediation of the concept of sustainability as long as we consider that, as a homeostatic reference, it can and should play a regulatory role in tourism practice while mobilising people and resources to achieve tourism purposes. Firstly, the question of defining what will be the purposes, resources and people who are present in a certain project or in a certain tourist reality arises.

To the UNWTO (2005) sustainable tourism should:

1. "Make optimal use of environmental resources that constitute a key element in tourism development, maintaining essential ecological processes and helping to conserve natural heritage and biodiversity.
2. Respect the socio-cultural authenticity of host communities, conserve their built and living cultural heritage and traditional values, and contribute to inter-cultural understanding and tolerance.
3. Ensure viable, long-term economic operations, providing socio-economic benefits to all stakeholders that are fairly distributed, including stable employment and income-earning opportunities and social services to host communities, and contributing to poverty alleviation.

Sustainable tourism development requires the informed participation of all relevant stakeholders, as well as strong political leadership to ensure wide participation and consensus building. Achieving sustainable tourism is a continuous process and requires constant monitoring of impacts, introducing the necessary preventive and/or corrective measures whenever necessary. Sustainable tourism should also maintain a high level of tourist satisfaction and ensure a meaningful experience to the tourists, raising their awareness about sustainability issues and promoting sustainable tourism practices amongst them."

To Becken and Carmignani (2016, p. 64), "peace is the process of reducing violence and achieving a balance of powers within overarching norms and institutions". The authors also point out the concept of "peacekeeping", which, in its complex scope, involves "activities that contribute to ending the incidence of conflict, as well as preventing the onset of new conflict". These activities are of a political, economic, social and legal nature, public or private. Now, it is precisely in this multidimensional context that the issue of the contribution of tourism to peace, whether in global or national terms, has to be considered since there are multiple interactions that tourism maintains with those activities. Only then can tourism contribute to the development of societies, an inalienable assumption for the construction and maintenance of peace. Within this framework, Becken and Carmignani (pp. 77-78) present, among others, the following conclusions: "stimulation of tourism in a post-conflict situation is important"; "rebuilding tourism in a post-conflict county presents opportunities, most prominently for poverty alleviation and social equity", namely when it privileges "long-lasting, sustainable, inclusive and resilient outcomes", for example, in job creation; the contribution to peacebuilding and development is best harnessed as part of a broader government-led strategy"; finally, the collaboration between public and private sectors is vital to any tourism development strategy, especially in post-war communities.

Therefore, it is important to assume the multidimensional character of sustainability and the necessary involvement of all the players, including stakeholders, policymakers and tourists themselves in all that has to do with the natural heritage and its biodiversity, cultural heritage and its values, as well as the justice in the distribution of socio-economic benefits. This is a requirement for active participation in a process that requires the monitoring and intervention of the actors involved, who need to be aware of the sustainability issues and their impact on the meaning it assumes in the search for the wellbeing of tourists.

If tourism can contribute to peace, it can also trigger disturbing situations of the harmony between host communities and visitors. In fact, physical proximity is not in itself a guarantee of personal proximity. It may even create conflict by incompatibilities of different nature. Firstly, there are the risks of deterioration of natural and social contexts, as well as the incompatibilities between representations, purposes and values and the clash between economic conditions.

Hence the growing importance of the notion of sustainability whenever the assumptions and consequences of tourism are problematised as a practice that mobilises people and resources.

According to Nuwan Herath (2010), who echoes what is a generalised understanding, to promote tourism as a sustainable industry, there are three sectors that must function in articulation:

First, 'host communities' should focus more on the quality of service that they deliver to incoming tourists and must be responsible for protecting their surrounding environment (eco systems). Secondly, governments should make available the necessary infrastructure expansion and must implement essential 'regulatory mechanisms' to protect the industry from exploitation and environmental degradation. Finally, foreign stakeholders should support local initiatives through investment or provide the necessary support that is required for the qualitative growth of the tourism industry."

In the name of promoting peace, tourism is required to be sustainable far beyond strict economic criteria, due to ethical demands on host communities, governments and foreign stakeholders, which should operate in a network. The sense of responsibility for the preservation, protection and quality of the respective ecosystems in their connection with the growth of tourist activity is essential. A harmonious relationship between visitors and hosts and between both and the surrounding environment is, in these circumstances, essential for the recognition of tourism as a socially credible practice.

Sustainability becomes even more sensitive as the natural, heritage and human resources that underpin the attractiveness of tourist destinations are vulnerable and therefore very sensitive as to how they are exploited. All of these resources need protection, and the people who form the host communities must be active agents in their structuring and operationalisation, including those that directly concern them. Firstly, because of their resilience in the face of the possible negative implications of the tourism products to which they end up being part. These implications undermine the sustainable development of the territories, now and for the future.

In this regard, it is worth remembering the definition of the strategy for sustainable tourism given by the European Charter for Sustainable Tourism in Protected Areas:

Any form of tourism development, management or activity which ensures the long-term protection and preservation of natural, cultural and social resources and contributes in a positive and equitable manner to the economic development and wellbeing of individuals living, working, or staying in protected areas.

Also according to this Charter, there are ten principles that should support sustainability in tourism:

1. "To involve all those implicated by tourism in and around the protected area
2. To prepare and implement a Sustainable Tourism Strategy and Action Plan for the protected area
3. To protect and enhance the area's natural and cultural heritage
4. To provide visitors with a high quality experience in all aspects of their visit
5. To communicate effectively to visitors, tourism businesses and residents about the special qualities of the area
6. To encourage specific tourism products which enable discovery and understanding of the area
7. To increase knowledge of the protected area and sustainability issues
8. To ensure that tourism supports and does not reduce the quality of life of local residents
9. To increase benefits from tourism to the local economy
10. To monitor and influence visitor flows to reduce negative impacts."

These principles underline operational concepts such as "protected areas", "natural and cultural heritage" and "quality of life", which, being more evident here, must be taken into account not only in the narrow territories that are formally considered to be protected areas. In addition to being intrinsically related to each other, their nature and development imply the assumption of their senses by all actors, as the objectivity of the situations targeted in terms of their emergence and natural, cultural, economic and social consequences always has as a corollary the representations of these same actors on each one of them. Being not just mechanical phenomena, presuppose the formation of the consciences of those who are their protagonists precisely in favour of sustainability.

In turn, to help the tourist destinations measure their performance in relation to sustainability, the European Commission has developed a 'European Tourism Indicators System' (ETIS). ETIS has been implemented in two pilot phases since 2013, by over 100 destinations.

Here are the results of some case studies that highlight the importance of the people involved and their training:

- Looking back at his experience, Vasil Marinov, head of the Geography of Tourism department of Sofia's St. Kliment Ohridski University and Project Leader during the ETIS testing phase in the Danube region, says:

We immediately saw that ETIS provided a good basis to assess and then support tourism development in Bulgaria. The benefits of such comprehensive testing were not obvious to some local stakeholders though, so we ended up providing guidance and training along the way. The awareness-raising aspect has become a priority because we want to show that monitoring of sustainable tourism is important over the long term.

- Marina Bresciani, Regional Project Coordinator at Italia Executive, says:

Territorial design and labelling are concepts that local authorities were largely unfamiliar with, but in order to determine the sustainability and accessibility levels of the provinces, we analysed the results of the indicator testing together with the participants, which got them much more engaged and improved the understanding of why it is important to have a strong network of destinations united under the Terrae Anio Iubensanae label."

- Jana Apih, Managing Director of the sustainable tourism factory GoodPlace, in cooperation with the Slovenian Tourism Board, wrote:

We wanted to use the ETIS indicators not just as an evaluation tool, so we broadened the formula using the Global Sustainable Tourism Review tool and introduced scores that tourism coordinators could apply at their destination to obtain the Slovenia Green certification. The scheme is quite complex, but in the first edition of our training in June 2015, we had 15 destination coordinators motivated and ready to enter the race for the greenest destination – with success! So far, 7 destinations and 6 hotels have been awarded Slovenia Green certificates.

In all these cases, the importance of those in charge of implementing policies and strategies for the promotion of sustainable tourism is evident, as well as the knowledge and critical capacity that they must possess in order to properly fulfil their roles and functions. Indeed, if there is an activity in which the notion of the ecosystem is important, it is certainly tourism. In addition to its potential for wealth creation and wellbeing, the sensitivity of the contexts in which it develops cannot be neglected, which are particularly vulnerable when disturbed by the construction of hotels, by space sharing with new groups of visitors and by a redefinition of objectives and practices of all kinds.

FROM ETHICAL ASSUMPTIONS TO HOSPITALITY AS A HUMANISATION OF THE ENCOUNTER IN THE TOURIST RELATIONSHIP

Fennell (2018, p. 330) considers that "tourism problems are the result of many complex issues, which are diverse and multifaceted given the social, economic and ecological basis of tourism". The interplay of interests that runs through this issue raises ethical questions due to the need to ensure that there is justice in the decisions and measures that are taken. The governance then seems to be the most appropriate strategy for this purpose today, as Fennell says (p. 209), quoting Rosenau: "In contemporary times, governance has been defined as a 'set of regulation mechanisms in a sphere of activity, which function effectively even though they are not endowed with formal authority'". This is because, having a broader and more informal scope than the government, it promotes consensus based on values, ethics, love and cooperation, which means that resources are shared mainly by ethical criteria and not so much through institutional structures of power. The latter, tending to produce winners and losers, create a social climate that favours misunderstandings and confrontations.

Presenting sex tourism as the most reprehensible type of tourism because it is done in favour of the circumstantial pleasure of some against the dignity of the victims, Fennell also points out that at least five ethical theories condemn it:

1. "Utilitarianism in showing that its social costs are greater than its benefits;
2. The ethics of rights questions whether those who commit such acts have the right to do so;
3. The distributive justice ethics warns that the morally correct action must produce 'a fair distribution of benefits and costs for everyone';
4. Virtue ethics considers that the morally correct action is the one which 'displays good moral virtues' like compassion, fairness, benevolence, etc.;
5. Finally, the ethic of care emphasises the importance of the quality of each other's relationships with others, including here the impacts of human action on nature".

These ethical approaches must always be considered because the issue of the relationship between tourism and peace necessarily involves relationships between people and here through the ability to accept the other as a resilient and dignified being.

In fact, it is important to bear in mind that the issue of the connections between tourism and peace, as it implies a critical and civic awareness of the importance of respect for personal and natural otherness, in terms of their dignity, resilience and eventual fragility, is, above all, a question of citizenship. Only then can it be consequently assumed in its complexity, demanding the education of all those directly involved in the tourist activity as its promoters or consumers, making it an authentically political issue as it concerns society as a whole. Therefore, it can no longer be seen as a responsibility for strictly technical training of the respective professionals.

Thus, the principle that there is no direct, causal, linear and univocal relationship between tourism and peace is accepted, although it is understood that such a restriction should not and cannot prevent the appreciation of the existence of important connections between tourism projects and the construction or maintenance of the peace. Thus having prior knowledge to acquire, ethical and civic requirements to be met and sensitivities to be formed, it is crucial to recognise, therefore, that education assumes, in tourism as an industry and in the societies from which tourists come as well as in the communities of reception, a decisive importance as a mediating practice. Practice that can help to take advantage of the meeting possibilities offered by tourism and, at the same time, avoid the serious conflicts that the stay of visitors in the territories can cause on both sides.

Reflecting on the relationship between tourism and peace implies, as we saw, taking into account the notions and practices of hospitality and sustainability. Sometimes this theme ends up resulting in almost militant speeches that are only concerned with praise the virtues of tourist activity with regard to peaceful coexistence through the cross-cultural relations it provides. In this context, the conditions under which this happens are not questioned. We know, however, that tourism can, on the contrary, favour conflict situations, namely due to cultural and social shocks between visitors and hosts or the reactions of host communities when the tourist contingents exceed certain limits or the economic benefits for local people are not real.

Hospitality, as a dimension of the tourist activity, requires, from the outset, a clarification of the meaning of the term. In a restricted sense, hospitality can be seen as confined to the services provided by the hotel industry, with regard to the organisation of accommodation and restaurants, or cover aspects of the personal relationship, even if professional and commercial, between hosts and guests. This seems to us to encompass and surpass the other and to be anthropologically more relevant, mainly in a study on the assumptions of the connections between tourism and peace that are neither evident nor mechanical, although, in a positive sense, potential and desirable.

In close connection with hospitality, sustainability refers to a framework of requirements regarding the viability and effective interest of tourism for human wellbeing in general and its stakeholders in particular. Once again, we favour a broad concept of sustainability that covers the conditions and consequences of tourism in all its complexity. It is, therefore, a matter of considering not only the economic contours of this activity but its multidimensionality, where social, cultural and identity dynamics are present, along with those of a personal nature, that will necessarily have to be present in the hospitality relations, as above we define them.

Thus, we come to the issue of peace, to demand, complementarily, the contemplation and appreciation of the importance of education of all the actors involved, since peace is not only the absence of war, but it implies, above all, a harmonious coexistence between people and societies. It is essentially the result of social systems and organisations in which hospitality and sustainability are understood in their axiological importance and practised as realities, namely for what interests us directly here, in the dynamics of tourism.

In the book "La fin de l'hospitalité", by G. Le Blanc and F. Brugère (2017, p. 13), a work critically reviewing the dramas experienced in Europe by migrants from Africa, comes a disturbing conclusion at some point:

When we stop welcoming, we push away worlds that are still there, we make invisibility and we believe we can erase human lives that knock on our door. 'We get rid of these humans. Does anyone have an objection? No. Walls are raised and everything is settled. When you stop seeing someone, there is no one else'. Doing as if there is no one where there is someone, is the opposite of hospitality, is in hospitality. Now inhospitality becomes hostility or hatred.

When we consider the practices in which hospitality proves to be an important dimension that requires, therefore, the clarification of its epistemological and anthropological conceptualisation, we must value its status not only in a rational aspect but also in what it has to do with feelings and emotions. As a rational practice, strategic and pragmatic aspects prevail, seeking a balance between the personal investment made with others and the achievement of results. Results that, in the specific case of tourism, have to translate into their profitability obtained through the wellbeing and satisfaction of the customer.

The relational scopes of tourism as an activity that can provide the encounter with the other, namely as a "cross-cultural understanding", have to go through hospitality conceived as a practice of empathy while reaching an "intersubjective understanding level through which we can recognise and respect the subjectivity and agency of others" (Tucker, 2016, p. 41). Therefore, we have to realise that there are contingencies resulting from the differences between "'unsetled' and 'confrontational' empathy on the one hand, and 'lazy empathy' on the other, which attempts to feel hope or rage on behalf of others whilst remaining unquestioned and unreflective". It is interesting to welcome here the ethnographic perspective of the interpersonal relationship in tourism, even when it comes to voluntary tourism, with the purpose of promoting solidarity.

In this regard, Frazer and Waitt (2016, pp. 177, 182), about a little-studied aspect as is the case of embodied pain, highlight the contingencies of "empathic pain arising from the living conditions of those they aimed to help were felt through their bodies", as "an economy of empathic pain circulating between bodies can operate to both conceal and repeat asymmetric colonial power relations" or, on the contrary, "encounters of empathic pain may fuel the passion of some volunteer tourists and evoque an ethics of a responsible hope". This highlights the importance of the emotional dimension as "embodied and affective sense of being and engagement". Still in the context of interpersonal relationships in tourism, the importance of the concepts of gratitude and kindness should be considered, as highlighted by Filep, Macnaughton and Glover (2017, p. 35). Through a study with Canadian tourists in New Zealand, the authors conclude that, for the most part, "a resounding issue participants highlight in receiving kindness was the growth of their own personal confidence, and the sense that, while the encounter was brief, the personal impact was profound and eudemonic in character". These feelings naturally contribute to a peaceful relationship, contrary to Derrida's convictions, cited by these authors (p. 28), for whom "tourism and hospitality present the risky possibility that guests could become adversaries or freeloaders of their host's generosity".

In order not to remain indifferent, we have to value justice as it implies the openness to accept different and new points of view without closing ourselves in our convictions, thus not being willing to listen and critically question everything that can transform us or coexist dialogically, even with the otherness that inhabits us. In fact, if hospitality institutes humanity because plurality is specifically human, we cannot consider it spontaneous. Hence the importance of an ethics of hospitality by the assumption of responsibility, by the exercise of solicitude that, according to the values of a humanisation of humanity, highlights its inherent principles, namely the respect for human vulnerability through the openness and solidary acceptance, instead of the blind and imperative exercise of power. We have to encourage, with others, the emergence of humanity from the people we are and counteract the inhumanity – and inherent propensity for violence – that also lives in us. This inhumanity prevents coexistence and fuels violence, precisely by trying to suppress otherness, as it leads us to reject others as strangers – and, thus, the risks of surprises caused by unforeseen events.

As Innerarity (2001, p. 20) stated, "an ethics of hospitality, insofar as it implies generosity, openness and a favourable disposition towards what is complex, is necessarily an ethics of instability, as the host's situation is vulnerable whose forecasts are always threatened by the inopportune of a visit". It is in the face of these challenges that, in the absence of reflection and civic training, personal affirmation may result in a manifestation of self-sufficiency that, quite simply, ignores the fragility of each one and the personal and social wealth that may arise from a relationship open to others, subject to surprises and the questioning of projects of sovereign self-determination.

Van Den Dungen (2018, p. 1) talks about "peace tourism" clarifying that this expression "involves visits to places, at home and abroad, which are significant because of their association with such notions as peace-making, peaceful conflict resolution, prevention of war, resistance to war, protesting war, nonviolence and reconciliation". In our opinion, this notion will benefit if it is extended beyond cities, museums and other places that, in one way or another, witness to peace, as the author's scope seems to be. Without prejudice to considering this dimension, it should also serve to express tourist practices that carry a broad sense of hospitality, recognise and implement through them the solidarity and positive tolerance that can now build transcultural cores that aggregate diversity and detonate not conflicts – evident or hidden – but rather of social spaces of conviviality. This is what Ravi Kumar (2018, p. 1) means when he states: "Tourism provides an effective tool for promoting peace which in turn leads to cross-cultural interaction resulting in tolerance, national integration, compassion, goodwill, educational standards, justice and mutual respect for each other".

H. Arendt considers that "plurality" constitutes human specificity. This idea is also addressed by G. Vincent (2015, pp. 163-164) that argues that it is hospitality that establishes this plurality encouraging meetings between people, a symbolic field of "our common co-humanisation" rather than "unpredictable accidents" or "collisions due to the evil they inflict on each other". Thus, "to practice hospitality is to show that we know not to be indifferent to the suffering of others who lack roof and food."

THE IMPORTANCE OF EDUCATION FOR THE PROMOTION OF TOURISM FOR PEACE

Several studies, considering possible positive and/or negative impacts of tourism on societies, draw attention to such a purpose for the relevance that training and education can have to ensure the prevalence of the former. Nayomi and Gnanapala (2015), for example, based on several authors, find that more than 80% of travelers' fees benefit airlines, hotels and other international companies and not so much local businesses and workers. In order to mitigate this negative impact for the populations of the host communities, they argue in particular that "sustainable tourism development requires the informed participation of all relevant stakeholders, as well as strong political leadership to ensure wide participation and consensus building" (p. 60). As part of this concern, the authors report the results of an investigation into the process of setting up a hotel in Sri Lanka. Among the several conclusions, it seems to us especially important those that they propose as solutions for job level friction and for changing living styles and behaviour of the young: to provide skill development and professional training for newcomers of the industry, to motivate the local students for high education and to advice young generation by showing the impact of changing the behaviour, respectively.

Dewey's (1983) perspectives on the formative role of democratic values and the need to deconstruct ideologemes provide us with the most appropriate entry point for education intervention in this field. In fact, Dewey considers that philosophy should be seen as cultural criticism, that is, as an activity committed to overcoming the obstacles to the wellbeing of people and communities. To pursue with this purpose, a critique of beliefs and representations, which appear to be dominant in culture. Thus, philosophy should provide, as a critical exercise shared by all, an analysis of the present in search of opportunities for collective action and personal development for each one, being conceived essentially as a method of inquiry and political intervention – or intelligent experience –, philosophy coincide with democracy itself, understood mainly as a continuous process of democratisation.

For Dewey, being contemporary implies being able to not only have access to the goods of a given time, but also to have a critical conscience. Consciousness that presupposes to have the ability to question the situation to better understand it, as well as looking at it with the perplexity that its contradictions, antinomies and enigmas raise.

In this context, questions arise and necessarily lead to problematisation that is at the heart of the "reflective experience", with the thought being experience and experience being educational. The perplexity ensues, especially when the connections that, in search of solutions, we try to establish between what is known and the unknown surprise us.

Thus, the experience implicit in the inquiry is of interest not so much for the apprehension of a given object of knowledge, but rather for the opportunities for reorientation it provides. Observation is, for this educator and philosopher, a diagnosis that should bring opportunities for anticipation and preparation for a better future. This is the greatest sense of philosophy as a method of problematisation in which ideas emerge as leading ideas, as instruments to be used as ideals or projections that equate and integrate ends and means in the form of successively qualitatively more and more successful experiences.

Dewey's conviction that human problems can be solved, first of all, by education and here namely by the philosophy conceived with the general theory of education, was assumed and developed by the interactive constructivism of the Cologne program founded by Kersten Reich. As Stefan Neubert summarises (2003, p. 1):

The interactive constructivist understanding of education involves, among other things, theoretical perspectives on (1) observers-participants-agents in cultural practices, routines, and institutions, (2) processes of communication with particular focus on the dimension of lived relationships, (3) the interplay between the symbolic resources of a life-world, the imaginative desire of subjects, and the occurrence of real events, (4) the connections between processes of construction, reconstruction, and deconstruction in the cultural production of realities, (5) involvements of discourse and power, (6) cultural diversity, otherness, and incommensurability in multicultural contexts.

Comprehensive pedagogy can and must then foster and explore the relationship between learning and the experiences of the real, always valuing pragmatic and contextualist approaches, given their importance for the education and training of people. Thus, taking into account their cultural and social constraints and interests, which translate into a diversity of interpretations that, being irreconcilable and ambivalent, have to be understood in a vision that perceives them in the plurality of discourses that express them.

These speeches result from the constructions of the actors involved that produce them as observers based on their cultural assumptions and in interaction with others.

In this way, education is always a constructive activity carried out by those in training, based on their own life experiences, under the guidance of a trainer. Also, according to Neubert (2003, p. 6), "learning begins when learners use and expand their constructive agencies to solve problems and create meanings in the concrete situations they find themselves in". It is precisely here that the main proposals of interactive constructivism emerge, becoming unavoidable challenges for pedagogy. We refer to the notions of "incommensurability" and "otherness". That brings with it the realisation of the impossibility, in our global societies, to deny the heterogeneity of cultures, values, norms and traditions that, however, live a proximity induced by the current means of communication. If the otherness is enshrined in this way, there are also difficulties to demand that, through education, without dissolving the diversity of identities of groups and people, sharing and understanding platforms are also built, albeit fragile.

As Hein summarised (1991), "we do not learn isolated facts and theories in some abstract ethereal land of the mind separate from the rest of our lives: we learn in relationship to what else we know, what we believe, our prejudices and our fears". The author noted that "unless we know 'the reasons why', we may not be very involved in using the knowledge that may be instilled in us, even by the most severe and direct teaching".

Without these anthropological and epistemological perspectives that inspire and encourage training projects with a democratic profile, it will be impossible, in our understanding, to approach and understand, with the necessary depth, the complex issue of the relationship between tourism and peace. Once the nature and meaning of this relationship have been identified, we will then have the conditions to plan how education, understood as the process of training all involved actors – namely professionals, tourists and host communities – can and should exercise its training function of the human in the human being, the only way that can favour the construction and maintenance of peace. With the awareness that this is necessarily a very demanding purpose, Dewey and interactive constructivism emerge, as we have seen, as being able to sustain the most appropriate pedagogical proposals. These proposals are essential to implement in order to form visions of the world and others, visions that are prior to any acquisition of specialised technical skills or the exploitation of any travel itinerary.

CONCLUSION

In abstract terms, peace will be a purpose that can be easily accepted by a vast majority of people and societies, being the violent confrontation between them its antithesis. On the other hand, these people and societies can see tourism positively due to the ability it has to generate a rapprochement with regard to the exchange of knowledge and cultural representations. However, with this approach, the eventual confrontation of threatened identities and interests can also emerge as a perverse effect. In other words, an important connection can be found between the wishes of peaceful coexistence and the wellbeing of the populations. In one way or another, including because of objective constraints of a political, social and economic nature, outside the tourism-peace dyad, tourist activity can face situations of war or post-war and terrorism that hinder its development. The challenges to a critical positioning of the researcher then arise through the basic question: does tourism favour peace, or does tourism need peace?

On the other hand, we realise that the issue cannot be confined to the logic of linear causalism because there are many factors at play, which will stop us from maintaining a euphoric attitude regarding the existence of a directly positive relationship through which tourism would necessarily favor peace or a radical scepticism of denying such causality. The proposal of authors such as Wintersteiner and Wuohlmuther, mentioned before, when defending the adoption of the concept of "peace-sensitive tourism" instead of "peace-through-tourism" is particularly important, as they consider essential to acquire and cultivate a predisposition to accept that we can learn from others.

In fact, by valuing the idea of a "positive peace" for the importance given to this opening aimed at promoting a more harmonious coexistence, we will be expanding the notion of local, regional or universal sustainability. As a result, along with the aspects related to the natural and socio-economic balance, taken in the strict sense, hospitality and empathy also intervene.

In fact, the quality of human relationships, an aspect closely linked to the sphere of emotions and feelings in which tourism plays an important role in relation to people's expectations and needs, fills an essential dimension of the human condition. Providing the true encounter and meaningful life with it, that is, a life with renewed meaning, in this way favours wellbeing and with it the drive for a good social relationship, respect for others and individual and family happiness. collective. The positive psychology (Peterson, 2006) may have a relevant importance here in the training of tourism professionals, precisely because it combines happiness with the requirement of a meaningful life.

In a way or another, the issue of peace versus tourism gains a renewed interest and thus enters the sphere of the full exercise of citizenship as an anthropological practice. It is in this context that human rights, in particular, assume special relevance in all that they imply in the coherent exercise of respect for the identity of others, for tolerance and for the perception of the potential advantages of tourism as a promoter of the encounter between people. It is about promoting the humanisation of the relations between guests and host communities in everything that these relations assume a dimension of cross-cultural knowledge, an awareness of otherness and an ethical requirement intrinsically contingent by the human vulnerability that underlies them.

In this chapter, we intend to demonstrate that, finally, the issue of relations between tourism and peace leads to a problematisation through the recognition of the importance of the central role of education in the training of all actors in the tourist activity, not only professionals but also tourists and hosts. Some studies focus on the tourism practice as a learning place (cf. Brougère, 2012), that is, on the educational role that tourism can assume in the broader scope of social tourism and not so much on education – civic and cultural – to be a tourist. Other works identify this need (Ceriani et al., 2005) but, as Brougère concludes, without ever entering the "black box" about the learning necessary for someone to "become a tourist". It is in fact, above all, the formation of civic consciences not by any moral indoctrination but by the acquisition of rational, emotional and ethical skills that allow each one to understand his idiosyncrasy while accepting the plurality of views of the world in their own identities and resilience. Interactive constructivism, when considering as fundamental aims of education the recognition of incommensurability and otherness as key conditions for understanding and taking advantage of the heterogeneity of our societies, emerges as the educational orientation that can best support, either informally or in schools, the training of tourists and host communities and, in training centres, support the preparation of professionals in the sector.

REFERENCES

Algarve Regional Coordination and Development Commission. (2020). *RIS3 - Algarve 2014-2020, Estratégia Regional de Investigação e Inovação para a Especialização Inteligente, Cresc Algarve.* http://s3platform.jrc.ec.europa.eu/s3-platform-registered-regions

Becken, S., & Carmignani, F. (2016). Does tourism lead to peace. *Annals of Tourism Research, 61*, 63–79. doi:10.1016/j.annals.2016.09.002

Brougère, G. (2012). Pratiques touristiques et apprentissages. *Mondes du tourisme, 5*, 62-75. doi:10.4000/tourisme.415

Caneday, L. (1991). Chapter. In J. Zeigler & L. Caneday (Eds.), Tourism and Leisure: Dynamics and Diversity (pp. 83–91). National Recreation and Parks Association.

Ceriani, G., Duahamel, Ph., Knafou R. & Stock, M. (2005). Le tourisme et la rencontre de l'autre. Voyage au pays des idées reçues. *L'autre, cliniques, cultures et sociétés, 6*(1), 71-82.

Dario, T. (2019). *Atlas of Migration – 2019.* Publications Office of the European Union. https://publications.jrc.ec.europa.eu/repository/handle/JRC118415

Dewey, J. (1916). *Democracy and Education: An Introduction to the Philosophy of Education*. Macmillan.

Dewey, J. (1983). The need for for a recovery of philosophy. In J. A. Boydston (Ed.), *Middle Works.* Southern Illinois University Press.

Dewey, J. (1991). Creative Democracy - The Task Before Us. In The Later Works, (vol. 14). Carbondale: Southern Illinois University Press.

European Union. (2019). *Atlas of Migration 2019.* Luxembourg: Publications Office of the European Union. https://ec.europa.eu›...›EU Science Hub

Fennell, D. A. (2018). *Tourism Ethic.* Channel View Publications.

Filep, S., Macnaughton, J., & Glover, T. (2017). Tourism and gratitude: Valuing acts of kindness. *Annals of Tourism Research, 66*, 26–36. doi:10.1016/j.annals.2017.05.015

Frazer, R., & Waitt, G. (2016). Pain, politics and volunteering in tourism studies. *Annals of Tourism Research, 57*, 176–189. doi:10.1016/j.annals.2016.01.001

Geethika, N., & Gnanapala, W.K. (2015). Socio-economic impacts on local community through tourism development with special reference to heritance Kandalama. *Tourism, Leisure and Global Change, 2.* creativecommons.org/licenses

Hein, G. (1991). *Constructivist Learning Theory.* https://www.exploratorium.edu/education/ifi

Herath, N. (2010). Peace Through Tourism. In *Beyond Intractability Conflict Information Consortium.* University of Colorado Boulder. http://www.beyondintractability.org/essay/peace-through-tourism

Holland, S. (1991). Recreation and Tourism: Evolution of the Social Mission. In J. Zeigler & L. Caneday (Eds.), *Tourism and Leisure: Dynamics and Diversity* (pp. 66–81). National Recreation and Parks Association.

Innerarity, D. (2001). *Ética de la hospitalidad*. Ed. Península.

Kelly, I. (2012). Peace Through Tourism. An Implementation Guide. *The Journal of Tourism and Peace Research*, *2*(2), 32–49.

Khamouna, M., & Zeiger, J. (1995). Peace Through Tourism. *Parks & Recreation*, *30*(9), 80–86.

Knopf, R. (1991). Harmony and Convergence Between Recreation and Tourism. In J. Zeigler & L. Caneday (Eds.), *Tourism and Leisure: Dynamics and Diversity* (pp. 53–66). National Recreation and Parks Association.

Kumar, R. (2018). *Sustainable Tourism Development and Peace: a Local Community Approach*. Retrieved January 15, 2020 from https://www.researchgate.net/publication/326302785_sustainable_tourism_development_and_peace_a_local_community_approach

Le Blanc, G., & Brugère, F. (2017). *La fin de l'hospitalité*. Flammarion.

Leclerc, G. (2015). *Le désir de voyage et la quête de l'autre*. L'Harmattan.

Litvin, S. (1998). *Tourism: The World's Peace Industry?* Retrieved January 15, 2020 from https://www.researchgate.net/publication/258161060_Tourism_The_World's_Peace_industry

McIntosh, R., Goeldner, C., & Richtie, J. (1995). *Principles, Practices and Philosophies* (7th ed.). John Wiley.

Milano, C. (2019). Overtourism and Tourismphobia: A Journey Through Four Decades of Tourism Development, Planning and Local Concerns. *Tourism Planning & Development*, *16*(4), 353-357.

Moser. (2019). *Public narratives and attitudes towards refugees and other migrants Germany country profile*. ODI Evidence. Ideas. Change. www.odi.org›sites›files›resource-documents

Neubert, S. (2003). *Some Perspectives of Interactive Constructivism on the Theory of Education*. University of Cologne. Retrieved January 28, 2020, from https://www.hf.uni-koeln.de/data/dewey/File/Neubert_Introduction.pdf

Peterson, C. (2006). *A Primer in Positive Psychology*. Oxford University Press.

Pratt S. & Liu A. (2015). Does Tourism Really Lead to Peace? A Global View. *International Journal of Tourism Research*.

Rogerson, C. M. (2015). Tourism-agriculture linkages in rural South Africa: evidence from the accommodation sector. In A. Spenceley & D. Meyer (Eds.), *Tourism and Poverty Reduction: Principles and impacts in developing countries*. Routledge.

Sharma, A., & Thapar, M. (2016). Development of tourism in the Third World nations: a comparative analysis. *Proceedings of the 7 International Conference on Sustainable Tourism*. www.witconferences.com

UNWTO. (2005). *Making Tourism More Sustainable - A Guide for Policy Makers*. UNEP and UNWTO.

UNWTO ST-EP. (2011). *Foundation, Sustainable Tourism for Eliminating Poverty*. http://unwtostep.org/

Van Den Dungen, P. (2018). *Peace Tourism*. Retrieved January 15, 2020, from https://worldbeyondwar.org/peace-tourism/

Vincent, G. (2015). *Hospitalité: la naissance symbolique de l'Humain*. Presses Universitaires de Strasbourg.

Wintersteiner, C., & Wuohlmuther, W. (2014). Peace Sensitive Tourism: How Tourism Can Contribute to Peace. In C. Wuohlmuther & W. Wintersteiner (Ed.), International Handbook on Tourism and Peace. Academic Press.

WTTC. (2016). *Tourism as a driver of peace*. Retrieved January 15, 2020, from https://www.wttc.org/publications/2016/tourism-as-a-driver-of-peace/

Chapter 16
Academic Mobility:
Bridging Tourism and Peace Education

Anabela Monteiro
https://orcid.org/0000-0001-8506-6073
Universidade Europeia, Portugal

Sofia Lopes
https://orcid.org/0000-0001-7803-513X
Universidade Europeia, Portugal

Fabio Carbone
Coventry University, UK

ABSTRACT

Education has played an influential role over time in economic, social, and cultural development and, hence, in the formation of individuals. The considerable increase in academic mobility, in recent decades, shows that it is an essential tool in students' education. Due to its specific characteristics, academic mobility has a strong link to multiculturalism, and one of its missions is to be a transmitter of values and is an important element for educational tourism. In this sense, this chapter intends to demonstrate the pedagogical potential of academic mobility, its individual and social effects in terms of peace education, and to explore the relationship between academic mobility and tourism for an environment of intercultural and peace sharing.

INTRODUCTION

In the broad context of education and its transformative power, it is important to understand the pedagogical potential of academic mobility to boost intercultural competencies and as a form of educational diplomacy. The interest in academic mobility, as well as the study of its motivations and effects, have increased over the past few decades. At the base of academic mobility programs, the multicultural contact at both social and educational level is one of the intrinsic elements of this experience, if not the most

DOI: 10.4018/978-1-7998-5053-3.ch016

important. As Banks (2009, p. 14) stated, "the major goal of multicultural education is to restructure schools so that all students acquire the knowledge, attitudes and skills needed to function in ethnically and racially diverse communities and nations, and in the world".

The different educational systems are promoters for the acceptance of cultures. According to Silva and Rebolo (2017, p. 189),

Education is an important contributor to make possible a change for a society that discusses cultural differences, taking into account the necessary recognition of the legitimacy of different cultures [...] There are, implicated in intercultural education processes, not only strategic, structural, technological, political and control elements but also, elements of the human and cultural dimension.

He who is different from me does not impoverish me – he enriches me [...] (Saint-Exupéry, 1942).

Academic mobility and educational tourism are promoters of global citizenship, and this strand is a pillar for the peacebuilding project. This study focuses its attention on the relationship between academic mobility and educational tourism and their involvement in the peacebuilding process and in the resolution of conflicts. Therefore, this chapter is organised according to the main objectives. For the understanding of academic mobility, it was necessary to approach the role of education as a formative agent for peacebuilding and conflict transformation, the educational curriculum and the role of academic institutions in promoting intercultural understanding. The definition and importance of academic mobility, as well as its programs, the main motivational factors and their influence on conflict transformation were also approached. For the understanding of what literature has studied about this theme was done a systematic literature review with the main concepts, last, the relation between academic mobility and educational tourism and their contribution to the peacebuilding process was also highlighted.

EDUCATION, A FORMATIVE AGENT FOR PEACEBUILDING AND CONFLICT TRANSFORMATION

Throughout the centuries, education has always been an actor of economic, social and cultural development, having as its functions the correlation between the individual and society, the individual and the state and the individual and the globalised world. In 1946, the United Nations Educational, Scientific, and Cultural Organization was established to contribute to world peace and security through education, science, culture, and communications.

Nowadays, the new communication challenge focuses mainly on the need for better understanding, new ways of accepting and interacting with other cultures, the presence of multiculturalism. Bourdieu (1996) states, in this context, that "culture is the substantial content of education, its source and its final justification [...] one cannot be thought without the other". For this reason, it is urgent that institutions, namely educational ones, have an innovative capacity to think about culture with the spectrum that the global world allows the development of inter-global competences based on international knowledge (UNESCO, 2017).

The cornerstone of academic institutions is education. It is crucial not to fail to point out that education was and will always be a tool for educational diplomacy, because it has an invisible built-in function of the construction of moral values, which, in turn, will reflect the students' personality (MGIEP,

2014). Education is a formative agent, which, through its means, aims to provide the individual with tools to deal with his surroundings, namely other individuals. This ability is, mainly, due to education allows the cohabitation of various values, as is the case of the academic mobility that can lead to "brain transformation" (Scott, 2015).

Nowadays, we are witnessing the progress to a dichotomous era, the need for authentic cultural affirmation, but at the same time dictating differences, placing an identity crisis on the agenda because "mass culture integrates and incorporates at the same time into a polyculture reality" (Morin, 2002, p. 16). The academic institutions have an important role in promoting an intercultural understanding to address shortcomings and to be a socialising agent, thus "experiences and practices that come along with international mobility of students and staff need to be negotiated in their everyday lives" (Bilecen & Van Mol, 2017, p. 1247).

The construction of the contemporary individual is based above all on a transformative education that leads the individual to have to adapt and constantly renew because, according to Durkheim (1978, p. 42), "society finds itself, with each new generation, as if facing a blank slate, on which it is necessary to build almost everything again. Here is the work of education. It creates in man a new being".

THE ROLE OF ACADEMIC INSTITUTIONS IN PROMOTING INTERCULTURAL UNDERSTANDING

Academic institutions have to be integrators of pluralism, making it their ally and not a factor of differentiation. As stated by Araujo and Ramos (2014, p. 92), "a need for communication between different cultures intensifies, due to increasingly complex conflicts and social interactions [...] education, and intercultural are related to address this plurality".

Higher education has to adopt cultural pluralism as a strategic axis because "adopting the doctrine of cultural pluralism, diversity is understood as a value and, therefore, is also understood as a good to preserve. The action is oriented to identify and monetise diversity as an educational resource" (Costa, 2019, p. 13). The coexistence of diversity is a fact, institutions should not deny it, we know that there are positive and negative points, but these points will strengthen the knowledge and awareness of the other's existence and the "acquisition of intercultural competences" (Moreira & Antão, 2019).

Another important point of educational institutions is the synergy between teaching and the surrounding community, as pointed out in the book "The student city strategic planning for student communities in EU cities" by the European Institute for Comparative Urban Research. "Universities and higher research institutions have relevant regional impacts [...] economic and cultural links between universities and cities translate into strategic common management of all the areas of contact between them" (Van Den Berg & Russo, 2017, p. 1).

This scientific capital, in addition to helping in the development of the community, allows for an interconnection between effective members of the institution (students, researchers, entrepreneurs, etc.) and local residents as a force of contemporaneity. According to Strandberg (2017, p. 9),

Social infrastructure enables innovations that constitute a new order of things – interacting elements that, when they work together, create new forms of value. These builds and creates the future, rather than reacting to it, and brings new resilient systems into being to enable, scale and accelerate social progress.

This will put the various elements in interrelation with all kinds of religion, culture, social habits, political parties, and these exchanges will create a sociocultural balance at the moment and in the future for the locals and especially for these students. However, there may also be cultural shocks if there has not been a back study by local political, social and cultural bodies (Leimgruber, 2018). There should be a sharing in the participatory strategy to be adopted, involving all stakeholders, fostering sustainable formal and informal sharing (Figure 1).

Figure 1. Higher education as an engine of sustainable urban development
Source: Van Den Berg and Russo, 2017

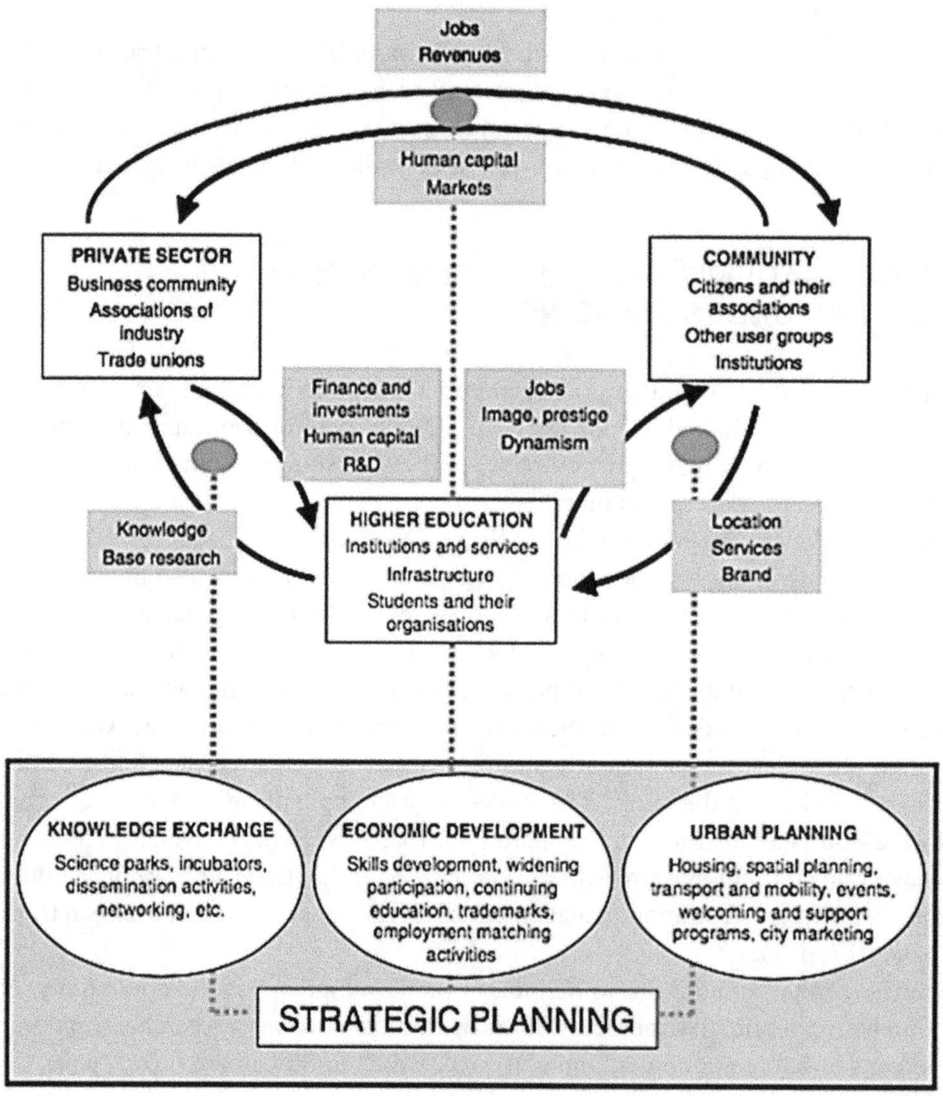

Being aware of the growth of the practices of international mobility, it is essential to raise awareness of educational institutions about their role in the process of educational diplomacy. International relations in educational institutions will be a dynamic and reflective instrument for an increasingly complex world, having as its main function to teach students skills and a reflective attitude to face the new demands.

THE EDUCATIONAL CURRICULUM AS AN INSTRUMENT FOR PEACEBUILDING

In the curriculum of courses in general, there are always some curricular units that are fundamental for understanding the historical process of the world and consequently understanding the current context of the globalised world. For example, in the study "The Role of Higher Education Institutions in Building a Culture of Peace: An Albanian Case" by Shehi, Ozcan and Hagen (2018), the purpose of the study was to explore the status of peace education and its correlation with student perceptions about peace. The study concludes that social science programs offer a range of topics related to peace education, particularly in social sciences, where three sciences have been identified, namely sociology, history and economics, where students show that positive human-centred topics on peace are taught in the course units.

The study, mentioned above, demonstrates that students consider the topics covered in curricular units – cultural studies, human rights, international relations, democracy and religious tolerance – important factors for building and understanding the peace process (Shehi et al., 2018). Academic curricula can and should be an aid to educational diplomacy because "the conscious and systematic adoption of pedagogical practices, ultimately oriented by the principles and values of intercultural dialogue…" (Bizarro & Braga, 2006, p. 5).

Due to such a wide range of education and its current characteristics of multiculturalism, education has a potential role of messenger of peace and transmitter of values. In a study carried out on the importance of textbooks, it reflects what the authors intend when referring to the curriculum. Ide, Kirchheimer and Bentrovato (2018, pp. 288-289) mention three important points regarding the importance of writing the manuals and their use:

- "school textbooks are indicators of dominant – or hegemonic – political knowledge in each society";
- "school textbooks have a large coverage and audience as they typically serve as quasi-mandatory readings for young people. School textbooks are the first – and often the only – books on social, political and economic issues people read in their lifetime";
- the considerable coverage and impact of school textbooks make them important objects of peace and conflict processes themselves".

Over the years, there have been areas that somewhat touch on direct or indirect learning about injustices and past human rights damages such as gender education, civic education intercultural and interfaith education (Nucci, 2001). All these tools serve a common purpose, play an important role in understanding tolerance, diversity, individual and social responsibility in the process of active participation by the individual (Figure 2).

Figure 2. Example of 'Whole School Approach' 7 developed by the school in Philippines
Source: Agnihotri, 2017

In summary, it is necessary to base teaching on a curriculum that values group work, promoting cooperation among all. Curricular units should have a practical component, for example, using Program Based Learning (PBL), which could improve this strategy. In cases of resorting to examples, they should resort to real cases of different origins (ethnic, religious, political, etc.) enriching the teacher's discourse (UNESCO, 2004). The professor's attitude is critical, being a reference and a model to follow and so it should have a fair attitude and an open vision. One of the aims of educational institutions should be to promote authentic peace, student well-being and social peace with a holistic education (Miller, Nigh, Binder, Novak & Crowell, 2018).

An example of an institution struggling to place education as an essential instrument in the fight against injustice at all levels, but above all in building peace and developing a sustainable world is *The Mahatma Gandhi Institute*. One of its main objectives is to help governments renew education systems, building on UNESCO's long experience and Mahatma Gandhi heritage. UNESCO has always expressed that education is a powerful tool for promoting peace, upholding human dignity and achieving sustainable development (MGIEP, 2017).

Another pedagogical instrument of vital value for democracy and human rights, but unfortunately a little debated issue, is the United Nations Charter and the Universal Declaration of Human Rights, adopted and proclaimed by the United Nations General Assembly (resolution 217 A-III) on 10 December 1948. According to Article 26,

1. Everyone has the right to education. Instruction will be free, at least in elementary and fundamental grades. Elementary instruction will be mandatory. Professional-technical education will be accessible to all, as well as higher education based on merit;
2. Instruction will be directed towards the full development of the human personality and the strengthening of respect for human rights and fundamental freedoms. Instruction will promote understanding, tolerance, and friendship among all nations and racial or religious groups and will assist the United Nations in maintaining peace;
3. Parents have a prior right to choose the kind of instruction to be given to their children.

The point two meets what have been saying, education is a means to an organised society, to a democratic rule of law, to fight injustice, to sustainable development, to a culture of peace and a means to equality. The Universal Declaration of Human Rights is not a curricular unit, but a useful pedagogical tool to use as a means to engage the student in a sphere of critical analysis, understanding their rights and others, a human pedagogy.

It is notable how the topic of education, in 1793, is referred to as a fundamental element of a society in its various aspects, pointed out in the French National Convention in 1793. In its article 22 attested that "education is the need of all. Society must, with all its power, favour the advancement of public intelligence and make education available to all citizens". Another important point in education is its similarity to progress/development. Education has an idiosyncratic element in the development of the improvement of the human condition, whether socially, economically, scientifically, culturally or personally (Hanushek & Woessmann, 2008).

In 2015, the United Nations established the New Sustainable Development Agenda for 2030, where it is stereotyped a new model of international cooperation.

As we embark on this great collective journey, we pledge that no one will be left behind. Recognising that the dignity of the human person is fundamental, we wish to see the goals and targets met for all nations and peoples and for all segments of society. And we will endeavour to reach the furthest behind first. (p. 6)

In the same document, Sustainable Development Goal 4 for Education was said: "to ensure inclusive and quality education for all and promote lifelong learning" (United Nations, 2015). "Universities, as cultural, professional and political institutions, go beyond the educational system and are more involved in the processes of development and economic and social change than other educational organisations" (Schwartzman, 2018, p. 45). Education is involved in all branches of globalisation, being a key component in creating an international environment (Lemoine, Jenkins & Richardson, 2017). In the emergence of the century, education cannot merely stagnate; it must use its knowledge to create an environment of synergies because education produces direct and indirect benefits.

The questions that arise are how academic institutions should educate and, how to bring the terms of peace, security, human rights, dignity, equality through contemporary and engaging student education. All of these issues should be discussed with students, put them think, fostering open dialogue. The answers are not stereotyped, because they lie within each of us, if the teacher is unaware of these topics and does not take them into account with a pedagogical factor, there are no responses. It is necessary to try to sensitise the academic community.

According to Felgueiras, Siqueira, Lima and Oliveira (2018, p. 67),

Being the educational environment at any level (i.e. basic, medium or higher) a place of plurality, of differences, where dialogue is often not easy and disagreements and discrimination still occur, the human rights education can change the foundations of these relationships.

Education must impart knowledge, but above all, it must be a means of social, cultural and political transformation for a sustainable and peaceful future.

ACADEMIC MOBILITY PROGRAMS: WHAT IS THE SYMBIOSIS BETWEEN EDUCATION, CULTURE, TOURISM AND THE PEACEBUILDING PROCESS?

Education has to adapt to the new modalities of educational actions, which currently do not take place only in the school space. Education passes institutional barriers, the contemporary socialisation system imposes this change (Setton, 2005). These pluralities of actions must be limited to the construction of a full individual, with an education that allows him to be a formative agent for the construction of peace and a regenerator of conflicts.

Currently, there are several student exchange programs, such as Erasmus + in Europe; Fulbright in the United States of America; China Scholarship Council in China; Australia Awards Scholarships in Australia; Mexican Agency for International Cooperation and Development in Mexico, among others. The practice of academic mobility is an increasingly common resource for students. According to UNWTO (2020), it is estimated that there will be almost 300 million international youth trips per year.

The considerable increase in academic mobility over the last decades shows that this aspect is a very strong component of educational institutions. In 2017, there were more than 5.3 million international students, 3.3 million more than in 2000 (GMDAC, 2018). More than half of them were matriculated in educational programs in five countries: United States of America; United Kingdom; Australia; France and Germany. Major international students sending countries included China, India, Germany, South Korea, Nigeria, France, Saudi Arabia, and several Central Asian countries (GMDAC, n.d). Table 1 represents the number of total incoming students, both genders provided by UNESCO's Institute for Statistics, for the year 2017 (GMDAC, 2018).

In the table above, it is possible to verify that a large number of countries receive students from different parts of the world to perform academic mobility. There is a big bet on cross-border education, a way of imposing international agencies such as the OECD, the United Nations Educational and Scientific Organization (UNESCO) or the World Bank. The focus of this strand is mainly to "eliminate barriers and boundaries of knowledge between countries" (Pinto & Larrechea, 2018).

Regarding the number of students that go to the United States of America, the five countries that send the most students are China (321,625); India (142,618); Korea (56,186); Saudi Arabia (54,709) and Canada (27,800) (Figure 3). Nonetheless, North American students commonly choose to study in the United Kingdom (16,178); Mexico (11,109); Canada (8,799); Grenada (6,109) and Germany (5,710) (Figure 4). The United States receives, significantly, more students than those it sends.

Table 1. Total inbound internationally mobile students

Country	2017	Country	2017
World	5309240,396	Greece	25067
High income countries	3886512,063	Finland	24168
Northern America and Europe	3302325,287	Portugal	22194
North America and Western Europe	2765854,953	Brazil	20671
Europe	2107298,678	Morocco	20121
Middle income countries	1364986,333	Ireland	19983
Northern America	1195026,608	Lebanon	19670
Upper middle-income countries	1098834,333	Belarus	17570
East Asia and the Pacific	1084896,11	China, Macao Special Administrative Region	14821
United States of America	984897,6084	Kyrgyzstan	14796
Central and Eastern Europe	661383,3333	Kazakhstan	13850
Asia (Eastern and South-eastern)	644617,642	Senegal	13536
Western Asia and Northern Africa	472708	Ghana	12978
Asia (Eastern)	443411,642	Serbia	11624
Oceania	441798,4676	Qatar	11034
United Kingdom of Great Britain and Northern Ireland	435734	Slovakia	10764
Oceania (Australia/New Zealand)	433880,4676	Dominican Republic	9607
Asia (Western)	386509	Norway	8955
Australia	381202	Benin	8867
Arab States	331144	Algeria	8546
Lower middle-income countries	266152	Georgia	7994
Germany	258873	Indonesia	7763
France	258380	Bosnia and Herzegovina	7275
Russian Federation	250658	Grenada	6587
Canada	209979	Tunisia	6370
Latin America and the Caribbean	209676	Latvia	6130
Latin America and the Caribbean	209526	Lithuania	5806
Asia (South-eastern)	201206	Bahrain	5616
Japan	164338	Kenya	5254
China	157108	Côte d'Ivoire	4736
Africa (Sub-Saharan)	137892	Chile	4708
Sub-Saharan Africa	137613	Colombia	4550
Small Island Developing States	109850	Armenia	4512
Small Island Developing States	109504	Azerbaijan	4274
Turkey	108076	Viet Nam	4162
Malaysia	100765	Republic of Moldova	3964
Asia (Central and Southern)	100373	Cameroon	3941
Italy	97563	Estonia	3917
Netherlands	96289	Namibia	3422
Argentina	88873	Luxembourg	3298
Africa (Northern)	86199	Burkina Faso	3283
Landlocked Developing Countries	79622	Slovenia	3090
Saudi Arabia	78344	Oman	3044
Austria	73964,3448	Niger	2793
Republic of Korea	70796	Madagascar	2342
Asia (Southern)	68799	Tajikistan	2238
South and West Asia	68799	Mauritius	2087
Spain	64927	Burundi	1789
United Arab Emirates	64447	Mongolia	1520
Poland	63925	Rwanda	1397
Switzerland	53368	Sri Lanka	1313
Singapore	53204	Iceland	1231
Ukraine	52768	Botswana	1180
New Zealand	52678,46756	El Salvador	658
Central Asia	49874	Uzbekistan	603
India	46703	Mozambique	590
South Africa	45334	Saint Lúcia	479
Belgium	44978	Lao People's Democratic Republic	472
Czechia	44261	Brunei Darussalam	436
Jordan	39577	Mauritania	279
China, Hong Kong Special Administrative Region	34348,642	Andorra	182
Denmark	33602	Bermuda	150
Asia (Central)	31574	Cabo Verde	102
Sweden	28747	Lesotho	74
Hungary	28628	Monaco	3
Mexico	25125		

Source: GMDAC, 2018

Figure 3. The flow of student studying in the United States of America
Source: UNESCO, 2017

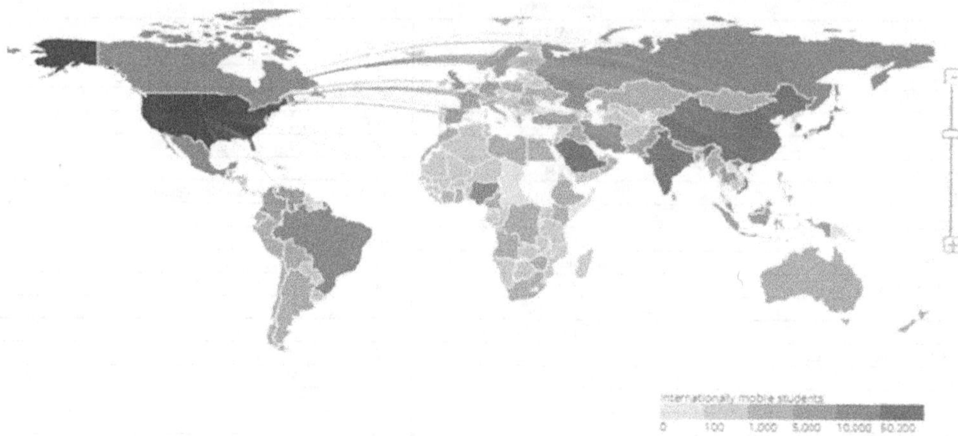

Figure 4. The flow of student of the United States of America studying abroad
Source: UNESCO, 2017

Academic mobility is stimulated as a form of cultural, social, political and economic integration (Lima, 2009). The Erasmus+ program has an unprecedented, intercultural and educational aspect of enrichment (van't Klooster, van Wijk, Go & van Rekom, 2008). According to the results of Salamanca's Convention (2001), the student and the staff mobility fosters the ability to cope with a new cultural and teaching environment, as well as, the ability to understand other cultures. It is a requirement in the light of globalisation, but it also promotes coherence at European level and improves the scientific perspective (Osterwalder, 2001). This practice represents a must-have experience among the student community in higher education, aiming to be global citizens. According to the Portuguese Secretary of State for Education, "Erasmus is one of the main tools for peace in Europe. It can bring young people from across

Europe in contact with others by showing that there is cultural diversity and also that we are more equal than different" (Costa, 2018).

Furthermore, to the growing cultural diversity of European societies, Erasmus+ has a number of objectives, but what concerns is the support for policy reform, whose outlined steps are to prepare young people for the new intercultural reality that they are in daily contact to promote: "democratic values, fundamental rights, intercultural understanding and active citizenship" (European Commission, 2018, p. 2). The projects also aim to promote the social inclusion of disadvantaged and migrant individuals, while preventing and combating discriminatory practices"

The cultural element is strongly connected to this kind of experience. As culture is one of the most referred points, it is important to remember what the United Nations Educational, Scientific and Cultural Organization (UNESCO), in the international conference held in Mexico City, consolidated about the importance of the role of culture in conflict mediation.

In its broadest sense, culture can now be understood as the integral complex of distinct spiritual, material, intellectual and emotional traits that characterise a society or social group. It includes not only the arts and letters, but also ways of life, fundamental human rights, value systems, traditions and beliefs. (UNESCO, 1982, p. 1)

In 1959, Hall, in his work, *The Silent Language*, drew attention to what he called "intercultural communication" the author analysed the need to know the other as a way to promote a change in attitude towards understanding:

The aim of intercultural communication is to promote changes in behavior through understanding, above all, a better understanding of yourself, the development of self-awareness, the understanding of the cultural self, of your own identity compared to the identity of the other. This phenomenon differentiates one person's adaptability from another with a different culture (Hall, 1959, p. 186).

The understanding of different cultures allows, as Proust (2016) stated, that "the real voyage of discovery consists, not in seeking new landscapes, but in having new eyes" and thus assuming a new position before the other.

Regarding the question about the connection between education, culture and tourism in the peacebuilding process, the answer lies in the simplicity that the academic mobility and educational tourism experiences achieve in their educational aspect as a form of personal, social and cultural growth because "every tourist learns about different aspects that the destination offers" (OMT, 2003) and "initiatives of tourism that contribute to international understanding, improving the quality of the environment, preserving heritage and, through these initiatives, helping to build a peaceful and sustainable world" (Salazar, 2006, p. 3). Getting to know other people, cultures, countries is a way of being a "Citizenship Diplomacy", a more effective tool for building a peaceful understanding between people and countries (D'amore, 2014).

MOTIVATIONAL FACTORS FOR ACADEMIC MOBILITY AND THEIR RELATIONSHIP TO CONFLICT TRANSFORMATION

More than ever, it is crucial to understand the main reasons that lead the students to undertake academic mobility. With regards to the motivations, a study "European Academic Mobility and Educational and Cultural Tourism: Decision and Motivation Factors" (Monteiro, 2014) revealed that one of the priority factors in choosing to pursue academic mobility has been linked to decision making as the main but not the only reason, thus, there is no single motif but an aggregation of motifs that co-exist synchronously (Monteiro, 2014).

According to the results of Monteiro (2017) study, there are three major groups of students: one group includes all who favour culture-related activities; a second group, students who carry out academic mobility to gather new experiences mainly at the personal level; a third group consisting mainly of students performing academic mobility in Europe and selecting cultural and academic development as the main reasons.

According to the report "The International Higher Education Report: Student Trends 2018", more than 30 million students from 221 countries have achieved academic mobility. More than 20,000 surveys of international students (42% male and 58% female) at 18-29 years old were conducted answering why they are pursuing academic mobility. The reasons are to explore new cultures and experiences, making friends and expanding their career opportunities (Table 2).

Table 2. Reasons for doing an academic mobility

Educations.com's International Higher Education Report: Student Trends 2018 (Prestage, Hopton-Stewart&the educations.com team)	Current students
Experiencing a new culture	59%
Expanding career opportunities	37%
Making friends and expanding professional network	30%

Looking at these results, contact with another culture and personal growth are the main decisive factors. These factors are also revealed in other studies (Table 3).

Through the words presented in table 3, a *WorldCloud* analysis was also conducted to summarise the most commonly used words in the answers to the reasons that led students to achieve academic mobility (Figure 5). The most prominent word is "cultural", confirming the desire to know other cultures, to be inserted in another environment and to be able to be integrated into another reality. This reinforces "the role of academic mobility as an instrument of educational and cultural diplomacy". This willingness to know, and desire to understand each other, allows these young people to have a diplomatic mindset.

According to the motivational factors, it is possible to verify the importance of academic mobility on the transformation of conflicts, namely, because the basis of the reasons is the desire to get to know other cultures. Students stay, from three months to a year, in another country, consequently, they will live with the natives of the region and also with other international students. This intercultural relationship will allow us to meet the other and put ourselves in the other's place, otherness.

According to Laplantine (2000, p. 21),

Table 3. Summaries of previous studies conducted about the reasons for academic mobility

Study	1st reason	2nd reason	3rd reason
Murphy-Lejeune (2000)	Language and cognitive development	Acquisition of social skills	Intercultural learning
Richards and Wilson (2003)	Knowledge of other cultures	Emotion of the experience	Increase knowledge
Richards and Wilson (2004)	Explore other cultures	Accomplish a trip to enthusiasm	Increase knowledge
Krzaklewska and Krupnik (2006)	Contact with another language	Have new experiences	Career opportunity
Caudery, Petersen and Shaw (2007)	Improve the british accent	Experience, emotion, fun	Finding a new culture
Li and Bray (2007)	Academic benefits	Social and cultural	Economic benefits
Kasravi (2009)	Personal growth	Cultural Experience	Increase knowledge
Mendes, Pedradas and Pereira (2009)	Cultural experience	Professional career	Different environment
Jašková and Heczková (2010)	Contact with another language	Life experience	Better future perspective
Alfranseder, Fellinger, Taivere and Krzaklewska (2011)	Knowledge of other persons	Knowledge of other cultures	Personal development
Mészáros (2011)	Learn or improve other languages	Life experience	Professional career
Dias (2013)	Academic training	Learn or improve other languages	Tourism
Fombona, Rodríguez and Sevillano (2013)	Learn or improve other languages	European experience	Academic benefits
Monteiro (2014)	New experiences	Socialisation	Cultural reasons
Lima and Riegel (2015)	Intercultural experience	Learn or improve other languages	Professional skills
Oliveira and Freitas (2016)	Knowledge of other cultures	Learn or improve other languages	International career
Monteiro (2018)	The social experience of living in a different country	Meet new people and make new friends	Desire to travel

The experience of otherness (and the elaboration of that experience) leads us to see what we would not have been able to imagine, given our difficulty in fixing our attention on what is usual, familiar, daily, and which we consider 'evident'. Gradually, we noticed that the smallest of our behaviors (gestures, mimics, postures, affective reactions) are not really 'natural'. Then we start to be surprised by what concerns ourselves, to spy on us. The (anthropological) knowledge of our culture inevitably passes through the knowledge of other cultures and we must especially recognise that we are a possible culture among many others, but not the only one.

Figure 5. WorldCloud of the answer to the survey
Source: Educations.com, 2019

Figure 6. Number of publications per country

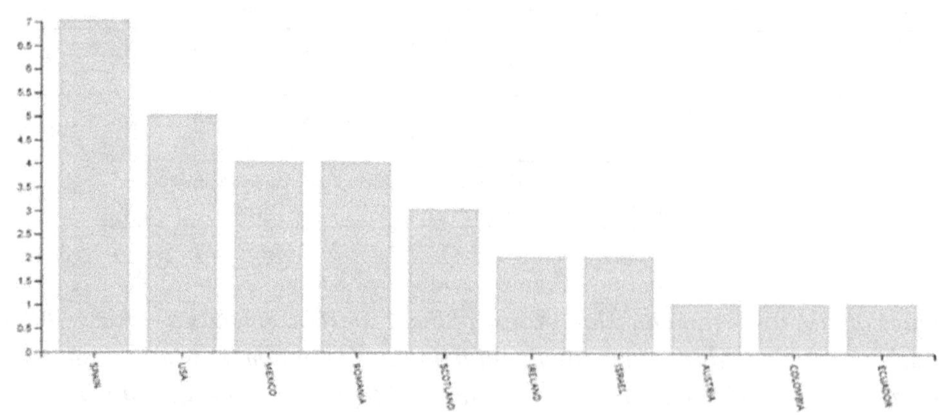

INTERCULTURAL EDUCATION AND PEACE: A SYSTEMATIC LITERATURE REVIEW

In order to understand to what extent this topic - the relationship between education, academic mobility and educational diplomacy in the field of tourism and peace – has been approached by the literature and its importance for the development of future studies, pointing out the trends (Araújo, 2006; Oliveira, Barbosa, Rezende, Silva & Albuquerque, 2013) was elaborated a bibliometric analysis using VOSviewer software which allows "creating maps based on network data and for visualising and exploring these maps" (Eck & Waltman, 2019).

To analyse the literature approach on these topics, it was elaborated: i) a bibliometric analysis about the management WoS Core Collection[1] publications that have as topic "intercultural education", as well as "peace"; and ii) a bibliometric analysis about the management WoS Core Collection publications that as a topic "peace", "education" and "tourism".

The terms "intercultural education" and "peace" are present in 40 articles from different countries published between 2015 and 2019 (Figure 6), the highest percentage is published in the category of Education Educational Research (Figure 7).

Figure 7. Category of Web of Science

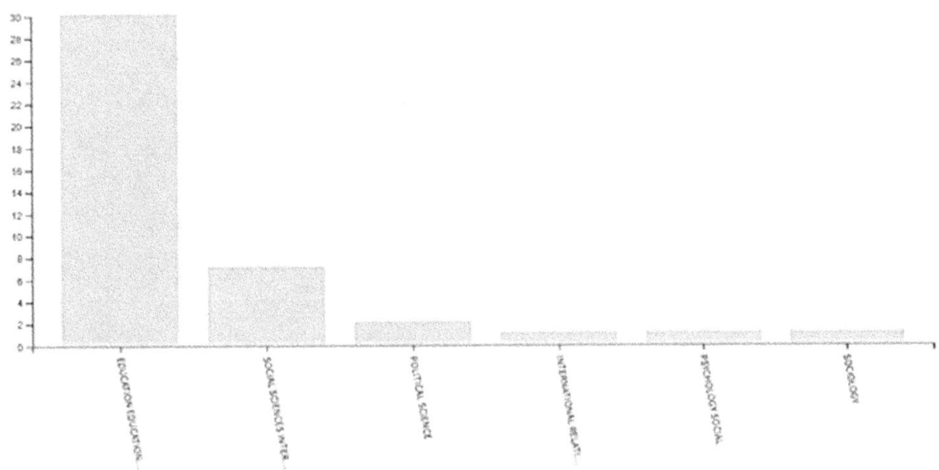

Using the grouping technique, it is possible to verify that there are 5 major groups identified by 5 different colours (red, green, blue, yellow and purple) that demonstrate the approximation of words, based on the number of times the word is used. The closer two terms are, the stronger is their relationship in terms of co-quote links (Figure 8).

Regarding the five groups with a greater relationship, these are: (i) intercultural peace & integration, (ii) knowledge & world, (iii) university & diversity, (iv) woman & peace education, (v) interreligious education & dialogue. The major keywords that stand out are: (i) education, (ii) intercultural education, (iii) peace; (iv) experience, and (v) culture.

Concerning the green cluster, its main publications contemplate aspects such as education, experience and culture. In the study elaborated by Kotite (2012), for example, education can contribute to the acceptance of a different culture:

[...] education can contribute to mitigating conflict along the three primary aspects of conflict: structural, behavioural and attitudinal. Education can alter societal contradictions (structural), improve relations and interactions (behavioural) and encourage changes in attitudes (attitudinal) in ways that can reduce the risk of conflict and help build a sustainable peace. (p. 13)

Figure 8. Terms' clusters of general literature

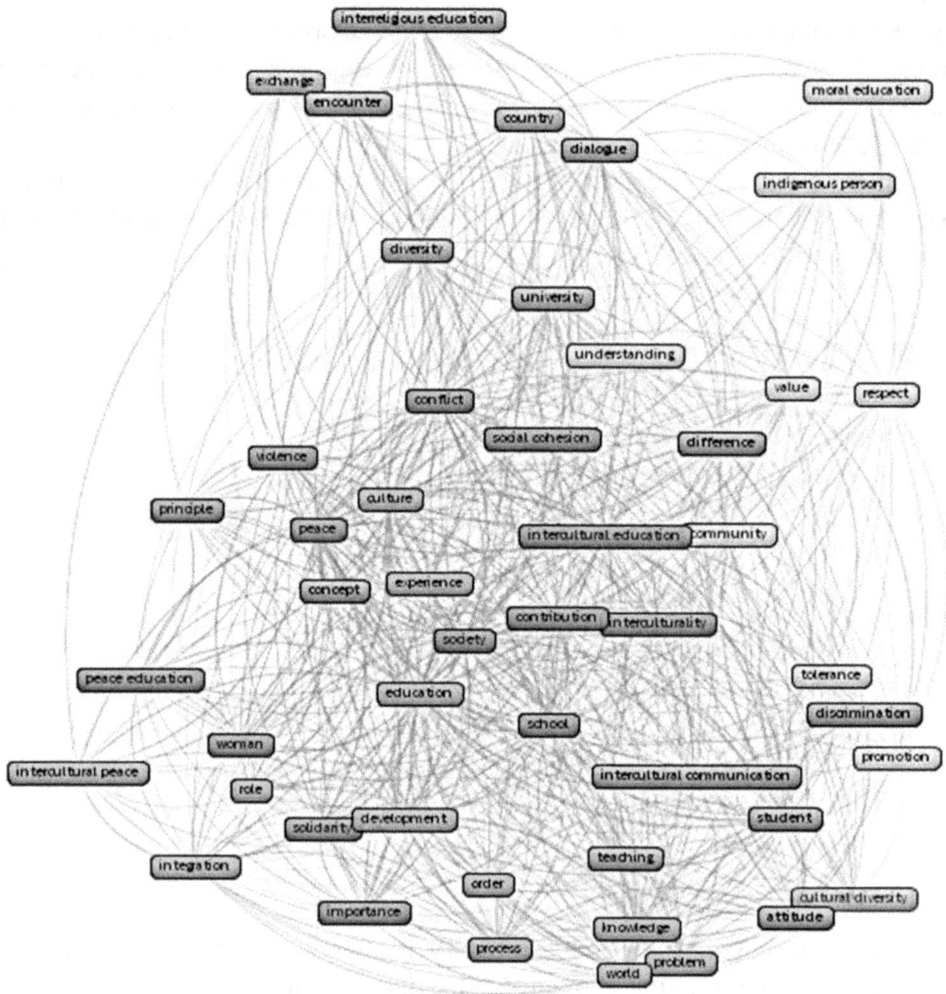

In the red cluster, in which the main studies contemplate aspects related to intercultural education and peace, as stated by Heleta and Deardorff (2017 p. 55), "intercultural competencies can assist in building trust and promoting peace and reconciliation among students, academics, and formerly separated communities".

EDUCATION, TOURISM AND PEACE: A SYSTEMATIC LITERATURE REVIEW

The search for the terms "education", "tourism" and "peace" revealed that they are consistent in 64 articles published between 2015 and 2019, from 10 different countries (Figure 9), the highest percentage is published in the category of *Hospitality Leisure Sport Tourism* (Figure 10).

Figure 9. Number of countries' publications

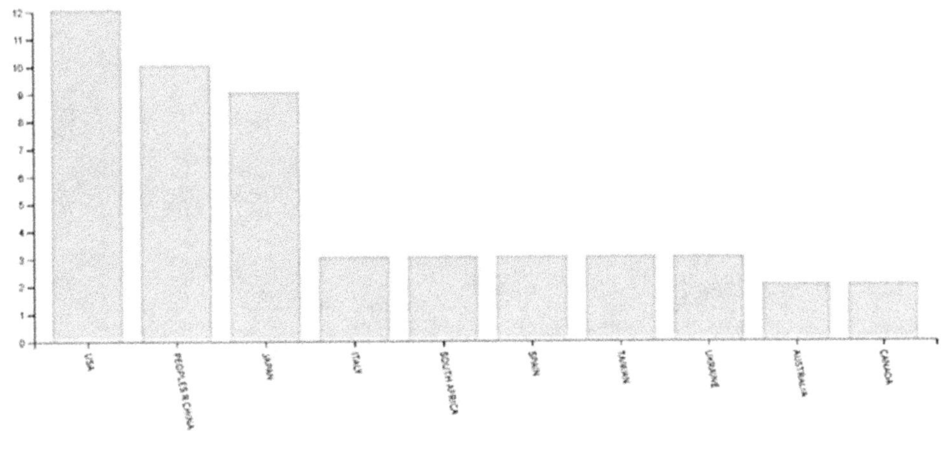

Figure 10. Web of Science categories

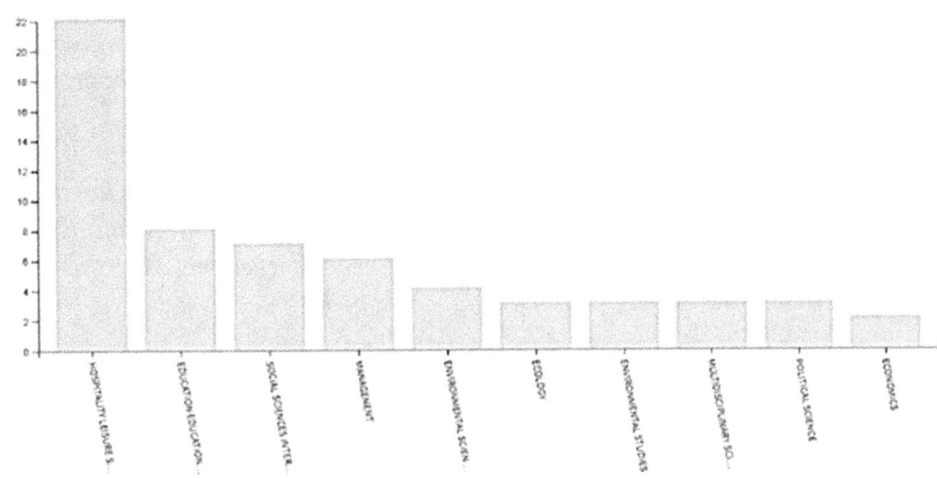

According to the usage of the grouping technique, there are five groups identified by five different colours (blue, green, yellow, purple and red) that demonstrate the approximation of words (Figure 11). The three main groups with greater relationship are: (i) student & experience; (ii) conflict & region & perception; and (iii) problem & cultural landscape & development. The four major keywords are: (i) development, (ii) peace, (iii) tourists, and (iv) experience.

Figure 11. Terms' clusters of general literature

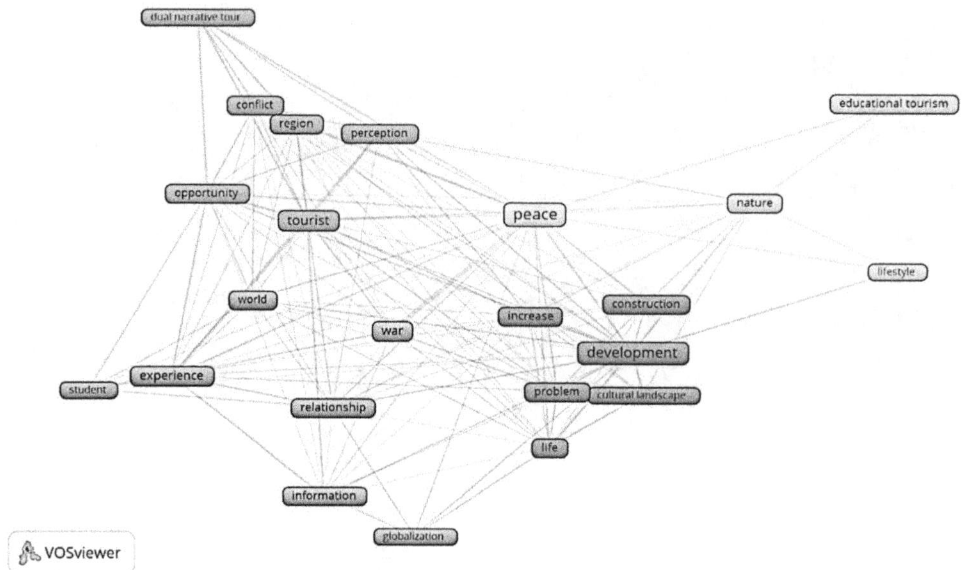

Concerning the red cluster, its main publications contemplate aspects such as development. In the study elaborated by González, Raesfeld and Ponce (2019), was verified the importance of education development for peace:

In the context of globalisation where linguistic, religious and cultural differences rise, the education must have as a main objective that all people without any distinction can develop themselves to its fullest, to exercise their human rights to use the full extent of the fundamental freedoms. With this premise, it will be plausible a model of society that strengthens its social net in peaceful coexistence and of respect. (p. 117)

The yellow cluster is strongly related to the green one, namely, there is a strong connection between peace and tourists. According to CPRPE (2013):

[...] we need both a complex understanding of tourism and a complex understanding of peace, when it comes to discussing the impact that tourism may have on peace processes. We have to study tourism in the broader context of human mobility while developing a concept of peace that goes far beyond the political dimension or negative peace (absence of war). (p. 19)

The analysis based on published articles shows that there is not much occurrence of these terms in publications, probably because this theme has been carried out only in recent studies, namely, since 2015. As stated by Mathews-Aydinli (2016), "even today, in an era of both increased numbers of participants in international educational exchanges and a growing appreciation of the potential for such exchanges to contribute to intercultural understanding, and thus public diplomacy, research on the topic is limited" (p. 4).

The observation of these data allows us to conclude that little has been published on this topic, but it reveals a strong presence of interdisciplinary between education, academic mobility and educational tourism, mainly after the phenomenon of globalisation, generating new challenges (Makarova, Makarova & Korsakova, 2019). Therefore, this chapter contributes to a better understanding of this topic.

EDUCATIONAL TOURISM: A BRIDGE BETWEEN EDUCATION AND PEACEBUILDING PROCESS

Educational tourism is based on an educational program and attempts to enhance the students' knowledge, skills and behaviour. This type of tourism is related to any "program in which participants travel to a location as a group with the primary purpose of engaging in a learning experience directly related to the location" Roger (1998, cited in Bhuiyan, Islam, Siwar & Ismail, 2010, p. 19).

Academic mobility programs are one of the vehicles for this type of tourism. Educational tourism, according to Jafari (2002, p. 170), allows the "train and develop the knowledge, skill, mind and character of students through a process formal schooling and teaching… who bring with them a diversity of skills, knowledge and attitudes". Jafari also emphasises that this form of tourism allows the construction of the individual with a multifaceted formation and that, in some way, this experience as stated by Malta and Carneiro (2005, p. 116), "is one of the more visible and strident faces of globalisation processes". This link represents one of the main contributions to the creation of "a society based on social justice through equal opportunities, the fair distribution of power and resources, equal protection and impartial application of the law and, above of everything, of mutual cultural understanding and respect" (Carbone, 2017).

The academic mobility and educational tourism provide the students with the opportunity to develop cultural awareness by experiencing first-hand differences and similarities of hosting country's culture and – on the other hand - to become representatives of their own cultural values abroad. This last characteristic is outlined by Jafari and Way (1994, p. 76), "despite the fact that the most tourist share certain behavioural aspects, most still retain elements of their home culture".

According to the European Higher Education Area (EHEA), the quality of mobility is as an added value enabling personal improvement in the training of skills, tolerance, pure intercultural skills and democratic competences (European Commission, 2015). Teichler and Jahr (2001) also reinforce the same idea by stressing that mobility is a voluntary action "of a horizontal basis and of cultural, social and economic enhancement both for the individual and for society" (p. 444). Therefore, academic mobility and educational tourism allow growth at different levels and enables mutual understanding and improvement of international cooperation.

The mobility for educational purposes reflects the need for an "experience of alterity", as "the (anthropological) knowledge of our culture inevitably passes through knowledge of other cultures, and we must especially recognise that we are one culture among many, but not the only one" (Laplantine, 2000, pp. 12-13). The interest towards this topic is thus given by the potential that academic mobility and, consequently, educational tourism (Ritchie, Carr, & Cooper, 2003) have to prepare the students to create a better future of global understanding.

Many terms within academic mobility bring us closer to terms essential to building peace, because "education is one of the fundamental pillars of human rights, democracy, sustainable development and peace" (UNESCO, 1988).

International exchange within higher education has several missions. There is great interest from governments and higher education institutions in fostering the learning of cultural, social, academic and political differences between states (Abreu, 2018). Linked directly to this is, one of the main reasons expressed by students to have such an experience is the desire to know other cultures, and this is the basis of the peace process. Governments must increasingly invest in this transcendental synergy, the transmission of peace through the knowledge society and public diplomacy, which "encompasses the dimensions of international relations beyond traditional diplomacy, the interaction of groups and interests of one country with another, and public opinion in other countries, among other aspects" (Abreu, 2018). There is another point that can and should be used in this relationship, the paradiplomacy (Senhoras, 2013).

Paradiplomacy is based on cooperation at multilateral level, involving national, regional and foreign agents and public or private (Dias, 2010). This relationship allows a rapprochement between different actors and civil society because "subnational entities, with their tentacles and ramifications, go where central governments cannot reach, breaking the distance between external political and the real desires of the population" (Santana, 2009, p. 42). As higher institutions increasingly value cultural diversity and are an agent providing academic mobility programmes, there is, therefore, cooperation between different agents of different cultures (Prieto, 2004). Paradiplomacy is a factor that also enables the management of a development strategy between local and global institutions, basically a "development cooperation" and the "art of meeting" (Jesus, 2017). Here is the core of cooperation between education and the peace process.

The Intra-Africa Pax Lusófona (IAPL) project is an example of this cooperation. It is the result of a partnership between the European Commission (managed by the Education, Audio-visual and Culture Executive Agency – EACEA), with the support of the African Union. The IAPL aims at 67 academic and professional mobility's directed solely at Law and Peace Education (PaxLusófona, 2017) by 2022. The project is supported by several Portuguese, Angolan, Cape Verde, Mozambique and Sao Tome and Principe educational institutions. The IAPL's main objective, through academic mobility, is to develop skills that allow "the creation of synergies and boost postgraduate education in Africa, based on the values of respect for the historical and cultural identity of the countries involved, the exchange of experiences and knowledge, the adaptation or introduction of academic programmes at postgraduate level to promote, build and maintain peace" (PaxLusófona, 2017).

It is necessary to revalue existing resources among educational institutions, local, national and international governments because the culture of wellbeing has a direct relationship with peace, development, social justice and democracy (Mayor, 2010). The students are about to enter into the professional world, the real world, where they will face challenges and will have to overcome barriers. If these young people were well prepared for cultural, social and economic diversity, the reality of a globalised world accentuated by a peace process will be much more possible. The education is a tool "through which the individual takes history into his own hands, in order to change its course. How? Believing in the student, in his ability to learn, discover, create solutions, challenge, face, propose, choose and assume the consequences of his choice" (Fuck, 2001, pp. 14-15). Young people with this experience have more potential to become a social actor with awareness in society and in the world.

CONCLUSION

Education has played an important role over time in the economic, social and cultural development and consequently in the training of individuals. Academic institutions have a significant function in promoting intercultural understanding and, academic mobility emerges as an experience across borders, where the students have the opportunity to know and interact with different cultures. Academic mobility is stimulated, especially, by the desire to get to know other environments and allows the development of empathy, solidarity, respect for differences and diversity. All the "Ensamblage" between academic mobility, educational diplomacy and educational tourism form the perfect recipe for a symbiosis of cooperative understanding and an intense cultural renewal.

Education is the gene for a world of mutual understanding, for a dialogue between values and ethical knowledge. Education should not be a single tool, but rather a multifunctional tool that can be used through pedagogical instruments. Educational institutions are a vehicle to convey this message through a pedagogical way, the *status quo* of the contemporary world. Education is the system that can support young people to understand this world of uncertainty and conflict because "the transformation of social systems (conflicts) requires secondary learning (second-order learning), a willingness and ability to challenge the underlying assumptions" (Sousa, 2017, pp. 64-65).

According to Marx (n.d., p. 232):

In essence, therefore, like all rights, is the right to inequality. The right can only consist, by nature, in the application of an equal measure; but unequal individuals [...] can only be measured by a measure as long as they are considered from an equal point of view.

We need to work together, looking for the best ways to interpret the new assumptions of this world to make globalisation a tool for unity and not a disadvantage. Academic mobility is a way for the exchange of ideas between different individuals that allow the dialogue concerning all the points that create a society and its implied social functions. Academic mobility and educational tourism contribute to an open mind, for acceptance of the differences, contributing to peace.

REFERENCES

Abreu, J. R. (2018). *Mobilidade acadêmica como instrumento soft power: A experiência dos USA, Brasil e México* (PhD Thesis). Universidade de Brasília, Brasília.

Agnihotri, S. (2017). Critical reflection on the role of education as a catalyst of peace-building and peaceful coexistence. *Universal Journal of Educational Research*, *5*(6), 911–917. doi:10.13189/ujer.2017.050601

Alfranseder, E., Fellinger, J., Taivere, M., & Krzaklewska, E. (2011). *E-Value-ate your exchange: Research report of the ESNSurvey 2010*. Erasmus Student Network AISBL.

Araújo, C. A. (2006). Bibliometria: Evolução histórica e questões atuais. *Questão*, *12*, 12–32.

Araujo, D. F., & Ramos, M. C. P. (2014). *Um olhar sobre a educação superior do turismo para uma nova cidadania Intercultural*. Cairu em Revista - Sociedade, Educação, Gestão e Sustentabilidade.

Banks, J. A. (2009). Multicultural education: Dimensions and paradigms. In *The Routledge international companion to multicultural education* (pp. 29–52). Routledge. doi:10.4324/9780203881514

Bhuiyan, M. A. H., Islam, R., Siwar, C., & Ismail, S. M. (2010). Educational tourism and forest conservation: Diversification for child education. *Procedia: Social and Behavioral Sciences*, *7*, 19–23. doi:10.1016/j.sbspro.2010.10.003

Bilecen, B., & Van Mol, C. (2017). Introduction: International academic mobility and inequalities. *Journal of Ethnic and Migration Studies*, *43*(8), 1241–1255. doi:10.1080/1369183X.2017.1300225

Bizarro, R., & Braga, F. (2006). *Formação de professores de línguas estrangeiras: Reflexões, Estudos e Experiências*. Porto Editora.

Bourdieu, P. (1996). *Razões práticas: Sobre a teoria da ação*. Papirus Editora.

Carbone, F. (2017). International tourism and cultural diplomacy: A new conceptual approach towards global mutual understanding and peace through tourism. *Tourism: An International Interdisciplinary Journal*, *65*, 61–74.

Caudery, T., Petersen, M., & Shaw, P. (2007). The language environments of exchange students at Scandinavian universities. *Researching Content and Language Integration in Higher Education*, 233-250.

Costa, D. (2019). Diversidade cultural na universidade em Portugal: Tendências e desafios. In S. Gonçalves & J. J. Costa (Eds.), Diversidade no Ensino Superior (pp. 11-37). Estratégias de Ensino e Sucesso Académico: Boas Práticas no Ensino Superior.

Costa, J. (2018, May 10). Erasmus. Um dos principais instrumentos de paz na Europa. *Renascença*. Retrieved from: https://rr.sapo.pt/2018/05/10/mundo/erasmus-um-dos-principais-instrumentos-de-paz-na-europa/noticia/112913/

CPRPE. (2013). *International handbook on tourism and peace*. Klagenfurt/Celovec.

D'Amore, L. (2014). Peace through tourism: An historical and future perspective. In C. Wohlmuter & W. Wintersteiner (Eds.), *International handbook on tourism and peace* (pp. 355–370). Drava Verlag.

Dias, R. (2010). Um tema emergente nas relações internacionais: A paradiplomacia das cidades e municípios. Âmbito Jurídico, Rio Grande, 13(79).

Dias, R. V. A. (2013). *Conhecimento e interesse do estudante de bacharelado em Educação Física da UFSC em intercâmbio acadêmico internacional*.

Durkheim, E. (1978). *Educação e sociedade*. Melhoramentos.

Eck, N. J., & Waltman., L. (2019). VOSviewer Manual. *Manual for VOSviewer version 1.6.13*.

Educations.com. (2019). International higher education report: Student trends 2018. *Educations.com*. https://www.educations.com/higher-education-news/international-higher-education-report-18-13755

European Commission. (2015). *The European higher education area in 2015: Bologna process implementation report*. Author.

European Commission. (2018). *Erasmus+ annual report 2016*. Activity report, Brussels: EU Publications.

Felgueiras, A. C., Siqueira, A. S., Lima, J. D., & Oliveira, R. C. (2018). Educação em direitos humanos: História e conceitos. *Revista Augustus*, *23*(45), 47–70. doi:10.15202/1981896.2018v23n45p47

Fombona, J., Rodríguez, C., & Sevillano, Á. P. (2013). The motivational factor of Erasmus students at the university. *International Education Studies*, *6*(4), 1–9. doi:10.5539/ies.v6n4p1

Fuck, I. T. (2001). Alfabetização de adultos: Relato de uma experiência construtivista. Petrópolis: GE-EMPA.

GMDAC. (2018). *International students - IOM's Global Migration Data Analysis Centre.* Retrieved January 1, 2020, from: https://migrationdataportal.org/themes/international-students

GMDAC. (n.d.). *Migration data portal: International students.* Retrieved January 1, 2020, from: https://migrationdataportal.org/themes/international-students

González, R. E. D., Raesfeld, L., & Ponce, B. A. (2019). Interculturality: Contributions and proposals from an interdisciplinary focus in the institute of social sciences and humanities of the Autonomous University of the State of Hidalgo. *Universidad y Sociedad, 11*(1), 116–127.

Hall, E. T. (1959). *The Silent language*. Doubleday.

Hanushek, E. A., & Woessmann, L. (2008). The role of cognitive skills in economic development. *Journal of Economic Literature, 46*(3), 607–668. doi:10.1257/jel.46.3.607

Heleta, S., & Deardorff, D. K. (2017). The role of higher education instituitions in developing intercutural competence in peace-building in the aftermath of violent conflit. In D. K. Deardorff & L. A. Arasaratnam-Smith (Eds.), *Intercultural competence in higher education: international approaches, assessment and application*. Routledge. doi:10.4324/9781315529257-6

Ide, T., Kirchheimer, J., & Bentrovato, D. (2018). School textbooks, peace and conflict: An introduction. *Global Change, Peace & Security, 30*(3), 287–294. doi:10.1080/14781158.2018.1505717

Jafari, J. (Ed.). (2002). *Encyclopedia of tourism*. Routledge. doi:10.4324/9780203195673

Jafari, J., & Way, W. (1994). Multicultural strategics in tourism. *The Cornell Hotel and Restaurant Administration Quarterly, 35*(6), 72–79. doi:10.1177/001088049403500614

Jašková, M., & Heczková, L. (2010, July). ERASMUS mobility: Students motivation and satisfaction. *ERACON Congress*, 1-5.

Jesus, D. S. V. (2017). A arte do encontro: A paradiplomacia e a internacionalização das cidades criativas. *Revista de Sociologia e Politica, 25*(61), 51–76. doi:10.1590/1678-987317256104

Kasravi, J. (2009). *Factors influencing the decision to study abroad for students of color: Moving beyond the barriers* (PhD Thesis). University of Minnesota, Minnesota.

Kotite, P. (2012). Education for conflict prevention and peacebuilding. In *Meeting the global challenges of the 21st century*. International Institute for Educational Planning.

Krzaklewska, E., & Krupnik, S. (2006). *Research report. The experience of studying abroad for exchange students in Europe*. Brussels: Erasmus Student Network in cooperation with Petrus Communications.

Laplantine, F. (2000). *Aprender Antropologia*. Brasiliense.

Leimgruber, W. (2018). *Between global and local*. Routledge. doi:10.4324/9781351162722

Lemoine, P. A., Jenkins, W. M., & Richardson, M. D. (2017). Global higher education: Development and implications. *Journal of Education and Development, 1*(1), 58–71. doi:10.20849/jed.v1i1.253

Li, M., & Bray, M. (2007). Cross-border flows of students for higher education: Push–pull factors and motivations of mainland Chinese students in Hong Kong and Macau. *Higher Education*, *53*(6), 791–818. doi:10.100710734-005-5423-3

Lima, M. C. (2009). Mobilidade mundial de estudantes. *GV EXECUTIVO*, *8*(2), 66–69. doi:10.12660/gvexec.v8n2.2009.23837

Lima, M. C., & Riegel, V. (2015). Mobilidade acadêmica made in south: Refletindo sobre as motivações de estudantes brasileiros e colombianos. *Magis. Revista Internacional de Investigación en Educación*, *8*(16), 109–132. doi:10.11144/Javeriana.m8-16.mams

Makarova, E. A., Makarova, E. L., & Korsakova, T. V. (2019). The role of globalisation and integration in interdisciplinary research, culture and education development. *Journal of History Culture and Art Research*, *8*(1), 111–127. doi:10.7596/taksad.v8i1.1957

Malta, P. A., & Carneiro, M. J. (2005). Turismo e educação, mobilidades e ensino superior: Outras Teias de Globalização. *Revista Turismo & Desenvolvimento*, *3*, 115–128.

Marx, K. (n.d.). Crítica ao programa de Gotha. In K. Marx (Ed.), *ENGELS, Friedrich: Obras Escolhidas* (pp. 203–234). Alfa-Omega.

Mathews-Aydinli, J. (2016). *International education exchanges and intercultural understanding: Promoting peace and global relations*. Springer.

Mayor, F. (2010). Nutrindo uma cultura de paz. *Comitê Paulista para a década da cultura de paz: Um programa UNESCO: 2000-2010.*

Mendes, R., Pedradas, S., & Pereira, C. (2009). *Caracterização do Programa Erasmus: Análise das motivações, condições de frequência e expectativas dos estudantes.*

Mészáros, A. (2011). The motivation to study abroad: What motivates students to apply for Erasmus grant? Technical University of Cartagena, Department of Economics.

MGIEP. (2014). *Education for peace and sustainable development: Concepts, clarity and cohesion.* MGIEP.

MGIEP. (2017). *Rethinking schooling for the 21st century: The state of education for peace, sustainable development and global citizenship*. Author.

Miller, J. P., Nigh, K., Binder, M. J., Novak, B., & Crowell, S. (2018). Sustainable peace for the Planet. International Handbook of Holistic Education, 79-88.

Monteiro, A. (2014) *A mobilidade académica europeia e o turismo educativo e cultural: factores de decisão e motivação* (PhD thesis). University ofCoimbra, Coimbra.

Monteiro, A. (2017). *Turismo educativo e cultural: Factores de decisão e de motivação.* Alemanha: Novas Edições Académicas.

Monteiro, R. M. A. (2018). *Universidade do Porto: Fatores de decisão dos estudantes internacionais.*

Moreira, D., & Antão, G. (2019). Diversidade e mobilidade no ensino superior. In S. Gonçalves & J. J. Costa (Eds.), *Diversidade no ensino superior*. CINEP/IPC.

Morin, E. (2002). Cultura de massas no século XX: O espírito do tempo: neurose. *Forense Universitária*, 1-102.

Murphy-Lejeune, E. (2000). Mobilité internationale et adaptation interculturelle: Les étudiants voyageurs européens. *Recherche & Formation*, *33*(1), 11–26. doi:10.3406/refor.2000.1614

Nucci, L. (2001). Morality in context: Issues of development. In L. Nucci (Ed.), *Education in the moral domain*. Cambridge University Press. doi:10.1017/CBO9780511605987.006

Oliveira, A. L. D., & Freitas, M. E. D. (2016). Motivações para Mobilidade Acadêmica Internacional: A visão de alunos e professores universitários. *Educação em Revista*, *32*(3), 217–246. doi:10.1590/0102-4698148237

Oliveira, S. C. M., Barbosa, E. S., Rezende, I. C. C., Silva, R. P. A., & Albuquerque, L. S. (2013). Bibliometria em artigos de contabilidade aplicada ao setor público. Anais do Congresso Brasileiro de Custos-ABC.

Organização Mundial do Turismo. (2003). *Turismo internacional: Uma perspectiva global*. Bookman.

Osterwalder, K. (2001). Convention of the work of thematic groups: Convention of European higher education institutions. Salamanca, 29-30.

PaxLusófona. (2017). *Intra-África Pax Lusófona*. Retrieved January 3, 2020, from: http://www.intraafricapaxlusofona.org/index.php/apresentacao-do-projeto

Pinto, M. M., & Larrechea, E. M. (2018). Internacionalização da educação superior: Uma análise das tendências de mobilidade dos estudantes entre países do norte e do sul global. *Avaliação: Revista da Avaliação da Educação Superior*, *23*(3), 718–735. doi:10.15901414-40772018000300009

Prieto, N. C. (2004). *O outro lado do novo regionalismo pós-soviético e da Ásia-Pacífico: A diplomacia federativa além das fronteiras do mundo ocidental. A dimensão subnacional e as relações internacionais*. EDUC.

Proust, M. (2016). *A prisioneira: Em busca do tempo perdido* (Vol. 5). Relógio D'Água.

Richards, G., & Wilson, J. (2003). Today's youth travellers: Tomorrow's global nomads. *New Horizons in Independent Youth and Student Travel*, 1-43.

Richards, G., & Wilson, J. (2004). The international student travel market: Travelstyle, motivations, and activities. *Tourism Review International*, *8*(2), 57–67. doi:10.3727/1544272042782183

Ritchie, B. W., Carr, N., & Cooper, C. P. (2003). *Managing educational tourism* (Vol. 10). Channel View Publications. doi:10.21832/9781873150528

Salamanca Convention. (2001). *The Bologna Process and the European higher education area*. EUA Genève.

Salazar, N. B. (2006). Building a "culture of peace" through tourism: Reflexive and analytical notes and queries. *Universitas Humanistica, 62*, 319–336.

Santana, J. L. (2009). *O papel da paradiplomacia nas relações internacionais: A ascensão das unidades subnacionais num contexto mundial globalizado* (Dissertation). State University of Santa Cruz, Bahia, Brasil.

Schwartzman, S. (2018). A contribuição da educação para o progresso social. *Ciência & Trópico, 42*(1), 1–76.

Scott, P. (2015). Dynamics of academic mobility: Hegemonic internationalisation or fluid globalisation. *European Review (Chichester, England), 23*(1), 55–69. doi:10.1017/S1062798714000775

Senhoras, E. M. (2013). Múltiplas camadas das relações internacionais entre a diplomacia e a paradiplomacia. *Revista Intellector, 9*(18), 1–14.

Setton, M. D. G. J. (2005). A particularidade do processo de socialização contemporâneo. *Tempo Social, 17*(2), 335–350. doi:10.1590/S0103-20702005000200015

Shehi, R. Z., Ozcan, S., & Hagen, T. (2018). The role of higher education institutions in building a culture of peace: An Albanian case. *Journal of Peacebuilding & Development, 13*(1), 46–61. doi:10.1080/15423166.2018.1427136

Silva, V. A. D., & Rebolo, F. (2017). A educação intercultural e os desafios para a escola e para o professor. *Interações (Campo Grande), 18*(1), 179–190. doi:10.20435/1984-042X-2017-v.18-n.1(14)

Sousa, R. R. P. (2017). Relações internacionais e o estudo da paz e conflito: O contexto da resolução de conflitos. *Incursões na Teoria da Resolução de Conflitos*, 49-98.

Strandberg, C. (2017). *Maximising the capacities of advanced education institutions to build social infrastructure for canadian communities*. JW McConnell Family Foundation.

Teichler, U., & Jahr, V. (2001). Mobility during the course of study and after graduation. *European Journal of Education, 36*(4), 443–458. doi:10.1111/1467-3435.00081

UNESCO. (1998). *Declaração mundial sobre educação superior no século XXI: Visão e ação, marco referencial de ação prioritária para a mudança e o desenvolvimento da educação*. Universidade Metodista de Piracicaba.

UNESCO. (2004). Embracing diversity: toolkit for creating inclusive, learning-friendly environments. *UNESCO Digital Library website.* Retrieved February 1, 2020, from: https://unesdoc.unesco.org/ark:/48223/pf0000137522

UNESCO. (2017). *Global flow of tertiary-level students* Retrieved January 3, 2020, from: http://uis.unesco.org/en/uis-student-flow#slideoutmenu

United Nations. (2015). *17 Goals to transform our world.* Retrieved December 21, 2019, from: https://www.un.org/sustainabledevelopment/education/

Van Den Berg, L., & Russo, A. (2017). *The student city: Strategic planning for student communities in EU cities*. Routledge.

van't Klooster, E., Van Wijk, J., Go, F., & Van Rekom, J. (2008). Educational travel: The overseas internship. *Annals of Tourism Research*, *35*(3), 690–711.

ENDNOTE

1 The search was completed on matters available in journals/proceedings indexed in the Clarivate Analytics Web of Science Core Collection database (www.webofknowledge.com).

Chapter 17
Genetic Tourism:
Key to Multicultural Understanding and Peacebuilding?

Carla Villagra Álvarez
Laboratory of Multidisciplinary Research in Tourism (LMRT), University of Girona, Spain

Neus Crous-Costa
Laboratory of Multidisciplinary Research in Tourism (LMRT), University of Girona, Spain

ABSTRACT

This chapter introduces the birth of a new tourism modality, genetic tourism, by comparing it with genealogical tourism. Once this comparison was carried out, it was possible to determine the characteristics of genetic tourism and to study its possible positive effects concerning cultural sensitivity and transculturality, related to multicultural understanding and peacebuilding. The exploratory study was done by analysing travel blogs and online news related to genetic tourism, which results were used to identify a series of topics related to cultural sensitivity and transculturality. Among these topics, the author highlights the predisposition of the studied subjects to define themselves as citizens of the world, enabling people-to-people interactions and the belief of a positive relationship between genetic tourism, tolerance, cultural ethnorelativism, and reductions of negative stereotypes.

INTRODUCTION

The chapter presents the birth of a new tourism modality, genetic tourism. This type of tourism is based on a trip created thanks to the results of genetic tests that determine the ethnic origin of the subjects and that nowadays are carried out by companies such as MyHeritage, 23andMe, 24Genetics or OneFam.

As a consequence of genetic tourism's short life cycle, the author proposes a comparison between this type of tourism and genealogical tourism, a type of tourism widely studied by the academic community. The comparison aims to construct a theoretical framework which contains the typical characteristics of genetic tourism, which will allow studying the foundations of the apparent positive relationship between genetic tourism, cultural sensitivity and transculturality.

DOI: 10.4018/978-1-7998-5053-3.ch017

For the creation of the conceptual framework concerning genetic tourism, it was first necessary to present the characteristics of genealogical tourism, with the intention of the subsequent comparison between the two mentioned tourism modalities and the detection of genetic tourism's characteristics. The study of the relationship between genetic tourism, cultural sensitivity and transculturality includes the definition of the aforementioned cultural concepts and the connections between them and genetic tourism.

The study of connections between genetic tourism, cultural sensitivity and transculturality has been continued in an analysis of travel blogs and online news explaining the experience of subjects who have previously undergone a genetic test. From this analysis, issues related to the cultural concepts found in this study have been detected, such as an increase in tolerance and cultural ethnorelativism and a reduction in negative stereotypes, thus expanding the studies related to tourism typologies, multicultural understanding and world peace consolidation.

Through the analysis of a number of personal experiences, this chapter contributes to the advancement of knowledge in the domain of travel and tourism by introducing a new tourism modality: genetic tourism, which has become possible thanks to technological and scientific advances of recent years and to demonstrate its positive relationship with cultural sensitivity and transculturality, and by default, with the enhancement of multicultural understanding and the consolidation of world peace. As it will be argued, genetic analysis, which is seen as a type of reliable proof of identity, contribute to expand taker's cultural horizons and to foster, given individual capacity of expenditure, travel to regions that were not always considered previously. In this regard, genetic analysis foster intercultural contact through tourism, even when it becomes hard for the tourism industry (especially DMOs) to capitalise its potential.

METHODOLOGY

This study is based on a qualitative perspective: the analysis of travel blogs to identify the most recurrent topics in genetic tourism. Thus, the authors conducted a thematic analysis to describe the stages and implications in the pre-trip stage: from the moment participants chose to get tested to the moment they receive the results.

Blog analysis presents some noticeable limitations. Researchers do not have the opportunity to ask for clarification or to dig deeper into specific information. Yet, it also offers a clear advantage: bias is reduced, precisely because the writings reflect what was most important to each individual.

Travel blogs containing posts about genetic tourism experiences were looked for using standard Internet search engines. The selection was narrowed using the following keywords: travel blog DNA (in English, Spanish and Catalan), the DNA journey blog, DNA ancestors blog, travel blog the DNA journey (in English, Spanish and Catalan). All queries were carried out in three languages: English (as a *lingua franca*), Spanish (as another of the most widely spoken languages on the Internet) and Catalan (as a mother tongue of the researchers). Saturation of information was the criterium used to stop adding more blogs to the pool of data.

BACKGROUND

Genealogy tourism can be recognised in a wide variety of terms: ancestral tourism, roots tourism, legacy tourism, family or diaspora tourism (Alexander, Bryce, & Murdy, 2017); but each and all of these may be defined as a branch of cultural tourism which aims to enable people to learn more about the culture of their ancestors (Russell, 2008) and which responds to our human desire to discover our own identity, a concern that has been increasing in line with globalisation and the attendant international movements of people (Meethan, 2004).

The appeal of this type of tourism has to do with the personalisation of history as being intimately linked to our life experience and that of our ancestors (Alexander et al., 2017), all of which is perceived as shaping our identity by way of the biographical stories of those who went before us (Meethan, 2004). Thanks to the connections thus established between human history and personal history, "few tourism experiences may be more emotional and nostalgic than those involving tourists' own personal connection to their heritage" (Ray & McCain, 2012, p. 982). The powerful emotional charge of this form of tourism is reflected in the fact that genealogy is now one of the most popular hobbies in the United States (Josiam & Frazier, 2008), a pastime characterised by a high level of personalisation, prolonged length of stay and the forging of deep connections between individuals (Alexander et al., 2017).

Genealogy tourism is born of our human need to recognise a given geographical space as "home"; a need that is found in all of the world's cultures and responds to people's curiosity about and wish to recognise their identity and that of their own kind (Meethan, 2004). Genealogy tourism is also linked to the need for answers to questions such as "Who am I? Where did I come from? or Who were my ancestors?" (Josiam and Frazier, 2008, p. 51). In addition, this type of tourism is related to the search for personal identity, the desire for connection with a specific geographical space, the return to "home", an intellectual challenge, the closure of a stage in life, a mission, an obligation to one's ancestors, a way of closing a circle, and a sacred journey or journey of the community (Ray & McCain, 2009).

The existence of countries with a major migratory tradition, such as Italy, the United Kingdom, Ireland, India or China, makes genealogy tourism a great opportunity for the tourism industry (Alexander et al., 2017) as a market capable of generating a considerable volume of revenue (Ray & McCain, 2009). Indeed, India is set to become one of the countries most deeply involved in genealogy thanks to a government programme to assist those who wish to do so in tracing their genealogical roots (Newland & Taylor, 2010).

However, even though certain countries have a significant migratory tradition and the launch of government programmes to facilitate genealogy research, this type of tourism has yet to attract more than a small number of participants (Alexander *et al.*, 2017). One explanation for this is the difficulty of obtaining accurate information, which Meethan (2004) illustrates as follows: "genealogists have to rely in great part on the construction of a logical and consistent pattern of events from an often incomplete and fragmentary record" (p. 142).

The methodology employed by participants in genealogy tourism can be divided into a public component, which is based on the bureaucratic systems of institutions, and a private component, which draws on the living memory of the participant's family (Meethan, 2004). Experts point out that the official archive documents provided by public bodies and institutions may contain errors and that uncritical trust in such information is likely to be counterproductive (Alexander et al., 2017). In addition, these institutions usually charge a fee and restrict access to their databases to a specific time, to some extent hindering the search for information by making it expensive and time-consuming (Alexander et al., 2017). As a result, the practice of this type of tourism is perceived as being confined to individuals with a high level of purchasing power (Ray & McCain, 2009). While it may be the case that the advent of the Internet and new technologies have made it easier to obtain relevant data, such information as is available is by no means complete and often difficult to access (Meethan, 2004). In light of the obstacles to obtaining accurate information through such channels, one of the reasons for participating in genealogy tourism not mentioned above is, in many cases, to look for information in archives and other sources far from the participant's place of residence (Meethan, 2004).

The private component of the methodology, based on the living memory of the family, is also unlikely to provide much relevant information, as memories rarely go back more than a couple of generations (Meethan, 2004). It should also be borne in mind that not all of us have extensive detailed knowledge about our ancestors and our family history, making it difficult for us to research our genealogy by way of family memory alone (Ray & McCain, 2012).

Genealogy tourism is characterised by the tendency to visit places which are not conventional tourist destinations (Alexander *et al.*, 2017), with the priority shifting from places of tourist importance to those of personal importance. Given this concentration on sites of personal importance, genealogy tourists are more likely to spend on small businesses and thus benefit the local community (Newland & Taylor, 2010) in their search for an "authentic experience of the local way of living" (p. 5).

This type of tourism is related to the incorporation of a personal narrative into a narrative that was previously felt to be alien and tends to result in a reduction in the person's scepticism and in their rejection of certain cultures (Russell, 2008) and an increase in intercultural tolerance: "many tourists learn to understand history through their own family history" (Ray & McCain, 2009, p. 2).

With regard to the presentation of the cultural aspects discussed in this chapter, Oberg (1960) argues that culture can be defined as a social construct that envelops us as individuals in every aspect of our lives. It can be said that "one's culture imperceptibly forms a mental framework through which individuals define their ontology, motivate and select their behaviours, and judge and evaluate the actions of others" (Sussman, 2000, p. 356). It is at this point that the concept of ethnocentrism, understood as the tendency of individuals to "[accept] their culture as both the best and the only way of doing things", comes to the fore (Oberg, 1960, p. 144). D'Antoni (2010) criticises ethnocentrism on the grounds that the concept of culture only makes sense "in a context of interculturality" (p. 3) in which we recognise our cultural identity through contact with other cultural realities. Contact between cultural realities requires "culture to be analysed no longer as a static element, existing within any society or group of individuals, but to consider it as an active element describing interaction between different societies" (Lingeberzins, 2011, p. 72).

In a globalised environment in which contact between cultures is the order of the day, the ability to adapt to new cultural realities presents itself as a key factor of coexistence between individuals with different cultural backgrounds. Anderson (1994) defines the concept of cultural adaptation as: "a commonplace process of learning to live with change and difference – in this instance, a changed environment and different people, different norms, different standards, and different customs" (p. 299). In order to achieve successful cultural adaptation, a great deal of importance attaches to cultural relativism, defined as the capacity of individuals to set aside the primary reactions of their culture of origin in order to accept and to be more tolerant of the traditions, behaviours and beliefs of another culture (Winkelman, 1994).

At the same time, cultural sensitivity and intelligence have a directly proportional relationship to the ease of adaptation of the individual: the greater the degree of cultural intelligence, the greater the ease of adaptation, and the lower the degree of cultural intelligence, the less ease of adaptation (Presbitero, 2016; D'Souza, Singaraju, Halimi, & Sillivan, 2016). Earley and Ang (2003) continue this line of reflection and define cultural intelligence as the capacity that facilitates adaptation in spaces of cultural exchange, emphasising the importance of an active attitude, which enhances cultural adaptation, for the ability to enter a new culture quickly and efficiently (Belford, 2017). Sussman (2000) detects two additional variables that influence the individual's ability to adapt to culture: cultural centrality, understood as the rooting of the subject in relation to his or her cultural reality, and cultural flexibility, understood as the ability to model one's own cultural reality to adapt to another cultural reality.

The aforementioned cultural concepts are of key importance in relation to multicultural understanding and world peace. However, Welsch (1999) argues that understanding and peace will only be possible if society abandons the emphasis on interculturality and multiculturalism, which he defines as concepts that can be visualised as spheres and which therefore "[...] can, according to the logic of this conception, do nothing other than ccolliding with one another" (p. 196). Mathé (2018) argues that the concept of interculturality does not facilitate multicultural understanding or the consolidation of world peace, in that, while recognising the existence of diverse cultural realities, it does not promote deep contact between them. Welsch (1999) and Mathé (2018) have similar views of interculturalism but diverge in their views on multiculturalism. Mathé (2018) defines multiculturalism as "[...] integrated, deep cooperation with other cultures, synthesis and empathy, with reflexive experiences in performing activities" (p. 42); the stage in which society arrives at multicultural understanding and the consolidation of world peace. For Welsch (1999), the concept of transculturality is the optimum stage for achieving these goals in so far as it represents cultures as networks which communicate with and nourish one another in a symbiotic manner, while at the same time manifesting their similarities and differences.

Contact and knowledge between cultures are presented as "a necessary step to get to recognise the nature of these differences and to relate to others through attitudes of understanding and tolerance which are premises for genuine intercultural dialogue, especially in the tourism industry" (Albu, 2015, p. 7). Tourism is the industry most exposed to cultural contacts, and for this reason is a good field in which to work for the achievement of intercultural harmony, multicultural understanding, tolerance and the consolidation of world peace (Ward, Furnham and Bochner, 2005). Albu (2015) also highlights the importance of the tourism industry in terms of inter-cultural contact and sees it as contributing to the reduction of negative stereotypes and the improvement of international cooperation.

GENETIC TOURISM

The construction of the theoretical framework in relation to genetic tourism is realised through a comparison of this type of tourism with genealogy tourism. In terms of similarities, genetic tourism shares Russell's (2008) definition of genealogy tourism as a branch of cultural tourism which seeks to learn about the culture of the individual's ancestors. It also responds to our desire to find out more about our identity (Meethan, 2004).

Genetic tourism has the potential to shift the focus away from history as an external, alien concept towards a personal concept that is intimately linked to our life experience as individuals and that of our ancestors (Alexander *et al.*, 2017). It also has a high emotional charge thanks to the association of lived experience and the presence of personal bonds (Ray and McCain, 2012), which tends to foster the creation of deep connections between individuals and is reflected in a high degree of personalisation of the tourism phenomenon (Alexander *et al.*, 2017).

Genetic tourism shares with genealogical tourism a concern with our human need to recognise a geographical space as "home"; a need that responds to the curiosity of human beings to recognise our identity and that of our own kind (Meethan, 2004). It is also linked to the wish to provide answers to questions such as "Who am I? Where did I connection with the search for personal identity, the desire for a bond with a specific geographical space, a return to "home", an intellectual challenge, the closure of a stage or phase, a mission or an obligation to one's ancestors, a sense of coming full circle, and a pilgrimage or journey of community (Ray & McCain, 2009).

The existence of countries with a strong migratory tradition (Alexander *et al.*, 2017) must be seen as a business opportunity for genetic tourism, promising a market capable of generating a considerable volume of revenue (Ray & McCain, 2009).

Here, too, locations which are not conventional tourist destinations continue to gain ground (Alexander et al., 2017), as places of personal importance are ranked higher on the preference scale than places of tourist importance, and it is also the case that genetic tourists are more likely to spend on businesses run by the local community (Newland & Taylor, 2010) and on the search for "an authentic experience of the local way of living" (p. 5).

Genetic tourism also has in common with genealogical tourism the incorporation of a personal narrative into a narrative previously felt to be alien and tends to result in a reduction in scepticism and in the rejection of certain cultures (Russell, 2008) and an increase in intercultural tolerance: "many tourists learn to understand history through their own family history" (Ray & McCain, 2009, p. 2).

With regard to the dissimilarities between genetic tourism and genealogy tourism, it is worth noting the different methodologies involved in participation in one and the other. Genetic tourism is underpinned by DNA analysis of saliva samples to reveal the ancestral origin of the individual, whereas genealogy tourism is based on public-private research in institutions and on the testimony of relatives (Meethan, 2004). As a result, genetic tourism does not suffer from the difficulty of obtaining information that besets genealogy tourism (Meethan, 2004), with a concomitant rise in the number of participants in this form of tourism (Alexander et al., 2017). Furthermore, the elimination of research processes in institutions makes genetic tourism less expensive and less time-consuming (Alexander et al., 2017), thereby extending the potential market for this type of tourism to individuals with an average level of purchasing power (Ray & McCain, 2009).

Genetic tourism affords a new perspective on the need to identify a geographical space as 'home' (Meethan, 2004): while genealogical tourism tends to focus on one particular geographical location, genetic tourism is capable of mapping onto several geographical spaces at once thanks to its distinctive methodology. Individuals who do not possess extensive knowledge about their ancestors and their family history (Ray and McCain, 2012) are likely to perceive genetic tourism as an attractive alternative to genealogy tourism, in that it is not subject to the limitations of a fragmented family history. In addition, since the purpose of the trip is not to search for information in archives or institutions far from the traveller's place of residence (Meethan, 2004) but to engage with their ancestral origins, the stay is unlikely to be as prolonged as in the case of genealogy tourism (Alexander et al., 2017), because there is no need to spend time searching out data in bureaucratic systems.

In short, genetic tourism and genealogical, heritage or family tourism are part of what experts classify as cultural tourism. Both have high emotional baggage due to the presence of personal bonds and both share the same motivations (such as the search for personal identity, to recognise a geographical space as "home" or a mission to one's ancestors). Regarding the differences, methodologies used should be remarked: while genetic tourism is based on the collection of samples of genetic material; genealogical, family or heritage tourism needs previous historical and family research.

GENETIC TOURISM, MULTICULTURAL UNDERSTANDING AND PEACEBUILDING

The methodology employed in the practice of genetic tourism is likely to facilitate the discovery of unexpected ethnic connections that may lead those who engage in it to reflect more deeply on their cultural identity, so that this type of tourism is perceived as tending to reduce cultural ethnocentrism (Oberg, 1960; D'Antoni, 2010) and enhance cultural relativism (Winkelman, 1994). At the same time, genetic tourism ties in with the vision of cultural dynamism outlined by Lingeberzins (2011) and draws attention to the interactions between different cultural realities that have taken place again and again over the course of history.

Genetic tourism is associated with a process of enhancing the subject's capacity for cultural adaptation (Anderson, 1994; Winkelman, 1994) through the promotion of an active attitude (Belford, 2017) thanks to the high degree of personalisation of the experience (Alexander et al., 2017), which facilitates his or her introduction into and engagement with the new cultural reality. The enhancement of the capacity for cultural adaptation through genetic tourism also has a positive effect on the level of cultural intelligence and sensitivity (Presbitero, 2016; D'Souza, Singaraju, Halimi, & Sillivan Mort, 2016; Earley & Ang, 2003) due to the directly proportional relationship between the concepts. As a result, this type of tourism facilitates the adoption of a position characterised by a low degree of cultural centrality and a high degree of cultural flexibility (Sussman, 2000).

Mathé's (2018) model of multiculturalism and Welsch's (1999) model of transculturality both relate to genetic tourism, in so far as this form of tourism is highly compatible with the view of cultures as interrelated networks and can be seen as an active step towards multicultural understanding and the consolidation of world peace. Thus, all the cultural aspects mentioned in relation to genetic tourism have a direct bearing on the improvement of intercultural dialogue (Albu, 2015) in the tourism context and tend to propitiate harmony between cultures, tolerance, the reduction of negative stereotyping and an improvement in international cooperation (Albu, 2015; Ward, Furnham, & Bochner, 2005; Winkelman, 1994; Ray & McCain, 2009, Russell, 2008).

ANALYSIS OF TRAVEL BLOGS AND ONLINE NEWS

The exploratory study in this chapter takes the form of an analysis of travel blogs and online news related to genetic tourism, and more specifically with the personal experience of certain individuals at the moment of taking an ancestral origin test. Both travel blogs and online news are considered primary sources, and analysis has focused exclusively on the first-person accounts of the protagonists.

The identification of travel blogs and online news took the form of a Google search using the following keywords: "blog de viajes adn", "travel blog dna", "el viaje del adn blog", "the dna journey blog", "adn ancestros blog", "test adn blog de viajes", "travel blog the dna journey", "blog de viajes el viaje del adn" and "blog de voyage adn". The analysis was carried over a period of one week and continued until the data was saturated.

The purpose of this analysis was to expand the conceptual framework of genetic tourism as a new category and to detect recurring themes in relation to this type of tourism as such and in relation to the cultural aspects discussed in this chapter: cultural sensitivity, transculturality, multicultural understanding and the consolidation of world peace.

The author goes on to present the fourteen themes identified in order of frequency, from highest to lowest, together with a brief description of each and quoted comments on the theme in question from the protagonists of the experience. First of all, most subjects expressed feeling a greater or lesser degree of surprise at the test results:

"This test came as a huge surprise to me because I had thought that all my ancestors were from Spain, but my haplogroup – a group of people with similar genetic profiles who share a common ancestor – turns out to be in southern Sweden, encompassing 40% of the current population" (*El Confidencial*, 2017).

The surprise and the uncertainty occasioned by unexpected results caused the individuals studied to wonder where the test data come from: "Because, okay, we were not surprised to find out that I am, ethnically, 54% from the Iberian Peninsula and 10% from Eastern Europe. But where does that 16% of British blood come from?" (*Mi Mundo Sin Fronteras*, 2017). In many cases these feelings of surprise led the subjects to share the results with family members to shed light on the findings: "Great Britain? I mean, no one in my family knows anything about having relatives from there, and 11% is actually quite much" (*Swedish Nomad*, 2017). Furthermore, subjects tended not only to feel surprised at their results, but also appreciated that this response is common to most individuals who take the test and discover unexpected ancestral genetic linkages: "Though the DNA test confirmed a lot of what I already suspected about my heritage, I did – like many people – find some surprises" (*Around The World "L"*, 2017). However, despite the unexpected results, it is noteworthy that the surprises were mostly produced by a small percentage of the individual's total genetic heritage: "Needless to say, I flipped a little with the last three ethnicities, which are the ones least present in my DNA, but that doesn't make them any less surprising." (*Viaja Con Libertad*, s.d.).

Second, the test results generally prompt the individual to want to travel to the geographical regions they indicate; in other words, they play a significant part in determining the destination for the next trip: "Has this inspired us to travel more? Definitely: it has not only fueled us with a great deal of wanderlust to see more of the Middle East and Europe [...]" (*Nomadic Boys*, 2018). The test not only encourages the subject to travel, but introduces into their list of preferences destinations they had overlooked or looked down on in the past – "I'm excited to have a clearer picture of my lineage and have more motivation to travel to certain regions" (*Y Travel Blog*, 2018) – as well as destinations previously deferred for some indefinite future that are now filled with content by the test results:

"It goes without saying that now I have one more reason to visit all the mentioned places. A trip to the Balkans, for example, has always been high on my list. But now it got even more meaning [...] Do you see what I mean?" (*Travel Reportage*, 2017).

Genetic tourism also presents as a multifaceted tool, in that it motivates travel not only to destinations never previously visited – "I'm pretty sure if some country or region that I have not visited before shows up strongly then I may just have to pack my bags again" (*J 5 Travel*, 2018) – but also trips to destinations already visited, with the return visit to the destination being approached from a different perspective:

"And the funny thing is that before knowing the results I had already planned another trip there some time soon, so now I'll be going there with all the more reason, with other sensations and knowing that a part of the place lives in me" (*Dónde Te Metes*, 2017).

One of the distinguishing features of this form of tourism is that the trip is perceived as a special experience – "My curiosity has marked out my most special journey, the one that charts its course through the story of my life" (*A Tomar Por Mundo*, 2018) – and this sets genetic tourism apart from over other types of tourism. The subjects also point out that the meaning of the experience is richer for those who consider themselves travellers: "Knowing where you come from takes on an even greater significance, especially when you travel the world and feel that you have an affinity for a certain place without knowing why" (*Dónde Te Metes*, 2017).

Third, the experience is clearly related to a powerful emotional charge, the high point of which is the moment when the results are received: "The feeling that overwhelmed me must be similar to that of a someone who needs to know more about the father they never met or that of a father who wants to find out about the son or daughter he never knew" (*A Tomar Por Mundo*, 2018). Indeed, the subjects experience this powerful emotional charge even if they already had a certain amount of information about the origins of their ancestors: "After listening to family stories for years, I thought I had a pretty good idea of what the test would say. But actually seeing my DNA results for the first time was still thrilling …" (*Expert Vagabond*, 2018).

Fourth, the subjects report feeling nervous both in the process of waiting for the results – "And wait nervously for a couple of weeks, of course!" (*The Travel Tester*, 2016) – and at the moment of receiving the findings: "In all honesty, I was terrified opening the results. I have no idea why, and I hadn't expected to be so nervous, but suddenly all kind of thoughts shot through my mind" (*One Step 4 Ward*, 2017). In some cases, this nervousness is transmitted to other family members when the protagonists decide to share their experience with them: "Everyone in my house was really excited and intrigued and anxious to know the results" (*World Travel Feet*, 2016).

Fifth, there is a widespread tendency to make assumptions about the geographical areas that will appear in the test at the time of waiting for the results:

"They're going to tell me that I'm Iberian (Spanish/Portuguese) and Northern European (Scandinavian), besides having something from North Africa. Why? On my mother's side, they are very tall, with straight hair, and my grandmother is blonde with blue eyes. On my father's side, I'm in no doubt that there has to be a touch of Morocco or some neighbouring country: tall, dark, with curly black hair and big eyes" (*Una Vida Por Viajar*, s.d.).

Once the waiting is over and they have the results in their hands, subjects tend to claim that the assumptions they made before knowing their ancestral data were accurate: "I confess that some answers I already suspected" (*João Cajuda*, 2016).

Sixth, it is evident that one of the most frequent motives for taking the test is to gain fuller knowledge about oneself and to reflect on one's own identity:

"It's amazing how a simple saliva analysis makes you reappraise at the deepest level who you are, the sum of thousands of people who have existed on this planet, thousands of lives and stories we will never know what they were like, but thanks to their existence we are here" (*World Travel Feet*, 2016).

Seventh, the subjects' high degree of curiosity about genetics is clearly revealed, irrespective of whether the individual in question had a prior interest in these matters – "I've always been curious about my ancestry" (*Expert Vagabond*, 2018) – or their curiosity was first aroused when they heard about the test: "I started to become curious about what my DNA could tell me" (*Y Travel Blog*, 2018). As a rule, curiosity does not disappear when the results are obtained but is actually fuelled by the findings, leading the individual to come up with new questions that further stimulate their curiosity: "I enjoyed learning about my DNA because it has fed my curiosity even more" (*Y Travel Blog*, 2018). This renewal of curiosity causes some individuals to consider asking other family members to take the test in order to continue to explore the subject: "If I can get my parents or my grandmother to do it, I'll find out even more about my ancestors" (*El País*, 2017). Some subjects associate their curiosity about the ancestry test with the scant information about their family history and the difficulties of genealogy: "It was impossible for me to draw up a family tree and I had only managed to go back as far as my grandparents'

grandparents, just a century ago in the history of my family, with a very high degree of endogamy" (*La Cosmopolilla*, 2017). At the same time, the scarcity of information leads individuals to fill in gaps in the past with imagined family histories that are not necessarily true:

"The uncertainty has haunted me all my life. I always saw a long way beyond the photos of the first trips stuck in the family album and the drawer full of old scraps my grandmother kept in a rickety old cupboard in a beautiful courtyard full of flowers. I knew that behind the semblance of normality there were a lot of stories to discover [...] and I had long wanted to know the story of my life" (*A Tomar Por Mundo*, 2018).

Eighth, the subjects tend to believe that a person's identity is not delimited by the borders within which they were born or lived most of their life, but includes all the genetic information that appears in the test results; and this leads them to define themselves as citizens of the world:

"The conclusion, for me, is clear. Maybe also because I see what I want to see. But it's obvious that we are citizens of the world, that migration makes more sense than flags and borders, that racism is utterly ridiculous and that, in fact, human race is a single race" (*Mi Mundo Sin Fronteras*, 2017).

The test is also understood as a tool to overcome intolerance, revealing as it does the genetic connections that exist between individuals who apparently share no traits in common – "It is important that we all be more tolerant with each other and get the message: It is much more what united us than what keep us apart!" (*João Cajuda*, 2016) – and emphasising the concept of diversity as something positive that enriches us by casting off stereotypes:

"Thanks to this daring I now believe more than ever in the different. I want to be an ambassador for the world of your diversity. I'm ready to join your fight against prejudice. To discover new horizons. To meet the thousands of good people that populate the Earth and make it a little better. To live. To feel. To travel" (*A Tomar Por Mundo*, 2018).

Ninth, a number of the subjects are evidently aware of taking part in a new and increasingly democratised tendency characteristic of a globalised world: "I do believe, thanks to tools like this, travel based upon following your DNA and discovering more of who you are, is becoming an emerging trend" (*Y Travel Blog*, 2018). One major difference between this and other trends is that it is regarded as deeply meaningful, having none of the frivolity of other modes: "Of the trends we've seen, DNA travel is the most meaningful and satisfying of all" (*Classic Journeys*, 2018).

Tenth, the subjects define the test as a unique opportunity that ought not to be missed by anyone who really wants to know where they come from: "Destiny had placed a unique opportunity in front of me. I was going to be able to answer a question I had been asking myself for a long time" (*A Tomar Por Mundo*, 2018).

Eleventh, some individuals claim to have felt a prior affinity for destinations that appear in their test results: "And 3% from Western Asia! Turkey and Persia (present-day Iran), the first the other country that fascinates me most, with Istanbul being one of my top ten favourite cities, and the second a destination that appeals to me more and more ..." (*La Cosmopolilla*, 2017). This is reflected in the fact that some people wonder about the extent to which the presence in their DNA of characteristics from other geographical areas influences their perception of these regions: "How much of my DNA had already influenced my travels subconsciously? You know how sometimes you feel called to visit a certain place, or when you arrive you feel as if you've been there before?" (*Y Travel Blog*, 2018).

Twelfth, the analysis shows that subjects perceive the diversity in the test results as a positive factor that makes them more interesting as individuals – "Now, married to a Spanish man and having two children, we are making the DNA mix even more interesting" (*Menudos Viajeros*, 2017) – whereas a low level of diversity is perceived as a negative aspect, associated with dullness and a far from noteworthy genetic past: “All my ancestors are from the south of the province of Granada: the valley of Lecrín, the Alpujarras and the Costa Tropical. How boring!” (*La Cosmopolilla*, 2017).

Thirteenth, some individuals place considerable emphasis on the low level of specification in the results for certain geographical areas and attribute this to the fact that the database with which their genetic material is being compared is small, making it difficult to obtain the desired specificity:

“Unfortunately, Ancestry’s Asian database isn’t very exhaustive, so the result of my test was just that I am 100% East Asian. 100% East Asian. I kind of already guessed that and I was, therefore, a little disappointed when I received my results” (*Mitziemee*, 2017).

And, fourteenth, one of the subjects’ most frequent concerns is that the past may be forgotten, with the test being appreciated as a tool with which to keep the past alive from generation to generation:

“The uncertainty that dies away beyond the predictable ‘Italy and Spain’, for the possibility of unveiling the mystery, at how unfair and how stupid it has always seemed to me the thoughtlessness of letting the past die of neglect” (*Los Viajes de Nena*, 2017).

The subjects’ concern that the past should not be forgotten is also apparent in their wish to give their ancestors the place they deserve in their family history:

“That I was seeing that landscape that my grandpa Pepe had had to turn his back on, that with my arrival in those streets all the dreams he must have had were coming home. No one understood it in that way, but I felt I was connecting with my great-grandfather whom I never knew, with my genes, with my blood” (Los Viajes de Nena, 2017).

DISCUSSION

After presenting the analysis of travel blogs and online news and having identified the topics mentioned in the previous section, the author defines a clear pattern of behaviour that has been presented in four differentiated phases, and they refer to the perceptions of the subjects studied in the pre-mediation stage.

The first phase is the moment when the individuals plan to take the test. Some are attracted to this experience in order to satisfy their curiosity, while others admit to having had a prior interest in genetics for some time, which is evidenced by attitudes such as recurring thoughts about ancestors or inquiring about family members. Subjects who want to satisfy their curiosity are characterised by a low prior interest in the subject of genetics, an interest that begins to emerge when they are offered to take the test. Subjects with a high prior interest in genetics are characterised by being subjects who have already experimented with genetics in the past, or by subjects which have always been interested in genetics. They all agree on the idea that the test offers a unique opportunity, an exciting experience or a need to meet.

The second phase refers to the moment when the individuals decide to take the test and in which they passively wait for the results. It is a phase in which the subjects’ nervousness is emphasised, which is alleviated in two different ways: by searching for genetic information in family members or by occupying their time with other activities. All individuals associate results with low genetic diversity to failure, and results with high diversity to success. It is in this wait that the subjects generate lines of thought about the results that will appear in the test.

The third phase is characterised by the discovery of results and, therefore, it is a phase full of emotions in which individuals may find themselves in two situations that are not necessarily mutually exclusive. The first situation occurs when the subject is surprised by the results and the second when the subject is not surprised by the results. The two situations are broken down into two attitudes that are once again not mutually exclusive: the affinity that some emphasise between the destinations they are attracted to and the results and the creation of imaginary discourse in relation to their ancestors. Most claim to have met their need, but others claim to have generated even more curiosity about genetics. Some individuals do not feel their need to be satisfied as a result of a lack of specification in the results, which is due to a poor database.

Finally, the fourth phase refers to the conclusion that individuals draw from the experience. The subjects conclude that it is necessary to travel to the destinations that appear in the results even if they have been previously visited, as the trip is perceived as an experience that will be lived from a different perspective. Whether new destinations were recurrent in the individual's thinking in the past or not, a desire to visit them now appears. That is, in the case that the individual already had in mind to visit the destination, the genetic test is presented as a decisive tool (a key decision-making factor) in carrying out those journeys that are delayed in time. The test is also perceived as a tool against intolerance that brings individuals closer to those who were previously alien to them. As it has been argued at the beginning of this text, this is a key factor for tourism as a way for multicultural understanding and world peace.

As a personal experience, it has been said that related tourism niches are linked to the human need of perceiving geographical space as "home", this need is found all around the world's cultures due to a human need to recognise their identity and that of their own kind. The particular appeal of genetic tourism lies in the reliability of the test's results and the personalisation of history, which is inextricably linked to life experience and identity. Results of blog analysis show how deeply emotional genetic discovery is, featuring a particular (and almost accidental) methodological strength: stories were spontaneously shared, thus eliminating the bias that may have been introduced by the researcher.

Beyond the issue of intercultural understanding, which may be aroused in individuals that were not inclined to it previously to taking the test, genetic tourism can be linked to other topics that nourish tourism academic theories. Although it extends beyond the scope of the present text, the implications of genetic tourism described clearly link travelling to self-knowledge and the development of one's identity. The construction of the self has been lightly touched by tourism research so far, especially from the perspective of the traveller, probably due to the difficulties of getting comparable data of such deep, transformative experiences by using the methods usually employed in tourism research. Some contributions to this have been made in relation to specific niches, such as voluntourism (Uriely, Reichel and Ron, 2002) and creative tourism (Richards and Raymond, 2000). Broader approaches to this personal growth through travel have been undertaken by authors like McWha, Frost and Laing (2018) or Beerli, Díaz Meneses and Moreno Gil (2007). The study of this re-creation of the self by means of genetic-related destinations may shed some light on new, deeper issues connected not only to familiar oral traditions (as it is the case of ancestral tourism) but also to the issue of travel as a space for meaningful learning. In spite of the efforts already undertaken by academics in this field, there is a lack of abridging, consistent, corpus of theory.

Imaginaries play a crucial role in tourism in a variety of levels. As Vidal-Casellas (2020, in press) describes, an imaginary is made up of words, emotions, fragments of emotions, wishes and many other seemingly random components which are put together individually and socially. Imaginaries may coincide more or less accurately with facts, but they are always an opportunity, a possibility for future realisation. Thus, it is clear how it relates to the formation of desires for (or aversion towards) travelling. As it has been already discussed, the results of the tests are able to change the perception of places, even in the case of individuals with previous negative judgements.

Connecting to this idea, the possibility of using genetic tests for tourism promotion of product creation is a double-edged sword. On the one side, results obtained in this exploratory study seem to indicate that perceptions of places always change for the better, thus a genetic test generates a positive attitude towards a place. On the other hand, it is not possible for DMOs or tour operators to control the results of the test or, for their purposes, what destinations will be targeted.

CONCLUSION

This chapter presents the birth of a new tourism modality, genetic tourism, which has become possible through recent scientific and technological improvements, and which the author defines as a cultural tourism segment. The aim is to know the culture of the ancestors of the individual through a DNA test of ethnic origin.

The creation of the theoretical framework for the contextualisation of this type of tourism is based on a comparison between genetic tourism and family tourism, a tourism modality partly related to genetic tourism, which has also served to show the similarities and differences between the two types. One of the greatest differences is the methodology used: the genetic tourism is based on a salivary test of ethnic origin, while the family tourism is based on family research through public institutions and the family memory itself. Thus, it can be argued that genetic tourism stands out for a large amount of information stored generation after generation in the individual's DNA, while genealogy tourism stands out for its high-quality information based on names, birth dates or death.

Therefore, genetic tourism is characterised as a type of tourism that generates deep emotions in the subject due to the high level of personalisation of the experience, since it is difficult to think of a more personalised experience than that based on the tourist's own genetic material. The close bond between the experience and the individual causes the traveller to feel a special connection for the destinations that appear in the results of his test. This connection translates into a predisposition to move away from the tourist paths designed to achieve closer and more authentic contact with the local population and to approach the culture of their ancestors. The penetration of tourists into traditionally non-tourist spaces of the destination increases the possibility of spending on businesses run by the local community. Likewise, the predisposition to move away from the tourist routes predefined would reduce the internal overcrowding of the destination, from traditionally tourist areas to traditionally non-tourist areas, and the external overcrowding, understood more or less as the equitable distribution of genetic tourists globally because tourists do not have the ability to decide their destination.

The increase in migratory movements in world dynamics and the existence of countries with a great migratory tradition has made this tourist modality attractive to all those individuals who feel detached from their origins and ask questions related to their identity. personal; questions that are closely related to the motivations that tourists express in relation to this type of tourism: the search for personal identity, the emotional connection to a geographical space, the obligation to ancestors and the conclusion of a sacred journey.

In relation to the cultural aspects discussed in this chapter, it is noteworthy that genetic tourism facilitates the individual's identification with attitudes closer to ethnorelativism, which transforms history from alien to a personal concept with the incorporation of a narrative, and realise that cultures are formed of elements that communicate and nourish in a symbiotic way, also approaching the tourism typology to the concept of transculturality. Cross-culturality is perceived as a key element in reducing rejection to other cultures, reducing scepticism multicultural understanding, consolidating world peace, eradicating racism or increasing tolerance, for if the subject accepts his own transculturality, he is more likely to accept the transculturality of the other as well.

Regarding the detection of recurring themes in relation to the tourist typology itself and in relation to the cultural aspects discussed in this chapter, the author presents fourteen topics that are summarised in the following key phrases: surprises in the results of the test, results as motivation for travel, test as an exciting experience, nervousness about test results, assumptions about the results and expected results, test as a tool for knowing one's identity, curiosity as motivation, definition of individuals as citizens of the world, test understood as a fashion, test understood as an opportunity, prior interest in the destinations that appear in the results, the more diversity in the results the better, the less specification in the results and the feeling of compliance towards to the ancestors. It is noted that the analysis has allowed to add key points in the definition and conceptualisation of genetic tourism that were not detected in the construction of the theoretical framework, as curiosity as motivation, and allowed to finish confirming points that they have detected and that have been reflected in the personal experiences of the individuals, like test as exciting experience, test like tool for the knowledge of the own identity or feeling of fulfilment towards the ancestors.

Once these themes are detected, the author realises a clear pattern of behaviourthat is repeated in most individuals and reflects on the creation of four differentiated phases that refer to the perceptions of the subjects studied in the stage of the preceding. The four phases through which the individuals go are: approach of the experience, acquisition of the test and wait for the results, discovery of the results, and conclusion of the experience and outputs generated.

In conclusion, the purpose of this chapter is to define and identify the characteristics of genetic tourism as a new tourism modality, by comparing them with family tourism, to expand the following theoretical framework through the analysis of travel blogs and online news related to genetic tourism and linking tourism typology with the cultural aspects discussed in this chapter, such as cultural sensitivity or cross-culturality, and their consequent positive effects on multicultural understanding and the consolidation of world peace.

LIMITATIONS

The first noticeable limitation is the limited academic literature dealing specifically with genetic tourism. Thus, the theoretical background had to be built by combining information from related fields, as it often happens in tourism domains. So far, this seems to be embedded in the nature of tourism as a research field.

Results obtained from the empirical study are sound and already provide an insight into this new modality of tourism and its relation to interculturality, thus allowing to make an early contribution to theory. In spite of this, perspectives given in this text should be backed up, discussed or modified by further studies. Due to the intimate nature of the genetic analysis, authors would suggest approaching the topic from qualitative perspectives, which allow for a better understanding of personal perspectives and feelings, that can later be related to industry procedures.

A final limitation, from an academic perspective, is that this research has been designed to focus only on the pre-trip stage. Further research should be designed to study the trip itself – and how often it does happen – and perhaps, more importantly, the aftermath of the journey: how is it integrated by each individual and how does it contribute to reframing his/her life (if it does indeed happen).

FUTURE RESEARCH DIRECTIONS

Given the high degree of the experience personalisation, genetic tourism would seem to be related to solo travellers A future line of research could be identifying the market segments that are most attracted to genetic tourism as a tourism modality. Likewise, the present chapter focuses only on the perceptions that subjects have before the trip, and that is why another future line of research could be related to the study of tourist behaviour on the trip itself and the after-trip plus the detection of possible variations of the subjects' perceptions according to the phase in which they are immersed. Finally, the author proposes a study about the treatment of the subjects' genetic data carried out by the companies and labs.

REFERENCES

J5Travel. (2018). *Explore your DNA*. Retrieved 26 March 2019, from https://www.j5travel.com/blog/explore-your-dna

A tomar por mundo. (2018). *Estoy hecho de pedacitos de ti, mundo: Mi viaje del ADN*. Retrieved 26 March 2019, from: https://atomarpormundo.com/mi-viaje-del-adn/

Albu, C. E. (2015). Intercultural communication in tourism. *Transcultural Management Journal*, *17*(1), 7–14.

Alexander, M., Bryce, D., & Murdy, S. (2017). Delivering the past: Providing personalised ancestral tourism experiences. *Journal of Travel Research*, *56*(4), 543–555. doi:10.1177/0047287516643414

Anderson, L. E. (1994). A new look at an old construct: Transcultural adaptation. *International Journal of Intercultural Relations*, *18*(3), 293–328. doi:10.1016/0147-1767(94)90035-3

Bade, K. (2016). *El viaje más interesante de mi vida*. Retrieved 25 March 2019, from: https://worldtravelfeet.com/viaje-mas-interesante-vida-adn/

Beerli, A., Díaz Meneses, G., & Moreno Gil, S. (2007). Self-congruity and destination choice. *Annals of Tourism Research*, *34*(3), 571–587. doi:10.1016/j.annals.2007.01.005

Belford, N. (2017). International students from Melbourne describing their transcultural transition experiences: Culture shock, social interaction, and friendship development. *Journal of International Students*, *7*(3), 499–521. doi:10.32674/jis.v7i3.206

Cajuda, J. (2016). *The DNA journey – Momondo*. Retrieved 26 March 2019, from: http://www.joaocajuda.com/the-dna-journey-momondo/

Cimarosti, G. (2017). *The DNA journey: I took a DNA test and found about my origins*. Retrieved 28 March 2019, from: https://www.travelreportage.com/2017/02/09/the-dna-journey-i-took-a-dna-test-and-found-about-my-origins/

Classic Journeys. (2018). *DNA travel is a real trip*. Retrieved 26 March 2019, from: https://www.classicjourneys.com/blog/dna-travel-trip/

Cosmopolilla. (2017). *El test de ADN de Momondo: Viaje a mis orígenes*. Retrieved 26 March 2019, from: https://lacosmopolilla.com/viaje-a-mis-origenes-test-de-adn-de-momondo/cultures

D'Antoni, M. (2010). Tradición investigativa y constructos para comprender aspectos de la inserción profesional docente: Pluralismo cultural, aculturación, shock cultural. *Actualidades Investigativas en Educación, 10*(3), 1–18.

D'Souza, C., Singaraju, S., Halimi, T., & Sillivan Mort, G. (2016). Examination of cultural shock, inter-cultural sensitivity and willingness to adapt. *Education + Training, 58*(9), 906–925. doi:10.1108/ET-09-2015-0087

Dónde te metes. (2017). *Mis resultados del test de ADN #THEDNAJOURNEY*. Retrieved 25 March 2019, from https://dondetemetes.net/2017/08/05/mis-resultados-del-test-de-adn-thednajourney/

Earley, P. C., & Ang, S. (2003). *Cultural intelligence: Individual interactions across cultures*. Academic Press.

Gómez, A. (2017). *La fiebre del ADN llega a España: "Mi familia es murciana pero mis ancestros son vikingos"*. Retrieved 26 March 2019, from: https://www.elconfidencial.com/tecnologia/ciencia/2017-03-19/genealogia-adn-genetica-suecia-murcia-ancestros-vikingos_1350698/

Jiménez, R. (2017). *Mi prueba de ADN: 80% ibérica*. Retrieved 26 March 2019, from https://elpais.com/tecnologia/2017/02/19/soy_techie/1487540938_038219.html

Josiam, B. M., & Frazier, R. (2008). Who am I? Where did I Come from? Where do I go to find out? Genealogy, the Internet and Tourism. *Tourismos: An International Multidisciplinary Journal of Tourism, 3*(2), 35–56.

Karsten, M. (2018). *My personal DNA journey: learning where I come from*. Retrieved 27 March 2019, from: https://expertvagabond.com/ancestry-dna-test/

Krook, N. (2016). *Take a home DNA test and open your world*. Retrieved 28 March 2019, from: https://www.thetraveltester.com/home-dna-test/

Kvaternik, L. (2017). *Lo que un test de ADN desveló sobre mi origen étnico*. Retrieved 28 March 2019, from: https://www.mimundosinfronteras.com/test-adn-origen-etnico/

Lazzarino, L. (2017). *El viaje que me parió: crónica de un test de ADN*. Retrieved 25 March 2019, from: https://losviajesdenena.com/test-de-adn-momondo/

Libertad. (n.d.). *¿Merece la pena pagar por un kit de ADN para conocer tus orígenes étnicos?* Retrieved 27 March 2019, from: http://viajaconlibertad.com/merece-pena-kit-adn-origenes-etnicos/

Lingeberzins, E. (2011). Transcultural communication challenges of international tourism business environment. *Organizacijø Vadyba: Sisteminiai Tyrimai*, 69-79.

Makepeace, C. (2018). *Exploring your DNA ancestry trail around the world*. Retrieved 25 March 2019, from https://www.ytravelblog.com/dna-ancestry-travel/

Marshall, L. (2017). *DNA test fun: ancestry travel inspiration.* Retrieved 26 March 2019, from: https://www.aroundtheworldl.com/2017/07/12/dna-ancestry-travel/

Máthé, I. (2018). Multiculturality and Interculturality in Tourism. *Valahian Journal of Economic Studies*, *9*(23), 39–46. doi:10.2478/vjes-2018-0004

McWha, M., Frost, W., & Laing, J. (2018). Travel writers and the nature of self. Essentialism, transformation and (online) construction. *Annals of Tourism Research*, *70*, 14–24. doi:10.1016/j.annals.2018.02.007

Mee, M. (2017). *My DNA journey: Where I'm from.* Retrieved 28 March 2019, from: https://www.mitziemee.com/momondo-dna-journey-blog/

Meethan, K. (2004). To stand in the shoes of my ancestors. In T. Coles & D. J. Timothy (Eds.), *Tourism* (pp. 139–150). Diasporas and Space.

Menudos Viajeros. (2017). *Un viaje a travês del ADN.* Retrieved 25 March 2019, from https://www.menudosviajeros.com/blog/un-viaje-traves-del-adn

Morec, I. (n.d.). *Prueba del ADN: El viaje que nunca imaginé.* Retrieved 27 March 2019, from http://www.unavidaporviajar.com/reflexiones/test-adn/

Newland, K., & Taylor, C. (2010). *Heritage tourism and nostalgia trade: A diaspora niche in the development landscape.* Academic Press.

Nomadic Boys. (2018). *Ancestry DNA Review: A unique way to inspire travelling.* Retrieved 28 March 2019, from: https://nomadicboys.com/ancestrydna-review/

Oberg, K. (1960). Cultural shock: Adjustment to new cultural environments. *Practical Anthropology*, *7*(4), 177–182. doi:10.1177/009182966000700405

Presbitero, A. (2016). Culture shock and reverse culture shock: The moderating role of cultural intelligence in international students' adaptation. *International Journal of Intercultural Relations*, *53*, 28-38.

Ray, N. M., & McCain, G. (2009). Guiding tourists to their ancestral homes. *International Journal of Culture, Tourism and Hospitality Research*, *3*(4), 296–305. doi:10.1108/17506180910994514

Ray, N. M., & McCain, G. (2012). Personal identity and nostalgia for the distant land of past: Legacy tourism. *International Business & Economics Research Journal*, *11*(9), 977–989. doi:10.19030/iber.v11i9.7181

Richards, G., & Raymond, C. (2000). Creative Tourism. *ATLAS News 23.* Retrieved 12 December 2019, from: http://www.atlas-euro.org/archives.aspx#events_archive

Russell, D. W. (2008). Nostalgic tourism. *Journal of Travel & Tourism Marketing*, *25*(2), 103–116. doi:10.1080/10548400802402271

Sussman, N. M. (2000). The dynamic nature of cultural identity throughout cultural transitions: Why home is not so sweet. *Personality and Social Psychology Review*, *4*(4), 355–373. doi:10.1207/S15327957PSPR0404_5

Uriely, N., Reichel, A., & Ron, A. (2002). Volunteering in tourism: Additional thinking. *Tourism Recreation Research, 28*(3), 57–62. doi:10.1080/02508281.2003.11081417

Vidal-Casellas, D. (2020. (in press). "Los imaginarios del turismo" in Vidal-Casellas, D. and Crous-Costa, N. Dimensión Socio-cultural del turismo. Manual docente. Girona. *Documenta Universitaria.*

Waltner, A. (2017). *My DNA journey*. Retrieved 26 March 2019, from https://www.swedishnomad.com/my-momondo-dna-test/

Ward, C., Furnham, A., & Bochner, S. (2005). Varieties of culture travellers: Tourists. In *The psychology of culture shock* (2nd ed., pp. 127–141). Routledge. doi:10.4324/9780203992258

Ward, J. (2017). *Motivational Monday: My DNA journey, where I come from*. Retrieved 28 March 2019, from: https://onestep4ward.com/motivational-monday-dna-journey-i-come/

Welsch, W. (1999). Transculturality: the puzzling form of cultures today. *Spaces of Culture: City, Nation. WORLD (Oakland, Calif.)*, 194–213.

Winkelman, M. (1994). Cultural shock and adaptation. *Journal of Counseling and Development, 73*(2), 121–126. doi:10.1002/j.1556-6676.1994.tb01723.x

Chapter 18
The Role of Visiting Professionals in Peacebuilding

Katerina Antoniou
https://orcid.org/0000-0003-2685-1500
University of Central Lancashire, Cyprus

ABSTRACT

Seeking to inform the peace-through-tourism scholarly debate, this chapter revisits this complex relationship from the angle of peacebuilding professionals. International peacebuilding settings habitually welcome peace professionals as visiting contributors, consultants, and freelancers. These visiting peacebuilders are characterised by a form of geographical hybridity that identifies them as a distinct audience in international peacebuilding and a niche audience of international travellers. In order to examine their role and contribution in international peacebuilding, the present study employs a qualitative methodological approach that combines historical narrative and field interviews to the case of the Cyprus peacebuilding discourse. This chapter argues that visiting peacebuilders have a significant role in formulating optimal encounters between peacebuilder audiences and in increasing the effectiveness of international peacebuilding projects.

INTRODUCTION

Research on the relationship between tourism and peace is not new. The majority of this research has focused on verifying or refuting the argument that leisure tourism can contribute significantly to peace, either by acting as an indicator of restored relations between former rivals (Bar-Tal & Bennink, 2004), or by encouraging their positive interaction in post-conflict settings (Anastasopoulos, 1992; Sonmez and Apostolopoulos, 2000; Moufakkir and Kelly, 2010). Nonetheless, in the contemporary globalised world, leisure tourism does not fully capture the nature of international travel activity; travel is no longer simply an expression of escapism, but is often an entrenched component of daily routine, with both domestic and international work travel occurring for many on a weekly and often on a daily basis. With this observation in mind, the scholarly focus on 'peace through leisure tourism' has left non-leisure travellers, such as conflict resolution experts and the niche audience of peace professionals in particular, underexplored.

DOI: 10.4018/978-1-7998-5053-3.ch018

More specifically, visiting peacebuilders such as consultants, international facilitators and peace activists are scarcely identified as a unique audience of international travellers, while they are also rarely viewed as an independent stakeholder within the study of international peacebuilding initiatives. Peacebuilding research has thus far categorised peacebuilders as either local or international (Mac Ginty, 2015; Richmond, 2007), often overseeing a third category of travelling peace professionals that were characterised by their short-term involvement in peacebuilding projects. As a result, the contribution of these professionals and the impact of their travel activity on establishing and sustaining peace have been largely overlooked, not only in Tourism literature, but also within Conflict Resolution scholarship. This chapter seeks to shed light on the travel component within peacebuilding settings by acknowledging the contribution of short-term, hybrid professionals in international peacebuilding; an audience defined as *Visiting Peacebuilders*.

The underexplored audience of traveling professionals in peacebuilding settings does not fully conform to the conventional categorisation of peacebuilder audiences into either local or international. Instead, it is characterised by both geographical and temporal hybridity due to the traveller status of its members. This observation suggests that the binary local-international categorisation of peacebuilders might be simplistic or inaccurate. It is therefore imperative to acknowledge Visiting Peacebuilders as a distinct stakeholder in international peacebuilding and at the same time revisit the restrictive present categorisation of peacebuilding professionals.

In consideration of the hybrid nature of peace professionals (Mac Ginty, 2015), this chapter re-conceptualises the categorisation of peacebuilders beyond their geographical affiliation. In doing so, it adopts a single-case-study model and focuses on the Cyprus conflict, which has been characterised by numerous failed attempts at reaching a peace settlement between the country's partitioned Greek Cypriot and Turkish Cypriot communities. Cyprus showcases a decades-long discourse of international peacebuilding initiatives (Jarraud et al., 2013; Ladini, 2009), allowing for the identification of diverse peacebuilder audiences and at the same time examining the evolution of their interaction through time. The chapter draws upon the island's peacebuilding interventions and the role of peacebuilder audiences from the early years of communal partition, to the authorisation of movement across the Buffer Zone, and today's normalisation of intercommunal cooperation – within the confines of Nicosia's peacebuilding community.

To address the identified scholarly gap on the study of visiting professionals in international peacebuilding initiatives, this chapter proceeds to examine the categorisation of peacebuilding professionals and its evolution towards what is referred to as fourth-generation peacebuilding (Richmond *et al.*, 2011). It proceeds to identify traveller audiences that are relevant to the peacebuilding scene through a review of peace-oriented forms of tourism. The case study of the Cyprus peacebuilding discourse is then introduced to closely examine the interactions of Visiting Peacebuilders with long-term professionals within the island's peacebuilding projects. Findings on the contribution of Visiting Peacebuilders are discussed in reference to Allport's (1954) Contact Hypothesis, which suggests four conditions for optimal and meaningful intergroup contact.

The research aim of the chapter is to examine the ways in which Visiting Peacebuilders, as a distinct audience in international peacebuilding, contribute to the effectiveness of international peacebuilding projects. Allport's Contact Hypothesis plays a central role in achieving this aim, by outlining the nature of interactions among the categories of professionals collaborating under international peacebuilding initiatives – including Visiting Peacebuilders. These categories are further discussed below.

CATEGORIES OF PEACEBUILDING PROFESSIONALS

Tschirgi (2004: 2) argues that the 'original definition' of peacebuilding speaks of 'non-military interventions by external actors to help war-torn societies not only to avoid a relapse into conflict, but more importantly, to establish the conditions for sustainable peace'. This initial conceptualisation, however, has translated into what scholarship defines as four generations of peacebuilding. According to Richmond *et al.* (2011), first-generation peacebuilding was a limited intervention for conflict management, offering little beyond the capacities of peacekeeping. The second generation was a more integrated approach to peacebuilding that managed to acknowledge human needs and bottom-up opportunities for conflict resolution (*ibid.*). The third peacebuilding generation and the one most operated over the late-twentieth century was an approach primarily concerned with the structural reconstruction of states into liberal democracies (*ibid.*), a primarily state-centric approach to peacebuilding that was also referred to as statebuilding (Barnett and Zurcher, 2009; Woodward, 2017).

In recent years, critical and post-structuralist peacebuilding accounts have acknowledged a fourth generation in peacebuilding (Richmond, 2010; Richmond *et al.*, 2011; Roberts, 2011, 2011b; Richmond, 2012; Jabri, 2013), one that rejects third-party interventions for peacebuilding and instead emphasises the input of local populations in peacebuilding processes, with their inclusion as equal partners ensuring their empowerment and emancipation. Roberts (2011) refers to fourth-generation peacebuilding as emancipatory peacebuilding, while Jabri (2013) differentiates between colonial and post-colonial peacebuilding; the former grants agency to international experts through external peacebuilding intervention, while the latter supports self-determination and shifts the focus to local agency.

The transition from third-generation peacebuilding to a fourth-generation that enabled local populations to have increased agency in peacebuilding processes emphasised the dichotomy between local and international peacebuilding stakeholders. Doyle and Sambanis (2000) embrace the local-international differentiation through their *Peacebuilding Triangle,* which introduces three factors that are critical to the success of peacebuilding initiatives. These factors are first, the local source(s) of dispute, second, the local capacities for achieving reconciliation, and third, international support, which indicate a state-centric philosophy of third-generation peacebuilding, and at the same time a supportive role for international experts.

Richmond (2009) and Acharya (2004) highlight that third-generation local-international distinctions in peacebuilding settings have become increasingly hybridised. Mac Ginty (2015) has argued that the evolution of peacebuilding incorporates the geographically hybrid nature of peacebuilding professionals. It is vague – to the least – to discuss the geographical affiliation of peacebuilders as a constant, when witnessing local peacebuilders representing and being employed by international agencies, or peacebuilders of foreign origin working within local NGOs for years. The national affiliation of these professionals is, in many cases irrelevant, as their professional development within the field of peacebuilding work evolved along hybrid geographical and national boundaries. The emergence of fourth-generation peacebuilding has come to address the phenomenon of hybridity across categories of peacebuilding professionals, and in doing so, revisit the agency of local, international, and hybrid peacebuilder audiences.

The hybridity of the local-international differentiation across peacebuilders is not the only factor challenging its accuracy. A second factor is the assumption of a binary differentiation, which excludes the classification of peacebuilders in more than two categories. Mathieu (2018) acknowledges that the binary classification of local versus international stakeholders in peacebuilding research is not only restrictive, but it can even hinder the attempt for achieving equality across the two stakeholder categories, and stigmatise the local contributors as beneficiaries to the process. Paffenholz (2015) adopts a similar stance by stating that the attempt to elevate the role of local peacebuilders as equal contributors is jeopardised by the binary understanding of local versus international stakeholders, as it simplifies the stakeholder scene to such an extent that vital stakeholder categories such as local elites are overlooked. In other words, the assumed classification of peacebuilding professionals into either locals or internationals establishes a binary categorisation of an audience much more diverse in nature. Paffenholz suggests that a multidisciplinary approach to reconceptualising peacebuilding stakeholders beyond the restrictive local-international classification will address this scholarly deficit, a position that this chapter advances on.

In order to address the literature gap of a limiting binary classification for peacebuilding professionals from a multidisciplinary approach, as Paffenholz (2015) advocates, this chapter expands peacebuilder categorisation through the lens of tourism, and reviews the involvement of visiting professionals in peacebuilding. Looking at peacebuilder classification more closely, both the local and international categories refer to the professionals involved in a peacebuilding initiative throughout its duration, but fail to differentiate between long-term and short-term involvement in a project. This excludes short-term contributors such as visiting experts, consultants and analysts, who can be local, international, or have a hybrid geographical affiliation to the conflict. Short-term contributors could include sociologists, economists, conflict resolution experts, heritage and urban planning consultants, facilitators, and gender experts, depending on the themes and needs of each peacebuilding project.

Visiting contributors to international peacebuilding projects feature unique characteristics that differentiate them from locally-stationed and local peacebuilders. These are the (1) short-term nature of their contributions to the projects, (2) their short-term interactions with other peacebuilder audiences, and (3) their hybrid geographical and institutional affiliations. Examining these characteristics as potential factors of influence on international peacebuilding can thus be done more effectively by identifying Visiting Peacebuilders as a distinct audience. Employing Tourism literature to revisit and reconceptualise peacebuilding stakeholders enables the scholarly relationship between tourism and peace to move beyond the causality argument of peace through tourism (D'Amore, 2009; Higgins-Desbiolles & Blanchard, 2010; Moufakkir & Kelly, 2010) and assess the contribution of tourism to peace from an underexplored, niche perspective.

PEACE TOURISM AND THE AUDIENCE OF VISITING PEACEBUILDERS

Examining short-term contributors to international peacebuilding initiatives allows for the expansion of the limiting local-international peacebuilder classification. In doing so, it is important to review tourism literature on international and domestic visitors that could be identified as short-term contributors to peacebuilding projects, or Visiting Peacebuilders (VPs).

Van den Dungen (2014) states the obvious oversight that literature on peace and tourism has neglected to identify peace tourism, or travel that is specifically motivated by and associated with conflict resolution practices and war prevention. The author clarifies that this type of tourism can be valid for both national and international contexts, reaffirming the geographical hybridity of the travellers undertaking it. Van den Dungen introduces examples of peace tourism destinations, including peace museums, city peace trails, and peace cities – cities that symbolise peace through their historical discourse or strive for peace through their institutions. Van den Dungen also identifies himself as a member of the international peace movement, and a peace tourist (*ibid*), reflectively acknowledging the gap in examining this audience.

Peace tourism can refer to both leisure and non-leisure travel. Leisure travellers can be motivated to experience destinations associated with a past war or ongoing conflict, an activity that falls within the realm of dark tourism (Stone, 2006), or to actively engage with peacebuilding settings and societies transitioning from conflict. The latter can appeal to peace activists seeking to observe or volunteer in peace-oriented initiatives.

Israel and Palestine are popular destinations for peace activist tourism, or what has also been referred to as political tourism and justice tourism. Blanchard and Higgins-Desbiolles (2013) present political tourism as a tool of parties in conflict for sharing their perspective internationally and advocate for their demands. In the case of the Israeli-Palestinian conflict, Palestinians have, according to the authors, utilised political tourism to raise international awareness on the conflict and seek justice for the Palestinian community. Chaitin (2011) echoes the use of political tourism as an awareness-raising tool, and suggests that political tourism in Israel and Palestine can support peace and social justice by creating better informed international peace activists. Brin (2006) identifies activist tourists to Israel-Palestine as politically-oriented tourists, who arrange their visit through locally-stationed peace-oriented organisations, and seek to increase their understanding of the conflict through reality tours and direct experience on the ground. The author finds these politically-oriented visitor experiences problematic, as they can be less supportive to peace and more vested in promoting a biased political agenda.

Higgins-Desbiolles (2008) moves beyond the category of political tourism and defines justice tourism as travel activity that acts in line with international social justice objectives and seeks to address and reform global inequalities. Isaac (2014) places justice tourism under the umbrella term *responsible tourism*, and argues that it can contribute to peacebuilding through intercultural exchange and economic activity that supports vulnerable and marginalised communities in conflict, giving the example of justice travel to Palestine.

The above examples can be placed within leisure-based peace tourism, with the multiple examples of international travel for peace activism providing an informed understanding of its diverse applicability. Despite so, there is an evident scarcity when it comes to identifying non-leisure peace travellers. To differentiate this category from the broader peace tourist category, non-leisure peace travellers are the ones travelling for professional purposes to a destination in order to contribute, through their professional expertise, to the prevention of war, the stable transition of a post-conflict setting and for laying the ground for effective reconciliation and sustainable peace. To this end, non-leisure peacebuilding travellers, or Visiting Peacebuilders (VPs) as defined earlier, are the business tourists, or traveling professionals, directly engaged in peacebuilding projects due to their professional capacity and expertise. Differentiating this category from the broader leisure-oriented peace tourism category will allow for the output of these travelling professionals to be examined in a more detailed manner as a distinct audience of peacebuilding professionals in international peacebuilding settings. The category of VPs is thus further explored through the case study section below, pertaining to the peacebuilding discourse in Cyprus.

METHODOLOGY AND RESEARCH DESIGN

The study applies an exploratory qualitative research approach towards the single-case-study model of Cyprus. This approach is employed to achieve the de-categorisation of peacebuilding professionals from their existing binary classification as either local or international, and their re-categorisation in consideration of each professional's level and duration of involvement. More specifically, the examination of

Visiting Peacebuilders as a distinct stakeholder in peacebuilding is achieved through a combination of narrative history and direct interviews taken from peacebuilders involved with the Cyprus reconciliation efforts. Interviewees include locally-stationed peacebuilders, representatives of international organisations on the ground, as well as experts with short-term involvement and visiting consultants. Narrative history is employed to contextualise the analysis of the identified case study and provide the background information for the emergence and evolution of the Cyprus peacebuilding community. The research approach adopted achieves both the detailed encapsulation of the Cyprus peacebuilding discourse, as well as the empowerment of the population under study, whose input is featured in the form of direct interview quotes (Foley, 2002).

Structured, open-ended interviews were conducted with 21 members of the Cyprus peacebuilding community, including three international organisation representatives (IOR), eleven peacebuilders working as local civil society experts (local stakeholders, or LS), and seven Visiting Peacebuilders (VP). The structured interviews lasted for approximately one hour each and were carried out on separate occasions on a one-on-one basis; all feedback from interviewees was anonymised, ensuring privacy and confidentiality of the information shared. The interviews were conducted between November 2014 and May 2016, aligning with the completion of the United Nations Development Programme: Action for Cooperation and Trust (UNDP-ACT) project in 2015, which entailed the majority of the peacebuilding activity in Cyprus during that period. Complementary remarks – that were collected through non-structured, informal discussions with additional members of the peacebuilding community – are also included, to inform of the status of the Nicosia-based peacebuilding community after 2016.

Data collected from the structured interviews and statements from informal conversations were analysed using thematic content analysis, to identify prevailing themes and relevant patterns to be examined. With regards to the discussion of findings, the chapter explores the audience of Visiting Peacebuilders in relation to other peacebuilding professionals, and employs the Contact Hypothesis to examine their interaction based on the Hypothesis' four optimal conditions of intergroup contact. This research design is adopted to address the study's aims as stated earlier: to assess whether Visiting Peacebuilders have positive encounters with the peacebuilding community on the ground and whether these encounters can be further optimised to achieve the implementation of more effective peacebuilding initiatives.

PEACEBUILDER AUDIENCES AND INTERACTIONS: FINDINGS FROM CYPRUS

Over the course of the twentieth century, Cyprus became the home of steep ethnic conflict between its two major ethnic communities, Greek and Turkish Cypriots. Civil unrest following the island's independence in 1960 resulted in a foreign military intervention by Turkey in 1974 and to the geographical partition of the two communities. UN-mediated negotiations between the communities' political leadership have yet to result in a comprehensive settlement of what has come to be known as the Cyprus Problem, with the conflict in Cyprus to be widely referred to as protracted, intractable and frozen (Azar, 1990; Fisher, 1997; Broome, 1998; Sozen, 2006; Makriyianni and Psaltis, 2007). Despite the lack of a comprehensive settlement, Cyprus counts decades of international peacebuilding efforts, involving a vast array of both local and global peacebuilding agents.

Following the geographical partition of the Cypriot population in 1974, contact between the two communities was only enabled by authority permission and, thus, remained at negligible levels. Commencing in 1979 and receiving support from international agencies UNDP[1] and USAID[2], the Nicosia

Master Plan was the first of many intercommunal initiatives to follow, featuring a joint project involving the two municipalities of the partitioned capital city of Nicosia (UNDP, 2015).

The Cyprus peacebuilding discourse demonstrates an evolving peace-oriented, local civil society that involves a considerable contribution of Visiting Peacebuilders. UNDP's archives (Mahallae, 2015) illustrate a 34-year timeline of intercommunal engagement in peacebuilding initiatives that aimed, among others, to support capacity development for a local peacebuilding civil society. Visiting Peacebuilders to the early UNDP projects included architects, urban planners, sociologists, and economists, who were invited as international experts to provide their input – initially, through the Nicosia Master Plan, this input was on a common plan for the divided city of Nicosia, and thereafter to establish additional intercommunal structures.

Other VP audiences included academics and conflict resolution researchers, who engaged with the Cyprus peacebuilding discourse by implementing conflict resolution and problem-solving workshops for members of the two communities during the 1980s and up until the late 1990s[3]. As Rothman (1999: 177) stated, Cyprus offered 'an incubator for conflict resolution scholars as they apply their skills to a relatively non-volatile but none the less deeply intransigent conflict'. Cyprus was an intriguing case study for conflict professionals and appealed to a wide range of VPs for examining the conflict and parties involved. Rothman himself is an example of a conflict resolution expert visiting Cyprus on a short-term, professional basis, in order to participate in the 1994 Cyprus Conflict Resolution Consortium as an action-researcher.

Between 1998 and 2003, 200 additional peacebuilding initiatives were supported by the Bicommunal Development Program (BDP), including primarily cultural heritage, environmental and public health projects, again involving international experts, who shared techniques and best practices (UNDP, 2015). The BDP was replaced, upon its completion, with the UNDP-ACT project, which commenced in 2005 and aimed at restoring intercommunal efforts for peace after the backlash of the 2004 Annan Plan[4] rejection (Jarraud *et al.*, 2013). Local civil society grew in expertise and, with the financial support of international funding schemes, was able to achieve public outreach and initiate joint advocacy efforts aimed at influencing the stagnated political negotiation process (Jarraud *et al.*, 2013). Despite being criticised for only impacting Nicosia's elites (Autesserre, 2014; Ladini, 2009), one of the breakthroughs achieved through international peacebuilding funding was the establishment of an intercommunal, peace-oriented civil society, which enabled multiple non-governmental organisations to implement peacebuilding activities across the divide (Jarraud *et al.*, 2013).

Intercommunal cooperation and social development evolved exponentially in Nicosia after the opening of crossing checkpoints and the ability of locals to access both sides of the UN-administered Buffer Zone. Intercommunal advocacy for peace continued to grow and culminate in a diversity of forms, from joint political movements to intercommunal think-tanks and forums. A project that is undoubtedly considered a milestone for intercommunal contact and local peacebuilding efforts is the restoration of a building within a publicly accessible part of Nicosia's Buffer Zone and its transformation into a multi-functional intercommunal venue called the Home for Cooperation (H4C, 2014).

Build Peace (2015) and the Power of One (2012) are examples of peacebuilding conferences that attracted tens of Visiting Peacebuilders to the island through their professional capacity. Additionally, for the period January 2012 – December 2014, events at the Home for Cooperation featured 23 VPs (AHDR, 2015) – mainly for conference lectures, book presentations, exhibitions and talks – while freelance journalists, photographers, authors and film-makers who have visited the Home during this period are additional to this number. The figures suggest that the Home for Cooperation, an innovation that institutionalised intercommunal cooperation and physical contact, has evolved into a hub for the island's peacebuilding initiatives, attracting peace-oriented visitors from the two communities, but also foreign Visiting Peacebuilders. While the UNDP-ACT programme ended in 2015, a number of the peacebuilding civil society organisations active in Cyprus have continued their activities through other funding opportunities, arguably of smaller scale. In 2019, the Home continues to attract funding and international visitors, while hosting a variety of events and festivals, featuring art, creative expression, dance and athletic activities, as well as academic and scholarly initiatives. As an administrator of the Home states, international visitors continue to visit, often in more frequency than local audiences (LS40, 2019).

Members of the Cyprus peacebuilding community acknowledge the participation and contribution of Visiting Peacebuilders in the Cyprus peacebuilding discourse – from the early years of its establishment until the more recent UNDP-ACT initiatives – as significant. USAID representatives IOR2 and IOR3 expressed the significance of the presence of international experts in the initial projects; they did not only benefit locals by providing technical expertise on the issues discussed, but were also able to assist the debate on more emotional or controversial issues that were found primarily in culture and heritage-related projects. In the following words:

Sometimes we would identify notable international scholars and try to link them up with locals. For example, youth camps were initially organised by finding the international experts first and then locating the children that would participate. Intercommunal initiatives were developed by identifying a local shared need of wide interest and matching it with international expertise: so the international expert would come in contact with the locals and share their expertise on an intercommunal level, to build bridges (IOR2, 2015).

Cultural heritage is an area that without international participation, no intercommunal contact and success would have been possible. The international focus became the turning point for overcoming difficulties on a topic like cultural heritage that was an emotional attachment for the public (IOR3, 2015).

While international agency representatives perceived the involvement of short-term foreign experts as successful, this position was not always echoed by local peacebuilders:

Some foreign experts came in because they had similar experiences and they shared their best practices. Their role was mainly consultative. Were they helpful? To be honest, it felt that the agenda was already set. It was supposed to be a discussion, interactive, but that wasn't the case.

I don't feel that VPs gave us something that we couldn't have figured out on our own. VPs were not that informed over the bi-communal dynamic. We created strong [intercommunal] relations within the team, we had a strong bond, and then these external professionals came in to write a short report and we would have to take that report and incorporate their feedback into our work. But that wasn't always feasible or practical (LS11, 2016).

The international peacebuilding structures enabled Greek Cypriots and Turkish Cypriots to establish genuinely positive relationships and effective synergies, yet this was not the case for early visiting consultants. LS11 admitted that the structures in place did not enable the VP audience to build effective partnerships with the locals, despite their often promising and catalytic contributions:

I think [VPs] have surely had an impact in Cyprus' post-conflict transition. I found out about them while doing research for a peacebuilding history infographic. I think the early mediators and conflict resolution trainers who came and trained the first wave of civil society professionals on the island were definitely influential, not just one or two professionals but many more (LS1, 2015).

LS1 made reference to early researchers for their influential role as conflict and peacebuilding experts in Cyprus' – then embryonic – peacebuilding movement. IOR3 clarified that peacebuilding programmes were designed to provide knowledge to locals, distinguishing between the professional capacities of locals and foreigners. More specifically, locals were defined as *beneficiaries* and international experts as the neutral agents assisting technical barriers to intercommunal collaboration.

The premature status of the local peacebuilding movement notwithstanding, there was an evident imbalance between foreign professionals that were introduced as experts and local professionals that were labelled as beneficiaries. Looking at the work of Fisher, Burton and Doob in Cyprus (Fisher, 1997), the foreign experts treated local audiences as research participants rather than equal partners, effectively establishing a colonial and outdated – as Yanow (2009), Jabri (2013), and Gobo and Marciniak (2016) would argue – relationship of researchers versus natives.

The year 2003 marked a new era for intercommunal relations in Cyprus, as both Greek and Turkish Cypriots were able to cross to the other community and access the island in its entirety. LS6 argues that the checkpoint opening was of historic significance, as it enabled intercommunal contact, the visualisation of a potential solution, and possibly economic balance between the partitioned communities. LS9 reiterated that the opening of the checkpoints was a highpoint for the island's peacebuilding efforts:

We had meetings within the Ledra Palace and everything was very structured, but now you just wave at the guards and pass. The Home for Cooperation provides a relaxed atmosphere and the difference is huge in terms of accessibility for intercommunal contact, in terms of perceptions and in terms of the environment and setting (LS9, 2016).

In the months following April 2004 and the rejection of the Annan Plan the decrease in intercommunal activity was notable. LS21 (2016), civil society practitioner in youth projects, indicatively mentioned a halt in youth activities after 2004 because 'there was no point', no end goal of reunification the youth would hope for. This was reaffirmed by LS3:

The momentum our organisation and our work had created reached a peak in 2004 but the political elite and 2004 referendum results managed to reverse it. The social and political levels affect one another. If the political elite make a change then they can drive the social level and vice versa (LS3, 2015).

LS30 highlighted the importance of VPs in a challenging period like the post-2004 years:

It was actually a visiting expert that brought us together [Greek Cypriots and Turkish Cypriots] right after 2004 to discuss about the economy in the north and next steps, at a time when nothing was really happening, intercommunal-wise (LS30, 2016).

Commencing in 2005, the funding schemes provided by UNDP-ACT, USAID, and the European Union were received positively by local civil society practitioners, as they enabled their work to resume and evolve. UNDP-ACT representative IOR1 acknowledged the sustainability challenges faced by the projects implemented:

There has been positive feedback from local stakeholders who received grants, but traditional grant-giving projects cannot ensure their sustainability. It's a risk we take and therefore we try to select the best to implement these grants. The important thing is to create solid, sustainable partnerships, and perhaps it is the most important element of our work, even more than the production of knowledge products which also stay on and are available to people at any time (IOR1, 2015).

Following the opening of the checkpoints, intercommunal synergies and organisations sought the opportunity to establish shared working spaces, a vision that gave way to the establishment of the Home for Cooperation in 2011.

Then there was the need for a space. [Our organisation] had already been intercommunal and the first conferences we organised took place at Ledra Palace and the Fulbright Commission. We were looking for a more permanent space somewhere in the area. PRIO and the UN supported us by providing a room in the Ledra Palace hotel.

An abandoned building [on the same street] was soon identified, and we sought financial support to restore it. A number of international organisations supported the restoration of the building and the operation of the Home [thereafter].

International experts visiting the Home include political elites, such as UN Special Advisers and Heads of Mission, as well as various Ambassadors stationed in Cyprus and interested in our perspective. With regards to Visiting Peacebuilders, we have often welcomed researchers on education and conflict, journalists, authors, academics and others. I would say that today, approximately half of our visitors are invited by our local organisation and the other half visit at their own initiative (LS3, 2015).

LS5 acknowledged the importance of the physical illustrations achieved, speaking in reference to the delivery of intercommunal theatre workshops and their transformative capacities. These workshops are particularly impactful, but perhaps so is the 'mere presence of people in the Buffer Zone' and the manifestation of diverse audiences interacting, both locals and foreigners. Regarding Visiting Peacebuilders:

Academic field work [by VPs] has come to fit into the local culture, for example field work illustrated at the Home has had a great impact; it made its way back. The fact that work by VPs can potentially return back [in practice] is really important and impactful (LS5, 2015).

Peacebuilding efforts in Cyprus were actively shared on online platforms and social media, which often enabled VPs to choose Cyprus as a destination in their peace-related work. Swedish film-maker, writer and peace activist, VP1, who was eager to visit Cyprus and discuss possible future collaborations that would include film-making workshops for diverse youth audiences, shared her experience with the Cyprus Buffer Zone peacebuilding scene:

I am amazed that in the middle of a horrible place there are still people fighting to make it better. Their courage and desire for peace and a better future makes me believe in humankind and in a way reminds me of my mother's life work to fight [South African] apartheid with good deeds.

During the past 8 years I have been to Cyprus maybe 6-7 times. When I met with a Cypriot girl at a youth conference she invited me to come and see her organisation's work in the Buffer Zone. I went there to show a movie against hate-crime but it ended up being also a meeting to discuss how we can work together (VP1, 2014).

A number of local peacebuilders agree that organic interactions with Visiting Peacebuilders were promising and enthusiastic, but unfortunately local civil society didn't always have the capacity to accommodate transnational synergies beyond their funded project outputs. On the other hand, the UNDP-ACT projects continued to recruit international experts and consultants from a top-down approach, whom local professionals dishearteningly admitted did not effectively contribute to their work and professional development:

I don't see any vested interests [by recruited international experts] because they share best practices from analogous conflict and reconciliation models. However there was no significant capacity building offered. Capacity building is important but there is expertise available locally and that was not fully acknowledged. Personally, I developed my expertise by visiting foreign conferences, not as much by the visits of VPs here (LS6, 2015).

UNDP did some very random projects with [invited] VPs. Out of the blue UNDP decided that gender was the flavour of the month and they wanted to bring together women to discuss gender. But they brought a woman that was completely ignorant of the Cyprus context. And if the trainer was more appropriate in terms of context, things would have been more successful. She was a gender expert but she wasn't impactful. UNDP had a good position as an international agency to make a great impact, but they didn't do a good job. They could have contributed to a powerful women's movement across the divide. The mistake is that they copy-pasted methods and projects that worked elsewhere but they weren't adjusted appropriately (LS9, 2016).

In terms of the bi-communal structure, [invited] VPs had zero impact, because they couldn't address the bi-communal dynamics present, especially the practical challenges that came up, for example advocacy towards the leaders could not be implemented in the north – international standards were not applied in the north, only in the south (LS10, 2016).

Nicosia's Buffer Zone structures between 2011 and 2015 – which were primarily the institutional structures of international peacebuilding projects and the physical structure of the Home – accommodated VPs very differently, with international projects failing to effectively engage invited experts and local professionals, while the Home provided a space for organic relations to form. Local professionals talk positively of organic interactions that allowed them to interact with VPs on an expert-to-expert basis, with LS7 calling VPs the 'neutral voice' between locals and international agencies that shared new perspectives. The peacebuilder continues to acknowledge that VPs could have helped even more if international agencies were not 'in the way':

Most VPs came through the [UNDP-ACT] projects for trainings and during the [interregional conference] the Power of One. It was liberating to talk to VPs, but no in-depth discussions and connections were made, because the project design did not allow more time for more in-depth work. International agencies seemed to be 'in the way' because they created a setting that applied to the project objectives but did not necessarily achieve the best knowledge exchange possible (LS7, 2016).

LS3 elaborates on VP contribution:

Foreign audiences show more interest in the Home than local audiences, due to ideological discrepancies and a lack of active locals, active within civil society. Some VPs promote awareness for the Home [for Cooperation] and this can even reach local audiences. There is a level of legitimacy when a story comes from foreign journalists, for example, by Brownbook and the Guardian. Promoting the Home and its activity is the greatest thing, the best impact achieved [by VPs]. They share our vision and bring other VPs closer. And they even reach those who can potentially provide financial support. Another impact that is important is to host VPs that worked in other conflict and post-conflict areas and [can] prove to local audiences that reconciliation is possible (LS3, 2015).

It can be understood that local stakeholders from the two communities differentiated between foreign experts who were short-term visitors (VPs) and long-term stationed representatives of international agencies. This was clearly highlighted by local professionals engaged with the UNDP-ACT programme:

If more foreign experts came in we would have had a bigger impact. The whole idea and philosophy behind our intercommunal civil society organisation was the outcome of a research made by an Irish practitioner. That research was adopted by UNDP and implemented as understood (LS9, 2016).

For our organisation, VPs did help us figure out our concept, our philosophy and strategy, although the timing and conditions were such that we didn't take full advantage of them (LS10, 2016).

Collaborating with foreign professionals gives legitimacy to a great extent and has been catalytic for the two civil society organisations I have been involved with on intercommunal projects (LS5, 2015).

[Visiting] conflict professionals have definitely had an impact in Cyprus' reconciliation process. They have raised awareness, provided feedback, and contributed to a change of narrative. They acknowledge the peace work done, and there have been opportunities for joint projects with a positive impact (LS2, 2015).

The positions recorded in the surveys indicatively suggest locals welcoming the input of VPs but at times finding that their influence had been restricted due to unfavourable conditions. These conditions are further examined in relation to Allport's (1954) optimal conditions for intergroup contact as outlined in the Contact Hypothesis, which is employed in the analysis and discussion below.

DISCUSSION OF FINDINGS

Peacebuilder Audiences and Interaction

Despite the limitations, local peacebuilding professionals in Cyprus indicatively admit that VPs have overall had a positive contribution to the local peacebuilding scene, especially in their ability to share best practices and reconciliation models, and add legitimacy to local initiatives. Nevertheless, it is important to examine the conditions under which this peacebuilder audience was involved in the Cyprus peacebuilding discourse, specifically in reference to their interaction with other peacebuilding professionals.

Allport's (1954) Contact Hypothesis states four conditions for optimal intergroup contact that can induce positive interactions and reduce prejudice. Allport's Hypothesis has often been employed in peacebuilding research to assess the relationships between rival communities, yet it has not been thoroughly applied to address the intergroup dynamics within the community of peacebuilding professionals. These four conditions for optimal intergroup contact are: (1) the equal status of participating parties, (2) their support by an institutional or social authority figure, (3) their intergroup cooperation, and (4) a common goal. This study employs these conditions to assess the interaction between International Organisation Representatives, Visiting Peacebuilders, and Local Peacebuilders within the Cyprus peacebuilding discourse, in an effort to identify the potential for their optimal collaboration.

Equal status: The early years of international peacebuilding in Cyprus had clear hierarchical structures with both international agency representatives and invited international experts assuming hegemonic roles over local partners. Their local counterparts were perceived as beneficiaries with little expertise and were only incorporated into peacebuilding projects as research participants, primarily to assist international professionals in assessing local dynamics and testing intergroup interaction. This discrepancy narrowed down throughout the years that followed, mainly as Visiting Peacebuilders assumed more neutral consultative roles. Equal status between local and visiting peacebuilders was effectively achieved through the organic interactions that were enabled between the two groups of professionals after the establishment of the Home, and through the increased international conference opportunities that emerged. Nevertheless, equality was up until the completion of the UNDP-ACT projects in 2015 never optimally achieved between international agency representatives and local peacebuilders, as the former continued to identify local counterparts as beneficiaries to grant-giving projects, despite endorsing their input more than the earlier years.

The discourse of international peacebuilding in Cyprus reflects the evolution of international peacebuilding from first generation basic peacekeeping to UNDP-ACT's third generation peacebuilding, which achieved partial local empowerment and capacity-building. Interestingly, the equal status of local-international counterparts and the emancipatory intent of fourth-generation peacebuilding were not achieved through the activities implemented under the grant-giving projects, but through the physical and institutional structures these projects helped to establish. Notably, the Home acted as a peacebuilding hub for Nicosia and the rest of the island, enabling ad-hoc organic interactions for a diversity of peace contributors, and dissolving hierarchical structures of peacebuilding agency.

Authority Support: the partitioned communities in Cyprus were highly securitised and suspicious in the years before 2003. Political leaders were therefore reluctant to support any peacebuilding efforts, which mainly remained marginalised. The opening of the checkpoints and the possibility of intercommunal interaction enabled a gradual turn to peace-oriented narratives, which were formally expressed by 2015, by then Greek Cypriot President Nicos Anastasiades and Turkish Cypriot community leader Mustafa Akinci.

Despite the change in political leadership attitudes, the peacebuilding community's credibility mainly came from the support they received by international organisations and experts, which local peacebuilders saw as neutral and credible voices in support of peace. As a result, the international agencies involved in the Cyprus peacebuilding scene, primarily UNDP and USAID, acted as the institutional authority under which local peacebuilding efforts were financially and politically sheltered, making it even harder for the agencies to achieve an equal status with their local collaborators.

Intergroup Cooperation: The testimonies of local peacebuilders reveal to a large extent that there was evident willingness from local stakeholders to create synergies with international visiting experts. Such synergies were in frequent occasions jeopardised by the rigid structure of grant-giving projects, however, they were often achieved in an organic and meaningful manner after the establishment of the Home. Intergroup collaboration was continuous for international agency representatives working with both the audience of local stakeholders, as well as visiting international experts.

As seen from the findings, intergroup cooperation was often achieved by engaging in activities of mutual interest, such as the theatre workshops LS5 mentioned. These activities covered a wide spectrum of themes, including other artistic and athletic initiatives, but also youth empowerment workshops, academic and practitioner conferences, film and documentary screenings. Local peacebuilders, as well as locally-stationed international agencies, acted primarily as hosts to Visiting Peacebuilders, whether the latter were seeking to feature their work publicly or simply to engage with the local peacebuilding momentum.

Common Goal: Similarly to the factor of intergroup cooperation, the common goal for the reunification of Cyprus, the reconciliation of its partitioned communities, and the establishment of viable peace was evidently and thoroughly shared by all three groups of peacebuilding professionals involved in the Cyprus-based initiatives. Nevertheless, as shown from peacebuilder testimonies, the approaches adopted by each group to achieve this goal varied significantly, leading to a lack of efficiency and a deficit of sustainability in the partnerships and practices implemented.

While the overarching common goal that brought the peacebuilding community together was undoubted, it was often overshadowed by diverging interim goals that each peacebuilder audience had, such as timely project completion, addressing budget limitations, and dealing with pending tasks for the grant-giving projects. The urgency of these interim goals influenced the decision-making for peacebuilders, particularly for local stakeholders, and often prevented the longevity and sustainability of partnerships with Visiting Peacebuilders. As a result, the common goal, although evident, was often lost between International Organisation Representatives and Local Stakeholders. At the same time, the hybridity and short-term involvement of Visiting Peacebuilders restricted many encounters with Local Stakeholders from going beyond superficial interactions, in the absence of effective structures that could assist the exchange between local expertise and international best practices – and potentially contribute to the continuance of these encounters.

Peacebuilder Interaction and the Peacebuilder Paradigm

The application of the Contact Hypothesis to assess the intergroup dynamics between the audiences of professionals involved in the Cyprus peacebuilding discourse reveal a set of interesting observations regarding peacebuilding effectiveness for the ethnically-partitioned island. First, the establishment of hierarchical relations across peacebuilder groups and the lack of emancipatory, fourth-generation peacebuilding structures hindered the optimal and organic interaction between the peacebuilder audiences involved. This led to superficial encounters between visiting and local peacebuilders that lacked sustainability, failing to achieve noteworthy peacebuilding impact. Moreover, the status of international agencies as the institutional authorities overseeing the projects implemented did not allow for their representatives to collaborate with local peacebuilders as equal partners. This observation suggests that the hybrid audience of Visiting Peacebuilders is more likely to achieve equal and organic partnerships with local peacebuilders than representatives of donor agencies.

The Contact Hypothesis suggests that peacebuilder dynamics across the examined audiences would optimise if the four conditions above are applied to a larger extent than they have been applied until now. For the case of the Cyprus peacebuilding scene, this can be achieved by establishing equal local-international partnerships, while maintaining some form of institutional authority support. Differentiating between international experts that are locally-stationed as international agency representatives and short-term visiting international experts allows for the reconceptualisation of peacebuilder encounters in a more optimal way.

As suggested by the findings, international agency representatives can maintain an authority position and refrain from direct implementation in peacebuilding projects, while supporting Visiting and Local Peacebuilders to create organic partnerships. These partnerships will foster equal and reciprocal encounters, and intergroup cooperation towards the common goal of sustainable, positive peace, satisfying Allport's optimal conditions for intergroup contact. This triangular structure for intergroup peacebuilder interaction, or what can be referred to as the *Peacebuilder Paradigm*, allows both local and international experts to contribute towards international peacebuilding initiatives in a balanced manner, while enabling international agencies to maintain their institutional authority as facilitative agents. The *Peacebuilder Paradigm* and the structures for optimal peacebuilder contact it entails are further demonstrated in the model below:

FUTURE RESEARCH DIRECTIONS

Conflict Resolution research has put forth two main observations regarding the nature and effective implementation of international peacebuilding initiatives. The first is the hybridity of peacebuilding professionals (Mac Ginty, 2015; Richmond, 2009), and the second is the overlooked agency of local peacebuilding stakeholders (Jabri, 2013; Sending, 2009). The *Peacebuilder Paradigm* incorporates both observations by introducing Visiting Peacebuilders as a hybrid audience of contributors to peacebuilding projects, and at the same time providing a structure of peacebuilder interaction that acknowledges the centrality of local professionals.

The hybridity of the Visiting Peacebuilder audience entails a wide diversity of members, each offering a unique perspective and contribution to a project's objectives. Looking at Visiting Peacebuilders from a Tourism perspective, one can refer to this audience as a subcategory of peace tourists that are characterised by their direct and professional involvement in international peacebuilding projects. International peacebuilding settings can therefore be conceived as prospective destinations for a niche audience of professional peace tourists.

Figure 1. The Peacebuilder Paradigm

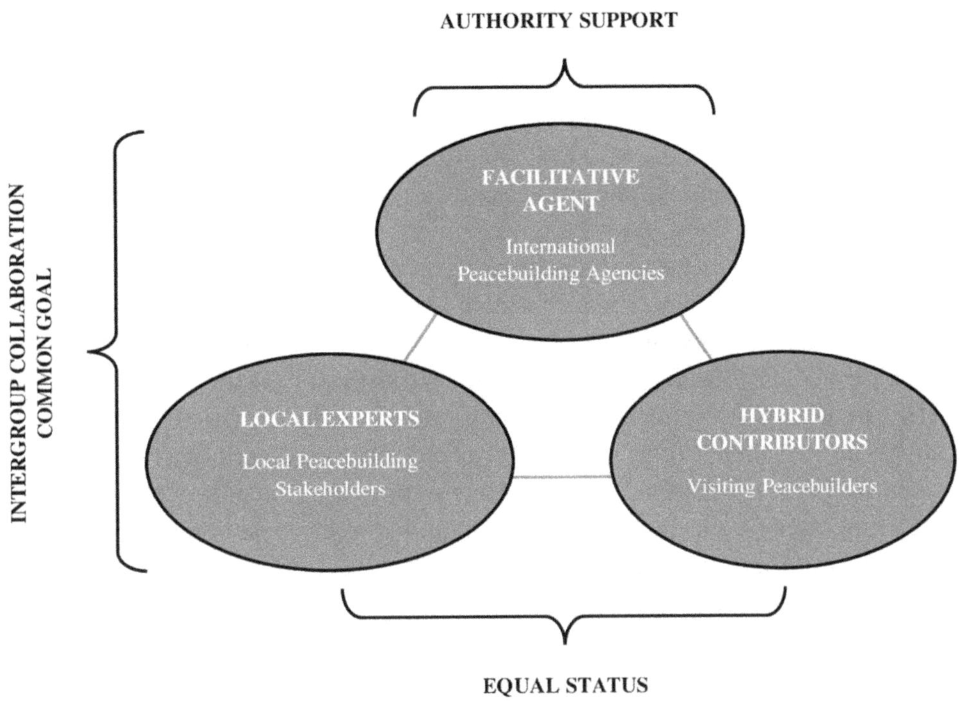

In addition to revisiting the categorisation of peacebuilding professionals, the *Peacebuidler Paradigm* adjusts the central role international peacebuilding agencies held within first, second, and third-generation peacebuilding structures to a less partial, facilitative role. This allows international organisations to support the interactions across Visiting and Local Peacebuilders and to act as the mediator in the possibility of a dispute. Mediation has been widely used in peacebuilding to act as a bottom-up approach to reconciling rival communities. Zehr (2008) discusses Victim-Offender Mediation (VOM) as a popular form of restorative justice in processes of reconciliation, while Sending (2009: 22) argues that mediation has the ability to overcome the 'arrogance of peacebuilding' by empowering locals and other direct stakeholders to take ownership of the reconciliation process. From an analogous approach, De Coning (2018) agrees that peacebuilders should only assume a facilitative role in order to assist the process of reconciliation by simply enabling self-organisation. These approaches are in line with fourth-generation peacebuilding (Richmond *et al.*, 2011), or what Abu-Nimer (2001) defines as *elicitive peacebuilding,* which emphasises local agency in peacebuilding processes and the vitality of local ownership in reconciliation initiatives. Nevertheless, it is important to apply emancipatory peacebuilding practices not only to the communities involved in conflict, but most importantly to the peacebuilders themselves.

The *Peacebuilder Paradigm* comes to suggest a novel structure for intergroup interaction across peacebuilder audiences, which enables their emancipation and balanced professional contribution by applying the optimal conditions for intergroup contact put forth by the Contact Hypothesis. This structure suggests the direct interaction between local peacebuilders and visiting experts, with locally-stationed international agencies assuming a facilitative role across the two groups of peacebuilding experts. The *Peacebuilder Paradigm* puts local peacebuilding professionals at the forefront of peacebuilding initiatives, while incorporating the international expertise of visiting contributors, and acknowledging the authority of international funding agencies. The latter can act to ensure reciprocal and emancipated interactions across local and international experts without taking control of the process, and acting as a mediator when local and international perspectives clash.

Mediation can, in the context of the *Peacebuilder Paradigm*, act as a facilitative tool for improving and encouraging positive relations across peacebuilding professionals, while encouraging the active participation of travelling peacebuilding professionals to engage in locally-stationed, internationally supported peacebuilding initiatives. Identifying visiting peacebuilders as a distinct stakeholder in international peacebuilding allows for the reconceptualisation of peacebuilding structures in a way that more effectively embraces fourth-generation peacebuilding and enables optimal intergroup contact.

Findings from Cyprus are not exclusive to the ethnically partitioned Mediterranean island. Cyprus and the Israeli-Palestinian conflict share the common characteristics of identity-based intractable conflict, for which conflict transformation efforts on the ground have been ineffective. It is additionally a conflict setting that attracts a diverse range of peace tourists and VPs on a continuous basis. As the peacebuilding course in Cyprus shows, socio-political developments such as the Annan Peace Plan failure can have a drastic effect on processes of conflict transformation and severely delay their progress. Nevertheless, applying the appropriate tools and establishing more solid structures for peacebuilding can gradually and more effectively lead to viable, positive peace.

The *Peacebuilder Paradigm* can be an appropriate tool for a diversity of conflict settings, as it does not dictate the approach through which peacebuilding activities will be implemented, but instead provides the guidelines for bringing the appropriate stakeholders together and enabling them to set their own dynamics. It is imperative, however, to assess the practical applicability of this structure and devote future research to reaffirming or disproving its validity and viability. Future research projects can focus on structuring designated international peacebuilding settings on the basis of the *Peacebuilder Paradigm* (see Figure 1), and as such assess its applicability and ability to establish optimal peacebuilder collaborations.

CONCLUSION

International and local peacebuilders are catalytic to the success of international peacebuilding projects. Yet acknowledging their contributions as the outcome of a dyadic collaboration is restrictive and inaccurate. By examining International Peacebuilding from a Tourism perspective, this study revisited the categorisation of peacebuilding professionals and identified an underexplored audience of hybrid, short-term contributors to peacebuilding projects – the Visiting Peacebuilders. Acknowledging this category of overlooked contributors can more accurately inform international peacebuilding projects on the diversity of stakeholders involved in their implementation, and on the significance of their contributions. In addition, identifying Visiting Peacebuilders as a distinct audience in international peacebuilding endorses the practical application of fourth-generation peacebuilding through the *Peacebuilder Paradigm*, a structure of peacebuilder interaction that revisits power imbalances in peacebuilding settings and embraces the emancipation of local stakeholders.

Findings from Cyprus present three key audiences of peacebuilding professionals: International Organisation Representatives, Local Stakeholders and Visiting Peacebuilders. The Cyprus peacebuilding discourse indicates that interactions between the three audiences evolved from a period of 'expert-native' relationships to a peace-oriented civil society that granted agency to both local and international peacebuilders. Early activities used locals as participants to the research studies of foreign experts, whereas more recent initiatives such as interregional conferences featured the work and input of both local and international experts. Local stakeholders reported power imbalances in their relations with international agencies despite this shift, indicating that their presence and work could have been more emancipatory to local audiences.

Relations between the three identified peacebuilder audiences can be further optimised based on the insights provided by the case study of Cyprus. Effective communication, reciprocal interaction, and positive encounters among peacebuilders can be achieved if Visiting Peacebuilders more actively assume the role of the international counterpart in international peacebuilding projects. This suggests establishing appropriate structures that put Visiting Peacebuilders at the forefront of international peacebuilding initiatives, such as interregional conferences and workshops. At the same time, locally-stationed representatives of international peacebuilding agencies should assume a less involved, mediating role that allows them to establish their institutional authority in the projects without creating power imbalances and unequal interactions among peacebuilder professionals. This would include less rigid structures to the projects, allowing local feedback to be considered throughout the design, implementation and evaluation of the projects. This structure of intergroup interaction between peacebuilding professionals, introduced here as the *Peacebuilder Paradigm*, allows for a more balanced and reciprocal interaction between local and global peacebuilding agents, providing international peacebuilding settings with optimal conditions for intergroup contact and as a result, increasing their effectiveness.

REFERENCES

H4C. (2015). *Home for Cooperation: About Us*. Retrieved from www.home4cooperation.info

Abu-Nimer, M. (2001). Conflict resolution, culture, and religion: Toward a training model of interreligious peacebuilding. *Journal of Peace Research*, *38*(6), 685–704. doi:10.1177/0022343301038006003

Acharya, A. (2004). How ideas spread: Whose norms matter? Norm localisation and institutional change in Asian regionalism. *International Organization*, *58*(2), 239–275. doi:10.1017/S0020818304582024

AHDR. (2015). *Association for Historical Dialogue and Research: Calendar.* Retrieved from www.ahdr.info

Allport, G. (1954). *The Nature of Prejudice*. Addison-Wesley Publishing Company, Inc.

Anastasopoulos, P. G. (1992). Tourism and attitude change: Greek tourists visiting Turkey. *Annals of Tourism Research*, *19*(4), 629–642. doi:10.1016/0160-7383(92)90058-W

Autesserre, S. (2014). *Peaceland: Conflict Resolution and the Everyday Politics of International Intervention*. Cambridge University Press. doi:10.1017/CBO9781107280366

Azar, E. E. (1990). Protracted International Conflicts: Ten Propositions. In J. Burton & F. Dukes (Eds.), *Conflict: Readings in Management and Resolution*. Springer.

Bar-Tal, D., & Bennink, G. H. (2004). The Nature of Reconciliation as an Outcome and as a Process. In From Conflict Resolution to Reconciliation. Oxford: Oxford University Press.

Barnett, M., & Zurcher, C. (2009). The peacebuilder's contract: how external statebuilding reinforces weak statehood. In R. Paris & T. D. Sisk (Eds.), *The Dilemmas of Statebuilding: Confronting the Contradictions of Postwar Peace Operations*. Routledge.

Bekerman, Z. (2007). Rethinking intergroup encounters: Rescuing praxis from theory, activity from education, and peace co-existence from identity and culture. *Journal of Peace Education*, *4*(1), 21–37. doi:10.1080/17400200601171198

Blanchard, L. A., & Higgins-Desbiolles, F. (Eds.). (2013). *Peace through tourism: promoting human security through international citizenship*. Routledge. doi:10.4324/9780203370384

Brin, E. (2006). Politically-oriented tourism in Jerusalem. *Tourist Studies*, *6*(3), 215–243. doi:10.1177/1468797607076672

Broome, B. J. (1998). Overview of conflict resolution activities in Cyprus: Their contribution to the peace process. *The Cyprus Review*, *10*(1), 47.

Chaitin, J. (2011). "Here's the Separation Wall": Political tourism in the Holy Land. *Conflict Resolution Quarterly*, *29*(1), 39–63. doi:10.1002/crq.21036

D'Amore, L. (2009). Peace through tourism: The birthing of a new socio-economic order. *Journal of Business Ethics*, *89*(4), 559–568. doi:10.100710551-010-0407-3

De Coning, C. (2018). Adaptive peacebuilding. *International Affairs*, *94*(2), 301–317. doi:10.1093/ia/iix251

Doyle, M. W., & Sambanis, N. (2000). International peacebuilding: A theoretical and quantitative analysis. *The American Political Science Review*, *94*(4), 779–801. doi:10.2307/2586208

Fisher, R. J. (1997). *Interactive Conflict Resolution*. Syracuse University Press.

Foley, D. E. (2002). Critical ethnography: The reflexive turn. *International Journal of Qualitative Studies in Education: QSE*, *15*(4), 469–490. doi:10.1080/09518390210145534

Gobo, G., & Marciniak, L. T. (2016). *What is Ethnography? In Qualitative Research*. Sage Publications.

Higgins-Desbiolles, F. (2008). Justice tourism and alternative globalisation. *Journal of Sustainable Tourism*, *16*(3), 345–364. doi:10.1080/09669580802154132

Higgins-Desbiolles, F., & Blanchard, L. (2010). *Challenging peace through tourism: placing tourism in the context of human rights, justice and peace*. CABI.

Isaac, R. A. M. I. (2014). *Responsible Tourism and Development in the Context of Peace-building. In International Handbook on Tourism and Peace*. Drava.

Jabri, V. (2013). Peacebuilding, the local and the international: A colonial or a postcolonial rationality? *Peacpebuilding*, *1*(1), 3–16. doi:10.1080/21647259.2013.756253

Jarraud, N., Louise, C., & Filippou, G. (2013). The Cypriot Civil Society movement: A legitimate player in the peace process? *Journal of Peacebuilding and Development*, *8*(1), 45–59. doi:10.1080/15423166.2013.791536

Ladini, G. (2009). Peacebuilding, United Nations and Civil Society: The Case of Cyprus. *The Cyprus Review*, *21*(2), 37.

Liutikas, D. (2012). Experiences of valuistic journeys: Motivation and behavior. In R. Sharpley & P. Stone (Eds.), *Contemporary Tourist Experience: Concepts and Consequences*. Routledge.

Mac Ginty, R. (2015). Where is the local? Critical localism and peacebuilding. *Third World Quarterly*, *36*(5), 840–856. doi:10.1080/01436597.2015.1045482

Mahallae. (2015). Retrieved from www.mahallae.org

Makriyianni, C., & Psaltis, C. (2007). The Teaching of History and Reconciliation. *The Cyprus Review*, *19*(1), 43–69.

Mathieu, X. (2019). Critical peacebuilding and the dilemma of difference: The stigma of the 'local'and the quest for equality. *Third World Quarterly*, *40*(1), 36–52. doi:10.1080/01436597.2018.1538732

Moufakkir, O., & Kelly, I. (2010). *Tourism, Progress and Peace*. Cabi. doi:10.1079/9781845936778.0000

Paffenholz, T. (2015). Unpacking the local turn in peacebuilding: A critical assessment towards an agenda for future research. *Third World Quarterly*, *36*(5), 857–874. doi:10.1080/01436597.2015.1029908

Richmond, O., Björkdahl, A., & Kappler, S. (2011). The emerging EU peacebuilding framework: Confirming or transcending liberal peacebuilding? *Cambridge Review of International Affairs*, *24*(3), 449–469. doi:10.1080/09557571.2011.586331

Richmond, O. P. (2007). Emancipatory forms of human security and liberal peacebuilding. *International Journal (Toronto, Ont.)*, *62*(3), 459–478. doi:10.1177/002070200706200302

Richmond, O. P. (2009). Becoming liberal, unbecoming liberalism: Liberal-local hybridity via the everyday as a response to the paradoxes of liberal peacebuilding. *Journal of Intervention and Statebuilding*, *3*(3), 324–344. doi:10.1080/17502970903086719

Richmond, O. P. (2010). A genealogy of peace and conflict theory. In Palgrave Advances in Peacebuilding (pp. 14-38). Palgrave Macmillan UK. doi:10.1057/9780230282681_2

Richmond, O. P. (2012). From peacebuilding as resistance to peacebuilding as liberation. *Rethinking peacebuilding: the quest for just peace in the Middle East and the Western Balkans*, 64-77.

Roberts, D. (2011). Beyond the metropolis? Popular Peace and Post-conflict peacebuilding. *Review of International Studies*, *37*(5), 2535–2556. doi:10.1017/S0260210511000234

Roberts, D. (2011b). *Liberal peacebuilding and global governance: Beyond the Metropolis*. Taylor & Francis. doi:10.4324/9780203825815

Rothman, J. (1999). Articulating Goals and Monitoring Progress in a Cyprus Conflict Resolution Training Workshop. In *Theory and Practice in Ethnic Conflict Management*. Palgrave Macmillan.

Sending, O. J. (2009). Why peacebuilders fail to secure ownership and be sensitive to context. *Security in Practice 1* (Working Paper). Norwegian Institute of International Affairs.

Sonmez, S. F., & Apostolopoulos, Y. (2000). Conflict resolution through tourism cooperation? The case of the partitioned island-state of Cyprus. *Journal of Travel & Tourism Marketing*, *9*(3), 35–48. doi:10.1300/J073v09n03_03

Sozen, A. (2006). *The Cyprus Impasse and Prospects for a Breakthrough. In Reflections on the Cyprus Problem: A Compilation of Recent Academic Contributions*. EMU Printing House.

Stone, P. R. (2006). A Dark Tourism Spectrum: Towards a typology of death and macabre related tourist sites, attractions and exhibitions. *Tourism: An Interdisciplinary International Journal*, *54*(2), 145–160.

Tschirgi, N. (2004). *Post-conflict Peacebuilding Revisited: Achievements, Limitations, Challenges*. International Peace Academy.

UNDP. (2015, December). Final Report. *Action for Cooperation and Trust in Cyprus*, *1*(October), 2005–2031.

Van den Dungen, P. (2014). *Peace Tourism. In International Handbook on Tourism and Peace*. Drava.

Woodward, S. L. (2017). *The Ideology of Failed States*. Cambridge University Press. doi:10.1017/9781316816936

Yanow, D. (2009). Organisational ethnography and methodological angst: Myths and challenges in the field. *Qualitative Research in Organisations and Management: An International Journal*, *4*(2), 186–199. doi:10.1108/17465640910978427

Zehr, H. (2008). Doing justice, healing trauma: The role of restorative justice in peacebuilding. *South Asian Journal of Peacebuilding*, *1*(1), 1–16.

KEY TERMS AND DEFINITIONS

Contact Hypothesis: The proposition developed by Gordon Allport in 1954 arguing that positive intergroup contact can reduce prejudice across members of different identity-based groups, and this can occur under four optimal conditions: equal status, support by a social/institutional authority, intergroup cooperation, and a common goal.

Intergroup Contact: Contact between people that affiliate with different social, ethnic or religious identities and classify themselves as members of different identity-based groups.

Mediation: In its non-legal application, mediation refers to a facilitated process—with the mediator being the facilitative agent—that can enhance the interaction and cooperation between two parties, in the attempt to prevent or resolve intergroup conflict.

Peace Tourism: Travel that is specifically motivated by and associated with conflict resolution practices and war prevention.

Peacebuilder Hybridity: The notion that geographical, institutional and other affiliations of professionals involved in international peacebuilding efforts is not static, but rather multifaceted and complex to pinpoint into the binary categorisation of local versus international peacebuilders.

Peacebuilder Paradigm: A structure of optimal interaction between the three main audiences of peacebuilding professionals—International Organisation Representatives, Local Stakeholders, and Visiting

Peacebuilders—under which International Organisation Representatives act as the institutional authority in international peacebuilding, and assume a facilitative role for enhancing the equal and reciprocal collaboration between Local Stakeholders and Visiting Peacebuilders.

Peacebuilding: The set of mechanisms, tools and processes implemented by conflict resolution experts, activists and volunteers in a society transitioning from conflict in order to prevent future conflict from escalating and to promote positive and trusted relations across former rivals.

Visiting Peacebuilders: Professionals, who are involved in the design, delivery or evaluation of international peacebuilding projects on an external, visiting, or ad-hoc basis; their involvement enables them to provide their expertise and consultancy to the project without joining it throughout its duration and without representing the agency funding the project. They can be identified as a niche audience of professional peace tourists.

ENDNOTES

1 United Nations Development Programme.

2 United States Agency for International Development.

3 In 1984, Harvard University professor Herbert Kelman organized the Interactive Problem-solving Workshop and in 1985, Yale University's Leonard Doob organized the Operation Locksmith Workshop. From 1988 to 1993, Ron Fisher (Canadian Institute for Peace and Security) organized four workshops on problem solving and conflict resolution. Analogous workshops were also carried out in 1992 and 1994 with Louise Diamond (Institute of Multi-Track Diplomacy) and Diana Chigas (Conflict Management Group), with the support of USAID and the Cyprus Fulbright Commission. Over the following three years, 32 foreign and 57 local trainers were involved in 47 workshops and 19 dialogue groups. During the same period, Fulbright scholar Benjamin Broome introduced the Interactive Management peace-building technique, while his initiative resulted in 15 new local projects involving both communities. Additional mediation trainings including role-play and dialogue were offered by Marco Turk and John Ungerlider, also Fulbright scholars (Mahallae, 2015; Ladini, 2009).

4 The Annan Plan was a comprehensive peace settlement for the reunification of Cyprus, drafted by UN Secretary General Kofi Annan and put forth to be accepted by both of the island's communities in separately-held referenda. In 2004, the Turkish Cypriot community accepted the plan by 65%, while the Greek Cypriot community yielded a staggering 76% against the Plan (Jarraud *et al.*, 2013).

Compilation of References

A tomar por mundo. (2018). *Estoy hecho de pedacitos de ti, mundo: Mi viaje del ADN*. Retrieved 26 March 2019, from: https://atomarpormundo.com/mi-viaje-del-adn/

Abreu, J. R. (2018). *Mobilidade acadêmica como instrumento soft power: A experiência dos USA, Brasil e México* (PhD Thesis). Universidade de Brasília, Brasília.

Abu-Nimer, M. (2001). Conflict resolution, culture, and religion: Toward a training model of interreligious peacebuilding. *Journal of Peace Research, 38*(6), 685–704. doi:10.1177/0022343301038006003

Acharya, A. (2004). How ideas spread: Whose norms matter? Norm localisation and institutional change in Asian regionalism. *International Organization, 58*(2), 239–275. doi:10.1017/S0020818304582024

Adams, N. (2005). Just like it used to be. In J. Davies (Ed.), A world in harmony: A celebration of the Llangollen International Eisteddfod (pp. 18-19). Llangollen: Friends of the Llangollen International Musical Eisteddfod.

Agencia para la Reincorporación y la Normalización. (2019). *La Reincorporación en cifras* (Publication 10 Dec 2019). Retrieved from http://www.reincorporacion.gov.co/es/reincorporacion/La%20Reincorporacion%20en%20cifras/Reincorporación%20en%20cifras%20-%20corte%2030112019.pdf

Agnihotri, S. (2017). Critical reflection on the role of education as a catalyst of peace-building and peaceful coexistence. *Universal Journal of Educational Research, 5*(6), 911–917. doi:10.13189/ujer.2017.050601

AHDR. (2015). *Association for Historical Dialogue and Research: Calendar.* Retrieved from www.ahdr.info

Aiken, N. T. (2008). Post-conflict peacebuilding and the politics of identity: Insights for restoration and reconciliation in transitional justice. *Peace Research, 40*(2), 9–38,111.

Ajanovic, E., Çizel, B., & Çizel, R. (2016). Effectiveness of Erasmus programme in prejudice reduction: Contact theory perspective. *Turisticko Poslovanje, 17*(17), 47–60. doi:10.5937/TurPos1617047A

Akerhielm, P., Dev, C. S., & Noden, A. A. (2003). Brand Europe: European integration and tourism development. Cornell Hotel and Restaurant Administration Quarterly, 88-93.

Alaoui, B. A. (2017). Morocco's security strategy: Preventing terrorism and countering extremism. *European View, 16*(1), 103–120. doi:10.100712290-017-0449-3

Albu, C. E. (2015). Intercultural communication in tourism. *Transcultural Management Journal, 17*(1), 7–14.

Alexander, M., Bryce, D., & Murdy, S. (2017). Delivering the past: Providing personalised ancestral tourism experiences. *Journal of Travel Research, 56*(4), 543–555. doi:10.1177/0047287516643414

Alfranseder, E., Fellinger, J., Taivere, M., & Krzaklewska, E. (2011). *E-Value-ate your exchange: Research report of the ESNSurvey 2010*. Erasmus Student Network AISBL.

Algarve Regional Coordination and Development Commission. (2020). *RIS3 - Algarve 2014-2020, Estratégia Regional de Investigação e Inovação para a Especialização Inteligente, Cresc Algarve.* http://s3platform.jrc.ec.europa.eu/s3-platform-registered-regions

Allport, G. (1954). *The Nature of Prejudice*. Addison-Wesley Publishing Company, Inc.

Allport, G. W. (1954). *The nature of prejudice*. Addison-Wesley.

Alluri, R. M. (2009). *The role of tourism in post-conflict peacebuilding in Rwanda.* Working Paper, 2. Bern: Swisspeace.

Alluri, R. M. (2009). *The role of tourism in post-conflict peacebuilding in Rwanda*. Swisspeace.

Alluri, R. M., Leicher, M., Palme, K., & Joras, U. (2013). Understanding economic effects of violent conflicts on tourism: Empirical refections from Croatia, Rwanda and Sri Lanka. In C. Wohlmuther & W. Wintersteiner (Eds.), *International Handbook on Tourism and Peace* (pp. 101–119). Drava Verlag.

Alzamora Salom, A. (2006). La espiritualidad del turismo (3rd ed.). Parròquia de Santa Ponça.

Amad, R., & Hertzog, A. (2016). Tourism, memory and place in a globalising world. *Cultural & Tourism Studies, 16*(3), 201–205.

Amir, Y., & Ben-Ari, R. (1985). International tourism, ethnic contact, and attitude change. *The Journal of Social Issues, 41*(3), 105–115. doi:10.1111/j.1540-4560.1985.tb01131.x

Anastasopoulos, P. G. (1992). Tourism and attitude change: Greek tourists visiting Turkey. *Annals of Tourism Research, 19*(4), 629–642. doi:10.1016/0160-7383(92)90058-W

Andersen, P. A., & Wang, H. (2009). Beyond language: Non-verbal communication across cultures. In L. A. Samovar, R. E. Porter, & E. R. McDaniel (Eds.), *Intercultural communication: A reader* (12th ed., pp. 264–281). Wadsworth Cengage Learning.

Anderson, L. E. (1994). A new look at an old construct: Transcultural adaptation. *International Journal of Intercultural Relations, 18*(3), 293–328. doi:10.1016/0147-1767(94)90035-3

Andersson, T. D., & Getz, D. (2008). Stakeholder management strategies of festivals. *Journal of Convention & Event Tourism, 9*(3), 199–220. doi:10.1080/15470140802323801

Andrews, H., & Leopold, T. (2013). *Events and the social sciences*. Routledge. doi:10.4324/9780203070741

Anson, C. (1999). Planning for Peace: The Role of Tourism in the Aftermath of Violence. *Journal of Travel Research, 38*(1), 57–61. doi:10.1177/004728759903800112

Anwar, A. (2019). *How China is using tourists to realise its geopolitical goals*. Retrieved June 4, 2020, from https://www.eastasiaforum.org/2019/09/19/how-china-is-using-tourists-to-realise-its-geopolitical-goals/

Araújo, C. A. (2006). Bibliometria: Evolução histórica e questões atuais. *Questão, 12*, 12–32.

Araujo, D. F., & Ramos, M. C. P. (2014). *Um olhar sobre a educação superior do turismo para uma nova cidadania Intercultural*. Cairu em Revista - Sociedade, Educação, Gestão e Sustentabilidade.

Archer, D., & Finger, K. (2018). *Walking in another's virtual shoes: Do 360-degree video news stories generate empathy in viewers?* Retrieved on 26 March 2020, from: https://www.semanticscholar.org/paper/Walking-in-Another's-Virtual-Shoes:-Do-360-Degree-Archer-Finger/c67c231eecd6b476e8e4e3e3031dae24a9de0e05/figure/0/0

Arendt, H. (2000). *Eichmann em Jerusalém*. Companhia das Letras.

Arita, S., Edmonds, C., Croix, S. L., & Mak, J. (2011). Impact of Approved Destination Status on Chinese travel abroad: An econometric analysis. 17 (5). *Tourism Economics*, *17*(5), 983–996. doi:10.5367/te.2011.0076

Aronson, E., Wilson, T. D., & Akert, R. M. (2012). *Sosyal psikoloji* [*Social psychology*] (O. Gündüz, Trans.). Kaknüs Yayınları.

Aulet, S. (2018). Spiritual Tourism in Europe: the SPIRIT-Youth project. In R. Butler & W. Suntikul (Eds.), *Tourism and Religion. Issues and implications* (pp. 200–219). Channel View Publications.

Autesserre, S. (2014). Going Micro: Emerging and Future Peacekeeping Research. *International Peacekeeping*, *21*(4), 492–500. doi:10.1080/13533312.2014.950884

Autesserre, S. (2014). *Peaceland: Conflict Resolution and the Everyday Politics of International Intervention*. Cambridge University Press. doi:10.1017/CBO9781107280366

Àvila, A. (2019, Jul 17). Los dos posibles caminos de las disidencias de las FARC. *Semana*. Retrieved from https://www.semana.com/opinion/articulo/los-dos-posibles-caminos-de-las-disidencias-de-las-farc-columna-de-ariel-avila/623873

Azar, E. (2003). Protracted social conflicts and second track diplomacy. In J. Davies & E. Kaufman (Eds.), *Second track/citizens's diplomacy: Concepts and techniques for conflict transformation* (pp. 15–30). Rowman & Littlefield.

Azar, E. E. (1990). Protracted International Conflicts: Ten Propositions. In J. Burton & F. Dukes (Eds.), *Conflict: Readings in Management and Resolution*. Springer.

B'tselem. (2019). Retrieved from https://www.btselem.org/

Bache, I., George, S., & Bulmer, S. (2011). *Politics in the European Union*. Oxford University Press.

Bade, K. (2016). *El viaje más interesante de mi vida*. Retrieved 25 March 2019, from: https://worldtravelfeet.com/viaje-mas-interesante-vida-adn/

Bagozzi, R. P. (1994). *Advanced methods of marketing research*. Blackwell Business.

Baker, D. (2014). The effects of terrorism on the travel and tourism industry. *International Journal of Religious Tourism and Pilgrimage*, *2*(1), 58–67. doi:10.21427/D7VX3D

Banda i Tarradellas, A. (1991). *Educació per la pau*. Biblioteca Cultural Barcanova.

Banerjee, T. (2019). One good tourist brings 10 more: ' Travel industry hopes J&K's new status will change fortunes. *ET Online*. Retrieved March 24, 2020, from: https://economictimes.indiatimes.com/magazines/panache/one-good-tourist-brings-10-more-travel-industry-hopes-jks-new-status-will-change-fortunes/articleshow/70586412.cms

Banks, J. A. (2009). Multicultural education: Dimensions and paradigms. In *The Routledge international companion to multicultural education* (pp. 29–52). Routledge. doi:10.4324/9780203881514

Barakat, S., & Chard, M. (2002). Theories, rhetoric and practice: Recovering the capacities of war-torn societies. *Third World Quarterly*, *23*(5), 817–835. doi:10.1080/0143659022000028639

Barlow, M. (1988). Tourism, peace, and conflict: A geographer's perspective. *Tourism-a Vital Force For Peace, 108.*

Barnett, M., & Zurcher, C. (2009). The peacebuilder's contract: how external statebuilding reinforces weak statehood. In R. Paris & T. D. Sisk (Eds.), *The Dilemmas of Statebuilding: Confronting the Contradictions of Postwar Peace Operations*. Routledge.

Barometer, C. (2020). *The Heidelberg Institute for Conflict Research (HIIK) and Institute of Political Science at Heidelberg University*. Retrieved March 18, 2020, from: https://hiik.de/conflict-barometer/current-version/?lang=en

Bar-On, D., & Kassem, F. (2004). Storytelling as a way to work through intractable conflicts: The German-Jewish experience and its relevance to the Palestinian-Israeli context. *The Journal of Social Issues*, *60*(2), 289–306. doi:10.1111/j.0022-4537.2004.00112.x

Bar-Siman-Tov, Y. (Ed.). (2004). *From conflict resolution to reconciliation*. Oxford University Press. doi:10.1093/acprof:oso/9780195166439.001.0001

Bar-Tal, D., & Bennink, G. H. (2004). The Nature of Reconciliation as an Outcome and as a Process. In From Conflict Resolution to Reconciliation. Oxford: Oxford University Press.

Basheer, K. A. (2019). *Psychology for B.SC. Nursing* (2nd ed.). PHI Learning Pvt. Ltd.

Bassas, A. (2008). *El matí amb Raimon Panikkar*. Proa.

Bauman, Z. (2007). *Tiempos líquidos: vivir en una época de incertidumbre*. Ed. Ensanyo TusQuets.

BBC News. (2012). *China travel agencies suspend Philippine tours*. Retrieved from BBC News: https://www.bbc.com/news/world-asia-18015538

Beaver, A. (2012). *A dictionary of travel and tourism*. Oxford University Press.

Becken, S., & Carmignani, F. (2016). Does tourism lead to peace? *Annals of Tourism Research*, *61*, 63–79. doi:10.1016/j.annals.2016.09.002

Beerli, A., Díaz Meneses, G., & Moreno Gil, S. (2007). Self-congruity and destination choice. *Annals of Tourism Research*, *34*(3), 571–587. doi:10.1016/j.annals.2007.01.005

Bekerman, Z. (2007). Rethinking intergroup encounters: Rescuing praxis from theory, activity from education, and peace co-existence from identity and culture. *Journal of Peace Education*, *4*(1), 21–37. doi:10.1080/17400200601171198

Belford, N. (2017). International students from Melbourne describing their transcultural transition experiences: Culture shock, social interaction, and friendship development. *Journal of International Students*, *7*(3), 499–521. doi:10.32674/jis.v7i3.206

Benfell, G. (2006). *Interview by D. Challinor: New Zealand National Army Museum*. Waiouru.

Bennett, M. J. (1984). *Towards ethnorelativism: A developmental model of intercultural sensitivity*. Paper presented at the Annual Conference of the Council on International Exchange, Minneapolis, MN.

Bennett, M. J. (1993). Towards ethnorelativism: A developmental model of intercultural sensitivity. In M. Paige (Ed.), Education for the intercultural experience. Yarmouth: Intercultural Press.

Bennett, M. J. (1986). A developmental approach to training for intercultural sensitivity. *International Journal of Intercultural Relations*, *10*(2), 179–195. doi:10.1016/0147-1767(86)90005-2

Bennett, M. J. (2004). Becoming interculturally competent. In J. Wurzel (Ed.), *Toward multiculturalism: A reader in multicultural education* (2nd ed., pp. 62–77). Intercultural Resource.

Bennett, M. J. (2013). *Basic concepts of intercultural communication: Paradigms, principles, & practices*. Intercultural Press.

Berdal, M. R. (1996). *Disarmament and demobilisation after civil wars: arms, soldiers and the termination of armed conflicts. Adelphi Paper 303*. Oxford University Press for the International Institute for Strategic Studies.

Berger, P. L. (1975). *Rumor de ángeles: La sociedad moderna y el descubrimiento de lo sobrenatural*. Herder.

Berno, T., & Ward, C. (2005). Innocence abroad: A pocket guide to psychological research on tourism. *The American Psychologist*, *60*(6), 593–600. doi:10.1037/0003-066X.60.6.593 PMID:16173892

Bhati, A., & Pearce, P. (2016). Vandalism and tourism settings: An integrative review. *Tourism Management*, *57*, 91–105. doi:10.1016/j.tourman.2016.05.005

Bhuiyan, M. A. H., Islam, R., Siwar, C., & Ismail, S. M. (2010). Educational tourism and forest conservation: Diversification for child education. *Procedia: Social and Behavioral Sciences*, *7*, 19–23. doi:10.1016/j.sbspro.2010.10.003

Biaett, V. (2015). Organic festivity: A missing element of community festival. In A. Jepson & A. Clark (Eds.), *Exploring community festivals and events* (pp. 7–30). Routledge.

Bilecen, B., & Van Mol, C. (2017). Introduction: International academic mobility and inequalities. *Journal of Ethnic and Migration Studies*, *43*(8), 1241–1255. doi:10.1080/1369183X.2017.1300225

Bimonte, S., & Faralla, V. (2012). Tourist types and happiness a comparative study in Maremma, Italy. *Annals of Tourism Research*, *39*(4), 1929–1950. doi:10.1016/j.annals.2012.05.026

Bizarro, R., & Braga, F. (2006). *Formação de professores de línguas estrangeiras: Reflexões, Estudos e Experiências*. Porto Editora.

Bizjak, B., Knežević, M., & Cvetrežnik, S. (2011). Attitude change towards guests with disabilities: Reflections from tourism students. *Annals of Tourism Research*, *38*(3), 842–857. doi:10.1016/j.annals.2010.11.017

Björkdahl, A., & Buckley-Zistel, S. (Eds.). (2016). *Spatialising Peace and Conflict: Mapping the Production of Places, Sites and Scales of Violence*. Palgrave Macmillan.

Blake, A., & Sinclair, M. T. (2003). Tourism crisis management: US response to September 11. *Annals of Tourism Research*, *30*(4), 813–832. doi:10.1016/S0160-7383(03)00056-2

Blanchard, J. F., & Hickey, D. V. (2012). More than two 'sides' to every story: An introduction to new thinking about the Taiwan Issue. In J.-M. Blanchard & D. V. Hickey (Eds.), *New thinking about the Taiwan issue: Theoretical insights into its origins, dynamics, and prospects* (pp. 1–27). Routedge. doi:10.4324/9780203131497

Blanchard, L.-A., & Desbiolles, H. F. (Eds.). (2013). *Peace through tourism: Promoting human security through international citizenship*. Routledge. doi:10.4324/9780203370384

Boada, I. (Ed.). (2004). *La filosofia intercultural de Raimon Panikkar*. Pòrtic.

Bonilla, J. (2018). *Un sendero para la construcción de paz en Choachí*. Blogs - El Tiempo. http://blogs.eltiempo.com/felicidad-desarrollo-turismo/2018/11/27/sendero-la-construccion-paz-choachi/

Bossuat, G. (2009). *Histoire de l'Union Européenne: Fondations, élargissements, avenir*. Paris: Belin.

Bourdieu, P. (1996). *Razões práticas: Sobre a teoria da ação*. Papirus Editora.

Boutros-Ghali, B. (1992). *An agenda for peace: Preventive diplomacy, peacemaking and peacekeeping*. United Nations-Security Council. doi:10.1177/004711789201100302

Bowd, R., & Özerdem, A. (2013). How to Assess Social Reintegration of Ex-Combatants. *Journal of Intervention and Statebuilding*, *7*(4), 453–475. doi:10.1080/17502977.2012.727537

Bowen, G. A. (2009). Document analysis as a qualitative research method (Report). *Qualitative Research Journal*, *9*(2), 27.

Bowen, J. (2020, January 29). Trump's Middle East peace plan: 'Deal of the century' is huge gamble. *The BBC*. Retrieved from https://www.bbc.com/news/world-middle-east-51263815

Braithwaite, D., & Lee, Y. L. (2006). Dark tourism, hate and reconciliation: The Sandakan experience. *IIPT Occasional Paper, 8*.

Branscombe, N. R., & Miron, A. M. (Eds.). (2004). *The social life of emotions*. Cambridge University Press.

Brewer, M. B. (1999). The psychology of prejudice: Ingroup love and outgroup hate? *The Journal of Social Issues, 55*(3), 429–444. doi:10.1111/0022-4537.00126

Brin, E. (2006). Politically-orientated tourism in Jerusalem. *Tourist Studies, 6*(3), 215–243. doi:10.1177/1468797607076672

Brin, E., & Noy, C. (2010). The said and the unsaid: Performative guiding in a Jerusalem neighbourhood. *Tourist Studies, 10*(1), 19–33. doi:10.1177/1468797610390982

Bronkhorst, S. van J., & Bob, U. (2010). Environmental conflicts: Key issues and management implications. *African Journal on Conflict Resolution*, (2). Retrieved March 20, 2019, from https://www.accord.org.za/ajcr-issues/%EF%BF%BCenvironmental-conflicts/

Broome, B. J. (1998). Overview of conflict resolution activities in Cyprus: Their contribution to the peace process. *The Cyprus Review, 10*(1), 47.

Brougère, G. (2012). Pratiques touristiques et apprentissages. *Mondes du tourisme, 5*, 62-75. doi:10.4000/tourisme.415

Brown, K. (2016). *CEO, China: The rise of Xi Jinping*. London: I.B.Tauris.

Brown, K. (2017). *What does China want? China's World*. London: I.B.Tauris.

Brown, L. (2008). Interviewer: B. Beach. Vietnam War Oral History Project (Vol OHInt-0959-07), VOH, Wellington, NZ.

Brown, F. (1989). Is tourism really a peacemaker? *Tourism Management, 10*(4), 270–271. doi:10.1016/0261-5177(89)90001-0

Brown, K. (2017). *China's world*. I.B. Tauris. doi:10.5040/9781350985704

Brown, R., & Hewstone, M. (2005). *An integrative theory of intergroup contact* (Vol. 37). Advances in Experimental Social Psychology. Academic Press.

Brundtland, I. (1987) *Informe: Nuestro futuro común*. Comisión Mundial sobre Medio Ambiente y Desarrollo.

Bruner, E. M. (1991). Transformation of self in tourism. *Annals of Tourism Research, 18*(2), 238–250. doi:10.1016/0160-7383(91)90007-X

Brunn, D. S., Bryne, S., McNamara, L., & Egan, A. (2010). Belfast landscapes: From religious schism to conflict tourism. *Focus on Geography, 53*(3), 81–123. doi:10.1111/j.1949-8535.2010.00011.x

Buda, D. M., d'Hauteserre, A.-M., & Johnston, L. (2014). Feeling and tourism studies. *Annals of Tourism Research, 46*, 102–114. doi:10.1016/j.annals.2014.03.005

Burnett, G. W. (1990). On the nature of peace in relationship to tourism: Three cases. *Tourism Review, 45*(1), 2–7.

Burns, P. M., & Novelli, M. (Eds.). (2007). *Tourism and politics*. Routledge. doi:10.4324/9780080467184

Butler, R. (2019). Tourism and conflict: A framework for examining risk versus satisfaction. In R. Isaac, E. Cakmak, & R. Butler (Eds.), *Tourism and hospitality in conflict-ridden destinations* (pp. 13–24). Routledge. doi:10.4324/9780429463235-2

Butler, R. W., & Mao, B. (2007). Conceptual and theoretical implications of tourism between partitioned states. *Asia Pacific Journal of Tourism Research*, 25–34.

ButlerR.SuntikulW. (2013). *Tourism and war* Retrieved from http://otago.eblib.com.au/patron/FullRecord.aspx?p=1092763

Cajuda, J. (2016). *The DNA journey – Momondo*. Retrieved 26 March 2019, from: http://www.joaocajuda.com/the-dna-journey-momondo/

Cameron, C. M., & Gatewood, J. B. (2000). Excursions into the un-remembered past: What people want from visits to historical sites. *The Public Historian, 22*(3), 107–127. doi:10.2307/3379582

Cameron, L., & Rutland, A. (2006). Extended contact through story reading in school: Reducing children's prejudice toward the disabled. *The Journal of Social Issues, 62*(3), 469–488. doi:10.1111/j.1540-4560.2006.00469.x

Campbell, T., Pickett, T., & Yoash-Gantz, R. (2010). Psychological rehabilitation for US Veterans. In E. Martz (Ed.), *Trauma rehabilitation after war and conflict* (pp. 159–176). Springer New York. doi:10.1007/978-1-4419-5722-1_8

Caneday, L. (1991). Chapter. In J. Zeigler & L. Caneday (Eds.), Tourism and Leisure: Dynamics and Diversity (pp. 83–91). National Recreation and Parks Association.

Carbone, F., Oosterbeek, L., & Costa, C. (2013). Paideia approach for heritage management: The tourist enhancement of archaeological heritage on behalf of local communities. *Pasos - Journal of Tourism and Cultural Heritage, 2*(2), 285-295.

Carbone, F. (2010). Valorização turística do património arqueológico em prol da sociedade: O Paideia approach. *Boletim ICOM, 2*(11), 2–10.

Carbone, F. (2016). Cultural heritage quality management: Analysis of archaeological heritage managers' perception. *European Journal of Tourism Research, 14*, 114–118.

Carbone, F. (2017). International tourism and cultural diplomacy: A new conceptual approach towards global mutual understanding and peace through tourism. *Tourism. An International Interdisciplinary Journal, 65*(1), 61–74.

Carbone, F. (2017). International tourism and cultural diplomacy: A new conceptual approach towards global mutual understanding and peace through tourism. *Tourism: An International Interdisciplinary Journal, 65*, 61–74.

Carbone, F. (2018a). A highest purpose of archaeology: Towards a future worthy of our children. *Global Journal of Archaeology and Antropology, 6*(4). Advance online publication. doi:10.19080/GJAA.2018.06.555693

Carbone, F. (2018b). Post-multicultural challenges for cultural heritage managers and museums in the age of migrations. *Museum Management and Curatorship*, 1–22.

Carbone, F. (2019). Tourism and peace: a tool for global integration and peace. In L. Oosterbeek & L. Caron (Eds.), *Resilience and Transformation in the territories of low demographic density: integrated methodologies of human and social sciences for integrated cultural landscape management* (pp. 149–165). Instituto Terra e Memoria.

Carbone, F., Oosterbeek, L., & Costa, C. (2012). Paideia approach: Bridging tourism and community development on the basis of heritage enhancement. *Journal of Tourism and Development, 17/18*, 165–166.

Carius, A. (2006). *Environmental Peacebuilding: Conditions for Success* (12th ed.). Woodrow Wilson International Center for Scholars.

Carneiro, M. J., Eusébio, C., & Caldeira, A. (2018). The influence of social contact in residents' perceptions of the tourism impact on their quality of life: A structural equation model. *Journal of Quality Assurance in Hospitality & Tourism, 19*(1), 1–30. doi:10.1080/1528008X.2017.1314798

Caudery, T., Petersen, M., & Shaw, P. (2007). The language environments of exchange students at Scandinavian universities. *Researching Content and Language Integration in Higher Education*, 233-250.

Causevi. S (2010). Tourism which erases borders: An introspection into Bosnia and Herzegovina. In O. Moufakkir & I. Kelly (Eds.), *Tourism, progress and peace* (pp. 48-64). Oxfordshire: Cab International.

Causevic, S., & Lynch, P. (2011). Phoenix tourism: Post-conflict tourism role. *Annals of Tourism Research*, *38*(3), 780–800. doi:10.1016/j.annals.2010.12.004

Causevic, S., & Lynch, P. (2013). Political (in) stability and its influence on tourism development. *Tourism Management*, *34*, 145–157. doi:10.1016/j.tourman.2012.04.006

CCTV 4. (2017). *CNTA publicou alerta sobre as viagens para a Coreia do Sul*. Retrieved from http://tv.cctv.com/2017/03/04/VIDERw2uBWJ4chDEBjFpxIva170304.shtml

CCTV News. (2012). *A China suspendeu as viagens para as Filipinas*. Retrieved from http://www.iqiyi.com/w_19rs0d7n5x.html#curid=3278338409_dc21c91e039390dfc405cce5d1a6fe2b

Çelik, S. (2018). Impacts of personality traits and tolerance levels of local tourists on perceived destination satisfaction. *Tourism Academic Journal*, *5*(1), 154–165.

Çelik, S. (2019a). Does tourism reduce social distance? A study on domestic tourists in Turkey. *Anatolia*, *30*(1), 115–126. Advance online publication. doi:10.1080/13032917.2018.1517267

Çelik, S. (2019b). Does tourism change tourist attitudes (prejudice and stereotype) towards local people? *Journal of Tourism and Services*, *10*(18), 35–46. doi:10.29036/jots.v10i18.89

Centeno, M. A. (2002). *Blood and debt: War and the nation-state in Latin America*. Park University Press, Penn State Press.

Centro Nacional de Memoria Histórica. (2012). *¡Basta ya! Colombia: Memorias de guerra y dignidad*. Retrieved April 2020, from http://www.centrodememoriahistorica.gov.co/micrositios/informeGeneral/descargas.html

Ceriani, G., Duahamel, Ph., Knafou R. & Stock, M. (2005). Le tourisme et la rencontre de l'autre. Voyage au pays des idées reçues. *L'autre, cliniques, cultures et sociétés*, *6*(1), 71-82.

Chaitin, J. (2011). "Here's the Separation Wall": Political tourism in the Holy Land. *Conflict Resolution Quarterly*, *29*(1), 39–63. doi:10.1002/crq.21036

Chalcraft, J., Magauadda, P., Solaroli, M., & Santoro, M. (2011). Music festivals as cosmopolitan spaces. In European Commission (Ed.), European arts festivals: Strengthening cultural diversity (pp. 25-36). Luxembourg: Publications Office of the European Union.

Challinor, D. (1998). *Talking the talk: New Zealanders remember the Vietnam War*. University of Waikato.

Challinor, D. (2009). *Grey ghosts: New Zealand Vietnam vets talk about their war*. HarperCollins.

Chan, T., & Rosario, L. (2017). *Pearl River Delta: From the world factory to global innovator*. Instituto Internacional de Macau.

Charmaz, K. (2006). Constructing grounded theory: A practical guide through qualitative analysis. *Sage (Atlanta, Ga.)*.

Charter, E. (2012-2016). *Earth Charter*. Recovered from: https://earthcharter.org/

Chauhan, V., & Khanna, S. (2009). Tourism: A tool for crafting peace process in Kashmir, J&K, India. *Tourismos*, *4*(2), 69–89.

Chen, G. M. (2009). *The impact of intercultural sensitivity on ethnocentrism and intercultural communication apprehension*. Intercultural Communication Studies XIX, University of Rhode Island. Retrieved on March 25 2020, from: https://web.uri.edu/iaics/files/01Guo-MingChen.pdf

Chen, D. P. (2017). *US-China rivalry and Taiwan's mainland policy: Security, nationalism, and the 1992 consensus*. Palgrave Macmillan. doi:10.1007/978-3-319-47599-8

Chen, G. M., & Starosta, W. J. (1996). Intercultural communication competence: A synthesis. *Communication Yearbook*, *19*, 353–384.

Chen, G. M., & Starosta, W. J. (1997). A review of the concept of intercultural sensitivity. *Human Communication*, *1*, 1–16.

Chen, G. M., & Starosta, W. J. (2000). The development and validation of the intercultural sensitivity scale. *Human Communication*, *3*, 1–15.

Chen, G. M., & Starosta, W. J. (2003). A review of the concept of intercultural awareness. In L. A. Samovar & R. E. Porter (Eds.), *Intercultural communication: A reader* (pp. 344–353). Wadsworth.

Chen, G. M., & Starosta, W. J. (2005). *Foundations of intercultural communication*. University Press of America.

Cheng, M., & Wong, A. I. (2014). Tourism and Chinese popular nationalism. *Journal of Tourism and Cultural Change*, *12*(4), 307–319. doi:10.1080/14766825.2014.914948

Cheng, Y.-H. (2010). Exploring passenger anxiety associated with train travel. *Transportation*, *37*(6), 875–896. doi:10.100711116-010-9267-z

Chew, E. Y. T., & Jahari, S. A. (2014). Destination image as a mediator between perceived risks and revisit intention: A case of post-disaster Japan. *Tourism Management*, *40*, 382–393. doi:10.1016/j.tourman.2013.07.008

Chhabra, D., Healy, R., & Sills, E. (2003). Staged authenticity and heritage tourism. *Annals of Tourism Research*, *30*(3), 702–719. doi:10.1016/S0160-7383(03)00044-6

Chiang, M.-H. (2018). *China-Taiwan rapprochement: The political economy of cross-straits relations*. Routledge.

Chilo Gorge Safari Lodge. (2019). *The Lodge*. Retrieved April 7, 2020 from the Chilo Gorge Image Gallery.

Chilo Gorge Safari Lodge. (2020). *About*. Retrieved January 5, 2020, from http://chilogorge.com/about/

China, D. (2019). *Decoding the outline development plan for Guangdong-Hong Kong-Macao Greater Bay Area (GBA)*. Retrieved on March 27 2020, from: https://www2.deloitte.com/content/dam/Deloitte/cn/Documents/international-business-support/deloitte-cn-csg-decoding-greater-bay-area-plan-en-190415.pdf

Chinese, V. O. A. (2012). *A China vai retirar todos os turistas chineses nas Filipinas*. Retrieved from VOAChinese: https://www.voachinese.com/a/article-20120514-tourists-return-151385825/958490.html

Chiutsi, S., & Saarinen, J. (2017). Local participation in transfrontier tourism: Case of Sengwe community in Great Limpopo Transfrontier Conservation Area, Zimbabwe. *Development Southern Africa*, *34*(3), 260–275. doi:10.1080/0376835X.2016.1259987

Cho, M. (2007). A re-examination of tourism and peace: The case of the Mt. Gumgang tourism development on the Korean Peninsula. *Tourism Management*, *28*(2), 556–569. doi:10.1016/j.tourman.2006.04.019 PMID:32287720

Chow, P. C. Y. (2008). An overview on the dilemma of 'One China' myth versus reality. In P. Chow (Ed.), *The 'One China' dilemma* (pp. 3–16). Palgrave Macmillan. doi:10.1057/9780230611931_1

Christensen, L. B., Johnson, R. B., & Turner, L. A. (2015). *Araştırma yöntemleri-desen ve analiz* [Research methods-pattern and analysis] (A. Aypay & E. Gumus, Eds. & Trans.). Anı Yayıncılık.

Cimarosti, G. (2017). *The DNA journey: I took a DNA test and found about my origins*. Retrieved 28 March 2019, from: https://www.travelreportage.com/2017/02/09/the-dna-journey-i-took-a-dna-test-and-found-about-my-origins/

Classic Journeys. (2018). *DNA travel is a real trip*. Retrieved 26 March 2019, from: https://www.classicjourneys.com/blog/dna-travel-trip/

Clough, N. R. (1993). *Reaching across the Taiwan Strait: People-to-people diplomacy*. Westview Press.

Clubb, G. (2014). "From Terrorists to Peacekeepers": The IRA's Disengagement and the Role of Community Networks. *Studies in Conflict and Terrorism*, *37*(10), 842–861. doi:10.1080/1057610X.2014.941434

Cohen, E. (1972). Toward a sociology of international tourism. *Social Research*, *39*(1), 164–182. doi:10.2307/40970087

Cole, J. M. (2015). *Black Island: Two years of activism in Taiwan*. Scotts Valley: CreateSpace Independent Publishing Platform.

Collins-Kreiner, N. (2019). Pilgrimage tourism-past, present and future rejuvenation: A perspective article. *Tourism Review*, *75*(1), 145–148. doi:10.1108/TR-04-2019-0130

Collins-Kreiner, N., Mansfeld, Y., & Kliot, N. (2006). The reflection of a political conflict in mapping: The case of Israel's borders and frontiers. *Middle Eastern Studies*, *42*(3), 381–408. doi:10.1080/00263200500521230

Colman, A. M. (2008). *A dictionary of psychology* (3rd ed.). Oxford University Press.

Colombia, P. N. U. D. (2019a, Feb 14). *El príncipe heredero Haakon de Noruega, embajador de buena voluntad del PNUD visita Mesetas para inaugurar Paraísos Ocultos*. Retrieved from https://www.co.undp.org/content/colombia/es/home/presscenter/pressreleases/2019/02/el-principe-heredero-haakon-de-noruega--embajador-de-buena-volun.html

Colombia, P. N. U. D. (2019b, Mar 25). *La casa de la vida. Recorrido turístico de naturaleza y reconciliación en el norte y nordeste de Antioquia*. Retrieved from https://pnudcolombia.exposure.co/la-casa-de-la-vida

Comaroff, J. L., & Comaroff, J. (2009). *Ethnicity, Inc*. University of Chicago Press. doi:10.7208/chicago/9780226114736.001.0001

Comunian, R. (2015). Festivals as communities of practice: Learning by doing and knowledge networks among artists. In C. Newbold, C. Maughan, J. Jordan, & F. Bianchini (Eds.), *Focus on festivals: Contemporary European case studies and perspectives* (pp. 53–65). Goodfellows Publishers. doi:10.23912/978-1-910158-15-9-2640

Convention on Biological Diversity. (2008). Protected areas in today's world: Their values and benefits for the welfare of the planet (CBD Technical Series No. 36). Montreal: Secretariat of the Convention on Biological Diversity.

Copper, J. F. (2010). *The A to Z of Taiwan (Republic of China)*. Scarecrow Press.

Cosmopolilla. (2017). *El test de ADN de Momondo: Viaje a mis orígenes*. Retrieved 26 March 2019, from: https://lacosmopolilla.com/viaje-a-mis-origenes-test-de-adn-de-momondo/cultures

Costa, D. (2019). Diversidade cultural na universidade em Portugal: Tendências e desafios. In S. Gonçalves & J. J. Costa (Eds.), Diversidade no Ensino Superior (pp. 11-37). Estratégias de Ensino e Sucesso Académico: Boas Práticas no Ensino Superior.

Costa, J. (2018, May 10). Erasmus. Um dos principais instrumentos de paz na Europa. *Renascença*. Retrieved from: https://rr.sapo.pt/2018/05/10/mundo/erasmus-um-dos-principais-instrumentos-de-paz-na-europa/noticia/112913/

Council (Council of the EC). (1984). *Council Resolution of 10 April 1984 on a Community Policy on tourism.* Retrieved from: https://eur-lex.europa.eu/legal-content/EN/TXT/?uri=CELEX%3A31984Y0430%2802%29

Council-of-Europe. (2005). *Council of Europe Framework Convention on the Value of Cultural Heritage for Society.*

Covane Community Lodge. (2020a). *Galleries.* Retrieved January 5, 2020, from http://www.covanecommunitylodge.com/gallery.html

Covane Community Lodge. (2020b). *About Covane Community Lodge.* Retrieved January 7, 2020, from https://www.covanecommunitylodge.com/about.html

CPRPE. (2013). *International handbook on tourism and peace.* Klagenfurt/Celovec.

Creswell, J. W. (2014). *Research design: qualitative, quantitative, and mixed methods approaches* (4th ed.). SAGE Publications.

Creswell, J. W. (2016). *30 Essential skills for the qualitative researcher.* SAGE Publications Ltd.

Crookes, P., & Davies, S. (2004). *Research into practice:essential skills for reading and applying research in nursing and health care.* Baillière Tindall.

Crouch, G. I., & Ritchie, J. B. (1999). Tourism, competitiveness, and societal prosperity. *Journal of Business Research, 44*(3), 137–152. doi:10.1016/S0148-2963(97)00196-3

D'Amore, L. (1988). Tourism: The world's peace industry. *Journal of Travel Research, 27*(1), 35–40. doi:10.1177/004728758802700107

D'Amore, L. (2010). Peace through tourism: The birthing of a new socio-economic order. *Journal of Business Ethics, 89*(S4), 559–568. doi:10.100710551-010-0407-3

D'Amore, L. (2014). Peace through tourism: An historical and future perspective. In C. Wohlmuter & W. Wintersteiner (Eds.), *International handbook on tourism and peace* (pp. 355–370). Drava Verlag.

D'Amore, L. J. (1988). Tourism: A vital force for peace. *Annals of Tourism Research, 15*(2), 269–283. doi:10.1016/0160-7383(88)90087-4

D'Amore, L. J. (1988). Tourism: A vital force for peace. *Tourism Management, 9*(2), 151–154. doi:10.1016/0261-5177(88)90025-8

D'Antoni, M. (2010). Tradición investigativa y constructos para comprender aspectos de la inserción profesional docente: Pluralismo cultural, aculturación, shock cultural. *Actualidades Investigativas en Educación, 10*(3), 1–18.

d'Hauteserre, A. M. (2004). Postcolonialism, colonialism, and tourism. A companion to Tourism. In A. Lew, M. Hall, & A. Willimas (Eds.), *A companion to tourism* (pp. 235–245). Bkacwell Publishing. doi:10.1002/9780470752272.ch19

D'Souza, C., Singaraju, S., Halimi, T., & Sillivan Mort, G. (2016). Examination of cultural shock, inter-cultural sensitivity and willingness to adapt. *Education + Training, 58*(9), 906–925. doi:10.1108/ET-09-2015-0087

Dario, T. (2019). *Atlas of Migration – 2019.* Publications Office of the European Union. https://publications.jrc.ec.europa.eu/repository/handle/JRC118415

Davison, W. D., & M., J. V. (1982). Foreign policy according to Freud. *Foreign Policy, 45*, 145–157. PMID:11632029

Dayton International Peace Museum. (n.d.). *About us: A brief history of America's only peace museum.* Retrieved January 5, 2020, from: https://daytonpeacemuseum.org/about/

de Andrade, J. V. (2006). *Turismo: Fundamentos e Dimensões*. Ática.

De Certeau, M. (1986). *Heterologies: Discourses on the other*. Manchester University Press.

de Coning, C. (2016). From peacebuilding to sustaining peace: Implications of complexity for resilience and sustainability. *Resilience*, *4*(3), 166–181. doi:10.1080/21693293.2016.1153773

De Coning, C. (2018). Adaptive peacebuilding. *International Affairs*, *94*(2), 301–317. doi:10.1093/ia/iix251

de Cultura de Pau, E. (2010). *Idees per a construir la pau. Diplomàcia municipal i altres iniciatives dels governs locals per construir la pau*. Xarxa de Municipis de la Diputació de Barcelona. Recovered from https://escolapau.uab.cat/municipisipau/municipis/idees2_cultura_de_pau.pdf

De Izertoren. (n.d.). *The Yser Tower*. Retrieved January 7, 2020, from: http://www.museumaandeijzer.be/ijzertoren/en/ijzertoren-diksmuide/yser-tower

de Jongh, A., Holmshaw, M., Carswell, W., & van Wijk, A. (2011). Usefulness of a trauma-focused treatment approach for travel phobia. *Clinical Psychology & Psychotherapy*, *18*(2), 124–137. doi:10.1002/cpp.680 PMID:20146201

De Kadt, E. (1979). *Tourism: Passport to development: Perspectives on the social and cultural effects of tourism in developing countries*. World Bank Press.

de Prat, C. R. (2013). *Euroscepticism, Europhobia and Eurocriticism: The radical parties of the right and left vis-à-vis the European Union*. Peter Lang. doi:10.3726/978-3-0352-6347-3

de Villiers, D. (2013). Cornerstones for a better world: Peace, tourism and sustainable development. In C. Wohlmuther & W. Wintersteiner (Eds.), *International handbook on tourism and peace* (pp. 78–86). Drava Verlag.

De Vries, H., & Wiegink, N. (2011). Breaking up and Going Home? Contesting Two Assumptions in the Demobilization and Reintegration of Former Combatants. *International Peacekeeping*, *18*(1), 38–51. doi:10.1080/13533312.2011.527506

Deventer, K. (2015). The public festival: Inspiration and interconnectivity at the heart of festivals. In C. Newbold, C. Maughan J., Jordan, & F. Bianchini (Eds.), Focus on festivals: Contemporary European case studies and perspectives (pp. 255-264). Oxford: Goodfellows Publishers.

Dewey, J. (1991). Creative Democracy - The Task Before Us. In The Later Works, (vol. 14). Carbondale: Southern Illinois University Press.

Dewey, J. (1916). *Democracy and Education: An Introduction to the Philosophy of Education*. Macmillan.

Dewey, J. (1983). The need for for a recovery of philosophy. In J. A. Boydston (Ed.), *Middle Works*. Southern Illinois University Press.

Diamond, L., & McDonald, J. (1996). *Multi-track diplomacy: A system approach to peace*. Kumarian Press.

Dias, R. (2010). Um tema emergente nas relações internacionais: A paradiplomacia das cidades e municípios. Âmbito Jurídico, Rio Grande, 13(79).

Dias, R. V. A. (2013). *Conhecimento e interesse do estudante de bacharelado em Educação Física da UFSC em intercâmbio acadêmico internacional.*

Diego, E. (2003). Una Percepción de la idea de Europa en España durante el periodo de entreguerras 1918-1939. *Cuadernos de Historia Contemporâne,* 311-324.

Dionysopoulou, P. (2012). European tourism policy. *Knowledge Management in Tourism: Policy and Governance Applications*, *4*, 283–300.

Dixon, J. (2001). Contact and boundaries: 'locating' the social psychology of intergroup relations. *Theory & Psychology, 11*(5), 587–608. doi:10.1177/0959354301115001

Donais, T. (2009). Empowerment or Imposition? Dilemmas of Local Ownership in Post-Conflict Peacebuilding Processes. *Peace and Change, 34*(1), 3–26. doi:10.1111/j.1468-0130.2009.00531.x

Dónde te metes. (2017). *Mis resultados del test de ADN #THEDNAJOURNEY*. Retrieved 25 March 2019, from https://dondetemetes.net/2017/08/05/mis-resultados-del-test-de-adn-thednajourney/

Doyle, M. W., & Sambanis, N. (2000). International peacebuilding: A theoretical and quantitative analysis. *The American Political Science Review, 94*(4), 779–801. doi:10.2307/2586208

Dreyer, J. T. (2016). *Middle Kingdom & Empire of The Rising Sun: Sino- Japanese Relations, Past and Present*. Oxford University Press.

Drifte, R. (2014). The Japan-China confrontation over the Senkaku/Diaoyu Islands: Between "shelving" and "dispute escalation". *The Asia-Pacific Journal, 12*(30 (3)), 1–61.

Duggan, P. (2011). *What have they done to the rain?* Kukupa Press.

Durkheim, E. (1978). *Educação e sociedade*. Melhoramentos.

Dwyer, C. D. (2012). Expanding DDR: The Transformative Role of Former Prisoners in Community-Based Reintegration in Northern Ireland. *The International Journal of Transitional Justice, 6*(2), 274–295. doi:10.1093/ijtj/ijs014

Dwyer, S. (2003). Reconciliation for realists. In C. Prager & T. Govier (Eds.), *Dilemmas of reconciliation: Cases and concepts*. Wilfrid Laurier University Press.

Eagly, A. H., & Chaiken, S. (1993). *The psychology of attitudes*. Harcourt Brace Jovanovich College Publishers.

Earley, P. C., & Ang, S. (2003). *Cultural intelligence: Individual interactions across cultures*. Academic Press.

EC. (1982). *Initial Guidelines for a Community Policy on Tourism* (COM/82/385 final). Retrieved from: https://op.europa.eu/en/publication-detail/-/publication/2c46eb35-049b-4fba-a8ae-2529521aa7fd/language-en

EC. (1985). *COM (85) 310 final.* White paper from the Commission to the European Council. Retrieved from: https://op.europa.eu/en/publication-detail/-/publication/4ff490f3-dbb6-4331-a2ea-a3ca59f974a8/language-en

EC. (1995). *COM (95) 97 final. The role of the Union in the field of tourism.* Commission Green Paper. Retrieved from https://op.europa.eu/en/publication-detail/-/publication/ed46c83e-95fc-4c5c-9dbc-da288647d7c2/language-en

EC. (1999). *COM 178/ Enhancing tourism's potential for employment Follow-up to the conclusions and recommendations of the High Level Group on Tourism and Employment.* Communication from the Commission to the Council, the European Parliament, the Economic and Social Committee and the Committee of the Regions. Brussels. Retrieved from https://eur-lex.europa.eu/legal-content/EN/TXT/?uri=CELEX:31999Y0623(01)

EC. (2001). *COM (2001) 665 final. Working together for the future of European tourism.* Commission communication to the Council, the European Parliament, the Economic and Social Committee and the Committee of the Regions. Brussels. Retrieved from https://eur-lex.europa.eu/LexUriServ/LexUriServ.do?uri=COM:2001:0665:FIN:en:PDF

EC. (2003). *COM (2003) 716 final. Basic orientations for the sustainability of European tourism.* Communication from the Commission to the Council, the European Parliament, the European Economic and Social Committee and the Committee of the Regions. Brussels. Retrieved from https://eur-lex.europa.eu/LexUriServ/LexUriServ.do?uri=COM:2003:0716:FIN:en:PDF

EC. (2006). *COM (2006) 134 final. A renewed EU Tourism Policy: Towards a stronger partnership for European Tourism*. Communication from the Commission. Brussels. Retrieved from https://eur-lex.europa.eu/LexUriServ/LexUriServ.do?uri=COM:2006:0134:FIN:en:PDF

EC. (2010). *COM (2010) 352 final. Europe: the world's No 1 tourist destination – a new political framework for tourism in Europe*. Communication from the Commission to the European Parliament, the Council, the European Economic and Social Committee and the Committee of the Regions. Retrieved from https://eur-lex.europa.eu/LexUriServ/LexUriServ.do?uri=COM:2010:0352:FIN:EN:PDF

EC. (2012). *COM (2012) 649 final. Implementation and development of the common visa policy to spur growth in the EU*. Communication from the Commission to the Council and the European Parliament. Brussels. Retrieved from https://ec.europa.eu/home-affairs/sites/homeaffairs/files/what-is-new/news/pdf/communication_on_visa_policy_to_spur_growth_com_2012_649_en.pdf

EC. (2012a). *Docs room*. Retrieved from http://ec.europa.eu/DocsRoom/documents?locale=en&tags=projects_on_cultural_tourism_2012

EC. (2014). *COM (2014) 86 final. A European Strategy for more Growth and Jobs in Coastal and Maritime Tourism*. Communication from the Commission to the European Parliament, the Council, the European Economic and Social Committee and the Committee of the Regions. Brussels. Retrieved from https://eur-lex.europa.eu/LexUriServ/LexUriServ.do?uri=COM:2014:0086:FIN:EN:PDF

EC. (2017). *Cultural Tourism – projects supported in 2017*. Retrieved from https://ec.europa.eu/docsroom/documents/29223

EC. (2018). *Tourism and Cultural and Creative Industries (CCIs) – projects supported in 2018*. Retrieved from https://ec.europa.eu/docsroom/documents/35247

Echandía, C. (2006). *Dos décadas de escalamiento del conflicto armado en Colombia 1986-2006*. Universidad Externado de Colombia.

Eck, N. J., & Waltman., L. (2019). VOSviewer Manual. *Manual for VOSviewer version 1.6.13*.

Economics, T. (2019a). *Macau tourist arrivals*. Retrieved on March 23 2020, from https://tradingeconomics.com/macau/tourist-arrivals

Economics, T. (2019b). *Macau GDP*. Retrieved on March 23 2020, from https://tradingeconomics.com/macau/gdp

Edensor, T. (1998). *Tourists at the Taj: performance and meaning at a symbolic site*. Routledge.

Edensor, T. (2007). Mundane mobilities, performances and spaces of tourism. *Social & Cultural Geography*, *8*(2), 199–215. doi:10.1080/14649360701360089

Edgell, D. L. (1990). *International tourism policy*. Van Nostrand Reinhold.

Edgell, D. L. (2014). Travel and tourism: The language of peace. *Journal of Hospitality and Tourism*, *12*(2), 30–41.

Educations.com. (2019). International higher education report: Student trends 2018. *Educations.com*. https://www.educations.com/higher-education-news/international-higher-education-report-18-13755

Ehrlich, S. R. (2014). *Hiroshima, Japan was devastated by an atomic bomb but tourists can't keep away*. CNN Press.

Elder, J. (1998). Expanding our options: The challenge of forgiveness. *Exploring Forgiveness*, 150-163.

Elias, N. (1978). *The civilising process* (Vol. 1). Blackwell.

Elliott, D. W. P. (2012). *Changing worlds: Vietnam's transition from Cold War to globalization*. Oxford University Press. doi:10.1093/acprof:oso/9780195383348.001.0001

Elliott, J. (2002). *Tourism: Politics and public sector management*. Routledge. doi:10.4324/9780203416136

Ellis, C., Adams, T. E., & Bochner, A. P. (2011). Autoethnography: An overview. *Forum. Qualitative Research, 12*(1).

Ellis, D. G. (2006). *Transforming conflict: Communication and ethnopolitical conflict*. Rowman & Littlefield Publishers.

Ellison, C. G., & Powers, D. A. (1994). The contact hypothesis and racial attitudes among Black Americans. *Social Science Quarterly, 75*(2), 385–400.

Environmental Peacebuilding. (2019). *About Environmental Peacebuilding*. Retrieved December 3, 2019, from https://environmentalpeacebuilding.org/about/

EP. (1996). *Resolution on the Commission report on the evaluation of the Community Action Plan to Assist Tourism 1993-1995 - Council Decision 92/421/EEC (COM(96)0166 - C4-0266/96)*. Retrieved from https://eur-lex.europa.eu/LexUriServ/LexUriServ.do?uri=CELEX:51996IP0299:EN:HTML

EP. (2020). *Tourism. Factsheets of the European Union*. Retrieved from www.europarl.europa.eu/factsheets/en

Ertamay, S. I., & Akbaba, A. (2019). European Union tourism policies. *International Eastern Mediterranean Tourism Symposium*.

Estol, J., Camilleri, M. A., & Font, X. (2018). European Union tourism policy: An institutional theory critical discourse analysis. *Tourism Review, 73*(3), 421–431. doi:10.1108/TR-11-2017-0167

Estol, J., & Font, X. (2016). European tourism policy: Its evolution and structure. *Tourism Management, 52*, 230–241. doi:10.1016/j.tourman.2015.06.007

European Commission. (2015). *The European higher education area in 2015: Bologna process implementation report*. Author.

European Commission. (2018). *Erasmus+ annual report 2016*. Activity report, Brussels: EU Publications.

European Union. (2019). *Atlas of Migration 2019*. Luxembourg: Publications Office of the European Union. https://ec.europa.eu›...›EU Science Hub

Fabiani, J. (2011). Festivals local and global: Critical interventions and the cultural public sphere. In C. Newbold, C. Maughan J., Jordan, & F. Bianchini (Eds.), Focus on festivals: Contemporary European case studies and perspectives (pp. 92-107). Oxford: Goodfellows Publishers.

Faltas, S. (2004). *DDR without Camps*. 15.

Fan, D., Zhang, H. Q., Jenkins, C. L., & Lin, P. (2017). Does tourist–host social contact reduce perceived cultural distance? *Journal of Travel Research, 56*(8), 998–1010. doi:10.1177/0047287517696979

Farmaki, A. (2017). The tourism and peace nexus. *Tourism Management, 59*, 528–540. doi:10.1016/j.tourman.2016.09.012

Farmaki, A. (2019). Visitor-host encounters in post-conflict destinations. The case of Cyprus. In R. Isaac, E. Çakmak, & R. Butler (Eds.), *Tourism and Hospitality in Conflict-Ridden Destinations*. Routlege. doi:10.4324/9780429463235-17

Farmaki, A., Khalilzadeh, J., & Altinay, L. (2019). Travel motivation and demotivation within politically unstable nations. *Tourism Management Perspectives, 29*, 118–130. doi:10.1016/j.tmp.2018.11.004

Feddes, A. R., Mann, L., & Doosje, B. (2012). From extreme emotions to extreme actions: Explaining non-normative collective action and reconciliation. *Behavioral and Brain Sciences*, *35*(06), 432–433. doi:10.1017/S0140525X12001197 PMID:23164364

Felgueiras, A. C., Siqueira, A. S., Lima, J. D., & Oliveira, R. C. (2018). Educação em direitos humanos: História e conceitos. *Revista Augustus*, *23*(45), 47–70. doi:10.15202/1981896.2018v23n45p47

Feng, X. (2012). Who are the "hosts"? Village tours in Fenghuang County, China. *Human Organization*, *71*(4), 383–394. doi:10.17730/humo.71.4.f738385u0206j866

Fennell, D. A. (2018). *Tourism Ethic*. Channel View Publications.

Filep, S., Macnaughton, J., & Glover, T. (2017). Tourism and gratitude: Valuing acts of kindness. *Annals of Tourism Research*, *66*, 26–36. doi:10.1016/j.annals.2017.05.015

Financial Times. (2019). *China suspends individual tourist permits to Taiwan before election*. Retrieved June 4, 2020, from: https://www.ft.com/content/6ba14934-b35e-11e9-8cb2-799a3a8cf37b

Finkel, R. (2009). A picture of the contemporary combined arts festival landscape. *Cultural Trends*, *18*(1), 3–21. doi:10.1080/09548960802651195

Fisas, V. (2002). *La paz es possible. Una agenda para la paz del siglo XXI*. Barcelona: de Bolsillo Plaza y Janés and Intermón Oxfam.

Fischer, S. (2013). Religious tourism: Business for peace in the Holy Land? In C. Wohlmuther & W. Wintersteiner (Eds.), *International handbook on tourism and peace* (pp. 120–142). Drava Verlag.

Fishbein, M., & Ajzen, I. (1975). *Belief, attitude, intention, and behavior: An introduction to theory and research*. Addison-Wesley.

Fisher, R. J. (1997). *Interactive Conflict Resolution*. Syracuse University Press.

Fisher, R. J., & Price, L. L. (1991). International pleasure travel motivations and post-vacation cultural attitude change. *Journal of Leisure Research*, *23*(3), 193–208. doi:10.1080/00222216.1991.11969853

Foley, D. E. (2002). Critical ethnography: The reflexive turn. *International Journal of Qualitative Studies in Education: QSE*, *15*(4), 469–490. doi:10.1080/09518390210145534

Fombona, J., Rodríguez, C., & Sevillano, Á. P. (2013). The motivational factor of Erasmus students at the university. *International Education Studies*, *6*(4), 1–9. doi:10.5539/ies.v6n4p1

Forero, S. (2019, Apr 30). Los tres productos más taquilleros de los excombatientes de las FARC. *El Espectador*. Retrieved from https://www.elespectador.com/colombia2020/pais/los-tres-productos-mas-taquilleros-de-los-excombatientes-de-las-farc-articulo-857922

Frank, A. (2005). What is dialogical research, and why should we do it? *Qualitative Health Research*, *15*(7), 964–974. doi:10.1177/1049732305279078 PMID:16093373

Franklin, A. (2003). The tourist syndrome: An Interview with Zygmunt Bauman. *Tourist Studies*, *3*(2), 205–217. doi:10.1177/1468797603041632

Frazer, O., & Ghettas, L. (Eds.). (2013). *Conflict Transformation in Practice: Approaches to Conflict Transformation*. Retrieved from https://css.ethz.ch/content/dam/ethz/special-interest/gess/cis/center-for-securities-studies/pdfs/Conflict_Transformation_in_Practice_2013.pdf

Frazer, R., & Waitt, G. (2016). Pain, politics and volunteering in tourism studies. *Annals of Tourism Research, 57*, 176–189. doi:10.1016/j.annals.2016.01.001

Freeman, C. M. L. (2012). The psychosocial need for intergroup contact:practical suggestions for reconciliation initiatives in Bosnia and Herzegovina and beyond. *Intervention (Amstelveen, Netherlands), 10*(1), 17–29. doi:10.1097/WTF.0b013e3283518de1

Frost, F. (1991). The Cambodia Conflict: The path towards peace. *Contemporary Southeast Asia, 13*(2), 119–163. doi:10.1355/CS13-2A

Fuck, I. T. (2001). Alfabetização de adultos: Relato de uma experiência construtivista. Petrópolis: GEEMPA.

Fundació, C. I. D. O. B. (2014). *El proceso de paz con las FARC-EP*. Retrieved April 2020 from https://www.cidob.org/publicaciones/documentacion/dossiers/dossier_proceso_de_paz_en_colombia/dossier_proceso_de_paz_en_colombia/el_proceso_de_paz_con_las_farc_ep

Fundación Ideas para la Paz. (2017). *La dejación de armas de las FARC y otros procesos de desarme en el mundo.* Retrieved from http://cdn.ideaspaz.org/media/website/document/598ca97c62763.pdf

Gajdošík, T., Sokolová, J., Gajdošíková, Z., & Pompurová, K. (2019). *Peace promotion through volunteer tourism. Marketing peace for social transformation and global prosperity*. IGI Global.

Galili, L. (2003, December 19). Long division. *Haaretz.* Retrieved from https://www.haaretz.com/1.4802471

Galtung, J. (1964). An editorial. *Journal of Peace Research, 1*(1), 1–4. doi:10.1177/002234336400100101

Galtung, J. (1969). Violence, peace, and peace research. *Journal of Peace Research, 6*(3), 167–191. doi:10.1177/002234336900600301

Galtung, J. (1976). Three approaches to peace: Peacekeeping, peacemaking, and peacebuilding. In J. Galtung (Ed.), *peace, war and defence: Essays in peace research* (pp. 282–304). Christian Ejlers.

Galtung, J. (1985). Twenty-Five Years of Peace Research: Ten Challenges and Some Responses. *Journal of Peace Research, 22*(2), 141–158. doi:10.1177/002234338502200205

Galtung, J. (1996). *Peace by peaceful means: Peace and conflict, development and civilisation.* International Peace Research Institute.

Gamper, J. A. (1981). Tourism in Austria a case study of the influence of tourism on ethnic relations. *Annals of Tourism Research, 8*(3), 432–446. doi:10.1016/0160-7383(81)90007-4

Garg, A. (2017). A study of tourist perception towards travel risk factors in tourist decision making. *Asian Journal of Tourism and Hospitality Research*, 47-57.

Garrison, J., & Philips, J. (1989). *The new diplomats: Citizens as ambassadors for peace.* Totnes: Green Books.

Garton, S. (2013). Longing for war:nostalgia and Australian returned soldiers after World War I. In T. G. Ashplant, G. Dawson, & M. Roper (Eds.), *The politics of war memory and commemoration.* Taylor and Francis.

Garzón, J. C., Prada, T., Silva, A., & Zárate, L. (2019). *Las trayectorias de reincorporación y seguridad de los excombatientes de las FARC.* Notas estratégicas 11. Fundación Ideas para la Paz. Retrieved from http://ideaspaz.org/media/website/FIP_NE_TrayectoriasFarc_Final_V02.pdf

Gates, H. L., & Appiah, A. (1999). *Africana: The encyclopedia of the African and African American experience.* Basic Civitas Books.

Geethika, N., & Gnanapala, W.K. (2015). Socio-economic impacts on local community through tourism development with special reference to heritance Kandalama. *Tourism, Leisure and Global Change, 2*. creativecommons.org/licenses

Gelbman, A., & Laven, D. (2016). Re-envisioning community-based heritage tourism in the old city of Nazareth. *Journal of Heritage Tourism, 11*(2), 105–125. doi:10.1080/1743873X.2015.1044993

Gelbman, A., & Maoz, D. (2012). Island of peace or island of war: Tourist guiding. *Annals of Tourism Research, 39*(1), 108–133. doi:10.1016/j.annals.2011.05.008

Gelézeau, V., De Ceuster, K., & Delissen, A. (2015). *De-bordering Korea: Tangible and intangible legacies of the sunshine policy*. Routledge.

George, R. (2003). Tourist's perceptions of safety and security while visiting Cape Town. *Tourism Management, 24*(5), 575–585. doi:10.1016/S0261-5177(03)00003-7

Gergen, K. J., McNamee, S., & Barrett, F. (2001). Toward transformative dialogue. *International Journal of Public Administration, 24*(7), 679–707. doi:10.1081/PAD-100104770

Getz, D. (2008). Event tourism: definition, evolution, and research, *Tourism Management, 29*(3), 403-428.

Getz, D., & Page, S. J. (2016). *Event studies: Theory, research and policy for planned events* (3rd ed.). Routledge. doi:10.4324/9781315708027

Glaesser, D. (2006). *Crisis management in the tourism industry*. Routledge. doi:10.4324/9780080464596

Gloppen, S. (2005). Roads to reconciliation: A conceptual framework. In S. G. E. Skaar & A. Suhrke (Eds.), *Roads to reconciliation*. Lexington Books.

GMDAC. (2018). *International students - IOM's Global Migration Data Analysis Centre*. Retrieved January 1, 2020, from: https://migrationdataportal.org/themes/international-students

GMDAC. (n.d.). *Migration data portal: International students*. Retrieved January 1, 2020, from: https://migrationdata-portal.org/themes/international-students

Gobo, G., & Marciniak, L. T. (2016). *What is Ethnography? In Qualitative Research*. Sage Publications.

Goffman, E. (1963). Stigma and social identity. *Understanding Deviance: Connecting Classical and Contemporary Perspectives*, 256-265.

Gómez, A. (2017). *La fiebre del ADN llega a España: "Mi familia es murciana pero mis ancestros son vikingos"*. Retrieved 26 March 2019, from: https://www.elconfidencial.com/tecnologia/ciencia/2017-03-19/genealogia-adn-genetica-suecia-murcia-ancestros-vikingos_1350698/

Gomez-Jacinto, L., Martin-Garcia, J. S., & Bertiche-Haud'Huyze, C. (1999). A model of tourism experience and attitude change. *Annals of Tourism Research, 26*(4), 1024–1027. doi:10.1016/S0160-7383(99)00063-8

González, R. E. D., Raesfeld, L., & Ponce, B. A. (2019). Interculturality: Contributions and proposals from an interdisciplinary focus in the institute of social sciences and humanities of the Autonomous University of the State of Hidalgo. *Universidad y Sociedad, 11*(1), 116–127.

Goodwin, H., Santilli, R. (2009). Community-based tourism: A success. *ICRT Occasional Paper, 11*(1), 37.

GOV.CN. (2012). *CNTA: Os turistas chineses estão a acelerar o seu retorno para a China*. Retrieved from Portal do Governo da RPC: http://www.gov.cn/jrzg/2012-05/13/content_2136105.htm

Graziano, P., & Vink, M. P. (2008). *Europeanization: New Research Agendas*. Palgrave Macmilan. doi:10.1057/9780230584525

Great Limpopo Transfrontier Park. (2019a). *About GLTP*. Retrieved August 7, 2019, from https://www.greatlimpopo.org/about-gltp/

Great Limpopo Transfrontier Park. (2019b). *Tourism*. Retrieved August 8, 2019, from https://www.greatlimpopo.org/tourism/

Great Limpopo Transfrontier Park. (2019c). *Tourism Opportunities*. Retrieved August 2, 2019, from https://www.greatlimpopo.org/tourism/tourism-opportunities/

Gruner, W. D. (1998a). Les idées de l'Europe politique au XVIIIe siècle. In K. Malettke (Ed.), *Imaginer L'Europe* (pp. 145–162). Éditions Belin.

Gruner, W. D. (1998b). Perceptions et conceptions de L'Europe au XIX siècle. In K. Malettke (Ed.), *Imaginer L'Europe* (pp. 163–190). Éditions Belin.

Grupo de Memoria Histórica. (Ed.). (2013). ¡Basta ya! Colombia, memorias de guerra y dignidad (Segunda edición corregida). Imprenta Nacional.

Günlü, E., Özgen, H. K. Ş., Dilek, S. E., Kaygalak, S., Türksoy, S. S., & Lale, C. (2015). Turkish visitors in Armenia: Any changes in attitudes and perceptions? *Journal of Tourism and Hospitality Management*, *3*(1), 29–43. doi:10.15640/jthm.v3n1a3

H4C. (2015). *Home for Cooperation: About Us*. Retrieved from www.home4cooperation.info

Habermas, J. (1962). The structural transformation of the public sphere: An inquiry into a category of bourgeois society. Cambridge: Polity.

Haessly, J. (2010). Tourism and a culture of peace. In O. Moufakkir & I. Kelly (Eds.), *Tourism, progress and peace* (pp. 1–16). CAB International. doi:10.1079/9781845936778.0001

Hall, C. M. (1998). *Tourism and politics: Policy, power and place*. Routledge.

Hall, C. M. (2010). Crisis events in tourism: Subjects of crisis in tourism. *Current Issues in Tourism*, *13*(5), 401–417. doi:10.1080/13683500.2010.491900

Hall, C. M. (2010). Politics and tourism: Interdependency and implications in understanding change. In R. Butler & W. Suntikul (Eds.), *Tourism and political change* (pp. 8–18). Goodfellow Publishers.

Hall, E. T. (1959). *The Silent language*. Doubleday.

Hall, M., & Tucker, L. (Eds.). (2004). *Tourism and postcolonialism*. Routledge. doi:10.4324/9780203392270

Halpern, J., & Weinstein, H. M. (2004). Rehumanizing the other: Empathy and reconciliation. *Human Rights Quarterly*, *26*(3), 561–583. doi:10.1353/hrq.2004.0036

Hammami, F., & Laven, D. (2017). Rethinking heritage from peace: Reflections from the Palestinian-Israeli conflict. In D. Walters, D. Laven, & P. Davis (Eds.), *heritage and peacebuilding* (pp. 137–148). The Boydell Press.

Hammer, M. R. (1989). Intercultural communication competence. In M. K. Asante & W. B. Gudykunst (Eds.), *Handbook of international and intercultural communication* (pp. 247–260). Sage.

Hamowy, R. (1990). Cato's Letters, John Locke, and the republican paradigm. *History of Political Thought*, *11*(2), 273–294.

Hannam, K., Sheller, M., & Urry, J. (2006). Mobilities, immobilities and moorings [Editorial]. *Mobilities*, *1*(1), 1–22. doi:10.1080/17450100500489189

Hanushek, E. A., & Woessmann, L. (2008). The role of cognitive skills in economic development. *Journal of Economic Literature*, *46*(3), 607–668. doi:10.1257/jel.46.3.607

Harlak, H. (2000). *Önyargılar: Psikososyal bir inceleme* [Prejudices: A psychosocial investigation]. Sistem Yayıncılık.

Harris, I. M. (2004). Peace education theory. *Journal of Peace Education*, *1*(1), 5–20. doi:10.1080/1740020032000178276

Hashimoto, A., & David, J. T. (2019). Contested geopolitical messages for tourists at the Okinawa Peace Park and memorials. *Japan Forum*, *31*(3), 378–407. doi:10.1080/09555803.2018.1451354

Hayes, B. C., & McAllister, I. (2001). Sowing Dragon's Teeth: Public support for political violence and paramilitarism in Northern Ireland. *Political Studies*, *49*(5), 901–922. doi:10.1111/1467-9248.00346

Haynes, J. (2005). World music and the search for difference. *Ethnicities*, *5*(3), 365–385. doi:10.1177/1468796805054961

Heater, D. (1992). *The idea of European Unity*. Leicester University Press.

Hedrick-Wong, Y., & Choong, D. (2014). *MasterCard 2014 Global Destination Cities Index*. MasterCard.

Hein, G. (1991). *Constructivist Learning Theory*. https://www.exploratorium.edu/education/ifi

Heleta, S., & Deardorff, D. K. (2017). The role of higher education instituitions in developing intercutural competence in peace-building in the aftermath of violent conflit. In D. K. Deardorff & L. A. Arasaratnam-Smith (Eds.), *Intercultural competence in higher education: international approaches, assessment and application*. Routledge. doi:10.4324/9781315529257-6

Herath, N. (2010). Peace Through Tourism. In *Beyond Intractability Conflict Information Consortium*. University of Colorado Boulder. http://www.beyondintractability.org/essay/peace-through-tourism

Heritage, M. C. a. (2008). *Fact Sheet 10: The Memorandum of Understanding and Tribute08*. Retrieved from http://www.nzhistory.net.nz/files/documents/vietnam-fact-sheet10.pdf

Heritage, M. C. a. (2013). *The Vietnam War*. Retrieved from http://www.nzhistory.net.nz/war/vietnam-war

Herrador-Valencia, D., & Riera, E. M. (2016). The Creation of a New Identity. *Defending the Forest in Cinquera in El Salvador.*, *19*, 25.

Hewstone, M. (1989). Changing stereotypes with disconfirming information. In D. Bar-Tal, C. F. Graumann, A. W. Kruglanski, & W. Stroebe (Eds.), *Stereotyping and Prejudice: Changing Conceptions* (pp. 207–223). Springer. doi:10.1007/978-1-4612-3582-8_10

Hewstone, M., Cairns, E., Voci, A., Hamberger, J., & Niens, U. (2006). Intergroup contact, forgiveness, and experience of "the troubles" in Northern Ireland. *The Journal of Social Issues*, *62*(1), 99–120. doi:10.1111/j.1540-4560.2006.00441.x

Higgins-Desbiolles, F., & Blanchard, L-A. (2012). Challenging peace through tourism: Placing tourism in the context of human rights, justice and peace. In O. Moufakkir & I. Kelly (Eds.), *Tourism, progress and peace* (pp. 35-47). Wallingford: CAB International.

Higgins-Desbiolles, F. (2003). Reconciliation tourism: Tourism healing divided societies. *Tourism Recreation Research*, *28*(3), 33–44. doi:10.1080/02508281.2003.11081415

Higgins-Desbiolles, F. (2008). Justice tourism and alternative globalisation. *Journal of Sustainable Tourism*, *16*(3), 345–364. doi:10.1080/09669580802154132

Higgins-Desbiolles, F., & Blanchard, L. (2010). *Challenging peace through tourism: placing tourism in the context of human rights, justice and peace*. CABI.

Hinch, T. (2010). Tourism as an instrument of foreign policy: The US trade embargo on Cuba. In R. Butler & W. Suntikul (Eds.), *Tourism and political change* (pp. 97–107). Goodfellow Publishers.

Hinds, R., & Carter, B. (2015). Indicators for conflict, stability, security, justice and peacebuilding. In *GSDRC6 Helpdesk Research Report* (Vol. 1230). University of Birmingham.

Hobsbawm, E. (2011). *A era dos extremos*. Editorial Presença.

Hobson, J. S. P., Heung, V. C. S., & Chon, K.-S. (1994). Vietnam's tourism industry: Can it be kept afloat? *The Cornell Hotel and Restaurant Administration Quarterly*, *35*(5), 42–49. doi:10.1177/001088049403500514

Hofstede, G. (1997). *Cultures and organisations: Software of the mind*. McGraw-Hill.

Hofstede, G. J., Hofstede, G., & Minkov, M. (2010). *Cultures and organisations: The software of the mind* (3rd ed.). McGraw-Hill.

Holland, S. (1991). Recreation and Tourism: Evolution of the Social Mission. In J. Zeigler & L. Caneday (Eds.), *Tourism and Leisure: Dynamics and Diversity* (pp. 66–81). National Recreation and Parks Association.

Hong Kong Poly, U. (2019). *Turning Macau into a smart city for tourism*. Retrieved on March 24 2020, from: https://www.polyu.edu.hk/cpa/excel/en/201905/connections/c2/index.html

Hooghe, L., & Marks, G. (2009). A postfuncionalist theory of European integration: From permissive consensus to constraining dissensus. *British Journal of Political Science*, *39*(1), 1–13. doi:10.1017/S0007123408000409

Hottola, P. (Ed.). (2009). *Tourism strategies and local responses in southern Africa*. CABI. doi:10.1079/9781845935085.0000

Hsiao, E. (2007). *Peace Parks: A Natural Alternative*. Retrieved February 13, 2019, from http://www.monitor.upeace.org/archive.cfm?id_article=456

Hsieh, J. F. (2016). Taiwan's general elections of 2016. In W. Lee (Ed.), *Taiwan's political re-alignment and diplomatic challenges: Politics and development of contemporary China* (pp. 49–69). Palgrave Macmillan.

Huang, E. (2017). *China is trying to punish South Korea by keeping its tourists away*. Retrieved from Quartz: https://qz.com/923890/china-retaliates-against-thaad-antimissile-system-and-bans-tourism-to-south-korea/

Humphreys, M., & Weinstein, J. M. (2007). Demobilization and Reintegration. *The Journal of Conflict Resolution*, *51*(4), 531–567. doi:10.1177/0022002707302790

Hunt, J. (2009). Wellington: VOH.

Ide, T. (2018). Does environmental peacemaking between states work? Insights on cooperative environmental agreements and reconciliation in international rivalries. *Journal of Peace Research*, *55*(3), 351–365. doi:10.1177/0022343317750216

Ide, T., Kirchheimer, J., & Bentrovato, D. (2018). School textbooks, peace and conflict: An introduction. *Global Change, Peace & Security*, *30*(3), 287–294. doi:10.1080/14781158.2018.1505717

Ikenberry, G., & Kupchan, C. (1990). Socialisation of hegemonic power. *International Organization*, *44*(3), 283–315. doi:10.1017/S002081830003530X

İnceoğlu, M. (2010). *Tutum algı iletişim* [Attitude perception communication] (5. Baskı). Beykent Üniversitesi Yayınları.

Innerarity, D. (2001). *Ética de la hospitalidad*. Ed. Península.

Institute for Economics and Peace. (2019). *Global Peace Index 2019: Measuring peace in a complex world, Sydney, June 2019*. Retrieved December 20, 2019, from: http://visionofhumanity.org/reports

International Institute for Peace through Tourism. (2003). *Tanzania Action Plan*. Retrieved March 16, 2020, from: https://www.iipt.org/conference/africanconference2003/presentations/TanzaniaActionPlan.pdf

International Institute for Peace through Tourism. (n.d.a). *About us: Mission statement*. Retrieved November 29, 2019, from: https://peacetourism.org/about-iipt/

International Institute for Peace through Tourism. (n.d.b). *Global Peace Parks Project*. Retrieved November 29, 2019, from: https://peacetourism.org/peace-parks-project/

International Institute for Peace through Tourism. (n.d.c). *Marketing peace through tourism: A 35 year history of IIPT*. Retrieved November 29, 2019, from: https://peacetourism.org/marketing-peace-through-tourism//

International Monetary Fund. (2019). *World Economic Outlook: GDP per capita*. Retrieved on March 24 2020, from: https://www.imf.org/external/datamapper/

International Peace Bureau. (n.d.). *International Network of Museums for Peace*. Retrieved December 15, 2019, from: http://www.ipb.org/members/international-network-of-museums-for-peace/

International Union for Conservation of Nature and Natural Resources, United Nations Environment Programme, World Wildlife Fund, Food and Agriculture Organisation of the United Nations, & UNESCO (Eds.). (1980). World conservation strategy: Living resource conservation for sustainable development. Gland, Switzerland: IUCN.

Isaac, R. (2014). Responsible Tourism and Development in the Context of Peace-Building. In C. Wohlmuther & W. Wintsersteiner (Eds.), *The International Handbook on Tourism and Peace* (pp. 87–100). Centre for Peace Research and Peace Eduaction of the Klagenfurt Univeristy and UNWTO.

Isaac, R. A. M. I. (2014). *Responsible Tourism and Development in the Context of Peace-building. In International Handbook on Tourism and Peace*. Drava.

Isaac, R., Cakmak, E., & Butler, R. (2019). Introduction. In R. Isaac, E. Cakmak, & R. Butler (Eds.), *Tourism and hospitality in conflict-ridden destinations* (pp. 11–12). Routledge. doi:10.4324/9780429463235-1

Isaac, R., & Platenkamp, V. (2016). Concrete U(dys)topia in Bethlehem: A city of two tales. *Journal of Tourism and Cultural Change*, *14*(2), 150–166. doi:10.1080/14766825.2015.1040413

Islam, M. R., & Hewstone, M. (1993). Dimensions of contact as predictors of intergroup anxiety, perceived out-group variability, and out-group attitude: An integrative model. *Personality and Social Psychology Bulletin*, *19*(6), 700–710. doi:10.1177/0146167293196005

Israeli Ministry of Foreign Affairs. (2014). *Christian tourism to Israel*. Retrieved from https://mfa.gov.il/MFA/AboutIsrael/Spotlight/Pages/Christian-tourism-to-Israel.aspx

Ivison, D. (2004). Locke, liberalism and empire. In *The Philosophy of John Locke* (pp. 99–118). Routledge.

Izotov, A., & Laine, J. (2013). Constructing (un)familiarity: Role of tourism in identity and region building at the Finnish-Russian Border. *European Planning Studies*, *21*(1), 95–110. doi:10.1080/09654313.2012.716241

J5Travel. (2018). *Explore your DNA*. Retrieved 26 March 2019, from https://www.j5travel.com/blog/explore-your-dna

Jaakson, R. (2004). Beyond the tourist bubble?: Cruiseship passengers in port. *Annals of Tourism Research*, *31*(1), 44–60. doi:10.1016/j.annals.2003.08.003

Jabri, V. (2013). Peacebuilding, the local and the international: A colonial or a postcolonial rationality? *Peacpebuilding, 1*(1), 3–16. doi:10.1080/21647259.2013.756253

Jafari, J. (1989). Tourism and peace. *Annals of Tourism Research, 16*(3), 439–443. doi:10.1016/0160-7383(89)90059-5

Jafari, J. (Ed.). (2002). *Encyclopedia of tourism*. Routledge. doi:10.4324/9780203195673

Jafari, J., & Way, W. (1994). Multicultural strategics in tourism. *The Cornell Hotel and Restaurant Administration Quarterly, 35*(6), 72–79. doi:10.1177/001088049403500614

Jalilvand, M. R., & Samiei, N. (2012). The effect of word of mouth on inbound tourists' decision for traveling to Islamic destinations (the case of Isfahan as a tourist destination in Iran). *Journal of Islamic Marketing, 3*(1), 12–21. doi:10.1108/17590831211206554

Jansen-Verbeke, M., & Go, F. (1995). Tourism development in Vietnam. *Tourism Management, 16*(4), 315–321. doi:10.1016/0261-5177(95)97356-U

Janzen, R. (2014). *Guatemalan Ex-Combatant Perspectives on Reintegration: A Grounded Theory*. 27.

Jarraud, N., Louise, C., & Filippou, G. (2013). The Cypriot Civil Society movement: A legitimate player in the peace process? *Journal of Peacebuilding and Development, 8*(1), 45–59. doi:10.1080/15423166.2013.791536

Jašková, M., & Heczková, L. (2010, July). ERASMUS mobility: Students motivation and satisfaction. *ERACON Congress*, 1-5.

Jencks, H. W. (1979). China's "punitive" war on Vietnam: A military assessment. *Asian Survey, 19*(8), 801–815. doi:10.2307/2643723

Jennings, K. M. (2007). The Struggle to Satisfy: DDR Through the Eyes of Ex-combatants in Liberia. *International Peacekeeping, 14*(2), 204–218. doi:10.1080/13533310601150800

Jesus, D. S. V. (2017). A arte do encontro: A paradiplomacia e a internacionalização das cidades criativas. *Revista de Sociologia e Politica, 25*(61), 51–76. doi:10.1590/1678-987317256104

Jiang, Y. (2013). *Disputa das Ilhas Diaoyu e perspectivas das relações sino-japonesas*. Retrieved from China Institute of International Studies: http://www.ciis.org.cn/chinese/2013-07/01/content_6074003.htm

Jiménez, R. (2017). *Mi prueba de ADN: 80% ibérica*. Retrieved 26 March 2019, from https://elpais.com/tecnologia/2017/02/19/soy_techie/1487540938_038219.html

Jimenez, C., & Kloeze, J. (2014). Analyzing the peace through tourism concept: The challenge for educators. *Sociology and Anthropology (Alhambra, Calif.), 2*(3), 63–70.

Jimenez, C., & te Kloeze, J. (2014). Analysing the peace through tourism concept: The challenge for educators. *Sociology and Anthropology (Alhambra, Calif.), 2*(3), 63–70. doi:10.13189a.2014.020301

Jin, X., Qu, M., & Bao, J. (2019). Impact of crisis events on Chinese outbound tourist flow: A framework for post-events growth. *Tourism Management, 74*(April), 334–344. doi:10.1016/j.tourman.2019.04.011 PMID:32287753

JNTO. (n.d.). *Estatísticas sobre visitantes estrangeiros no Japão*. Retrieved 11 2018, from JNTO: https://www.jnto.go.jp/jpn/statistics/data_info_listing/index.html?tab=block7

Johnson, P., & Thomas, B. (1992). *Perspectives on tourism policy*. Mansell.

Joint Management Board. (2002). *Joint Policy and Management Guidelines for the Great Limpopo Transfrontier Park*. Retrieved from https://pdf.usaid.gov/pdf_docs/PNADD330.pdf

Jones, J. (1954, July 7). Annual miracle of Llangollen. In 1954 International Eisteddfod Supplement. *Liverpool Daily Post.*

Joo, D., Tasci, A. D. A., Woosnam, K. M., Maruyama, N. U., Hollas, C. R., & Aleshinloye, K. D. (2018). Residents' attitude towards domestic tourists explained by contact, emotional solidarity and social distance. *Tourism Management, 64*, 245–257. doi:10.1016/j.tourman.2017.08.012

Josiam, B. M., & Frazier, R. (2008). Who am I? Where did I Come from? Where do I go to find out? Genealogy, the Internet and Tourism. *Tourismos: An International Multidisciplinary Journal of Tourism, 3*(2), 35–56.

Juan, Y., Choi, Y., An, S., Lee, C. K., & Lee, S. J. (2017). The constraints of Chinese tourists to visit Korea caused by THAAD using Q-methodology. *Asia Pacific Journal of Tourism Research, 22*(12), 1261–1273. doi:10.1080/10941665.2017.1391306

Juha's Guesthouse. (2020). *Home*. Retrieved from https://www.zarqabay.com/

Junaid, M. (2012). *Peace tourism and political games in Kashmir*. Retrieved January 10, 2020, from: https://www.aljazeera.com/amp/indepth/opinion/2012/07/201272612535508711.html

Kağıtçıbaşı, Ç., & Zeynep, C. (2014). *Günümüzde insan ve insanlar: Sosyal psikolojiye giriş* [Today, human and humans: Introduction to social psychology] (16th ed.). Evrim Yayınları.

Kang, M., & Moscardo, G. (2006). Exploring cross-cultural differences in attitudes towards responsible tourist behaviour: A comparison of Korean, British and Australian tourists. *Asia Pacific Journal of Tourism Research, 11*(4), 303–320. doi:10.1080/10941660600931143

Kant, I. (2015). *On perpetual peace*. Broadview Press.

Karsten, M. (2018). *My personal DNA journey: learning where I come from*. Retrieved 27 March 2019, from: https://expertvagabond.com/ancestry-dna-test/

Kasravi, J. (2009). *Factors influencing the decision to study abroad for students of color: Moving beyond the barriers* (PhD Thesis). University of Minnesota, Minnesota.

Kassis, R., Solomon, R., & Higgins-Desbiolles, F. (2017). Solidarity tourism in Palestine: The alternative tourism group of Palestine as a catalysing instrument of resistance. In R. Isaac, C. M. Hall, & F. Higgins Desbiolles (Eds.), *The politics and power of tourism in Palestine* (pp. 37–52). Routledge.

Kelly, I. (2012). Peace Through Tourism. An Implementation Guide. *The Journal of Tourism and Peace Research, 2*(2), 32–49.

Kelly, I. (2012). Peace through tourism: An implementation guide. *The Journal of Tourism and Peace Research, 2*, 32–49.

Kelly, I., & Nkabahona, A. (2010). Tourism and reconciliation. In O. Moufakkir & I. Kelly (Eds.), *Tourism, progress and peace* (pp. 228–241). CAB International. doi:10.1079/9781845936778.0228

Kelman, H. C., & Hamilton, V. L. (1989). *Crimes of obedience:Toward a social psychology of authority and responsibility*. Yale University Press.

Khamouna, M., & Zeiger, J. (1995). Peace Through Tourism. *Parks & Recreation, 30*(9), 80–86.

Kim, H. S. (2001). Korea's "Vietnam Question": War atrocities, national identity, and reconciliation in Asia. *Positions, 9*(3), 621-635. doi:10.1215/10679847-9-3-621

Kim, S. S., & Prideaux, B. (2003). Tourism, peace, politics and ideology: Impacts of the Mt. Gumgang tour project in the Korean Peninsula. *Tourism Management, 24*(6), 675–685. doi:10.1016/S0261-5177(03)00047-5

Kim, S. S., Prideaux, B., & Prideaux, J. (2007). Using tourism to promote peace on the Korean Peninsula. *Annals of Tourism Research, 34*(2), 291–309. doi:10.1016/j.annals.2006.09.002

Kim, S. S., Timothy, D. J., & Han, H.-C. (2007). Tourism and political ideologies: A case of tourism in North Korea. *Tourism Management, 28*(4), 1031–1043. doi:10.1016/j.tourman.2006.08.005

Kim, S., Prideaux, B., & Timothy, D. (2016). Factors affecting bilateral Chinese and Japanese travel. *Annals of Tourism Research, 61*, 80–95. doi:10.1016/j.annals.2016.08.001

Kim, Y., & Crompton, J. L. (1990). Role of tourism in unifying the Two Koreas. *Annals of Tourism Research, 17*(3), 353–366. doi:10.1016/0160-7383(90)90003-A

Kingma, K. (1997). *Demobilization of combatants after civil wars in Africa and their reintegration into civilian life*. 15.

Kingma, K. (2002). Demobilization, Reintegration and Peacebuilding in Africa. *International Peacekeeping, 9*(2), 181–201. doi:10.1080/113533312.2002.11978332

Klaic, D. (2006). "Festival", in Lexicon. *Performance Research, 4*(11), 54-55.

Knesset. (2018). *Basic Law: Israel: The Nation State of the Jewish People*. Retrieved from https://knesset.gov.il/laws/special/eng/BasicLawNationState.pdf

Knopf, R. (1991). Harmony and Convergence Between Recreation and Tourism. In J. Zeigler & L. Caneday (Eds.), *Tourism and Leisure: Dynamics and Diversity* (pp. 53–66). National Recreation and Parks Association.

Korstanje, M. E. (2011). Influence of history in the encounter of guests and hosts. *Anatolia, 22*(2), 282–285. doi:10.1080/13032917.2011.597991

Korstanje, M. E., & George, B. P. (2012). Falklands/Malvinas: A re-examination of the relationship between sacralisation and tourism development. *Current Issues in Tourism, 15*(3), 153–165. doi:10.1080/13683500.2011.569535

Korstanje, M. E., & Olsen, D. H. (2011). The discourse of risk in horror movies post 9/11: Hospitality and hostility in perspective. *International Journal of Tourism Anthropology, 1*(3), 304–317. doi:10.1504/IJTA.2011.043712

Korstanje, M. E., & Tarlow, P. (2012). Being lost: Tourism, risk and vulnerability in the post-'9/11' entertainment industry. *Journal of Tourism and Cultural Change, 10*(1), 22–33. doi:10.1080/14766825.2011.639455

Kosic, A., & Tauber, C. D. (2010). Promoting reconciliation through youth: Cross-community initiatives in Vukovar, Croatia. *Peace and Conflict, 16*(1), 81–95. doi:10.1080/10781910903487050

Kotite, P. (2012). Education for conflict prevention and peacebuilding. In *Meeting the global challenges of the 21st century*. International Institute for Educational Planning.

Kozak, M., & Decrop, A. (2009). *Handbook of tourist behavior theory & practice*. Routledge. doi:10.4324/9780203881804

Kriesi, H. (2016). The politicisation of European integration. *Journal of Common Market Studies, 54*, 32–47. doi:10.1111/jcms.12406

Krook, N. (2016). *Take a home DNA test and open your world*. Retrieved 28 March 2019, from: https://www.thetravel-tester.com/home-dna-test/

Krosnick, J. A., & Alwin, D. F. (1989). Aging and susceptibility to attitude change. *Journal of Personality and Social Psychology, 57*(3), 416–425. doi:10.1037/0022-3514.57.3.416 PMID:2778632

Krzaklewska, E., & Krupnik, S. (2006). *Research report. The experience of studying abroad for exchange students in Europe*. Brussels: Erasmus Student Network in cooperation with Petrus Communications.

Kumar, R. (2018). *Sustainable Tourism Development and Peace: a Local Community Approach.* Retrieved January 15, 2020 from https://www.researchgate.net/publication/326302785_sustainable_tourism_development_and_peace_a_local_community_approach

Kurze, D. (1998). La republica Christiana et l'Europe Médiévale. In K. Malettke (Ed.), *Imaginer L'Europe* (pp. 11–50). Éditions Belin.

Kvaternik, L. (2017). *Lo que un test de ADN desveló sobre mi origen étnico.* Retrieved 28 March 2019, from: https://www.mimundosinfronteras.com/test-adn-origen-etnico/

Ladini, G. (2009). Peacebuilding, United Nations and Civil Society: The Case of Cyprus. *The Cyprus Review, 21*(2), 37.

Ladrech, R. (2010). *Europeanization and national politics.* Palgrave Macmilan. doi:10.1007/978-1-137-06814-9

Lankford, V. S., Grybovych, O., & Lankford, K. J. (2008). Context for peace and community tourism: A worldview framework. *World Leisure Journal, 50*(3), 199–208. doi:10.1080/04419057.2008.9674553

Laplantine, F. (2000). *Aprender Antropologia.* Brasiliense.

Lavelle, A. (2006). Labor and Vietnam: A reappraisal. *Labour History,* 90.

Laws, E. (2005). *Tourism crises: Management responses and theoretical insight.* Psychology Press.

Lazzarino, L. (2017). *El viaje que me parió: crónica de un test de ADN.* Retrieved 25 March 2019, from: https://losviajesdenena.com/test-de-adn-momondo/

Le Blanc, G., & Brugère, F. (2017). *La fin de l'hospitalité.* Flammarion.

Lea, J. (2006). *Tourism and development in the Third World.* Routledge. doi:10.4324/9780203040393

Leandro, F. J. (2019a). The mesmerising journey from Gyeongju1 to Lisbon: The BRI as a mechanism of de-bordering, re-bordering, and co-bordering. *Baltic Journal of European Studies, 9*(2).

Leandro, F. J. (2019b). Macau Forum for three-continental dialogue: Perspectives and challenges for cooperation between Asia, Latin America and Central-Southeastern Europe. In S. S. Pajović, M. Andrijević, & S. C. Prado (Eds.), Europa Sudoriental, América Latina: Una mirada hacia la viabilidad de la cooperación triangular (pp. 39-40). Belgrade: Universidad Megatrend and International Federation of Study of Latin America and Caribbean (FIEALC).

Leclerc, G. (2015). *Le désir de voyage et la quête de l'autre.* L'Harmattan.

Lederach, J. P. (1997). *Sustainable reconciliation in divided societies.* USIP.

Lederach, J. P. (1999). *Building peace: Sustainable reconciliation in divided societies.* United States Institute of Peace Press.

Lederach, J. P. (2003). *Little Book of Conflict Transformation: Clear Articulation of the Guiding Principles by a Pioneer in the Field.* Good Books.

Lederach, J. P. (2005). *The moral imagination: The art and soul of building peace.* Oxford Press. doi:10.1093/0195174542.001.0001

Lee, I., Arcodia, C., & Lee, T. J. (2012). Benefits of attending a multicultural festival: The case of South Korea. *Tourism Management, 33*(2), 334–340. doi:10.1016/j.tourman.2011.04.001

Lee, Y. C., Lin, Y. C., Huang, C. L., & Fredrickson, B. L. (2013). The construct and measurement of peace of mind. *Journal of Happiness Studies, 14*(2), 571–590. doi:10.100710902-012-9343-5

Leimgruber, W. (2018). *Between global and local.* Routledge. doi:10.4324/9781351162722

Lemoine, P. A., Jenkins, W. M., & Richardson, M. D. (2017). Global higher education: Development and implications. *Journal of Education and Development, 1*(1), 58–71. doi:10.20849/jed.v1i1.253

Leong, A. M. W. (2008). International tourism and world peace. *Journal of Macau University of Science and Technology, 2*, 38–44.

Leopold, T. (2003). A provisional code of conduct for war heritage sites: Veterans' and managers' perceptions (Master of Tourism). University of Otago, Dunedin.

Leslie, D. (1996). Northern Ireland, tourism and peace. *Tourism Management, 17*(1), 51–55. doi:10.1016/0261-5177(96)83161-X

Levi, S., & Haekins, D. E. (2010). Peace through tourism – Commerce based principles and practices: A global view. *Journal of Business Ethics, 89*(S4), 569–585. doi:10.100710551-010-0408-2

Libertad. (n.d.). *¿Merece la pena pagar por un kit de ADN para conocer tus orígenes étnicos?* Retrieved 27 March 2019, from: http://viajaconlibertad.com/merece-pena-kit-adn-origenes-etnicos/

Lickorish, L., Jefferson, A., Bodlender, J., & Jenkins, C. L. (1991). Developing tourism destinations: policies and perspectives. Harlow: Longman.

Lim, Y. M. (n.d.). Imagine peace translated from Korean by Jessica Lee. *Transforming Tourism.* Retrieved March 15, 2020, from: http://www.transforming-tourism.org/goal-16b-peace.html

Li, M., & Bray, M. (2007). Cross-border flows of students for higher education: Push–pull factors and motivations of mainland Chinese students in Hong Kong and Macau. *Higher Education, 53*(6), 791–818. doi:10.100710734-005-5423-3

Lima, M. C. (2009). Mobilidade mundial de estudantes. *GV EXECUTIVO, 8*(2), 66–69. doi:10.12660/gvexec.v8n2.2009.23837

Lima, M. C., & Riegel, V. (2015). Mobilidade acadêmica made in south: Refletindo sobre as motivações de estudantes brasileiros e colombianos. *Magis. Revista Internacional de Investigación en Educación, 8*(16), 109–132. doi:10.11144/Javeriana.m8-16.mams

Lin, C., & Lee, M. (2018). Taiwan's opening policy to Chinese tourists and Cross-Strait relations: The impacts on inbound tourism into Taiwan. *Tourism Economics*, 1–18.

Lin, G. (2019). *Taiwan's party politics and Cross-Strait Relations in evolution (2008–2018).* Springer. doi:10.1007/978-981-13-5814-2

Lingeberzins, E. (2011). Transcultural communication challenges of international tourism business environment. *Organizacijø Vadyba: Sisteminiai Tyrimai*, 69-79.

Lin, S. S. (2016). *Taiwan's China's dilemma: Contested identities and multiple interests in Taiwan's Cross Strait economic policy.* Stanford University Press.

Litvin, S. (1998). *Tourism: The World's Peace Industry?* Retrieved January 15, 2020 from https://www.researchgate.net/publication/258161060_Tourism_The_World's_Peace_industry

Litvin, S. (1998). Tourism: The world's peace industry? *Journal of Travel Research, 37*(1), 63–66. doi:10.1177/004728759803700108

Liutikas, D. (2012). Experiences of valuistic journeys: Motivation and behavior. In R. Sharpley & P. Stone (Eds.), *Contemporary Tourist Experience: Concepts and Consequences*. Routledge.

Loth, W. (2014). *Building Europe. A history of European unification*. De Gruyter.

Lundin. (1998). *Mechanisms of Community Reception of Demobilised Soldiers in Mozambique*. 16.

Mac Ginty, R. (2013). Indicators+: A proposal for everyday peace indicators. *Evaluation and Program Planning*, *36*(1), 56–63. doi:10.1016/j.evalprogplan.2012.07.001 PMID:22868180

Mac Ginty, R. (2015). Where is the local? Critical localism and peacebuilding. *Third World Quarterly*, *36*(5), 840–856. doi:10.1080/01436597.2015.1045482

Mac Ginty, R., & Firchow, P. (2016). Top-down and bottom-up narratives of peace and conflict. *Politics*, *36*(3), 308–323. doi:10.1177/0263395715622967

Macau Government Tourism Office. (2019a). *Hotel's occupancy rate and average room rate*. Retrieved on March 24 2020, from: https://dataplus.macaotourism.gov.mo

Macau Government Tourism Office. (2019b). *Global indicators*. Retrieved on March 24 2020, from: https://dataplus.macaotourism.gov.mo/Publication/Report?lang=E

Macau Institute for Tourism Studies. (2019). *IFT in top 5 for Asian tertiary institutions teaching hospitality*. Retrieved on March 26 2020, from: https://www2.ift.edu.mo/NewsPortal/ift-in-top-5-for-asian-tertiary-institutions-teaching-hospitality/

MacDonald, J. (1993). Guidelines for newcomers to track two diplomacy. Occasional paper nº 2, *Institute of Multi-Track Diplomacy*, 1-10.

Mahallae. (2015). Retrieved from www.mahallae.org

Makarova, E. A., Makarova, E. L., & Korsakova, T. V. (2019). The role of globalisation and integration in interdisciplinary research, culture and education development. *Journal of History Culture and Art Research*, *8*(1), 111–127. doi:10.7596/taksad.v8i1.1957

Makepeace, C. (2018). *Exploring your DNA ancestry trail around the world*. Retrieved 25 March 2019, from https://www.ytravelblog.com/dna-ancestry-travel/

Makriyianni, C., & Psaltis, C. (2007). The Teaching of History and Reconciliation. *The Cyprus Review*, *19*(1), 43–69.

Malettke, K. (1998). L'Europe dans le 'Grand Dessein' de Sully. In K. Malettke (Ed.), *Imaginer L'Europe* (pp. 93–105). Éditions Belin.

Mallarach, J. M. (Ed.). (2012). *Spiritual Values of Protected Areas of Europe: Workshop 2011: Viim, Alemanya*. Federal Agency for Nature Conservation.

Malley, M. (2002). Letters/perspective. *Hotel & Motel Management*, 8.

Malta, P. A., & Carneiro, M. J. (2005). Turismo e educação, mobilidades e ensino superior: Outras Teias de Globalização. *Revista Turismo & Desenvolvimento*, *3*, 115–128.

Mansfeld, Y., & Pizam, A. (Eds.). (2006). *Tourism, security and safety*. Routledge.

Marshall, L. (2017). *DNA test fun: ancestry travel inspiration*. Retrieved 26 March 2019, from: https://www.aroundtheworldl.com/2017/07/12/dna-ancestry-travel/

Martínez, H., & Lefebvre, I. (2019). *La reincorporación económica de los excombatientes de las FARC. Retos y riesgos a futuro*. Notas estratégicas 09. Fundación Ideas para la Paz. Retrieved from http://ideaspaz.org/media/website/FIP_NE_Reincorporacion.pdf

Marx, K. (n.d.). Crítica ao programa de Gotha. In K. Marx (Ed.), *ENGELS, Friedrich: Obras Escolhidas* (pp. 203–234). Alfa-Omega.

Masalha, N. (2019). *Palestine: A four thousand year history*. Zed Books.

Mashike, L. (2004). "You are a time bomb…" Ex-combatants in post-conflict South Africa. *Society in Transition*, *35*(1), 87–104. doi:10.1080/21528586.2004.10419108

Maslow, A. H. (1971). *The farther reaches of human nature*. Viking.

Mateus, A. (2017). O papel do turismo na reinvenção do crescimento económico em Portugal. Relatório encomendado pela Associação Portuguesa das Agências de Viagens e Turismo (APAVT), *Augusto Mateus e Associados*, 1-38.

Máthé, I. (2018). Multiculturality and Interculturality in Tourism. *Valahian Journal of Economic Studies*, *9*(23), 39–46. doi:10.2478/vjes-2018-0004

Matheson, C. M. (2008). Music, emotion and authenticity: A study of celtic music festival consumers. *Journal of Tourism and Cultural Change*, *6*(1), 57–74. doi:10.1080/14766820802140448

Mathews-Aydinli, J. (2016). *International education exchanges and intercultural understanding: Promoting peace and global relations*. Springer.

Mathieson, A., & Wall, G. (1982). *Tourism: Economic, physical, and social impacts*. Longman.

Mathieu, X. (2019). Critical peacebuilding and the dilemma of difference: The stigma of the 'local'and the quest for equality. *Third World Quarterly*, *40*(1), 36–52. doi:10.1080/01436597.2018.1538732

Matteuccii, X. (2013). Photo elicitation: Exploring tourist experiences with researcher found images. *Tourism Management*, *35*, 190–197. doi:10.1016/j.tourman.2012.07.002

Mayor, F. (2010). Nutrindo uma cultura de paz. *Comitê Paulista para a década da cultura de paz: Um programa UNESCO: 2000-2010.*

McAdam, D., & Su, Y. (2002). The war at home: Antiwar protests and congressional voting, 1965 to 1973. *American Sociological Review*, *67*(5), 696–721. doi:10.2307/3088914

McClory, J. (2011). *The new persuaders II: A 2011 Global ranking of soft power*. Institute for Government.

McEvoy, K., & Shirlow, P. (2009). Re-imagining DDR: Ex-combatants, leadership and moral agency in conflict transformation. *Theoretical Criminology*, *13*(1), 31–59. doi:10.1177/1362480608100172

McGibbon, I. (2010). *New Zealand's Vietnam War: A history of combat, commitment and controversy*. Exisle Publishing Limited.

McGibbon, I., & Goldstone, P. (2000). *The Oxford companion to New Zealand military history*. Oxford Oxford University Press.

McIntosh, I. B., Swanson, V., Power, K. G., Raeside, F., & Dempster, C. (1998). Anxiety and health problems related to air travel. *Journal of Travel Medicine*, *5*(4), 198–198. doi:10.1111/j.1708-8305.1998.tb00507.x PMID:9876195

McIntosh, I., & Harman, L. (2017). *The many voices of pilgrimage and reconciliation*. CAB International. doi:10.1079/9781786393265.0000

McIntosh, R., Goeldner, C., & Richtie, J. (1995). *Principles, Practices and Philosophies* (7th ed.). John Wiley.

McKercher, B. (2008). The roots of stereotypes about tourists. *Society*, *45*(4), 345–347. doi:10.100712115-008-9113-5

McMullin, J. R. (2013a). *Ex-Combatants and the Post-Conflict State*. doi:10.1057/9781137312938

Mcmullin, J. R. (2013b). Integration or separation? The stigmatisation of ex-combatants after war. *Review of International Studies*, *39*(2), 385–414. doi:10.1017/S0260210512000228

McNamee, S. (2010). Research as social construction: Transformative inquiry. *Saúde & Transformação Social. Health and Social Change*, *1*(1), 9–19.

McNamee, S., & Gergen, K. J. (1999). *Relational responsibility: Resources for sustainable dialogue*. SAGE.

McWha, M., Frost, W., & Laing, J. (2018). Travel writers and the nature of self. Essentialism, transformation and (online) construction. *Annals of Tourism Research*, *70*, 14–24. doi:10.1016/j.annals.2018.02.007

Mearns, K. (2012). Community-based tourism and Peace Parks benefit local communities through conservation in Southern Africa. *Acta Academica*, *44*, 70–87.

Mee, M. (2017). *My DNA journey: Where I'm from*. Retrieved 28 March 2019, from: https://www.mitziemee.com/momondo-dna-journey-blog/

Meethan, K. (2004). To stand in the shoes of my ancestors. In T. Coles & D. J. Timothy (Eds.), *Tourism* (pp. 139–150). Diasporas and Space.

Menchero Sánchez, M. (2018). Colombia en posconflicto: ¿un turismo para la paz o paz para el turismo? *Revista Iberoamericana de Filosofía. Política y Humanidades*, *39*, 415–438.

Mendes, R., Pedradas, S., & Pereira, C. (2009). *Caracterização do Programa Erasmus: Análise das motivações, condições de frequência e expectativas dos estudantes*.

Menudos Viajeros. (2017). *Un viaje a través del ADN*. Retrieved 25 March 2019, from https://www.menudosviajeros.com/blog/un-viaje-traves-del-adn

Mény, Y. (2019). *Popolo ma non troppo. Il malinteso democratico*. Il Mulino.

Mészáros, A. (2011). The motivation to study abroad: What motivates students to apply for Erasmus grant? Technical University of Cartagena, Department of Economics.

Metsola, L. (2006). 'Reintegration' of Ex-combatants and Former Fighters: A lens into state formation and citizenship in Namibia. *Third World Quarterly*, *27*(6), 1119–1135. doi:10.1080/01436590600842407

MGIEP. (2014). *Education for peace and sustainable development: Concepts, clarity and cohesion*. MGIEP.

MGIEP. (2017). *Rethinking schooling for the 21st century: The state of education for peace, sustainable development and global citizenship*. Author.

Milano, C. (2019). Overtourism and Tourismphobia: A Journey Through Four Decades of Tourism Development, Planning and Local Concerns. *Tourism Planning & Development*, *16*(4), 353-357.

Miller, J. P., Nigh, K., Binder, M. J., Novak, B., & Crowell, S. (2018). Sustainable peace for the Planet. International Handbook of Holistic Education, 79-88.

Milman, A., Reichel, A., & Pizam, A. (1990). The impact of tourism on ethnic attitudes: The Israeli- Egyptian case. *Journal of Travel Research*, *29*(2), 45–49. doi:10.1177/004728759002900207

Mindful Travel Destinations. (2018). *Trails for Peace*. Retrived from: https://mindfultraveldestinations.com/trails-for-peace-2/

Ministério das Relações Exteriores do Japão. (2014). *Diplomatic Bluebook 2014 (period covered 2013 calendar year)*. Retrieved from Portal do Ministério das Relações Exteriores do Japão: https://www.mofa.go.jp/files/000055801.pdf

MLIT. (2012). *White Paper on Tourism in Japan 2012*. Retrieved from Japan Tourism Agency (MLIT): https://www.mlit.go.jp/common/001018364.pdf

MLIT. (2013). *White papers on tourism in Japan 2013*. Retrieved from Japan Tourism Agency (MLIT): https://www.mlit.go.jp/common/001211777.pdf

Money, R. B., & Crotts, J. C. (2003). The effect of uncertainty avoidance on information search, planning, and purchases of international travel vacations. *Tourism Management*, *24*(2), 191–202. doi:10.1016/S0261-5177(02)00057-2

Monteiro, A. (2014) *A mobilidade académica europeia e o turismo educativo e cultural: factores de decisão e motivação* (PhD thesis). University ofCoimbra, Coimbra.

Monteiro, A. (2017). *Turismo educativo e cultural: Factores de decisão e de motivação*. Alemanha: Novas Edições Académicas.

Monteiro, R. M. A. (2018). *Universidade do Porto: Fatores de decisão dos estudantes internacionais*.

Morec, I. (n.d.). *Prueba del ADN: El viaje que nunca imaginé*. Retrieved 27 March 2019, from http://www.unavidapor-viajar.com/reflexiones/test-adn/

Moreira, D., & Antão, G. (2019). Diversidade e mobilidade no ensino superior. In S. Gonçalves & J. J. Costa (Eds.), *Diversidade no ensino superior*. CINEP/IPC.

Morgan, M. (2008). What makes a good festival? Understanding the event experience. *Event Management*, *12*(2), 81–93. doi:10.3727/152599509787992562

Morin, E. (2002). Cultura de massas no século XX: O espírito do tempo: neurose. *Forense Universitária*, 1-102.

Moser. (2019). *Public narratives and attitudes towards refugees and other migrants Germany country profile*. ODI Evidence. Ideas. Change. www.odi.org›sites›files›resource-documents

Moufakkir, O., & Alnajem, M. (2017). Hostel front desk employees' perceptions of customers: A comparative analysis based on cultural background. *International Journal of Culture, Tourism and Hospitality Research*, *11*(3), 355–371. doi:10.1108/IJCTHR-07-2016-0068

Moufakkir, O., & Kelly, I. (2013). Peace through tourism: A sustainable development role for events. In T. Pernecky & M. Luck (Eds.), *Events, society and sustainability: Critical and contemporary approaches* (pp. 130–150). Routledge.

Moufakkir, O., & Kelly, I. (Eds.). (2010). *Tourism, progress and peace*. CABI. doi:10.1079/9781845936778.0000

Mouffakir, O., & Pernecky, T. (Eds.). (2015). *Ideological, social and cultural aspects of events*. CABI. doi:10.1079/9781780643526.0000

MREPRC. (2012). *Declaração do Ministério das Relações Exteriores da República Popular da China*. Retrieved from Portal do Ministério das Relações Exteriores da República Popular da China: https://www.fmprc.gov.cn/diaoyudao/chn/xwdt/t967820.htm

MREPRC. (2012). *Ilhas Diaoyu - Notícias*. Retrieved from Portal do Ministério das Relações Exteriores da República Popular da China: https://www.fmprc.gov.cn/diaoyudao/chn/xwdt/default_5.htm

MREPRC. (2012). *Ilhas Diaoyu*. Retrieved from Portal do Ministério das Relações Exteriores da República Popular da China: https://www.fmprc.gov.cn/diaoyudao/chn/xwdt/t910599.htm

Muggah, R. (2005). No Magic Bullet: A Critical Perspective on Disarmament, Demobilization and Reintegration (DDR) and Weapons Reduction in Post-conflict Contexts. *The Round Table*, *94*(379), 239–252. doi:10.1080/00358530500082684

Muggah, R. (Ed.). (2009). *Security and post-conflict reconstruction: dealing with fighters in the aftermath of war*. Routledge.

Mundo, E. (2009). *40 definiciones por la paz*. Recoverd from: http://fund-culturadepaz.org/doc/40_Definiciones_PAZ.pdf

Munro, L. K. (1953). New Zealand and the New Pacific. *Foreign Affairs*, *31*(4), 634–647. doi:10.2307/20030995

Murphy-Lejeune, E. (2000). Mobilité internationale et adaptation interculturelle: Les étudiants voyageurs européens. *Recherche & Formation*, *33*(1), 11–26. doi:10.3406/refor.2000.1614

Nakano, R. (2016). The Sino-Japanese territorial dispute and threat perception in power transition. *The Pacific Review*, *29*(2), 165–186. doi:10.1080/09512748.2015.1013493

Negrier, E. (2015). Festivalisation: Patterns and limits. In C. Newbold, C. Maughan, J. Jordan, & F. Bianchini (Eds.), *Focus on festivals: Contemporary European case studies and perspectives* (pp. 18–27). Goodfellows Publishers. doi:10.23912/978-1-910158-15-9-2668

Neubert, S. (2003). *Some Perspectives of Interactive Constructivism on the Theory of Education*. University of Cologne. Retrieved January 28, 2020, from https://www.hf.uni-koeln.de/data/dewey/File/Neubert_Introduction.pdf

Neulip, J. W. (2015). *Intercultural communication: A contextual approach* (6th ed.). Sage Publications.

Neumayer, E. (2004). The impact of political violence on tourism: Dynamic cross-national estimation. *The Journal of Conflict Resolution*, *48*(2), 259–281. doi:10.1177/0022002703262358

Newbold, C., Jordan, J., Bianchini, F., & Maughan, C. (2015). Introduction: Focusing on festivals. In C. Newbold, J. Jordan, F. Bianchini, & C. Maughan (Eds.), *Festivals in focus* (pp. xv–xxvi). Goodfellows Publishers. doi:10.23912/978-1-910158-15-9-2599

Newland, K., & Taylor, C. (2010). *Heritage tourism and nostalgia trade: A diaspora niche in the development landscape*. Academic Press.

Newman, D., & Passi, A. (1998). Fences and neighbours in the postmodern world: Boundary narratives in political geography. *Progress in Human Geography*, *22*(2), 186–207. doi:10.1191/030913298666039113

Nomadic Boys. (2018). *Ancestry DNA Review: A unique way to inspire travelling*. Retrieved 28 March 2019, from: https://nomadicboys.com/ancestrydna-review/

Nordvall, A., Pettersson, R., Svennson, B., & Brown, S. (2014). Designing events for social interaction. *Event Management*, *18*(2), 127–140. doi:10.3727/152599514X13947236947383

Northrup, T. (1989). The dynamic of identity in personal and social conflict. In L. N. Kriesberg, A. Terrell, & S. J. Thorson (Eds.), *Intractable conflicts and their transformation*. Syracuse University Press.

Novelli, M., Morgan, N., & Nibigira, C. (2012). Tourism in a post-conflict situation of fragility. *Annals of Tourism Research*, *39*(3), 1446–1469. doi:10.1016/j.annals.2012.03.003

Noy, C. (2004). This trip really changed me: Backpackers' narratives of self-change. *Annals of Tourism Research*, *31*(1), 78–102. doi:10.1016/j.annals.2003.08.004

Noy, C., & Cohen, E. (2005). *Israeli backpackers: From tourism to rite of passage*. SUNY Press.

Nucci, L. (2001). Morality in context: Issues of development. In L. Nucci (Ed.), *Education in the moral domain*. Cambridge University Press. doi:10.1017/CBO9780511605987.006

Nussio, E. (2018). Ex-combatants and violence in Colombia: Are yesterday's villains today's principal threat? *Third World Thematics: A TWQ Journal, 3*(1), 135–152.

Nyaupane, G. P., Teye, V., & Paris, C. (2016). Tourism and attitude change: The case of study abroad students (No. 64). Tourism Travel and Research Association: Advancing Tourism Research Globally.

Nye, J. (1999). Redefining the national interest. *Foreign Affairs, 78*(4), 22–35. doi:10.2307/20049361

Nye, J. Jr. (2008). Public diplomacy and soft power. *The Annals of the American Academy of Political and Social Science, 616*(1), 94–109. doi:10.1177/0002716207311699

NZPA. (2007). *NZ vets to get chance to visit Vietnam.* Retrieved from http://www.stuff.co.nz/national/48537/NZ-vets-to-get-chance-to-visit-Vietnam

Oberg, K. (1960). Cultural shock: Adjustment to new cultural environments. *Practical Anthropology, 7*(4), 177–182. doi:10.1177/009182966000700405

Okumus, F., Altinay, M., & Arasli, H. (2005). The impact of Turkey's economic crisis of February 2001 on the tourism industry in Northern Cyprus. *Tourism Management, 26*(1), 95–104. doi:10.1016/j.tourman.2003.08.013

Oliveira, S. C. M., Barbosa, E. S., Rezende, I. C. C., Silva, R. P. A., & Albuquerque, L. S. (2013). Bibliometria em artigos de contabilidade aplicada ao setor público. Anais do Congresso Brasileiro de Custos-ABC.

Oliveira, A. L. D., & Freitas, M. E. D. (2016). Motivações para Mobilidade Acadêmica Internacional: A visão de alunos e professores universitários. *Educação em Revista, 32*(3), 217–246. doi:10.1590/0102-4698148237

Ooi, N., & Lang, J. (2010). Backpacker tourism: Sustainable and purposeful? Investigating the overlap between backpacker tourism and volunteer tourism motivations. *Journal of Sustainable Tourism, 18*(2), 191–206. doi:10.1080/09669580903395030

Oosterbeek, L. (2016). Becoming human. New approaches for uncertain times. In L. Oosterbeek, M. Quagliuolo, & L. Caron (Eds.), Sustainability dilemmas. Transdisciplinary contributions to integrated cultural landscape management (Vol. 38-39, pp. 85-107). ITM.

Oosterbeek, L. (2019). Persona: From unrest to adaptation. *Plastir,* (56), 1-15.

Oosterbeek, L., Camara, A., & Martins, C. (2014). *Archaeology, societies and environments in Africa.* In *Proceedings of the XVI World Congress of the International Union of Prehistoric and Protohistoric Sciences.* Oxford: Archaeopress. 10.30861/9781407312972

Oosterbeek, L. (2017). Kóios and Phoïbe: knowledge through sociocultural matrices, in the framework of cultural integrated landscape management and sustainability science. In L. Oosterbeek, B. Werlen, & L. Caron (Eds.), *Sociocultural matrices: Transdisciplinary contributions to integrated cultural landscape management* (pp. 45–64). ITM.

Organização Mundial do Turismo. (2003). *Turismo internacional: Uma perspectiva global.* Bookman.

Osterwalder, K. (2001). Convention of the work of thematic groups: Convention of European higher education institutions. Salamanca, 29-30.

Otjes, S., & Katsanidou, A. (2017). Beyond Kriesiland: EU integration as a super issue after the Eurocrisis. *European Journal of Political Research, 56*(2), 301–319. doi:10.1111/1475-6765.12177

Özerdem, A. (2012). A re-conceptualisation of ex-combatant reintegration: 'Social reintegration' approach. *Conflict Security and Development*, *12*(1), 51–73. doi:10.1080/14678802.2012.667661

Pachankis, J. E. (2007). The psychological implications of concealing a stigma: A cognitive-affective-behavioral model. *Psychological Bulletin*, *133*(2), 328–345. doi:10.1037/0033-2909.133.2.328 PMID:17338603

Packer, J., Ballantyne, R., & Uzzell, D. (2019). Interpreting war heritage: Impacts of Anzac museum and battlefield visits on Australians' understanding of national identity. *Annals of Tourism Research*, *76*, 105–116. doi:10.1016/j.annals.2019.03.012

Pack, S. (2006). *Tourism and dictatorship: Europe's peaceful invasion of Franco's Spain*. Springer. doi:10.1057/9780230601161

Paffenholz, T. (2009). Understanding Peacebuilding Theory: Management, Resolution and Transformation. *New Routes*, *14*(2), 3–6.

Paffenholz, T. (2015). Unpacking the local turn in peacebuilding: A critical assessment towards an agenda for future research. *Third World Quarterly*, *36*(5), 857–874. doi:10.1080/01436597.2015.1029908

Pallaver, G., Gehler, M., & Cau, M. (2018). *Populism, populists, and the crisis of political parties: A comparison of Italy, Austria, and Germany 1990*-2015. Società Editrice il Mulino.

Palmer, C. A. (1994). Tourism and colonialism: The experience of the Bahamas. *Annals of Tourism Research*, *21*(4), 792–811. doi:10.1016/0160-7383(94)90084-1

Pan, H.-H. (2018). Does Cross-Strait tourism induce peace? Evidence from survey data on Chinese tourists and non-tourists. *International Relations of the Asia-Pacific*, *0*, 1–33.

Panikkar, R. (2004). *Pau i interculturalitat, una reflexió filosòfica*. Proa.

Pappé, I., & Chomsky, N. (2015). *On Palestine*. Haymarket Books.

Parkin, J. (2018, Dec 10). Colombian artist melts guns into tiles to make 'anti-monument' to conflict. *The Guardian: Americas Edition.* Retrieved from https://www.theguardian.com/world/2018/dec/10/doris-salcedo-colombian-artist-melts-guns-tiles-peace-monument

Parks Foundation. (2018). *Peace Parks Foundation Annual Review 2018*. Retrieved from https://www.peaceparks.org/wp-content/uploads/2019/07/PPF_Annual-Review-2018_Draft-15_web_smaller2.pdf

Patton, M. Q. (1989). A context and boundaries for a theory-driven approach to validity. *Evaluation and Program Planning*, *12*(4), 375–377. doi:10.1016/0149-7189(89)90054-2

PaxLusófona. (2017). *Intra-África Pax Lusófona.* Retrieved January 3, 2020, from: http://www.intraafricapaxlusofona.org/index.php/apresentacao-do-projeto

Peace Parks Foundation. (2015). *Great Limpopo Transfrontier Conservation Area Integrated Livelihood Diversification Strategy*. Retrieved from https://maps.ppf.org.za/arcgis/apps/MapJournal/index.html?appid=6a4ecb43005247b1a8adb88d7b79a917# Peace

Peace Parks Foundation. (2019a). *Great Limpopo*. Retrieved August 1, 2019, from https://www.peaceparks.org/tfcas/great-limpopo

Peace Parks Foundation. (2019b). *Our Journey*. Retrieved February 19, 2019, from https://www.peaceparks.org/about/our-journey/

Peace Parks Foundation. (2019c). *SA College for Tourism*. Retrieved May 4, 2019, from https://www.peaceparks.org/sact/

Pearce, P. L. (1980). A favorability-satisfaction model of tourists' evaluations. *Journal of Travel Research, 19*(1), 13–17. doi:10.1177/004728758001900103

Pécaut, D. (2008). *Las FARC: fuentes de su longevidad y de la conservación de su cohesión*. 29.

Peled-Elhanan, N. (2012). *Palestine in Israeli School Books: Ideology and Propaganda in Education*. I.B. Tauris. doi:10.5040/9780755608195

Pernecky, T. (2010). The being of tourism. *The Journal of Tourism and Peace Research, 1*(1), 1–15.

Peterson, C. (2006). *A Primer in Positive Psychology*. Oxford University Press.

Pettigrew, T. F. (1982). Prejudice. In T. F. Pettigrew, G. M. Frederickson, D. T. Knobel, N. Glazer, & R. Ueda (Eds.), Prejudice-dimensions of ethnicity (pp. 1–30). Harvard University Press.

Pettigrew, T. F. (1979). The ultimate attribution error: Extending Allport's cognitive analysis of prejudice. *Personality and Social Psychology Bulletin, 5*(4), 461–476. doi:10.1177/014616727900500407

Pettigrew, T. F. (1998). Intergroup contact. *Annual Review of Psychology*, 49. PMID:15012467

Pettigrew, T. F., & Tropp, L. R. (Eds.). (2000). *Does intergroup contact reduce prejudice? Recent meta-analytic findings*. Lawrence Erlbaum Associates.

Pettinger, T. (2018). *People who made a contribution to world peace*. Retrieved December 12, 2019, from: https://www.biographyonline.net/people/world-peace.html

Pforr, C., & Hosie, P. J. (2008). Crisis management in tourism preparing for recovery. *Journal of Travel & Tourism Marketing, 23*(2–4), 249–264. doi:10.1300/J073v23n02_19

Pigem, J. (2007). *El pensament de Raimon Panikkar. Interdependència, pluralisme, interculturalitat*. Institut d'Estudis Catalans.

Pigem, J. (2010). *Qüestió de valors. Del consumisme a la sostenibilitat*. Institut del Territori.

Pigem, J. (2015). *Intel·ligència Vital. Una visió postmaterialista de la vida i la consciència*. Kairós.

Pink, S. (2008). Analysing visual experience. In M. Pickering (Ed.), *Research methods for cultural studies* (pp. 125–149). Edinburgh University Press.

Pinto, M. M., & Larrechea, E. M. (2018). Internacionalização da educação superior: Uma análise das tendências de mobilidade dos estudantes entre países do norte e do sul global. *Avaliação: Revista da Avaliação da Educação Superior, 23*(3), 718–735. doi:10.15901414-40772018000300009

Pizam, A., Fleischer, A., & Mansfield, Y. (2002). Tourism and social change: The case of Israeli ecotourists visiting Jordan. *Journal of Travel Research, 41*(2), 177–184. doi:10.1177/004728702237423

Pizam, A., Jafari, J., & Milman, A. (1991). Inñuence of tourism on attitudes: US students visiting USSR. *Tourism Management, 12*, 47–54. doi:10.1016/0261-5177(91)90028-R

Porto, J. G., & Alden, C. (2016). *From soldiers to citizens: Demilitarization of conflict and society*. Routledge. doi:10.4324/9781315583525

PortSEurope. (2020). *Trade between China and Portuguese-speaking countries grew by 1.55% in 2019*. Retrieved on March 24 2020, from: https://www.portseurope.com/trade-between-china-and-portuguese-speaking-countries-grew-by-1-55-in-2019/

Pottorff, S. M., & Neal, D. D. M. (1994). Marketing implications for post-disaster tourism destinations. *Journal of Travel & Tourism Marketing*, *3*(1), 115–122. doi:10.1300/J073v03n01_08

Pratt S. & Liu A. (2015). Does Tourism Really Lead to Peace? A Global View. *International Journal of Tourism Research.*

Pratt, S., & Liu, A. (2016). Does tourism really lead to peace? A global view. *International Journal of Tourism Research*, *18*(1), 82–90. doi:10.1002/jtr.2035

Presbitero, A. (2016). Culture shock and reverse culture shock: The moderating role of cultural intelligence in international students' adaptation. *International Journal of Intercultural Relations*, *53*, 28-38.

Prideaux, B. (2005). Factors affecting bilateral tourism flows. *Annals of Tourism Research*, *32*(3), 780–801. Advance online publication. doi:10.1016/j.annals.2004.04.008

Prieto, N. C. (2004). *O outro lado do novo regionalismo pós-soviético e da Ásia-Pacífico: A diplomacia federativa além das fronteiras do mundo ocidental. A dimensão subnacional e as relações internacionais*. EDUC.

Prime Minister of Japan and His Cabinet. (2012). *Press Conference by Prime Minister Yoshihiko Noda*. Retrieved from Prime Minister of Japan and His Cabinet: https://japan.kantei.go.jp/noda/statement/201208/24kaiken_e.html

Proust, M. (2016). *A prisioneira: Em busca do tempo perdido* (Vol. 5). Relógio D'Água.

Pugel, J. (2006). *A Survey of Ex-combatants in Liberiai*. 74.

Putnam, R. (1993). *Making Democracy Work: Civic Traditions in Modern Italy*. Princeton University Press.

Putnam, R. (2000). *Bowling Alone: The Collapse and Revival of American Community*. Simon and Schuster.

Quinn, B. (2010). Arts festivals, urban tourism and cultural policy. *Journal of Policy Research in Tourism, Leisure & Events*, *2*(3), 264–279. doi:10.1080/19407963.2010.512207

Quinn, B. (2013). *Key concepts in event management*. Sage Publications. doi:10.4135/9781473914742

Quinn, B., & Wilks, L. (2013). Festival connections: People, place and social capital. In G. Richards, M. P. de Brito, & L. Wilks (Eds.), *Exploring the social impacts of events* (pp. 15–30). Routledge.

Ray, N. M., & McCain, G. (2009). Guiding tourists to their ancestral homes. *International Journal of Culture, Tourism and Hospitality Research*, *3*(4), 296–305. doi:10.1108/17506180910994514

Ray, N. M., & McCain, G. (2012). Personal identity and nostalgia for the distant land of past: Legacy tourism. *International Business & Economics Research Journal*, *11*(9), 977–989. doi:10.19030/iber.v11i9.7181

Reality Tours. (n.d.). *Reality tours*. Retrieved January 3, 2020, from: https://globalexchange.org/realitytours/

Reisinger, Y. (Ed.). (2013). *Transformational tourism: Tourist perspectives*. CAB International. doi:10.1079/9781780642093.0000

Reisinger, Y., & Mavondo, F. (2006). Cultural differences in travel risk perception. *Journal of Travel & Tourism Marketing*, *20*(1), 13–31. doi:10.1300/J073v20n01_02

Renfrew, C., & Bahn, P. (2012). *Archaeology. Theories, methods and practice* (6th ed.). Thames & Hudson.

Riba, N. (2019, March 10). Asked to strip by Security Guard, Palestinian artist refuses to fly to Israel. *Haaretz*. Retrieved from https://www.haaretz.com/israel-news/.premium-asked-to-strip-by-security-guard-palestinian-artist-refuses-to-fly-to-israel-1.7038673

Ribera Regull, R. (2007). *El Diàleg interreligiós*. Fragmenta.

Richards, G., & Raymond, C. (2000). Creative Tourism. *ATLAS News 23*. Retrieved 12 December 2019, from: http://www.atlas-euro.org/archives.aspx#events_archive

Richards, G., & Wilson, J. (2003). Today's youth travellers: Tomorrow's global nomads. *New Horizons in Independent Youth and Student Travel*, 1-43.

Richards, G. (2015). Imagineering events as interaction ritual chains. In G. Richards, L. Marques, & K. Mein (Eds.), *Event design: Social perspectives and practices* (pp. 14–24). Routledge.

Richards, G., de Brito, M. P., & Wilks, L. (Eds.). (2013). *Exploring the social impacts of events*. Routledge. doi:10.4324/9780203108123

Richards, G., & Hall, D. (Eds.). (2003). *Tourism and sustainable community development* (Vol. 7). Psychology Press.

Richards, G., & Wilson, J. (2004). The international student travel market: Travelstyle, motivations, and activities. *Tourism Review International*, *8*(2), 57–67. doi:10.3727/1544272042782183

Richmond, O. P. (2010). A genealogy of peace and conflict theory. In Palgrave Advances in Peacebuilding (pp. 14-38). Palgrave Macmillan UK. doi:10.1057/9780230282681_2

Richmond, O. P. (2012). From peacebuilding as resistance to peacebuilding as liberation. *Rethinking peacebuilding: the quest for just peace in the Middle East and the Western Balkans*, 64-77.

Richmond, O. P. (2005). *The transformation of peace*. Palgrave. doi:10.1057/9780230505070

Richmond, O. P. (2006). The problem of peace: Understanding the 'liberal peace'. *Conflict Security and Development*, *6*(3), 291–314. doi:10.1080/14678800600933480

Richmond, O. P. (2007). Emancipatory forms of human security and liberal peacebuilding. *International Journal (Toronto, Ont.)*, *62*(3), 459–478. doi:10.1177/002070200706200302

Richmond, O. P. (2009). Becoming liberal, unbecoming liberalism: Liberal-local hybridity via the everyday as a response to the paradoxes of liberal peacebuilding. *Journal of Intervention and Statebuilding*, *3*(3), 324–344. doi:10.1080/17502970903086719

Richmond, O. P., & Mitchell, A. (2011). *Hybrid forms of peace: from everyday agency to post-liberalism*. Springer.

Richmond, O., Björkdahl, A., & Kappler, S. (2011). The emerging EU peacebuilding framework: Confirming or transcending liberal peacebuilding? *Cambridge Review of International Affairs*, *24*(3), 449–469. doi:10.1080/09557571.2011.586331

Richter, L. K. (1989). The politics of tourism in Asia. Honolulu: University of Hawaii Press.

Richter, L. K. (1983). Tourism politics and political science. *Annals of Tourism Research*, *10*(3), 313–335. doi:10.1016/0160-7383(83)90060-9

Ritchie, B. W. (2004). Chaos, crises and disasters: A strategic approach to crisis management in the tourism industry. *Tourism Management*, *25*(6), 669–683. doi:10.1016/j.tourman.2003.09.004

Ritchie, B. W., Carr, N., & Cooper, C. P. (2003). *Managing educational tourism* (Vol. 10). Channel View Publications. doi:10.21832/9781873150528

Roberts, D. (2011). Beyond the metropolis? Popular Peace and Post-conflict peacebuilding. *Review of International Studies*, *37*(5), 2535–2556. doi:10.1017/S0260210511000234

Roberts, D. (2011b). *Liberal peacebuilding and global governance: Beyond the Metropolis*. Taylor & Francis. doi:10.4324/9780203825815

Rocha, A. (2003). Filosofia da Europa: Questões sobre a Europa. *Diacrítica, Filosofia e Cultura*, *17*(2), 7-22.

Rodríguez López, M., Andreouli, E., & Howarth, C. (2015). From Ex-Combatants to Citizens: Connecting Everyday Citizenship and Social Reintegration in Colombia. *Journal of Social and Political Psychology*, *3*(2), 171–191. doi:10.5964/jspp.v3i2.388

Rodríguez Rojo, M. (1995). *La educación para la paz y el interculturalismo como tema transversal*. Vilassar de Mar: Oikos-tau.

Rodríguez, P. L. M. (2011). *Los orígenes de la guerra civil española* (Vol. 154). Encuentro.

Rogerson, C. M. (2015). Tourism-agriculture linkages in rural South Africa: evidence from the accommodation sector. In A. Spenceley & D. Meyer (Eds.), *Tourism and Poverty Reduction: Principles and impacts in developing countries*. Routledge.

Rojo, V. (2017). *Historia de la guerra civil española*. RBA Libros.

Rolston, B. (2007). Demobilization and Reintegration of Ex-Combatants: The Irish Case in International Perspective. *Social & Legal Studies*, *16*(2), 259–280. doi:10.1177/0964663907076534

Romano, D. (2006). *The Kurdish National Movement: Opportunity, mobilisation, and identity*. Cambridge University Press. doi:10.1017/CBO9780511616440

Roser, M. (2020). *Tourism*. Retrieved January 13, 2020, from https://ourworldindata.org/tourism

Roszak, T. (1985). *Person / Planet. Towards a new ecological paradigm*. Kairós.

Rothman, J. (1999). Articulating Goals and Monitoring Progress in a Cyprus Conflict Resolution Training Workshop. In *Theory and Practice in Ethnic Conflict Management*. Palgrave Macmillan.

Rovelstad, J. M. (1988). *World awareness and perception search*. Paper presented at the first Global Conference Tourism: A vital force for peace. Vancouver, Canada.

Ruesch, M. (Ed.). (2011). *A peaceful Net? Intergroup contact and communicative conflict resolution of the Israel-Palestine conflict on Facebook*. Inter-Disciplinary Press.

Ruiz Monroy, H. E. (n.d.). *Cultura de paz. Un compromiso*. Retrived from: https://www.monografias.com/trabajos93/cultura-paz-compromiso/cultura-paz-compromiso.shtml

Russell, D. W. (2008). Nostalgic tourism. *Journal of Travel & Tourism Marketing*, *25*(2), 103–116. doi:10.1080/10548400802402271

Sack, R. D. (1986). *Human territoriality: its theory and history* (Vol. 7). CUP Archive.

Said, E. (1979). *The Question of Palestine*. Vintage Books.

Salamanca Convention. (2001). *The Bologna Process and the European higher education area*. EUA Genève.

Salazar, N. B. (2006). Building a 'culture of peace' through tourism: Reflexive and analytical notes and queries. *Universitas Humanistica*, (62), 319–336.

Salazar, N. B. (2006). Building a "culture of peace" through tourism: Reflexive and analytical notes and queries. *Universitas Humanistica*, *62*, 319–336.

Salazar, N. B. (2006). Building a culture of peace through tourism: Reflexive and analytical notes and queries. *Universitas Humanistica*, *2*, 319–333.

Salazar, N. B. (2010). *Envisioning Eden: Mobilising imaginaries in tourism and beyond* (Vol. 31). Berghahn Books.

Sánchez Sánchez, A., & Fernández Herreira, A. (Eds.). (1996). Dimensiones de la educación para la paz. Teoría y experiencia. Granada: Eirene.

Sandell, K. (2005). Access, tourism and democracy: A conceptual framework and the non-establishment of a proposed national park in Sweden. *Scandinavian Journal of Hospitality and Tourism*, *5*(1), 63–75. doi:10.1080/15022250510014291

Santana, J. L. (2009). *O papel da paradiplomacia nas relações internacionais: A ascensão das unidades subnacionais num contexto mundial globalizado* (Dissertation). State University of Santa Cruz, Bahia, Brasil.

Santos, L. R. (2001). Republicanismo e cosmopolitismo. A contribuição de Kant para a formação da ideia moderna de federalismo. In E. C. Leal (Ed.), *O federalismo europeu: História, política e utopia* (pp. 49–69). Edições Colibri. Instituto de História Contemporânea da Universidade de Lisboa.

Sassatelli, M. (2011). Urban festivals and the cultural public sphere: Cosmopolitanism between ethics and aesthetics. In L. Giorgio, M. Sassatelli, & G. Delanty (Eds.), *Festivals and the cultural public sphere* (pp. 12–28). Routledge.

Sasson, T., Shain, M., Hecht, S., Wright, G., & Saxe, L. (2014). Does Taglit-Birthright Israel Foster Long-Distance Nationalism? *Nationalism & Ethnic Politics*, *20*(4), 438–454. doi:10.1080/13537113.2014.969149

Sather-Wagstaff, J. (2016). *Heritage that hurts: Tourists in the memoryscapes of September 11*. Routledge. doi:10.4324/9781315427539

Scheyvens, R. (1999). Ecotourism and the empowerment of local communities. *Tourism Management*, *20*(2), 245–249. doi:10.1016/S0261-5177(98)00069-7

Scheyvens, R. (2002). Backpacker tourism and Third World development. *Annals of Tourism Research*, *29*(1), 144–164. doi:10.1016/S0160-7383(01)00030-5

Schmitz, C. L., Matyók, T., Sloan, L. M., & James, C. (2012). The relationship between social work and environmental sustainability: Implications for interdisciplinary practice. *International Journal of Social Welfare*, *21*(3), 278–286. doi:10.1111/j.1468-2397.2011.00855.x

Schneider, E. M. (2019). Touring for peace: The role of dual-narrative tours in creating transnational activists. *International Journal of Tourism Cities*, *5*(2), 200–218. doi:10.1108/IJTC-12-2017-0092

Schulenkorf, N., & Edwards, D. (2010). The role of sport events in peace tourism. In O. Moufakkir & I. Kelly (Eds.), *Tourism, progress and peace* (pp. 99–117). CABI. doi:10.1079/9781845936778.0099

Schulz, E. K. (2005). The meaning of spirituality for individuals with disabilities. *Disability and Rehabilitation*, *27*(21), 1283–12195. doi:10.1080/09638280500076319 PMID:16298931

Schwartzman, S. (2018). A contribuição da educação para o progresso social. *Ciência & Trópico*, *42*(1), 1–76.

Scott, J. W. (2013). Constructing familiarity in Finnish–Russian Karelia: Shifting uses of history and the re-interpretation of regions. *European Planning Studies*, *21*(1), 75–92. doi:10.1080/09654313.2012.716240

Scott, P. (2015). Dynamics of academic mobility: Hegemonic internationalisation or fluid globalisation. *European Review (Chichester, England)*, *23*(1), 55–69. doi:10.1017/S1062798714000775

Seaton, A. V., Smith, R. A., Inguanez, J., Pizam, A., Lebruto, S. M., Roper, A., & Boissevain, J. (2001). M. In J. Jafari (Ed.), *Encyclopedia of tourism* (pp. 366–403). Routledge.

Seattle, C. (1985). *Letter to the President of the United States*. Recovered from Carolina State University: http://www.csun.edu/~vcpsy00h/seattle.htm

Sebastião, D. (2019). Gregos e Troianos: A política de asilo na União Europeia, entre a politização e os direitos humanos. In I. Camisão & A. P. Brandão (Eds.), *O estado da União Europeia: Da(s) crise(s) à mudança?* (pp. 81–102). Editora Petrony.

Sebastião, D. (2020). Political parties in the EU political system. In R. Moldovan (Ed.), *The European Union: Policies, perspectives and politics* (pp. 131–152). Nova Science Publishers.

Selstad, L. (2007). The social anthropology of the tourist experience, exploring the 'middle role'. *Scandinavian Journal of Hospitality and Tourism*, *7*(1), 19–33. doi:10.1080/15022250701256771

Selwitz, R. (1996). Peace brings tourists to Northern Ireland. *Hotel and Motel Management*, *211*(7), 8, 40.

Sending, O. J. (2009). Why peacebuilders fail to secure ownership and be sensitive to context. *Security in Practice 1* (Working Paper). Norwegian Institute of International Affairs.

Senhoras, E. M. (2013). Múltiplas camadas das relações internacionais entre a diplomacia e a paradiplomacia. *Revista Intellector*, *9*(18), 1–14.

Séraphin, H., Butcher, J., & Korstanje, M. (2017). Challenging the negative images of Haiti at a pre-visit stage using visual online learning materials. *Journal of Policy Research in Tourism, Leisure & Events*, *9*(2), 169–181. doi:10.1080/19407963.2016.1261146

Seraphin, H., Korstanje, M., & Gowreesunkar, V. G. (2019). *Post-disaster and post-conflict tourism: Toward a new management approach*. Apple Academic Press.

Seth, C., & von Kulessa, R. (Eds.). (2011). *The idea of Europe: Enlightenment perspectives*. Open Book Publishers.

Setton, M. D. G. J. (2005). A particularidade do processo de socialização contemporâneo. *Tempo Social*, *17*(2), 335–350. doi:10.1590/S0103-20702005000200015

Shackley, M. (2001). *Managing sacred sites: service provision*. Continuum.

Shamsub, H., & Lebel, L. (2012). Identifying tourists with sustainable behaviour: A study of international tourists to Thailand. *Journal of Environmental Management & Tourism*, *3*(1). Advance online publication. doi:10.2478/v10260-012-0003-z

Sharma, A., & Thapar, M. (2016). Development of tourism in the Third World nations: a comparative analysis. *Proceedings of the 7 International Conference on Sustainable Tourism*. www.witconferences.com

Sharma, P., Charak, N. C., & Kumar, R. (2018). Sustainable tourism development and peace: A local community approach. *Journal of Hospitality Application & Research*, *13*(1), 37–56.

Sharma, P., Charak, N. S., & Kumar, R. (2018). Sustainaible tourism development and peace: A local community approach. *Journal of Hospitality Application & Reserach, 13*(1), 36–56.

Sharpley, R. (2009). Tourism, religion and spirituality. In T. Jamal & M. Robinson (Eds.), *The SAGE handbook of tourism studies* (pp. 237–253). SAGE. doi:10.4135/9780857021076.n14

Sharpley, R. (2018). *Tourism, tourists and society*. Routledge. doi:10.4324/9781315210407

Sharpley, R., & Jepson, D. (2011). Rural tourism: A spiritual experience? *Annals of Tourism Research, 38*(1), 52–71. doi:10.1016/j.annals.2010.05.002

Shavit, A. (2013). *My Promised Land*. Scribe.

Shehi, R. Z., Ozcan, S., & Hagen, T. (2018). The role of higher education institutions in building a culture of peace: An Albanian case. *Journal of Peacebuilding & Development, 13*(1), 46–61. doi:10.1080/15423166.2018.1427136

Shepherd, J., & Laven, D. (2019). Providing counter-narratives: The positive role of hostels in the Israeli-Palestinian context. *Tourism Geographies*, 1–24. Advance online publication. doi:10.1080/14616688.2019.1669215

Sherriff, L. (2020). Crossing divides: The hostel promoting tourism to ease Israel's tensions. *The BBC*. Retrieved from https://www.bbc.com/news/world-middle-east-51446690

Shin, Y. (2008). Peace and tourism vs. tourism and peace: Conceptual Issues. *International Journal of Hospitality and Tourism Systems, 1*(1), 25–35.

Shirley, G. (2018). The branding of post-conflict tourism destinations: Theoretical reflections and case studies. In E. Wylie, N. Andreas, & G. Jesse Hession (Eds.), *The tourism–disaster–conflict nexus* (Vol. 19, pp. 119–139). Emerald Publishing Limited. doi:10.1108/S2040-726220180000019007

Shiva, V. (2006). *Earth democracy: Justice, sustainability and peace*. Zed Books.

Sibley, D. (1995). *Geographies of exclusion*. Routledge.

Sidjanski, D. (1996). *O futuro federalista da Europa. A Comunidade Europeia, das origens ao Tratado de Maastricht*. Gradiva.

Sierra, L. (2019, Jul 17). Estos son los productos que ofrecen los exguerrileros en Agroexpo. *Pacifista*. Retrieved from https://pacifista.tv/notas/farc-agroexpo-reincorporacion-productos-etcr/

Sigelman, L., Bledsoe, T., Welch, S., & Combs, M. W. (1996). Making contact? Black-white social interaction in an urban setting. *American Journal of Sociology, 101*(5), 1306–1332. doi:10.1086/230824

Sigelman, L., & Welch, S. (1993). The contact hypothesis revisited: Black-White interaction and positive racial attitudes. *Social Forces, 71*(3), 781–795. doi:10.2307/2579895

Silva, J. T. (2012b). The non-governmental channels of intermediation in the renewed context of Cross Strait relations: The Taishang's economic dynamics. In B. Guo & C. Teng (Eds.), Taiwan and rise of China (pp. 145-163). Lanham: Lexington.

Silva, A. M. (2010). *História da Unificação Europeia: A integração comunitária (1945-2010)*. Imprensa da Universidade de Coimbra. doi:10.14195/978-989-26-0421-3

Silva, J. T. (2008). New trends in the Taiwan Strait conflict: The cooperative and quarrelsome framework. In L. Tomé (Ed.), *East Asia today* (pp. 321–340). Prefácio.

Silva, J. T. (2012a). Informal and nonofficial interactions in the new start of Cross Strait tRelations. In J. F. Blanchard & D. V. Hickey (Eds.), *New thinking about the Taiwan issue: Theoretical insights into its origins, dynamics, and prospects* (pp. 116–136). Routedge.

Silva, J. T. (2016). Chen Shui–bian. In K. Brown (Ed.), *Dictionary of Chinese biography* (Vol. 4, pp. 62–67). Massachuttts.

Silva, V. A. D., & Rebolo, F. (2017). A educação intercultural e os desafios para a escola e para o professor. *Interações (Campo Grande)*, *18*(1), 179–190. doi:10.20435/1984-042X-2017-v.18-n.1(14)

Simone-Charteris, M. T., & Boyd, S. W. (2010). Northern Ireland re-emerges from the ashes: The contribution of political tourism towards a more visited and peaceful environment. *Tourism, Progress, and Peace*, 179.

Simone-Charteris, M. T., & Boyd, S. W. (2010). *Northern Ireland re-emerges from the ashes: The contribution of political tourism towards a more visited and peaceful environment. In Tourism, progress and peace*. CABI.

Sin, N. (2018). *Macau turns to a hub*. Retrieved on March 24 2020, from: https://www.euromoney.com/article/b17fn-rfk46ln5m/macau-turns-to-a-hub

Sirakaya-Turk, E., Nyaupane, G., & Uysal, M. (2014). Guests and hosts revisited: Prejudicial attitudes of guests toward the host population. *Journal of Travel Research*, *53*(3), 336–352. doi:10.1177/0047287513500580

Siriphon, A., & Zhu, J. (2018). Outbound Tourists and policy reforms under the Chinese mobility regime. In China's Reforms in the New Era and Their Implications for East Asia (pp. 1–29). Academic Press.

Smith, S. A. (2018). Japan and China: A fresh start in a difficult era. *Asia-Pacific Review*, *25*(2), 75–88. doi:10.1080/13439006.2018.1551823

Smith, V. L. (1996). War and its tourist attractions. In A. Pizam & Y. Mansfield (Eds.), *Tourism, Crime, and International Security Issues* (pp. 247–264). Wiley.

Smith, V. L. (1998). War and tourism. An American ethnography. *Annals of Tourism Research*, *25*(1), 202–227. doi:10.1016/S0160-7383(97)00086-8

Smith, V. L. (Ed.). (1989). *Hosts and guests: The anthropology of tourism*. University of Pennsylvania Press. doi:10.9783/9780812208016

Snepenger, D., King, J., Marshall, E., & Uysal, M. (2006). Modeling Iso-Ahola's motivation theory in the tourism context. *Journal of Travel Research*, *45*(2), 140–149. doi:10.1177/0047287506291592

Sonmez, S. F., & Apostolopoulos, Y. (2000). Conflict Resolution Through Tourism Cooperation? The Case of the Partitioned Island-State of Cyprus. *Journal of Travel & Tourism Marketing*, *9*(3), 35–48. doi:10.1300/J073v09n03_03

Sorrells, K. (2013). *Intercultural communication, globalisation and social justice*. Sage Publications.

Soulier, G. (1997). *A Europa*. Instituto Piaget.

Sousa, R. R. P. (2017). Relações internacionais e o estudo da paz e conflito: O contexto da resolução de conflitos. *Incursões na Teoria da Resolução de Conflitos*, 49-98.

South African National Parks. (2000). *Bidding memorandum for the tender of concession sites*. South African National Parks.

South African National Parks. (2019). *Great Limpopo Transfrontier Park*. Retrieved August 9, 2019, from https://www.sanparks.org/conservation/transfrontier/great_limpopo.php

South China Morning Post. (2019). *Mainland China to boycott Golden Horse awards in latest attempt to squeeze Taiwan.* Retrieved March 30, 2020, from: https://www.scmp.com/news/china/politics/article/3021731/mainland-china-boycott-asias-oscars-latest-attempt-squeeze

Southern African Development Community. (1999). What governments can do for community tourism (Leaflet No. 2). Harare: Africa Resources Trust.

Southern African Development Community. (2012). *Transfrontier Conservation Areas.* Retrieved October 15, 2018, from https://www.sadc.int/themes/natural-resources/transfrontier-conservation-areas/

Sozen, A. (2006). *The Cyprus Impasse and Prospects for a Breakthrough. In Reflections on the Cyprus Problem: A Compilation of Recent Academic Contributions.* EMU Printing House.

Space for Giants, & United Nations Environment Program. (2019). *Developing Nature-Based Tourism in African State Protected Areas.* (Working Paper No 1). Space for Giants. Retrieved from https://spaceforgiants.org/wp-content/uploads/2019/06/Building-Africas-Wildlife-Economy-Space-for-Giants-Working-Paper-1.pdf

Spenceley, A. (2005). *Tourism investment in the Great Limpopo Transfrontier Conservation Area: Scoping report.* doi:10.13140/rg.2.1.3315.4407

Spenceley, A. (2006). Tourism in the Great Limpopo Transfrontier Park. *Development Southern Africa, 23*(5), 649–667. doi:10.1080/03768350601021897

Spracklen, K., & Lamond, I. R. (2016). *Critical event studies.* Routledge. doi:10.4324/9781315690414

Statistics and Census Service. (2019). *Statistics.* Retrieved on March 24 2020, from: https://www.dsec.gov.mo/

Stedman, S. J. (1997). Spoiler problems in peace processes. *International Security, 22*(2), 5–53. doi:10.1162/isec.22.2.5

Steger, I. (2020). *China tried to threaten Taiwan by weaponing tourism, but it didn't work.* Retrieved March 30, 2020, from: https://qz.com/1780709/taiwan-tourism-hits-record-high-despite-china-boycott/

Steiner, C. J., & Reisinger, Y. (2003). *Cross-cultural behaviour in tourism: Concepts and analysis.* Butterworth-Heinemann.

Stenvall, A., Laven, D., & Gelbman, A. (2017). The influence of social entrepreneurship in tourism on an Arab village in Israel. In P. J. Sheldon & R. Daniele (Eds.), *Social entrepreneurship in tourism: Philosophy and practice* (pp. 279–293). Springer. doi:10.1007/978-3-319-46518-0_17

Stephan, W. G., & Stephan, C. W. (1985). Intergroup Anxiety. *The Journal of Social Issues, 41*(3), 157–175. doi:10.1111/j.1540-4560.1985.tb01134.x

Stets, J. E., & Burke, P. J. (2000). Identity theory and social identity theory. *Social Psychology Quarterly, 63*(3), 224–237. doi:10.2307/2695870

Stewart, J., & Zediker, K. (2000). Dialogue as tensional, ethical practice. *Southern Journal of Communication, 65*(2-3), 224–242. doi:10.1080/10417940009373169

Stone, P. (2017). Preface. In D. Laven, D. Walters, & P. Davis (Eds.), Heritage and peacebuilding (pp. xii-xiv). Boydell & Brewer.

Stone, P. R. (2006). A Dark Tourism Spectrum: Towards a typology of death and macabre related tourist sites, attractions and exhibitions. *Tourism: An Interdisciplinary International Journal, 54*(2), 145–160.

Strandberg, C. (2017). *Maximising the capacities of advanced education institutions to build social infrastructure for canadian communities.* JW McConnell Family Foundation.

Straus, A., & Corbin, J. (1990). *Basics of qualitative research: Grounded theory procedures and techniques*. Sage.

Strömbom, L. (2019). Exploring Prospects for Agonistic Encounters in Conflict Zones: Investigating Dual Narrative Tourism in Israel /Palestine. *Alternatives*, *44*(2-4), 75–93. doi:10.1177/0304375419857421

Suansri, P. (2003). *Community based tourism handbook*. Responsible Ecological Social Tour-REST.

Sussman, N. M. (2000). The dynamic nature of cultural identity throughout cultural transitions: Why home is not so sweet. *Personality and Social Psychology Review*, *4*(4), 355–373. doi:10.1207/S15327957PSPR0404_5

Sveaass, N., & Castillo, M. (2000). From war hero to cripple: An interview study on psychosocial intervention and social reconstruction in Nicaragua. *Peace and Conflict*, *6*(2), 113–133. doi:10.1207/S15327949PAC0602_2

Swain, A., & Krampe, F. (2011). Stability and sustainability in peace building: Priority area for warfare ecology. In *Warfare Ecology* (pp. 199–210). Springer. doi:10.1007/978-94-007-1214-0_14

Szabo, J. S. (2015). Festivals, conformity and socialisation. In C. Newbold, C. Maughan, J. Jordan, & F. Bianchini (Eds.), *Focus on festivals: Contemporary European case studies and perspectives* (pp. 40–52). Goodfellows Publishers. doi:10.23912/978-1-910158-15-9-2662

Tamatea, P. P. (2007) Interview with Pepe Papa Tamatea/Interviewer: P. Diamond. Vietnam Oral History Project (Vol Reference Number OHDL-000513 digitial manuscript). Ministry of Culture and Heritage.

Tappe, O. (2011). Memory, tourism, and development: Changing sociocultural configurations and upland-lowland relations in Houaphan Province, Lao PD. *Journal of Social Issues in Southeast Asia*, *26*(2), 174–195. doi:10.1355j26-2b

Tarlow, P. (2014). *Tourism security: strategies for effectively managing travel risk and safety*. Elsevier.

Tashakkori, A., & Teddlie, C. (2003). *Handbook of mixed methods in social & behavioral research*. SAGE Publications.

Teddlie, C., & Yu, F. (2007). Mixed methods sampling. *Journal of Mixed Methods Research*, *1*(1), 77–100. doi:10.1177/1558689806292430

TedXTalks. (2017, September 17). *Me-Rational / Neta Hanien / TedxJaffa* [Video File]. Retrieved from https://www.youtube.com/watch?v=ZdNcOy_eAB0

Tehran Peace Museum. (n.d.). *Home*. Retrieved December 20, 2019, from: http://www.tehranpeacemuseum.org/index.php/en/

Teichler, U., & Jahr, V. (2001). Mobility during the course of study and after graduation. *European Journal of Education*, *36*(4), 443–458. doi:10.1111/1467-3435.00081

The Diplomat. (2019). *Taiwan slams China's decision to freeze individual travel to Taiwan*. Retrieved June 2, 2020, from: https://thediplomat.com/2019/08/taiwan-slams-chinas-decision-to-freeze-individual-travel-to-taiwan/

The International Ecotourism Society. (2019). *What Is Ecotourism?* Retrieved July 17, 2019, from https://ecotourism.org/what-is-ecotourism/

The North Africa Post. (2018). *Morocco's membership in Peace & Security Council boosts peace, stability in Africa*. Retrieved from https://northafricapost.com/23101-moroccos-membership-peace-security-council-boosts-peace-stability-africa.html

The Outpost Lodge. (2020). *The Outpost Lodge*. Facebook Homepage. Gallery. Retrieved on January 7, 2020, from https://www.facebook.com/pg/theoutpostlodge/photos/?ref=page_internal

Theidon, K. (2007). Transitional Subjects: The Disarmament, Demobilization and Reintegration of Former Combatants in Colombia. *The International Journal of Transitional Justice*, *1*(1), 66–90. doi:10.1093/ijtj/ijm011

Thiessen, C., & Darweish, M. (2018). Conflict resolution and asymmetric conflict: The contradictions of planned contact interventions in Israel and Palestine. *International Journal of Intercultural Relations*, *66*, 73–84. doi:10.1016/j.ijintrel.2018.06.006

Timothy, D. J. (2006). Safety and security issues in tourism. In *Tourism Management Dynamics* (pp. 43–51). Routledge. doi:10.1016/B978-0-7506-6378-6.50013-5

Timothy, D. J. (2018). Making sense of heritage tourism: Research trends in a maturing field of study. *Tourism Management Perspectives*, *25*, 177–180. doi:10.1016/j.tmp.2017.11.018

Timothy, D. J. (2019). Supranationalism and tourism: Free trade, customs unions, and single markets in an era of geopolitical change. In D. J. Timothy (Ed.), *Handbook of globalisation and tourism* (pp. 100–114). Edward Elgar Publishing. doi:10.4337/9781786431295.00018

Timothy, D. J., & Boyd, S. W. (2003). *Heritage tourism*. Pearson Education.

Timothy, D. J., & Kim, S. S. (2015). Understanding the tourism relationships between South Korea and China: A review of influential factors. *Current Issues in Tourism*, *18*(5), 413–432. doi:10.1080/13683500.2013.776019

Timothy, J. D. (2001). Tourism planning in borderlands. In *Tourism and political boundaries* (pp. 133–135). Routledge.

Ting-Toomey, S., & Chung, L. C. (2012). *Understanding intercultural communication* (2nd ed.). Oxford University Press.

Tomljenovic, R. (2010). Tourism and intercultural understanding or contact hypothesis revisited. In O. Moufakkir & I. Kelly (Eds.), *Tourism, progress and peace* (pp. 17–32). CABI. doi:10.1079/9781845936778.0017

Travel Palestine. (2019). Retrieved from http://travelpalestine.ps/

Tschirgi, N. (2004). *Post-conflict Peacebuilding Revisited: Achievements, Limitations, Challenges*. International Peace Academy.

Tse, T. S. (2013). Chinese outbound tourism as a form of diplomacy. *Tourism Planning & Development*, *10*(2), 149–158. doi:10.1080/21568316.2013.783738

Tse, T. S., & Hobson, J. S. P. (2008). The forces shaping China's outbound tourism. *Journal of China Tourism Research*, *4*(2), 136–155. doi:10.1080/19388160802279459

Tucker, H. (2001). Tourists and troglodytes: Negotiating for sustainability. *Annals of Tourism Research*, *28*(4), 868–891. doi:10.1016/S0160-7383(00)00084-0

Turner, V. (1969). *The ritual process*. Penguin.

Tzanelli, R. (2016). *Thanatourism and cinematic representations of risk: Screening the end of tourism*. Routledge. doi:10.4324/9781315624105

Tzanelli, R., & Korstanje, M. E. (2019). 6 On killing the 'toured object. In R. Isaac, E. Cakmak, & R. Butler (Eds.), *Tourism and hospitality in conflict-ridden destinations* (pp. 71–83). Routledge. doi:10.4324/9780429463235-6

U, C. (2019). *Macau | Average salary of local employees at MOP20,000 in Q1 2018*. Retrieved on March 24 2020, from: https://www.macaubusiness.com/macau-average-salary-local-employees-mop20000-q1-2018/

UNDP. (2015, December). Final Report. *Action for Cooperation and Trust in Cyprus*, *1*(October), 2005–2031.

UNESCO. (1998). *Declaração mundial sobre educação superior no século XXI: Visão e ação, marco referencial de ação prioritária para a mudança e o desenvolvimento da educação*. Universidade Metodista de Piracicaba.

UNESCO. (2004). Embracing diversity: toolkit for creating inclusive, learning-friendly environments. *UNESCO Digital Library website*. Retrieved February 1, 2020, from: https://unesdoc.unesco.org/ark:/48223/pf0000137522

UNESCO. (2017). *Global flow of tertiary-level students* Retrieved January 3, 2020, from: http://uis.unesco.org/en/uis-student-flow#slideoutmenu

UNESCO. (2019). *Diversidad de las expresiones culturales*. Retrieved from: https://es.unesco.org/creativity/

United Nations Environment Programme. (2004). *Environmental Management and Community Participation: Enhancing Local Programmes*. UNEP-IETC Urban Environment Series.

United Nations Environment Programme. (2018). *Environmental Cooperation for Peacebuilding*. Retrieved January 11, 2019, from https://web.unep.org/regions/rolac/environmental-cooperation-peacebuilding

United Nations Environment Programme. (Ed.). (2005). *Making tourism more sustainable: A guide for policy makers*. United Nations Environment Programme & World Tourism Organisation.

United Nations Verification Mission in Colombia - UNVMC. (2019a, Oct 1). *Report of the Secretary-General*. Retrieved from https://colombia.unmissions.org/sites/default/files/en_-_n1928697.pdf

United Nations Verification Mission in Colombia - UNVMC. (2019b, Dec 26). *Report of the Secretary-General*. Retrieved from https://colombia.unmissions.org/sites/default/files/n1942147.pdf

United Nations World Tourism Organisation. (2020). *Tourism in the 2030 Agenda*. Retrieved on March 23 2020, from https://www.unwto.org/tourism-in-2030-agenda

United Nations, Environment Programme, United Nations, & Development Programme. (2013). *The role of natural resources in disarmament, demobilization and reintegration: Addressing risks and seizing opportunities*. United Nations Environment Programme.

United Nations. (2014, May 9). *Integrated Disarmament, Demobilization and Reintegration Standards*. Retrieved from https://www.unddr.org/iddrs-framework.aspx

United Nations. (2015). *17 Goals to transform our world*. Retrieved December 21, 2019, from: https://www.un.org/sustainabledevelopment/education/

United Nations. (2017). *Global indicator framework for the Sustainable Development Goals and targets of the 2030 Agenda for Sustainable Development Goals and targets (from the 2030 Agenda for Sustainable Development) Indicators*. UN.

United Nations. (2017). *International Conference on the Question of Jerusalem*. Retrieved March 18, 2020, from: https://www.un.org/unispal/wp-content/uploads/2017/07/Baku-Report-Final.pdf

United Nations. (2019). *A/RES/73/338 Seventy-third session Agenda item 15*. Resolution adopted by the General Assembly on September 12, 2019. Retrieved January 4, 2020, from: https://undocs.org/en/A/RES/73/338

Universidad Nacional de Colombia. (2017, Jul 6). *Caracterización comunidad FARC-EP. Censo socioeconómico UN-CNR 2017*. Retrieved from http://pensamiento.unal.edu.co/fileadmin/recursos/focos/piensa-paz/docs/presentacion_censo_farc.pdf

UNWTO ST-EP. (2011). *Foundation, Sustainable Tourism for Eliminating Poverty*. http://unwtostep.org/

UNWTO. (2005). *Making Tourism More Sustainable - A Guide for Policy Makers*. UNEP and UNWTO.

UNWTO. (2011). *Tourism towards 2030: Global overview*. World Tourism Organization.

UNWTO. (2012). *International day of peace: Tourism as a force for development and peace*. Retrieved from https://www.unwto.org/archive/global/news/2012-09-21/international-day-peace-tourism-force-development-and-peace-0

UNWTO. (2016). *Tourism: A catalyst for peace and development*. Retrieved January 2, 2020, from: https://www.unwto.org/archive/asia/press-release/2016-07-14/tourism-catalyst-peace-and-development

UNWTO. (2018). *Tourism highlights*. World Tourism Organization.

UNWTO. (2019). *World tourism barometer*. Retrieved March 30, 2020, from: https://www.e-unwto.org/doi/pdf/10.18111/wtobarometereng.2019.17.1.1

UNWTO. (2020). *SDG 16 – Peace justice and strong institutions*. Retrieved April 1, 2010, from http://tourism4sdgs.org/sdg-16-peace-justice-institutions/

UNWTO. (n.d.). *About UNWTO*. Retrieved November 29, 2019, from: https://www.unwto.org/who-we-are

Upadhayaya, P. K. (2013). Peace sensitivity in tourism codes of conduct at destinations in conflict: A case study of partnership actions in Pokhara, Western Nepal. In C. Wohlmuther & W. Wintersteiner (Eds.), *International handbook on tourism and peace* (pp. 230–246). Drava Verlag.

Upadhayaya, P. K., Khatiwada, S. S., Upreti, B. R., & Sapkota, T. (2013). *Codes of conduct for peace responsive tourism in Pokhara: A manual*. Pokhara Tourism Council, Nepal Tourism Board, NCCR North-South.

Upadhayaya, P. K., Muller-Boker, U., & Sharma, S. R. (2011). Tourism amidst armed conflict: Consequences, copings, and creativity for peacebuilding through tourism in Nepal. *The Journal of Tourism and Peace Research*, *1*(2), 22–40.

Uriely, N., Reichel, A., & Ron, A. (2002). Volunteering in tourism: Additional thinking. *Tourism Recreation Research*, *28*(3), 57–62. doi:10.1080/02508281.2003.11081417

Urry, J. (1990). *The tourist gaze: Leisure and travel in contemporary societies*. Sage.

Urry, J. (2005). *The tourist gaze* (2nd ed.). Sage Publications.

Valecillo, Z. (2009). ¿Cómo acercar los bienes patrimoniales a los ciudadanos? Educación Patrimonial, un campo emergente en la gestión del patrimonio cultural. *PASOS: Revista de Turismo y Patrimonio Cultural, 7*.

Van Broeck, A. M., Guasca, M., & Vanneste, D. (2019). Tourism Initiatives as a Lever for Reintegration of Former FARC Combatants in Colombia. *Via. Tourism Review*, 15.

Van Den Berg, L., & Russo, A. (2017). *The student city: Strategic planning for student communities in EU cities*. Routledge.

Van Den Dungen, P. (2018). *Peace Tourism*. Retrieved January 15, 2020, from https://worldbeyondwar.org/peace-tourism/

Van den Dungen, P. (2014). *Peace Tourism. In International Handbook on Tourism and Peace*. Drava.

van't Klooster, E., Van Wijk, J., Go, F., & Van Rekom, J. (2008). Educational travel: The overseas internship. *Annals of Tourism Research*, *35*(3), 690–711.

VANZ News. (2011). Retrieved from https://www.veteransaffairs.mil.nz/about/whats-new/documents/vanznewsoct11.pdf

Var, T., & Ap, J. (1998). Tourism and world peace. In W. Theobald (Ed.), *Global tourism* (pp. 45–47). Routledge.

Vashishtha, P. (2016). *20 world organisations that strive to establish world peace*. Retrieved November 29, 2019, from: https://listsurge.com/20-world-organizations-strive-establish-world-peace/

Vidal-Casellas, D. (2020. (in press). "Los imaginarios del turismo" in Vidal-Casellas, D. and Crous-Costa, N. Dimensión Socio-cultural del turismo. Manual docente. Girona. *Documenta Universitaria.*

Vietnam visit puts ghosts to rest. (1998, July 20). *Waikato Times,* p. 10.

Vincent, G. (2015). *Hospitalité: la naissance symbolique de l'Humain.* Presses Universitaires de Strasbourg.

Vince, R., & Warren, S. (2012). Participatory visual methods. In G. Symon & C. Cassell (Eds.), *Qualitative organisational research: Core methods and current challenges* (pp. 275–295). SAGE. doi:10.4135/9781526435620.n16

Vision of Humanity. (2019). *Global peace index.* Retrieved March 23, 2020, from: http://visionofhumanity.org/indexes/global-peace-index/

Volkan, V. (1999a). Psychoanalysis and diplomacy. Part 1: Individual and large group identity. *Journal of Applied Psychoanalytic Studies, 1*(1), 29–55. doi:10.1023/A:1023026107157

Volkan, V. (1999b). Post-traumatic states: Beyond individual PTSD in societies ravaged by ethnic conflict. *Canadian Foreign Policy, 7*(1), 27–38. doi:10.1080/11926422.1999.9673198

Wachman, A. M. (2007). *Why Taiwan? Geoestrategic rationales for China's territorial integrity.* Stanford University Press.

Wachowiak, H. (Ed.). (2006). *Tourism and borders: Contemporary issues, policies and international research.* Ashgate.

Wallace, H., Pollack, M. A., & Young, A. R. (2010). *Policy-making in the European Union.* Oxford University Press.

Walmsley, B., & Franks, A. (2011). The audience experience: Changing roles and relationships. In Walmsley (Ed.), Key issues in the arts and entertainment industry (pp. 1-16). Oxford: Goodfellow Publishers.

Walters, D., Laven, D., & Davis, P. (2017). Introduction. In D. Laven, D. Walters, & P. Davis (Eds.), Heritage and peacebuilding (pp. 1-4). Boydell & Brewer.

Walters, G., & Mair, J. (2012). The effectiveness of post-disaster recovery marketing messages: The case of the 2009 Australian bushfires. *Journal of Travel & Tourism Marketing, 29*(1), 87–103. doi:10.1080/10548408.2012.638565

Waltner, A. (2017). *My DNA journey.* Retrieved 26 March 2019, from https://www.swedishnomad.com/my-momondo-dna-test/

Wang, N. (1999). Rethinking authenticity in tourism experience. *Annals of Tourism Research, 26*(2), 349–370. doi:10.1016/S0160-7383(98)00103-0

Ward, J. (2017). *Motivational Monday: My DNA journey, where I come from.* Retrieved 28 March 2019, from: https://onestep4ward.com/motivational-monday-dna-journey-i-come/

Ward, C., Furnham, A., & Bochner, S. (2005). Varieties of culture travellers: Tourists. In *The psychology of culture shock* (2nd ed., pp. 127–141). Routledge. doi:10.4324/9780203992258

Ward, E. V. (2009). Conñicts of interest: Plasticity of peace tourism and the 21st century nation. In J. Harris (Ed.), *The nation in the global era: Conflict and transformation* (pp. 304–316). Brill., doi:10.1163/ej.9789004176904.i-476.109

Weaver, D. B. (2000). The exploratory war-distorted destination life cycle. *International Journal of Tourism Research, 2*(3), 151–161. doi:10.1002/(SICI)1522-1970(200005/06)2:3<151::AID-JTR196>3.0.CO;2-G

Weber, M. (1947). *The theory of social and economic organisation.* Free Press.

Webster, C., & Ivanov, S. H. (2014). Tourism as a force for political stability. In C. Wohlmuther & W. Werner (Eds.), International handbook on tourism and peace (pp. 167-180). Celovec: Drava.

Webster, C., & Ivanov, S. (2013). Tourism as a Force for political stability. In C. Wohlmuther & W. Wintersteiner (Eds.), *International handbook on tourism and peace* (pp. 167–180). Drava Verlag.

Weizman, E. (2017). *Hollow Land: Israel's architecture of occupation.* Verso Books.

Welsch, W. (1999). Transculturality: the puzzling form of cultures today. *Spaces of Culture: City, Nation. WORLD (Oakland, Calif.)*, 194–213.

Wilks, L. (2011). Bridging and bonding: Social capital at music festivals. *Journal of Policy Research in Tourism, Leisure & Events*, *3*(3), 281–297. doi:10.1080/19407963.2011.576870

Williams, R. M. Jr. (1947). *The reduction of intergroup tensions*. Social Science Research Council.

Willibald, S. (2006). Does money work? Cash transfers to ex-combatants in disarmament, demobilisation and reintegration processes: Does Money Work? *Disasters*, *30*(3), 316–339. doi:10.1111/j.0361-3666.2005.00323.x PMID:16911431

Willson, G. B., McIntosh, A. J., & Zahra, A. L. (2013). Tourism and spirituality: A phenomenological analysis. *Annals of Tourism Research*, *42*, 150–168. doi:10.1016/j.annals.2013.01.016

Wilson, K., & van der Dussen, J. (1993). *The history of the idea of Europe*. Routledge.

Winkelman, M. (1994). Cultural shock and adaptation. *Journal of Counseling and Development*, *73*(2), 121–126. doi:10.1002/j.1556-6676.1994.tb01723.x

Wintersteiner, C., & Wuohlmuther, W. (2014). Peace Sensitive Tourism: How Tourism Can Contribute to Peace. In C. Wuohlmuther & W. Wintersteiner (Ed.), International Handbook on Tourism and Peace. Academic Press.

Wintersteiner, W., & Wohlmuther, C. (2013). Peace sensitive tourism: How tourism can contribute to peace. In C. Wohlmuther & W. Wintersteiner (Eds.), *International handbook on tourism and peace* (pp. 31–61). Drava Verlag.

Witersteiner, W., & Wohlmuther, C. (2014). Peace Sensitive Tourism: how Tourism Can Contribute to Peace. In C. Wohlmuther & W. Wintsersteiner (Eds.), *The International Handbook on Tourism and Peace* (pp. 36–61). Centre for Peace Research and Peace Eduaction of the Klagenfurt Univeristy and UNWTO.

Wohl, M. J. A., & Branscombe, N. R. (2008). Group threat, collective angst, and ingroup forgiveness for the war in Iraq. *Political Psychology*, *30*(2), 193–217. doi:10.1111/j.1467-9221.2008.00688.x

Wohlmuther, C., & Wintersteiner, W. (2014). *International handbook on tourism and peace*. Retrieved January 5, 2020, from: https://www.e-unwto.org/doi/pdf/10.18111/9783854357131

Wohlmuther, C., & Wintersteiner, W. (Eds.). (2014). International handbook on tourism and peace. Celovec: Drava.

Wohlmuther, C., Wintersteiner, W., & Wagner, M. (2013). Connected by the trails that used to divide us: Peace trails in the Alps-Adriatic Region. In C. Wohlmuther & W. Wintersteiner (Eds.), *International handbook on tourism and peace* (pp. 321–334). Drava Verlag.

Wohlmuther, C., & Wintsersteiner, W. (2014). Tourism and Peace – an emerging field of action and Research. In C. Wohlmuther & W. Wintsersteiner (Eds.), *The International Handbook on Tourism and Peace* (pp. 17–28). Centre for Peace Research and Peace Eduaction of the Klagenfurt Univeristy and UNWTO.

WOLA - Washington Office on Latin America. (n.d.). *El proceso de paz en Colombia*. Retrieved Arpil 2020, from https://www.wola.org/es/programa/colombia/el-proceso-de-paz-en-colombia/

Wond, T., & Macaulay, M. (2011). Extending time- extended benefit. *Public Management Review*, *3*(2), 309–320. doi:10.1080/14719037.2010.536059

Wood, M. E. (2002). Ecotourism: Principles, practices and policies for sustainability. Paris: United Nations Environment Programme.

Woodward, S. L. (2017). *The Ideology of Failed States*. Cambridge University Press. doi:10.1017/9781316816936

World Bank. (2019a). *International comparison program database*. Retrieved on March 24 2020, from https://www.worldbank.org/en/programs/icp

World Bank. (2019b). *GDP per capita, PPP (current international $)*. Retrieved on March 24 2020, from https://data.worldbank.org/indicator/NY.GDP.PCAP.PP.CD?locations=MO&view=chart

World Economic Forum. (2018). *The future of jobs 2018*. Retrieved on March 24 2020, from http://reports.weforum.org/future-of-jobs-2018/

World Tourism Organisation. (2019). *Tourism and the SDGs*. Retrieved July 17, 2019, from http://icr.unwto.org/content/tourism-and-sdgs

World Tourism Organization. (2018). *UNWTO tourism highlights* (2018 ed.). UNWTO.

World Travel & Tourism Council. (2019a). *City Travel & Tourism Impact Report*. Retrieved on March 24 2020, from: https://wttc.org/Research/Economic-Impact/Cities/

World Travel & Tourism Council. (2019b). *Macau: Country data*. Retrieved on March 24 2020, from: https://www.wttc.org/economic-impact/country-analysis/country-data/

World Travel & Tourism Council. (2020). *Economic impact reports*. Retrieved on March 24 2020, from: https://wttc.org/Research/Economic-Impact

World Travel and Tourism Council. (2016). *Tourism as a driver of peace: Quantitative analysis on the link between peace and tourism*. World Travel and Tourism Council.

WTTC. (2016). *Tourism as a driver of peace*. Retrieved January 15, 2020, from https://www.wttc.org/publications/2016/tourism-as-a-driver-of-peace/

Wu, S. (2013). *Solving disputes for regional cooperation and development in the South China Sea: A Chinese perspective*. Chandos Publishing. doi:10.1533/9781780633558

Xi, J. (2018). *Xi JinpingS: A grvernança da China*. Editora de Línguas Estrangeiras.

Yanow, D. (2009). Organisational ethnography and methodological angst: Myths and challenges in the field. *Qualitative Research in Organisations and Management: An International Journal*, *4*(2), 186–199. doi:10.1108/17465640910978427

Yılmaz, O., Cesur, S., & Bayad, A. (2018). Psychological correlates of negative attitudes of Turks and Kurds toward one another. *Turkish Psychological Articles*, *21*, 82–99.

Yin, M., & Zhao, S.-Z. (2006). Research on a dynamic model of trust-building within regional tourism alliances evidence from China. *Chinese Economy*, *39*(6), 5–18. doi:10.2753/CES1097-1475390601

Yoshida, K., Bui, H. T., & Lee, T. J. (2016). Does tourism illuminate the darkness of Hiroshima and Nagasaki. *Journal of Destination Marketing & Management*, *5*(4), 333–340. doi:10.1016/j.jdmm.2016.06.003

Zambrano Quintero, L. (2019). La reincorporación colectiva de las FARC-EP: Una apuesta estratégica en un entorno adverso. *Revista CIDOB d'Afers Internacionals*, *121*(121), 45–66. doi:10.24241/rcai.2019.121.1.45

Zebel, S., Zimmermann, A., Tendayi Viki, G., & Doosje, B. (2008). Dehumanization and guilt as distinct but related predictors of support for reparation policies. *Political Psychology*, *29*(2), 193–219. doi:10.1111/j.1467-9221.2008.00623.x

Zehr, H. (2008). Doing justice, healing trauma: The role of restorative justice in peacebuilding. *South Asian Journal of Peacebuilding*, *1*(1), 1–16.

Zhang, H., Cho, T., & Wang, H. (2019). The impact of a terminal high altitude area defense incident on tourism risk perception and attitude change of Chinese tourists traveling to South Korea. *Sustainability 2020, 12*(7), 1–13.

Zhang, C. X., Xiao, H., Morgan, N., & Ly, T. P. (2018). Politics of memories: Identity construction in museums. *Annals of Tourism Research*, *73*, 116–130. doi:10.1016/j.annals.2018.09.011

Zhang, H. Q. (2005). Impact of the tsunami on Chinese outbound tourism. *International Journal of Contemporary Hospitality Management*, *17*(5), 433–435. doi:10.1108/09596110510604850

Zhang, J. J. (2017). Rethinking 'heritage' in post-conflict tourism. *Annals of Tourism Research*, *66*(Supplement C), 194–196. doi:10.1016/j.annals.2017.06.005

Zhu, F. (2014). *China's first ADIZ decision: One year later.* Retrieved from https://amti.csis.org/chinas-first-adiz-decision-one-year-later/

About the Contributors

Jorge Tavares da Silva is an invited professor at the University of Aveiro (Portugal) and associate researcher at the Research Unit on Governance, Competitiveness and Public Policies (GOVCOPP). He holds a Ph.D. in International Relations from the Faculty of Economics at the University of Coimbra (Portugal), in the specific field of International Politics and Conflict Resolution. He is a founder member of the Observatory of China in Portugal. His research interests include the Chinese political economy, China foreign policy and Cross-Strait relations.

Zélia Breda holds a Ph.D. in Tourism, a Master's degree in Chinese Studies (Business and International Relations) and a degree in Tourism Management and Planning from the University of Aveiro, where she is Assistant Professor, at the Department of Economics, Management, Industrial Engineering, and Tourism. She is Director of the Master in Tourism Management and Planning, a full member of the Research Unit "Governance, Competitiveness and Public Policies" of the University of Aveiro, a founding member and vice-president of the Observatory of China, and a senior researcher of the spin-off company Idtour. She is also a member of the editorial and scientific boards of a few academic national and international journals, as well as a member of the organizing and scientific committees of international tourism conferences. She has authored and co-authored several national and international papers and communications on tourism, and has been taking part in several research projects in the tourism field, both as a member of the team and as a consultant.

Fabio Carbone is a Lecturer in International Tourism Management at Coventry University, and associate researcher at the Centre for Trust, Peace and Social Relations (Coventry University, UK). His research focuses mainly on the relations between cultural heritage management and tourism for the promotion of intercultural dialogue, global understanding and peace. He is Ambassador-at-large of the International Institute for Peace through Tourism (IIPT), and IIPT Special Envoy to Iran, where he founded the IIPT-Iran Chapter. His strong humanistic background is complemented by studies in the field of management and tourism development (PhD, University of Aveiro) and others such as International Relations & Cultural Diplomacy (Institute for Cultural Diplomacy, Berlin); Religion, Conflict and Peace; Humanitarian Response to Conflict and Disaster and Environmental Security; Sustaining Peace (EdX, Harvard University). Author of several academic and non-academic articles, he is also a practitioner and member of several boards and scientific committees worldwide. He belongs to the international UN Volunteers database, to serve in UN missions of Humanitarian Assistance and UN peacekeeping operation.

* * *

Katerina Antoniou is a lecturer at UCLan Cyprus, and the Course Leader for the BA Hospitality and Tourism Management course. She holds a PhD in International Peacebuilding from the University of Central Lancashire, an MSc in International Relations Theory from the London School of Economics, and a BA in Political Science and Economics from Clark University, Massachusetts. Her current research examines peacebuilding professionals in Cyprus, while recent publications have focused on processes of desecuritisation in peacebuilding. Additional research interests include social identity, intergroup contact, and travel behaviour in post-conflict settings. Dr Antoniou is a Fulbright Alumna.

Silvia Aulet is a Professor at the Faculty of Tourism of the University of Girona, teaching subjects in the Bachelor of Tourism, the Master of Cultural Tourism, the Erasmus Mundus in Tourism Management, and the Postgraduate Course "Tourism, Wellness and Spirituality". Her research line is cultural tourism focusing on religious and sacred sites. She has also worked in other areas related to heritage such as gastronomy or cultural accessibility. Apart from different publications on these topics, she is member of the editorial board of the International Journal on Religious Tourism and Pilgrimage, member of the association "Future for Religious Heritage", the UNITWIN UNESCO Chair "Culture Tourism and Development" and the Calonge-Sant Antoni Chair of "Gastronomy, Culture and Tourism" from the University of Girona. She has participated in several research studies both as an academic researcher and as a consultant. To mention some, she was an active researcher in the SPIRIT-Youth (a COSME project related to the creation of a spiritual tourism proposal trough Europe for young people) or INCOME Tourism (an Erasmus+ based on learning soft-skills in tourism through a new learning approach together with companies).

Sedat Çelik is an Associate Professor at the Department of Tourism and Hotel Management of Şırnak University. He completed his doctorate at Akdeniz University. His research interest areas contain destination, tourism management, organizational behaviour and social psychology of tourism. He is the editor-in-chief of the Journal of Hospitality and Tourism Issues (JOHTI).

Karen Davies is a Senior Lecturer in Events Management with a keen interest in festivals and their role in promoting social justice. Her PhD explored processes of intercultural communication at Llangollen International Musical Eisteddfod and the contribution that cultural events can make to the facilitation and encouragement of peace and global citizenship. Karen currently sits on the Executive Committee of the Association of Events Management Education.

Adalberto Dias de Carvalho is a Full Professor of the Faculty of Arts and Humanities of the University of Porto, where he is also an integrated researcher of the Research Group on Philosophy and Public Space of the Philosophy Institute. He was an Invited Professor at the University of Rouen (France), at the Portuguese Catholic University, and at the University Eduardo Mondlane (Mozambique), besides cooperating with the University of Paris 8. He is currently the main coordinator professor, director and president of the Technical-Scientific Council at ISCET – Higher Institute of Business Sciences and Tourism. He holds a PhD in Tourism by the University of Girona and in Philosophy by the University of Porto, has held several positions and important functions in the scope of his scientific activity, integrating several international scientific societies, and served as president of the Philosophy of Education Society of Portuguese Language. He is the coordinator of the tourism research area of the Center for Interdisciplinary Research and Community Intervention (CIIIC) of ISCET, where he also coordinates the postgraduate degree in Tourism and Hotel Management. He is author of scientific articles and communications, particularly in the area of island tourism, having recently published the book "Turismo Contemporâneo: Problemáticas e tendências" (2019).

Maria Dorsey received her MSc in Parks, Recreation, and Tourism from the University of Missouri-Columbia in 2006, and her PhD in Tourism and Peace and Conflict Studies from the University of Otago in 2015.

Mónica Guasca received an Erasmus Mundus Master in Sustainable Territorial Development from the Universities of Padova, University of Leuven and Paris 1 Panthéon-Sorbonne. She is currently a PhD candidate of the Division of Geography and Tourism at the University of Leuven (KU Leuven), Belgium. Her doctoral research explores the role of tourism in post-conflict areas in Colombia. Specifically, she is interested in analysing how local agency and spatial configurations related to tourism influence peace outcomes. Her broad research interests focus on development approaches applied in countries that are trying to recover from intra-state conflict, the physical and mental transformation of places marked by violence, and the study of peace at the micro-level.

Sweety Jamgade is an Assistant Professor at the Faculty of Hospitality and Catering Technology, in the Ramaiah University of Applied Sciences, Bangalore, India. She holds a PhD degree in the domain of Ecotourism in the faculty of Science and Technology, subject of Hotel Management and Catering Technology. She treasures 16 years of overall experience in the academics and hospitality industry. She has a keen interest in Tourism and Hospitality industry research, being passionate about Ecotourism and Sustainability projects. She has presented and published research papers in several national and international seminars and conferences. Also, published several research papers in various institutional journals, books and magazines. She is Editor-in-Chief of Hospitium (in-house) institute magazine, board member of other publications of the institute, member of Ecotourism Society of India, The International Ecotourism Society (TIES) and Indian Tourism and Hospitality Congress. She has a profound zeal to guide students in their professional and personal grooming. She is also an active travel review writer and blogger.

Maximiliano E. Korstanje is Editor-in-Chief of the International Journal of Safety and Security in Tourism (UP Argentina) and Editor-in-Chief Emeritus of the International Journal of Cyber Warfare and Terrorism (IGI-Global US). He is Senior Researcher in the Department of Economics at the University of Palermo, Argentina. In 2015 he was awarded a Visiting Research Fellow at the School of Sociology and Social Policy, University of Leeds, UK and the University of La Habana, Cuba. In 2017, he was elected as Foreign Faculty Member of AMIT, Mexican Academy in the study of Tourism, which is the most prominent institution dedicated to tourism research in Mexico. He has a vast experience in editorial projects, working as an advisory member of Elsevier, Routledge, Springer, IGI Global and Cambridge Scholar Publishing. He has visited and given seminars in many important universities worldwide. He has also recently been selected to take part in the 2018 Albert Nelson Marquis Lifetime Achievement Award, a great distinction given by Marquis Who´s Who in the world.

Daniel Laven is an Associate Professor of Human Geography in the Department of Economics, Geography, Law and Tourism at Mid Sweden University. Daniel's research is conducted under the auspices of the university's European Tourism Research Institute (ETOUR). Much of his research explores the intersection of heritage, tourism and peacebuilding activities. Daniel has worked actively in the Jordanian-Israeli-Palestinian context, Western Balkans, and at a number of sites throughout Europe. Prior to joining Mid Sweden University in 2010, Daniel spent over a decade working for the US National Park Service on issues of heritage stewardship and public memory.

Francisco Leandro received a PhD in Political Science and International Relations from the Catholic University of Portugal in 2010. From 2016 to 2017, he took part in a post-doctoral research programme on state monopolies in China - One Belt One Road studies. In 2014 and 2017, he was awarded the Institute of European Studies in Macau's (IEEM) Academic Research Grant, which is a major component of the Asia-Europe Comparative Studies Research Project. From 2014 to 2018, he was the Programme Coordinator at the Institute of Social and Legal Studies, Faculty of Humanities at the University of Saint Joseph in Macau, China. He is currently an Associate Professor and Assistant Dean of the Institute for Research on Portuguese-Speaking Countries at the City University of Macau, China. His most recent books are "The Challenges, Development, and Promise of Timor-Leste" (2019) and "The Belt and Road Initiative: An Old Archetype of a New Development Model" (2020).

Paula Duarte Lopes is a researcher on Peace Studies at the Center for Social Studies and an International Relations Professor at the School of Economics, University of Coimbra (Portugal). She received her PhD from The Johns Hopkins University, United States of America, and her Masters from the London School of Economics and Political Science, Great Britain. She has a Bachelor Degree (five years) in Economics from the School of Economics of the University of Coimbra. Her research interests include global interventionism, focusing on peace missions and development aid, and also including environmental governance, with a specific interest in international water conflicts. She has several publications on these issues.

Sofia Lopes holds a PhD in Tourism Management from ISCTE-IUL and Universidade Europeia, Lisbon, Portugal. She is currently Coordinator of Tourism Degree and Assistant Professor of Tourism in the Faculty of Tourism and Hospitality, Universidade Europeia, Lisbon. Her research focuses on the hospitality impact on tourism, namely, on health tourism and the convergence between hospitality in hotels and hospitality in hospitals. Her recent research has focused on the relation between tourism and peace, ecovillages and tourism, and hospitality and its relationship with absorptive capacity.

Anabela Monteiro holds a Doctorate degree in Tourism, Leisure and Culture from the University of Coimbra and has completed her initial training in Cultural Animation at ESAD.CR/IPL. She is currently Coordinator of Hotel Management Degree and Assistant Professor at the Faculty of Tourism and Hospitality of the Universidade Europeia of Lisbon. She is a researcher at the CEPESE, University of Porto; Researcher at Tourism and Hospitality Research, Faculty of Tourism and Hospitality, Universidade Europeia; columnist for Ponteditoramagazine; member of the Scientific Council - Tourism: Studies & Practices and reviewer - Journal of Entrepreneurship Education. Her area of interest and research focus mainly on the motives and decision factors that stimulate a decision and generate indicators of determining factors in the choice of destination.

Luiz Oosterbeek is graduated in History (1982) and holds a PhD in Archaeology (1994), He is the Secretary-General of the International Council for Philosophy and Human Sciences and Professor of the Polytechnic Institute of Tomar, being chairholder of UNESCO-IPT Chair in Humanities and Cultural Integrated Landscape Management. He conducted research in the fields of archaeology, heritage and landscape management in Portugal, Africa and Southern America. He is the author of over 300 papers and 70 books, having received prizes and awards from the European Commission, the Brazilian Order of Advocates, the Portuguese Ministry o Culture, Gulbenkian Foundation, Foundation for Science and Technology and several private sponsors. He is also President of the Instituto Terra e Memória (Portugal), Visiting Professor in several universities in Europe and Brazil, Director of the Museum of Prehistoric Art in Mação (PT), and Vice-President of HERITY International (IT). He is the Former Secretary-General of the International Union of Prehistoric and Protohistoric Sciences.

Caroline Ritchie was based in the Welsh Centre for Tourism Research at Cardiff Metropolitan University where she was appointed as Reader in Hospitality in 2012. Caroline taught at both undergraduate and post-graduate levels in her specialist subjects which included hospitality, (food and beverage) and alcoholic beverages, with particular emphasis upon related social and cultural issues. She also supervised several PhD students and was the Ethics Coordinator for her department, working with many international TNE partners in relation to research ethics. Caroline's main research area is socio-cultural interaction with wine and related consumer behaviour; she has also undertaken research into TNE student motivation. She has been involved in several national and international collaborative research projects in these areas.

Caroline Ritchie was based in the Welsh Centre for Tourism Research at Cardiff Metropolitan University where she was appointed as Reader in Hospitality in 2012. Caroline taught at both undergraduate and post graduate levels in her specialist subjects which included hospitality, (food and beverage) and alcoholic beverages, with particular emphasis upon related social and cultural issues. She also supervised several PhD students and was the Ethics Coordinator for her department, working with many international TNE partners in relation to research ethics. Caroline's main research area is socio-cultural interaction with wine and related consumer behaviour; she has also undertaken research into TNE student motivation. She has been involved in several national and international collaborative research projects in these areas.

Dina Sebastião is an Assistant Professor at the University of Coimbra, Faculty of Arts and Humanities, in the Department of History, European Studies, Archaeology and Arts, and a research member of the Centre of Interdisciplinary Studies of the 20th Century at the same university. She holds a PhD in European Studies, and has been researching on the relation of Iberian Peninsula and European integration, through the study of policies and parties, namely social-democracy, and has been developing studies about European public policies, the europeanization dynamics and problematisation of policies in the multilevel system of European Union. Within these subjects, she has been participating in international conferences and has published chapters and articles in peer-reviewed journals. She was awarded the Jacques Delors 2017 Prize/Portuguese Ministry of Foreign Affairs for the best academic study in EU issues.

Jack Shepherd is a doctoral candidate at the European Tourism Research Institute (ETOUR) at Mid Sweden University. He holds a BA in History and French from King's College London and an MSc in Tourism Studies from Mid Sweden University. His current research focuses on the relationship between toursism, peace and conflict in the Israeli-Palestinian context. Recently, he has explored such issues with forms of Creative Analytic Practice, including autoethnography. In addition, he has also published on gender in tourism research and on European Union funded projects in cross-border tourism cooperation.

Alina Stoica is an Associate Professor at the University of Oradea, Faculty of History, International Relations, Political Sciences and Communication Sciences, Department of International Relations and European Studies. She is member of the CEIS 20 (Centro de Estudos Interdisciplinares do Século XX) of the University of Coimbra, Portugal; Member of ISER (Institute for Euroregional Studies), Jean Monnet Centre of Excellence Oradea-Debrecen; Member of the Romanian Association of International Relations and European Studies; and Coordinator of the Jean Monnet Module "Cultural Diplomacy, at the East and West Borders at the European Union-an Essential Element in the Process of European Integration". She participated in numerous programs and national and international projects, lecturing at the University of Coimbra (Portugal), in Salamanca (Spain), Siena, Padua and Venice (Italy). She has published over 45 articles in specialised journals, volumes of international conferences or book chapters. She is the author of the books "Romanian-Portuguese Diplomatic Relations (1919-1933): Martinho de Brederode-Ambassador to Bucharest" (University of Oradea Press, 2011) and "History of the European Idea" (Lambert, 2015); co-editor of Volume IX and XVI of the Eurolimes Journal; coordinator of the volume "Education and Faith Towards a Reunification of the Values" (Marne, 2013); co-editor (with Ioan Horga and Maria Manuela Ribeiro) of the "Culture and Paradiplomatic Identity: Instruments in Sustaining EU Policy" (Cambridge Scholars Publishing, 2016); and co-editor of "The European Space: Borders and Issues" (with Mircea Brie and Florentina Chirodea) (University of Debrecen Press, 2016).

Edgar Tarrés is a Professor of Wellbeing, Happiness and Design of Life at the University of Girona and UNED Galicia, and Director of the Postgraduation in Tourism and Consciousness at the University of Girona. He is a mindful travel consultant, wellbeing into nature program trainer, and designer of wellbeing experiences. He is also a collaborator in different means of communication, lecturer, writer and composer.

Ana Margarida Teixeira holds a Bachelor's Degree in Tourism Planning and Management, from the University of Aveiro, and a Master's Degree in International Relations - Studies of Peace, Security and Development, from the University of Coimbra.

Anne Marie Van Broeck obtained her PhD in Social and Cultural Anthropology at the University of Leuven (KU Leuven) in 2000. Currently, she is a senior lecturer and researcher at the same university, Department of Earth & Environmental Sciences, Division of Geography and Tourism. She is involved in the Master of Tourism and the research group of Tourism. Her main research interests and lecturing topics are related to the Anthropology of Tourism (socio-cultural dimensions of tourism) and to Tourism and Conflict (conflicts caused by tourism; tourism in a context of political tensions; contested heritage). She has a strong interest in Colombia, where she lived for several years. At present, she has several research trips a year to this country and is involved in the research group of Tourism of the Institución Universitaria Colegio Mayor de Antioquia (Medellín, Colombia), where she used to teach.

Dominique Vanneste is a full professor in the Department of Earth and Environmental Sciences at the University of Leuven (KU Leuven), Belgium. She is involved in the research groups of Human Geography and Tourism and is the director of a research and development unit 'ASTOR' (Association for Tourism Research), as well as programme director for the KU Leuven part of an Erasmus Mundus Master in Sustainable Territorial Development. She represents KU Leuven with the UNITWIN-UNESCO network 'Culture, Tourism, Development'. Her main research interests and lecturing topics are economic geography (regional development and networking), historical geography (the relationship between landscape, heritage, identity and conservation) and tourism (sustainable tourism and social capital, heritage tourism, slow tourism, tourism as a regional development lever).

Carla Villagra Álvarez is graduated in Tourism from the University of Girona.

Jiawei Xing is a PhD student in Public Policies from the University of Aveiro, she has a MA in Literatures, Languages and Cultures from the University of Aveiro and a BA in Portuguese Language from the Dalian University of Foreign Languages. She was a Portuguese teacher in Colégio Guadalupe (Portugal), and presently she is member of the Research Unit 'Governance, Competitiveness and Public Policies (GOVCOPP)'.

Index

A

Academic Mobility 275-277, 282, 284-286, 288, 292-295
Agency 4, 6, 8, 67, 70, 75, 267, 282, 294, 322, 327, 330, 332-336, 341
Allport 110, 238-240, 249, 321, 331-332, 334, 340
Area A 196, 210, 217
Attitude Change 60, 220, 238-242, 248-249, 261

B

Barriers 36, 71, 92, 99, 109-110, 116, 119, 127, 137, 141, 143, 150, 153, 169, 184, 222, 230-231, 234, 239, 243, 250, 282, 294, 328

C

Catalyst 29, 53, 108, 118, 148, 185
Community-Based Tourism 76, 91, 95-96, 98, 107
Conflict Transformation 70, 87-90, 107, 276, 286, 336
Contact Hypothesis 238-240, 249-250, 321, 325, 331-335, 340
Contact Theory 109-110, 122, 125, 133, 240
Cooperation 16-17, 19, 24, 36-38, 57, 74, 89-92, 94, 98, 101, 111, 125, 138-141, 150-153, 159-160, 166, 173, 185, 187, 207-210, 213-214, 240, 259, 265, 280-282, 293-294, 306, 308, 321, 325-329, 331-334, 340
Cross-border Cooperation 213
Cross-Border Tourism 36, 45, 94, 96, 158
Cross-Strait Relations 136, 138, 140, 142, 144
Cultural De-bordering 148, 150, 152, 160
Cultural Ethnorelativism 302-303
Cultural Heritage Management 8, 10-11
Cultural Industries 148-152, 159-160
Cultural Sensitivity 152, 302-303, 305, 308, 315

D

Dark Consumption 17, 28
DDR 66-68, 70, 72, 79-80, 86
Demobilisation 67-68, 71-72, 74, 77, 86
Deteriorations 32-33, 45
Dewey 255, 269-270
Disarmament 66-68, 71-72, 86
Dual-Narrative Tour 217

E

Ecotourism 34-35, 41, 56, 58, 67, 91-92, 94-96, 100, 102-103, 107, 257
Empowerment 75-76, 98, 101, 322, 325, 332-333
Environmental Peacebuilding 88-90, 107
Equity 48, 259, 262
European Unity 164
Ex-Combatant 67, 69, 71, 73-78, 80, 86

F

FARC 56, 67, 70-72, 74-80, 86

G

Genealogical Tourism 302-303, 306-307
Genetic Test 303, 313-314
Genetic Tourism 302-303, 306-309, 313-316
Global Peace Index 19, 29, 31, 41, 45, 57
Great Limpopo 87-89, 91-92, 94-95, 100-101, 103
Green Line 209, 213, 217
Guerrilla 56, 68, 70, 74, 76, 78, 86

H

Hostels 195, 197, 200-202, 204-214, 217

I

IIPT 6, 35, 38-39, 48
Intercultural Communication 151, 219, 222-227, 231, 233, 285
Intercultural Dialogue 10-12, 50-51, 60, 279, 306, 308
Intercultural Exchange 221-222, 224, 226-227, 233-234, 324
Intergroup Contact 109-111, 121-126, 133, 239-240, 321, 325, 331-332, 334-337, 340
International and National Conflicts 41, 45
International Openness/Open Tourism 45
International Relations 8, 30, 137, 143-144, 165-166, 183-184, 278-279, 294
Israeli Arab 201-203, 217
Israeli Hostels Network (IHL) 217

J

Justice Tourism 324

M

Mahatma Gandhi 33, 280
Mass Tourism 6, 111, 133, 170-171, 257
Mediation 171, 285, 335-336, 340
Micro-level 67-68, 80
Militarisation 19, 29, 31, 36, 40, 45, 259
Mindful Travel Destinations 53, 55-56, 60
Motivational Factors 119, 276, 286
Multicultural Events 219, 222, 231, 233
Multiculturalism 148-149, 152-153, 275-276, 279, 306, 308

O

Open Tourism 34-36, 41, 45
Oslo Accords 197, 205, 217
Othering 195, 213
Otherness 197, 256, 265, 267-271, 286-287

P

P2P 148, 152-153, 160
Palestinian Authority (PA) 217
Peace Education 1, 11, 40, 48, 275, 279, 289, 294
Peace Parks 39, 45, 87-88, 90-92, 94-96, 101-103
Peace Tourism 29, 31, 34-37, 39-41, 261, 268, 323-324, 340
Peacebuilder Categorisation 323
Peacebuilder Hybridity 340
Peacebuilder Paradigm 333-337, 340
Peacebuilding 2, 5-6, 8-9, 29-31, 35, 40-41, 46-48, 53, 57, 61, 67-68, 79, 87-90, 101-103, 107, 149, 195, 213, 255, 262, 276, 279, 282, 285, 293, 302, 308, 320-330, 332-337, 340-341
Peacemakers 34, 45
People-to-People Diplomacy 137
Perpetual Peace 16, 166-167, 172
Phenomenology Research Design 238
Plurality 56, 256, 267-269, 271, 277, 282
Political Tourism 324
Positive Peace 87-92, 94, 96, 98, 100-103, 107, 220, 222, 259, 261, 271, 334, 336
Post-Conflict 6-7, 17, 20-25, 28, 66-68, 79-80, 86, 109, 122, 171, 262, 320, 324, 328, 331
Post-Conflict Destination 21-22, 28
Post-Conflict Tourism 6, 22-23, 67, 79-80, 86
Post-War Tourism 108, 114, 133
Prejudice 19, 110-111, 121, 123, 133, 238-240, 244, 248-249, 257, 261, 268, 311, 332, 340
Prosperity 23, 28, 52, 91, 160, 170
Protracted Conflict 71
Psychosocial 111, 116-117, 119, 121, 123, 127
Pull Factors 120, 133
Push Factors 133

R

Reconciliation 10, 39, 50, 66-67, 69, 74, 78, 80, 108-111, 115, 118-120, 122-127, 136-137, 219, 233, 261, 268, 290, 322, 324-325, 330-333, 335

S

Settlements 71-72, 78, 196, 207, 217
Social Exclusion 69
Social Psychology 239
Social Reintegration 66, 68-70, 72-73, 75, 77-80, 86
Societal Order 16
Socio-economic Development 91-92, 94-98, 100-101
Spiritual Tourism 59
Stereotypes 23, 80, 111, 169-170, 210, 223, 233, 240, 249, 302-303, 306, 311
Stigmatisation 69-70, 72, 76-79, 258
Structural Violence 76, 89, 107
Sustainability 1, 6, 11, 21, 29-30, 34-35, 37, 39-41, 45, 51-53, 60, 87-89, 91, 94, 96-98, 101-103, 149, 174-175, 177, 219, 255, 261-264, 266, 271, 329, 333-334
Sustainable Development Goals (SDGs) 1, 7
Sustainable Tourism 18, 30, 34-35, 39, 41, 57, 90, 94, 96, 98, 101, 261-265, 268

T

Tolerance 1, 7, 50-51, 59-61, 122, 125, 149, 152, 167, 185, 248, 250-251, 261, 268, 271, 279, 293, 302-303, 305-308, 315
Tourism Policy 164-166, 172, 174, 176-177
Tourist 6-7, 12, 17, 21, 23-24, 37, 39-40, 57, 60, 76, 86, 95-96, 98-101, 103, 110-111, 121-122, 136-137, 141, 143, 157-159, 169-171, 174, 176, 183-186, 189-190, 195, 197, 200-201, 212-213, 220, 222, 239-241, 243, 249, 251, 256-259, 261-266, 268, 270-271, 285, 293, 305, 307, 314-316, 323-324
Track Two Diplomacy 137
Transculturality 302-303, 306, 308, 314-315
Transfrontier Conservation Area 91, 94, 107

U

UNDP 325-326, 330-331, 333
UNWTO 6-7, 34, 39, 48, 137, 149, 151, 184, 219, 261, 282

V

Visiting Peacebuilders 320-321, 323-327, 329-330, 332-334, 336-337, 340-341
Visitor 60, 86, 90, 136, 151, 155-156, 185, 189, 324

W

West Bank 22, 196, 204-205, 207, 209-214, 217
World Peace 31, 55, 172, 219-220, 223, 243, 250, 259-260, 276, 303, 306, 308, 313, 315
WTTC 155, 259

www.ingramcontent.com/pod-product-compliance
Lightning Source LLC
LaVergne TN
LVHW060402300926
849162LV00006B/1326
* 9 7 8 1 7 9 9 8 5 0 5 3 3 *